Chemical Warfare Agents: Toxicity at Low Levels

Chemical Warfare Agents: Toxicity at Low Levels

Edited by
Satu M. Somani
James A. Romano, Jr.

CRC Press
Boca Raton London New York Washington, D.C.

Library of Congress Cataloging-in-Publication Data

Chemical warfare agents : toxicology at low levels / edited by Satu M. Somani, James A. Romano, Jr.

p. cm.

Includes bibliographical references and index.

ISBN 0-8493-0872-0 (alk. paper)

1. Chemical agents (Munitions) I. Somani, Satu M. II. Romano, James A.

RA648 .C546 2000

363.17'91—dc21 00-045426

CIP

Direct all inquiries to CRC Press LLC, 2000 N.W. Corporate Blvd., Boca Raton, Florida 33431.

International Standard Book Number 0-8493-0872-0

Library of Congress Card Number 00-045426

Printed in the United States of America 1 2 3 4 5 6 7 8 9 0

Printed on acid-free paper

Preface

We previously published a book on chemical warfare agents (Academic Press) in 1992. Since then, we have acquired considerable additional knowledge in this area. It is time to update our previous work, with particular emphasis on the low-level toxicology of chemical warfare (CW) agents. Chemical warfare agents are chemicals that have immediate, direct toxic effects on humans, animals, and plants and possible long-term, adverse effects on human health. Chlorine, phosgene, and mustard were CW agents used in World War I and in lesser conflicts thereafter. There was putative extensive use of CW agents in the Sino-Japanese War. Although CW agents were not used during World War II, much research was done in the development of toxicologic information and protective materials. However, mustard gas, defoliant, and nerve gases were used in localized wars in the 1960s, 1970s, and 1980s. Chemical warfare agents are primarily categorized as lethal and incapacitating agents. These agents also possess the attractive quality of being easy and inexpensive to synthesize on a large scale. A reasonable chemical-industrial set-up can be diverted to produce CW agents. Chemical warfare agents are particularly horrifying because their toxic effects are indiscriminate and thus affect not only military personnel but also the civilian population as a whole. Chemical warfare agents are becoming a major force in some of the militant developing countries. This is due to the fact that these agents can provide a substantial psychological edge to the military establishments of otherwise weak nations. Although acute toxicity and high-level dose toxicity were discussed in our previous volume, various review committees have suggested that there were data gaps in our information about the low-level toxicity of CW agents. The Gulf War of 1991 has raised our awareness of these gaps. Epidemiologic studies have indicated that more than 120,000 Gulf War veterans are suffering from many unexplained illnesses and are seeking medical care. Among the putative explanations for these illnesses include exposure to nerve agents or pretreatment drugs. Many United States and British troops were given pyridostigmine bromide as a pretreatment drug during 2 weeks of air and ground war to protect against the possible exposure to nerve gas. One of the notable nerve gases suspected to be present during the Gulf War was sarin. During war-time conditions, military personnel were under physical stress; some have argued for evidence of exposure to a low level of sarin. The toxicity of CW agents at low levels is a very special feature of this book. Certain factors such as stress, surroundings, and other chemical agents can interact with the toxicity of CW agents, and some of these interactions are described in this book.

There is a rapidly increasing interest in the low-level toxicology of CW agents. The National Institutes of Health, the Centers for Disease Control in Atlanta, the Veterans' Affairs Department, and the U.S. Army have a tremendous interest in this area, again stimulated by the aftermath of the Persian Gulf War. As a result of concern regarding a high incidence of undiagnosed illness among veterans of Operation Desert Shield/Storm, a Presidential Advisory Committee was formed to analyze the full range of the Federal Government's outreach, medical care, research, and coordinating activities pertinent to Gulf War Veterans' Illness (GWI). The Presidential Advisory Committee also looked at short- and long-term health effects of selected Gulf War risk factors, e.g., chemical/biological (C/B) weapons, depleted uranium, infectious diseases, anti-biological warfare agent (BWA) vaccines, pyridostigmine bromide (PB), etc. The Presidential Advisory Committee gave specific and serious attention to the question of health effects of low-level exposure to nerve CW agents. To close this gap in the current knowledge base, the Department of Defense (DoD) was urged to support additional research on the long-term health effects of low-level exposures to CW agents (nerve agents in particular). Such an increased level of research has already been initiated, and elements of it are discussed thoroughly in various chapters.

The chapter contributors are experts well-recognized for their contributions to the science of toxic chemicals. Their contributions are summarized as follows:

Romano, McDonough, Sheridan, and Sidell provide an overview of the health effects of low-level exposure to nerve agents. They begin with description of the biochemical and physiologic actions of these agents leading to their toxicity. The authors describe the catastrophic effects of the use of these agents and the resultant previous emphasis on lifesaving therapeutic interventions. The authors then discuss reasons for the current emphasis on long-term health effects of these agents, particularly with respect to the question of "low-level exposure." They attempt to provide workable definitions to the concepts of exposures and long-term health effects, review chronic health effects of acute exposures, review the contributions of *in vitro* studies to determine the health effects of low-level exposures and to provide a comprehensive, but perhaps not exhaustive, review of the literature surrounding chronic health effects of repeated low-level exposures, both animal and human. The authors close by expressing hope that the recent national investment into additional research will allow a more comprehensive assessment to unfold that will possibly contribute towards better treatment.

Benschop and DeJong provide a truly comprehensive review of the toxicokinetics of nerve agents. Their analysis includes toxicokinetics of G and V agents by inhalation or subcutaneous route, the influence of prophylaxis and therapy upon toxicokinetics of agents, and a chiral analysis of nerve agent stereoisomers. The development of this compendium of toxicologic data was partially dependent upon the development of improved methods of trace analysis in biological samples. Finally, the authors suggest that respiratory exposure for several hours to 20 ppb of nerve agent is near the lower limit of what can be reached with regard to toxicokinetics based on *in vivo* measurement of initial nerve agent. Further advances may enable reliable extrapolation of toxicokinetic results, even at low dosages, including extrapolation to man.

Somani and Husain described the low-dose toxicity of tabun, sarin, soman, and VX under normal as well as stressful conditions. These authors explained the interaction of environmental and physical stress on cholinergic as well as noncholinergic effects induced by low-dose exposure to nerve agents and their potential for additive or synergistic neuropathologic sequelae. Under certain conditions, nerve agents may also induce delayed neurotoxicity called organophosphate-induced delayed neurotoxicity (OPIDN), which is characterized by inhibition of the enzyme, neuropathy target esterase or neurotoxic esterase (NTE). The clinical symptoms of OPIDN are muscular weakness of the hind limb and ataxia. This chapter deals with the delayed neurotoxicity in terms of behavioral, biochemical, and histological changes. The enzyme NTE can be used as a marker for assessing delayed neurotoxicity in humans or animals exposed to neuropathic nerve agents. Physical stress seems to potentiate the delayed neurotoxicity caused by low-dose exposure to sarin.

Soreq, Kaufer, Friedman, and Glick point out that the complexity of the blood-brain barrier (BBB) has hampered research efforts to delineate its components and fully understand its mode of action. However, there have been recent significant advances for evaluating BBB integrity. These new techniques include *in vitro* approaches such as cell culture, organ systems, and imaging approaches. *In vivo* approaches include ischemia resulting from, say, carotid artery occlusion or cold injury in mice. Finally, transgenic and knockout animal models have been developed, which are helping to elucidate critical factors in BBB integrity.

Somani, Husain, and Jaganathan describe the pharmacokinetics and pharmacodynamics of carbamates (viz., pyridostigmine, physostigmine, or neostigmine) and several of the factors such as stress influencing them. Their extensive coverage of these compounds includes both human and animal studies. Among the potential uses for these compounds include their proposed use as pretreatments for nerve-agent poisoning by military personnel. Evidence supporting their effectiveness is presented and discussed. The pharmacokinetics of PB plays an important role in determining the pharmacodynamic effects in normal, disease, or stressful conditions, and in the presence of chemicals and low level nerve gas exposure. This chapter also discusses the pharmacokinetics and pharmacodynamics of physostigmine (PHY) under normal and stressful conditions. The influence of physical stress can at times be profound and these authors suggest that this area of research needs further exploration.

Doctor, Maxwell, Ashani, Saxena, and Gordon describe the progress made in exploring the use of enzymes to counteract the toxicity of organophophorus (OP) compounds. They describe the use of cholinesterase scavenging enzymes, comparing these to a number of pharmacologic antidotes whose actions and efficacy are well known. These studies have involved several animal species. Special emphasis is placed on the use of HuBChE as a scavenging enzyme. Strategies to improve the bioscavenging capability of cholinesterases are described. These include amplification of effectiveness of ChE using oximes, site-specific mutagensis of AChE, Huperazine A as a pretreatment drug, and the intriguing possibilities of immobilized cholinesterases to decontaminate and detoxify OP chemical warfare agents.

Lenz, Broomfield, Maxwell, and Cerasoli describe the use of scavenger enzymes as alternatives to conventional approaches to the management of nerve agent casualties. This new approach avoids side effects associated with current antidotal regimens. It also obviates the requirement, often difficult to achieve in a military setting, for rapid administration of pharmacologically sufficient drug to attain its therapeutic aim. Candidate bioscavenger proteins, which react quickly, specifically, and irreversibly with organophosphorus compounds are presented and discussed. This bond may be stoichiometric and sequester substrate or may be catalytic, hydrolyzing substrate into biologically inert products. Promising examples of each approach are presented and the advantages of the novel approach over conventional approaches are discussed.

Hurst and Smith discuss the clinical effects that may arise from chronic, sometimes symptomatic, low-dose exposure. They make the argument that long-term health effects deriving from acute, subclinical asymptomatic injury do not occur. They discuss the appearance of chronic health effects following a period of chronic, subclinical exposure. They also discuss the possibility of a "threshold" for these effects by describing the outcomes of more than 30 years of use of a sulfur mustard-containing petroleum formulation to treat psoriasis. They duly note the extensive evidence of a carcinogenic effect after repeated occupational exposure to sulfur mustard and summarize the *in vitro* findings of genotoxicity and metabolic disruption in several cell lines. The authors summarize the compilation of human, animal and *in vitro* data, and their implications for long-term health consequences are presented.

Weese provides a comprehensive review of measured association between putative environmental exposures during the Persian Gulf War and symptoms, reporting a clinical outcome, emphasizing the strength, if any, of measured relationships between solvents, smoke, pesticides, pyridostigmine bromide, and chemical warfare agents and specific conditions. Furthermore, this chapter provides an in-depth discussion of problems associated with the definition of cohorts, the use of data from Gulf War Registries, the problem of case definition, and the uncertain nature of putative exposures.

Borowitz, Isom, and Baskin describe key pathologic sequelae to acute and chronic exposure to cyanide. They provide exposure and risk assessment, with emphasis on the effects of cyanide on the neural tissue. These effects are primarily characterized as effects of cyanide on the metabolism of neurons, cyanide and oxidative stress in neuronal cells, cyanide-induced hyperpolarization, and neuronal activation by cyanide, processes which implicate abnormal sodium channel function in cyanide-induced neuronal damage. Endogenous generation of cyanide in neuronal tissue is also postulated as a causal mechanism in disease. Problems in metabolism of cyanide leading to chronic, low-level exposure are described and discussed.

Salem, Olajos, and Katz provide a historical overview of the testing and development of riot-control agents by the military forces of several nations, including the United States. They distinguish between riot-control agents as military chemicals vs. chemical warfare agents (such as nerve agents, blister agents, choking agents, blood agents, and incapacitating agents). Riot-control agents include three subclasses—lacrimators, sternutators, and vomiting agents—based on their salient physiological effects. Ocular, cutaneous, genotoxic, carcinogenic, and human toxicologic effects are provided for relevant instances of each of these classes of riot-control agents.

Adler, Oyler, Keller, and Lebeda provide an overview of botulinum neurotoxin action leading into a description of the syndrome known as botulism and a discussion of possible treatment options. Subsequently, Adler et al. develop purported terrorist or military anticipated use of botulinum neurotoxin and the threat thereof. This threat of use has led to investments in research that have achieved several major milestones and provided insights into mechanisms of action and a resolution of crystal structure. These authors suggest future promising areas of research into this problem. They end with brief discussions of some recent research success, viz., inhibitors of toxin binding, inhibitors of internalization, and inhibitors of translocation, providing examples in each case.

Romano and King suggest likely psychological, physiological, and neurobehavioral effects that may be encountered if chemical warfare agents are employed against U.S. forces, or even more troublesome, against U.S. citizens. They also describe the implications for health care if either these agents or their medical countermeasures are employed. Furthermore, because these pharmacologic and toxicologic actions could occur in the broad context of a nuclear, biological, or chemical environment with attendant confounding variables, they perhaps could lead to increased difficulty in the differential diagnosis of stress reaction vis-à-vis organophosphate-induced organic brain syndromes.

Moore and Alexander describe the organization and capabilities of the national response apparatus to a domestic or international terrorist use of a "weapon of mass destruction." This apparatus involves many federal agencies that support and complement local and state response systems which respond to such incidents. The review also discusses the implications of "low-level toxicity of chemical warfare agents" for the crisis and consequence management phases of the federal response. Finally, the authors provide a brief summary of how several federally funded research and development programs may enhance future response capabilities.

The editors wish to thank Ms. Patricia Little whose persistence, attention to detail, and sense of purpose kept the editors and many of the contributors on track. We also wish to thank her Springfield, IL counterpart, Ms. Judith M. Bryan. Without the efforts of these two individuals, this work would not have proceeded on schedule.

The editors wish to thank also Colonel James Little, Commander of the U.S. Army Medical Research Institute of Chemical Defense, for his support of the overall initiative, Dr. James King, who steadfastly pushed us in pursuit of scholarly excellence, and the contributors for submitting their work in a timely fashion and for making the necessary modifications. The Medical Research Institute of Chemical Defense is the Army's lead laboratory for the development of medical countermeasures to chemical warfare agents. It functions as a subordinate command of the U.S. Army Medical Research and Materiel Command, Ft. Detrick, MD. The editors thank Dr. Carl J. Getto, Dean and Provost, for his encouragement to publish this book. We also express gratitude to the reviewers who are identified in the acknowledgments in each chapter.

Finally, the editors wish to thank Candy Romano and Shipra Somani for their patience and encouragement. Without them, this task would have been onerous; with their support, it was an enriching experience.

Contributors

Dr. Michael Adler
Commander
U.S. Army Medical Research Institute of Chemical Defense
ATTN: MCMR-UV-PN (Dr. Adler)
3100 Ricketts Point Rd.
Aberdeen Proving Ground, MD 21010-5400

Steve M. Alexander
Program Manager
Domestic Preparedness
Battelle Memorial Institute
2012 Tollgate Rd., Suite 206
Bel Air, MD 21015

Yacov Ashani, Ph.D.
Israel Institute of Biological Research
Ness-Ziona
Israel

Steven I. Baskin, Ph.D., Pharm.D.
Commander
U.S. Army Medical Research Institute of Chemical Defense
ATTN: MCMR-UV-PB (Dr. Baskin)
3100 Ricketts Point Rd.
Aberdeen Proving Ground, MD 21010-5400

Hendrik P. Benschop, Ph.D.
Manager, Department of Chemical Toxicology
TNO Prins Maurits Laboratory
2280AA Rijswijk
The Netherlands

Joseph L. Borowitz, Ph.D.
Professor of Pharmacology and Toxicology
Department of Medicinal Chemistry and Molecular Biology
1021 Hovde Hall
Purdue University
West Lafayette, IN 47907-1021

Clarence A. Broomfield, Ph.D.
Commander
U.S. Army Medical Research Institute of Chemical Defense
ATTN: MCMR-UV-PB (Dr. Broomfield)
3100 Ricketts Point Rd.
Aberdeen Proving Ground, MD 21010-5400

Douglas M. Cerasoli, Ph.D.
Commander
U.S. Army Medical Research Institute of Chemical Defense
ATTN: MCMR-UV-PB (Dr. Cerasoli)
3100 Ricketts Point Rd.
Aberdeen Proving Ground, MD 21010-5400

Leo P. A. De Jong, Ph.D.
Department of Chemical Toxicology
TNO Prins Maurits Laboratory
2280AA Rijswijk
The Netherlands

Bhupendra P. Doctor, Ph.D.
Director, Division of Biochemistry
Walter Reed Army Institute of Research
503 Robert Grant Road
Silver Spring, MD 20910-7500

Alon Friedman, M.D., Ph.D.
Department of Neurosurgery
Ben-Gurion University
Beersheva 84105
Israel

David Glick, Ph.D.
Institute of Life Sciences
The Hebrew University of Jerusalem
Jerusalem 91904
Israel

Richard K. Gordon, Ph.D.
Division of Biochemistry
Walter Reed Army Institute of Research
503 Robert Grant Road
Silver Spring, MD 20910-7500

Charles G. Hurst, M.D.
Commander
U.S. Army Medical Research Institute of Chemical Defense
ATTN: MCMR-UV-ZM (Col. Hurst)
3100 Ricketts Point Rd.
Aberdeen Proving Ground, MD 21010-5400

Kazim Husain, Ph.D.
Department of Pharmacology
School of Medicine
Southern Illinois University
P.O. Box 19230
Springfield, IL 62794-1222

Gary E. Isom, Ph.D.
Associate Vice President for Research and Professor of Toxicology
School of Pharmacy and Pharmacal Sciences
1021 Hovde Hall
Purdue University
West Lafayette, IN 47907-1021

Ramesh Jaganathan, M.D., M.S.
Department of Pharmacology
School of Medicine
Southern Illinois University
Springfield, IL 62794-

Daniela Kaufer, Ph.D.
Institute of Life Sciences
The Hebrew University of Jerusalem
Jerusalem 91904
Israel

Sidney A. Katz, Ph.D.
Department of Chemistry
Rutgers University
3154 Penn St.
Camden, NJ 08102

James E. Keller, Ph.D.
Laboratory of Developmental Biology
National Institutes of Health
Bethesda, MD 20892

James M. King, Ph.D.
Deputy Director
Chemical and Biological Defense Information Analysis Center
P.O. Box 196
Gunpowder Branch
Aberdeen Proving Ground, MD 21010-0196

Frank J. Lebeda, Ph.D.
Toxinology Division
U.S. Army Medical Research Institute of Infectious Diseases
1425 Porter St.
Fort Detrick, MD 21702-5011

David E. Lenz, Ph.D.
Commander
U.S. Army Medical Research Institute
of Chemical Defense
ATTN: MCMR-UV-PB (Dr. Lenz)
3100 Ricketts Point Rd.
Aberdeen Proving Ground, MD
21010-5400

Donald M. Maxwell
U.S. Army Medical Research Institute
of Chemical Defense
ATTN: MCMR-UV-PB (Mr. Maxwell)
3100 Ricketts Point Rd.
Aberdeen Proving Ground, MD
21010-5400

John H. McDonough, Ph.D.
U.S. Army Medical Research Institute
of Chemical Defense
ATTN: MCMR-UV-PA (Dr. McDonough)
3100 Ricketts Point Rd.
Aberdeen Proving Ground, MD
21010-5400

David H. Moore, D.V.M., Ph.D.
Director
Medical Toxicology Programs
Battelle Memorial Institute
2012 Tollgate Rd., Suite 206
Bel Air, MD 21015

Eugene J. Olajos, Ph.D.
Director
U.S. Army Edgewood Chemical and
Biological Center
ATTN: AMSSB-RRT (Dr. Olajos)
5183 Blackhawk Rd.
Aberdeen Proving Ground, MD
21010-5424

George Oyler, Ph.D.
Department of Neurology
University of Maryland School of
Medicine
Baltimore, MD 21201

James A. Romano, Jr., Ph.D.
Commander
U.S. Army Medical Research Institute
of Chemical Defense
3100 Ricketts Point Rd.
Aberdeen Proving Ground, MD
21010-5400

Harry Salem, Ph.D.
Director
Edgewood Chemical and Biological
Center
ATTN: AMSSB-RRT (Dr. Salem)
5183 Blackhawk Rd.
Aberdeen Proving Ground, MD
21010-5424

Ashima Saxena, Ph.D.
Division of Biochemistry
Walter Reed Army Institute of Research
503 Robert Grant Road
Silver Spring, MD 20910-7500

Robert Sheridan, Ph.D.
Commander
U.S. Army Medical Research Institute
of Chemical Defense
ATTN: MCMR-UV-PN (Dr. Sheridan)
3100 Ricketts Point Rd.
Aberdeen Proving Ground, MD
21010-5400

Frederick R. Sidell, M.D.
14 Brooks Rd.
Bel Air, MD 21014

William J. Smith, Ph.D.
Commander
U.S. Army Medical Research Institute of Chemical Defense
ATTN: MCMR-UV-PB (Dr. Smith)
3100 Ricketts Point Rd.
Aberdeen Proving Ground, MD 21010-5400

Hermona Soreq, Ph.D.
Institute of Life Sciences
The Hebrew University of Jerusalem
Jerusalem, 91904
Israel

Satu M. Somani, Ph.D.
Professor of Pharmacology and Toxicology
Department of Pharmacology
School of Medicine
Southern Illinois University
P.O. Box 19629
Springfield, IL 62794-9629

Coleen Baird Weese, M.D.
Commander
U.S. Army Center for Health Promotion and Preventive Medicine
ATTN: MCHB-TS-COE (Dr. Weese)
5158 Blackhawk Rd.
Aberdeen Proving Ground, MD 21010-5403

Table of Contents

Chapter 1
Health Effects of Low-Level Exposure to Nerve Agents 1
James A. Romano, Jr., John H. McDonough, Robert Sheridan, and Frederick R. Sidell

Chapter 2
Toxicokinetics of Nerve Agents 25
Hendrik P. Benschop and Leo P. A. DeJong

Chapter 3
Low-Level Nerve Agent Toxicity under Normal and Stressful Conditions 83
Satu M. Somani and Kazim Husain

Chapter 4
Blood-Brain Barrier Modulations and Low-Level Exposure to Xenobiotics 121
Hermona Soreq, Daniela Kaufer, Alon Friedman, and David Glick

Chapter 5
Pharmacokinetics and Pharmacodynamics of Carbamates under Physical Stress...................................... 145
Satu M. Somani, Kazim Husain, and Ramesh Jaganathan

Chapter 6
New Approaches to Medical Protection against Chemical Warfare Nerve Agents...................................... 191
Bhupendra P. Doctor, Donald M. Maxwell, Yacov Ashani, Ashima Saxena, and Richard K. Gordon

Chapter 7
Nerve Agent Bioscavengers: Protection against High- and Low-Dose Organophosphorus Exposure...................................... 215
David E. Lenz, Clarence A. Broomfield, Donald M. Maxwell, and Douglas M. Cerasoli

Chapter 8
Chronic Effects of Acute, Low-Level Exposure to the Chemical Warfare Agent Sulfur Mustard 245
Charles G. Hurst and William J. Smith

Chapter 9
Gulf War Syndrome: Questions, Some Answers, and the
Future of Deployment Surveillance 261
Colleen Baird Weese

Chapter 10
Acute and Chronic Cyanide Toxicity 301
Joseph L. Borowitz, Gary E. Isom, and Steven I. Baskin

Chapter 11
Riot-Control Agents 321
Harry Salem, Eugene J. Olajos, and Sidney A. Katz

Chapter 12
Pharmacological Countermeasures for Botulinum Intoxication 373
Michael Adler, George A. Oyler, James E. Keller, and Frank J. Lebeda

Chapter 13
Psychological Factors in Chemical Warfare and Terrorism 393
James A. Romano, Jr. and James M. King

Chapter 14
Emergency Response to a Chemical Warfare Agent Incident: Domestic
Preparedness, First Response, and Public Health Considerations 409
David H. Moore and Steve M. Alexander

Index 437

1 Health Effects of Low-Level Exposure to Nerve Agents*

*James A. Romano, Jr., John H. McDonough, Robert Sheridan, and Frederick R. Sidell***

CONTENTS

I. Introduction 1
II. Chronic Health Effects of Acute Exposure.............................. 5
III. Chronic Health Effects of Repeated Low-Level Exposure................ 10
IV. The Contributions of *In Vitro* Studies to Determine Health Effects of Low-Level Exposure 14
V. Summary and Conclusions.. 17
References.. 18

I. INTRODUCTION

Nerve agents are highly toxic organophosphorous (OP) compounds that are chemically related to some insecticides (parathion, malathion). The four most common nerve agents are tabun (o-ethyl N,N-dimethyl phosphoramidocyanidate; military designation, GA), sarin (isopropyl methyl phosphonofluoridate; military designation, GB), soman (pinacolyl methyl phosphonofluoridate; military designation, GD), and VX (o-ethyl S-2-N,N-diisopropylaminoethyl methyl phosphonofluoridate). These compounds exist as colorless and relatively odorless liquids and are meant for use in weapon systems (shells, rockets, bombs) that are designed to deliver them as aerosols or fine sprays. They exert their toxic effects by inhibiting the cholinesterase (ChE) family of enzymes to include acetylcholinesterase (AChE; E.C.3.1.1.7), a critically important central nervous system (CNS) and peripheral nervous system (PNS) enzyme that hydrolyzes the neurotransmitter acetylcholine (ACh). Although the nerve agents can inhibit other esterases, their potency and specificity for inhibiting AChE account for their exceptionally high toxicity. For example, the rate constants for inhibition of AChE by soman, sarin, tabun, or VX are two to three orders of magnitude greater than for the more commonly known OP compounds such as DFP, paraoxon, or methylparaoxon.[1] Likewise, the rate constants for inhibition of AChE by

* The opinions or assertions contained herein are the private views of the authors, and are not to be construed as reflecting the view of the Department of the Army or the Department of Defense.

** All authors contributed equally.

the nerve agents are also two to five times greater than for trypsin (E.C.3.4.21.4), chymotrypsin (E.C.3.4.21.1), or carboxylesterase (E.C.3.1.1.1),[2] indicative of selective inhibition of this enzyme.

Nerve agents bind to the active site of the AChE enzymes, thus preventing them from hydrolyzing ACh. The enzyme is inhibited irreversibly, and the return of esterase activity depends on the synthesis of new enzyme (~1–3% per day in humans). All agents are highly lipophilic and readily penetrate the CNS. Acetylcholine is the neurotransmitter at the neuromuscular junction of skeletal muscle, the preganglionic nerves of the autonomic nervous system, the postganglionic parasympathetic nerves, as well as muscarinic and nicotinic cholinergic synapses within the CNS. Following nerve agent exposure and the inhibition of $\geqslant$ 60% of the AChE enzyme pool, levels of ACh rapidly increase at the various effector sites resulting in continuous overstimulation. It is this hyperstimulation of the cholinergic system at central and peripheral sites that leads to the toxic signs of poisoning with these compounds. The signs of poisoning include miosis (constriction of the pupils), increased tracheobronchial secretions, bronchial constriction, increased sweating, urinary and fecal incontinence, muscle fasciculations, tremor, and convulsions/seizures of CNS origin and loss of respiratory drive from the CNS. The relative prominence and severity of a given sign are highly dependent on the route and degree of exposure. Ocular and respiratory effects occur rapidly and are most prominent following vapor exposure, while localized sweating, muscle fasciculations, weakness, paralysis, and gastrointestinal disturbances are the predominant signs following percutaneous exposures and usually develop in a more protracted fashion. The acute lethal effects of the nerve agents are generally attributed to respiratory failure caused by a combination of effects at both central and peripheral levels and are further complicated by copious secretions, muscle fasciculations, and convulsions. There are several excellent reference sources that provide more detailed discussions of the history, chemistry, physiochemical properties, pharmacology and toxicology of nerve agents.[3–7]

Human estimates of nerve agent toxicity have been derived from animal studies. They range from 7 μg/kg (VX) to 80 μg/kg (tabun) as the LD_{50} for the i.v. route of administration,[7] while the percutaneous LD_{50} for tabun is estimated at 1000 mg, 1700 mg for sarin, 100 mg for soman, and 10 mg for VX, for a 70 kg person, respectively.[4] The rapid onset of effects and extreme toxicity have made these compounds eminently suitable for use as chemical warfare (CW) agents, and in some cases, many thousands of tons of these agents have been synthesized for military use. Exposure to lethal levels of nerve agents will produce toxicities that are precipitate in onset and catastrophic in effect.[8] For these reasons, major medical research efforts since the 1940s have focused on developing the best possible lifesaving therapeutic interventions, pretreatments, or, more recently, prevention of long-term changes in CNS function following a moderate to severe intoxication, using anticonvulsant drugs.[9]

Due to the focus on lifesaving interventions, it was not until the early 1980s that the question of chronic health effects of low-level exposure to nerve agents was subjected to its first major review. The Committee on Toxicology, National Academy of Sciences, studied the available literature reports from the soldier-volunteer test program of the former Army Chemical Center at the then Edgewood Arsenal, now a part of Aberdeen Proving Ground, MD.[10] Soldier-volunteers participated in this test program from 1958 to 1975. There were 15 anticholinesterases (anti-ChE) tested on approximately 1400 subjects during this timeframe, with the great majority of anti-ChE agents being tested during the 1950s and 1960s.

The National Academy of Sciences review found that mortality data compiled in 1981 did not indicate increased deaths among soldier-volunteers when compared to comparable soldiers outside the testing program. There was no clear-cut indication of long-lasting CNS effects and no evidence for mutagenicity, carcinogenicity, male reproductive, or cataractogenic effects.[10] The National Academy of Sciences review committee also reported confidence that its analyses would have had the power to detect any major health effects, had they been present. In general, that viewpoint was considered to be "state-of-the-art," with very little contention until the appearance of Persian Gulf War Illness in the early 1990s.

As a result of concern regarding a high incidence of undiagnosed illness among veterans of Operation Desert Shield/Storm, a Presidential Advisory Committee was formed to analyze the full range of the Federal Government's outreach, medical care, research, and coordinating activities pertinent to Gulf War Veterans' Illness (GWI). The Presidential Advisory Committee also looked at short- and long-term health effects of selected Gulf War risk factors, e.g., chemical/biological weapons, depleted uranium, infectious diseases, vaccines against potential biological warfare agents, pyridostigmine bromide, etc. The Presidential Advisory Committee gave specific and serious attention to the question of health effects of low-level exposure to nerve agents. Their conclusions could be summarized as follows:

1. Available scientific evidence does not indicate that long-term, subtle, neuropsychological and neurophysiological effects could occur in humans following low-level (asymptomatic) exposure.
2. The amount of data from either human or animal research on low-level exposures is minimal.
3. To close this gap in the current knowledge base, the Department of Defense was urged to support additional research on the long-term health effects of low-level exposures to CW agents, the nerve agents in particular.[11] Such an increased level of research has already been initiated, and some elements of it are discussed throughout this chapter.

Because of the great national interest, and perhaps because of technological advances in allowing public access to data, the current status of the federal portfolio of research in this area is readily available through the Internet. The current, annually updated summaries of progress in this research area of vital national interest can be found at http://www.va.gov/resdev/. This website is the Internet-based version of the Department of Veterans' Affairs Annual Report to Congress on Federally Sponsored

Research on Gulf War Veterans' Illness for 1998. At this site, one can find an "Overview of the Federal Research Program" (see Appendix D). Among the long-term research recommendations of that overview are the following:

- Development of exposure biomarkers for CW agents
- Development of a strategic research plan for investigating the long-term health effects of exposure to low concentrations of CW agents. The author of the overview notes that these recommendations have been guiding the selection of new research projects since November 1996. The Annual Report to Congress for 1998 lists 10 research projects whose primary focus relates to CW agent exposure and health effects.[12]

Another impetus for renewed investigations in this area is based on the question, "Is the current United States military medical treatment doctrine, as well as physical protective measures (protective masks, clothing, and support systems), adequate to protect soldiers in future deployments from effects of exposure to low levels of CW agents?"[13]

Two recent reviews of the literature on potential long-term health effects from low-level exposure to nerve CW agents have presented slightly different analyses of this issue and, not surprisingly, they have reached slightly different conclusions. Brown and Brix[14] argued that for nearly all accidental or wartime exposures to nerve agents or OP compounds, it is difficult to obtain reliable exposure data. Thus, they argued that exposures could be characterized as high, intermediate, or low, depending upon factors such as intensity of cholinergic signs (e.g., rhinorrhea, salivation, neuromuscular effects, etc.), level of ChE inhibition, and type of medical treatment required. Clearly identified long-term effects have been noted at or above their defined Intermediate Level Exposure. Long-term health effects, according to Brown and Brix,[14] are not reported in individuals experiencing repeated low-level exposure alone.

In his brief review of chronic effects of low-level exposure to anticholinesterases, Roy[15] concluded that "Concerns about major adverse health effects of low-level exposure to anticholinesterases in general seem entirely unwarranted on the basis of currently available literature, but the data are at present insufficient to reflect the possibilities of subtle, agent-specific effects." In the section labeled "Chronic Health Effects of Repeated Low-Level Exposure," we will also review the scientific basis for these health concerns.

It is common practice for toxicologists to differentiate exposure to chemicals based on the dose and the duration of exposure. Four timeframes have been used to define duration of exposures: acute, subacute, subchronic, and chronic. It is useful in light of today's interest in "long-term, low-level" exposures to clarify these terms. Acute exposure is defined as exposure to a chemical for less than 24 h. Subacute exposure refers to an exposure of 1 month or less, subchronic for 1 to 3 months, and chronic for more than 3 months. These exposures can be by any route; for most chemicals it is the oral route with the chemical given in the diet.[16] However, the limited animal studies using nerve agents have usually employed parenteral administration of the agent, and virtually all of them involve acute or subacute durations of exposure. All are intermittent, e.g., usually once a day. When referring to an inhalation exposure, the exposure duration most frequently used is 4h.

It is equally important to clearly define the term "low-level exposure." This term has seen many different usages in the papers reviewed by these authors. These appear to range from any non-lethal exposure through "subtoxic" (defined by DeMenti[17] as no clinical signs) to "subclinical" (defined by DeMenti as no clinical signs and no significant depression of ChE). Exposure, then, is any contact with a chemical that may induce a biochemical effect. Each definition suffers from arbitrariness and we see no way around this. For the purposes of this review, we will attempt to characterize each paper in terms of presence/absence of either clinical signs or symptoms (in the case of human studies), and level and type of ChE inhibition.

II. CHRONIC HEALTH EFFECTS OF ACUTE EXPOSURE

Much of the data regarding long-term neurological sequelae to exposures to cholinesterase inhibitors in man have been gathered following accidental exposures to organophosphate pesticides. While pertinent, extrapolation from these exposures to predictions of effects from nerve agents may be subject to risk. Several phenomena appear to differentiate nerve agent exposure from exposure to organophosphorus (OP) pesticides. These include:

1. The fact that the cholinergic crisis caused by acute, severe intoxication with the OP pesticides is generally much longer than that caused by OP nerve agents (days to weeks for pesticides vs. hours for nerve agents).
2. Many OP pesticides produce delayed peripheral neuropathy, a phenomenon known for more than 50 years, whereas nerve agents have caused polyneuropathy in animals only at doses manifold greater than the LD_{50}—a phenomenon only seen in the presence of massive pretreatment and therapy with atropine and oxime.[18]
3. The "intermediate syndrome," a delayed manifestation of OP poisoning seen in perhaps up to 100 accidentally poisoned patients,[19] has not been described after administration of nerve agents to animals, nor in the instances of nerve agent poisoning in man.[20]

Grob et al.[21] described the effects of acute to subacute short-term exposure of humans to DFP (1–2 mg, IM, daily for up to 7 days) on electroencephalographic (EEG) and psychological parameters. The changes produced by DFP included increases in EEG potential, frequency (especially noted was an increase in beta rhythm), more irregularities in rhythm, and by the intermittent appearance of abnormal waves similar to those seen in patients with grand mal epilepsy (high voltage waves of 3 to 6 Hz, usually most marked in frontal leads, and increased by hyperventilation). The CNS symptoms noted were excessive dreaming, insomnia, jitteriness and restlessness, increased tension, emotional lability, subjective tremulousness, nightmares, headache, increased libido, giddiness, drowsiness, paresthesias, mental confusion, and tremor. The EEG changes usually followed the onset of CNS symptoms. CNS symptoms and EEG changes were correlated with the depression of red blood cell ChE to 70 and 60% of original activity, respectively. Central nervous system symptoms disappeared within 1 to 4 days after exposure was stopped, while the EEG changes persisted in a diminishing degree from 8 to 42 days (average of 29 days). Essentially similar CNS symptoms and EEG changes were described by Holmes and Gaon[22] as occurring acutely in OP-pesticide-exposed workers. They also noted that the more severely exposed individuals or those with multiple exposures, tended to display persistent symptoms that included forgetfulness, irritability, and confused thinking, although the duration of these persistent symptoms was never clearly defined.

These CNS symptoms and EEG changes are virtually identical to those that have been reported to occur following symptomatic exposure to different nerve agents. Grob and Harvey[23,24] described extensive studies of the effects of sarin in man, to include effects on ChE, EEG, and behavior. They noted behavioral and EEG effects virtually identical to those reported for DFP. These effects began coincident with the depression of plasma and red blood cell ChE activity to approximately 60 and 50% of original activity, respectively, following a single i.v. dose, or 34 and 22% of original activity, respectively, following oral administration. These differences between i.v. and oral administration of sarin suggest that the rate of ChE inhibition, and consequently the rate of increase in CNS ACh, are important factors in the development of symptoms of exposure. Bowers et al.[25] studied the effects of the nerve agent VX in man and described behavioral symptoms of anxiety, psychomotor depression, a general intellectual impairment consisting of difficulties in concentration and retention, and sleep impairments generally involving insomnia due to excessive dreaming.

Psychological/behavioral effects were typically evident before the occurrence of physical symptoms. These effects were associated with whole blood ChE inhibitions of $> 60\%$. There have been descriptions of the acute toxic effects in humans that follow high-dose exposure ($\geq LD_{50}$) to the nerve agents soman, sarin, and VX.[8,26–29] The same cluster of behavioral symptoms that are reported following lower doses (anxiety, psychomotor depression, intellectual impairment, sleep disturbances) dominate the clinical picture in the immediate period following resolution of the acute toxic signs of intoxication and then slowly fade with time, sometimes taking months to fully resolve.

There have been a number of investigations as to the possible long-term consequences of an acute symptomatic exposure to OP compounds. For the nerve agents, Burchfiel et al.[30] evaluated the long-term effects of an acute high dose (5 μg/kg, i.v.) of sarin on the EEG of rhesus monkeys. The animals were paralyzed and artificially respirated during exposure since this dose of sarin produced generalized seizure activity on the EEG that lasted an average of 2.5 h. At both 24 h and 1 year following the exposure, there was a significant increase in the relative voltage in the beta frequency bands (13–22 Hz = beta-1; 22–50 Hz = beta-2) in the occipital-temporal EEG lead while the animals were awake in darkness. Similar EEG effects were seen in other animals in this study that were exposed to high doses of the chlorinated hydrocarbon, dieldrin. Functional behavioral tests of other rhesus monkeys exposed to sarin under identical conditions revealed no deficits in performance of a previously learned delayed response test 24 h after the exposure.[31] Duffy et al.[32] performed a similar analysis of EEG of munitions workers accidentally exposed to the nerve agent sarin at doses that produced clinical signs and symptoms of exposure and produced a reduction of erythrocyte ChE at least 25% below the individual's pre-exposure baseline. Within the exposed group, there was a maximally exposed subgroup that had experienced three or more such exposures. The study was performed at least one year after the last exposure. Univariate and multivariate analysis of the data show that the exposed group, especially the maximally exposed subgroup, displayed:

1. Elevated amounts of spectral energy in high-frequency beta activity
2. Visual inspection of the EEG showed decreased amounts of alpha (9–12 Hz) activity along with increased amounts of slow activity (0–8 Hz, delta and theta) and an increased amount of "nonspecific" abnormalities in the EEG background.
3. Increased amounts of rapid eye movement (REM) sleep.

The functional consequences of these EEG changes were not established, but this group reportedly had a high incidence of self-reported memory disturbances, difficulty maintaining alertness and appropriate focusing of attention.[33]

Several studies of the long-term effects of the sarin-exposure victims from Japan have been published. Yokoyama et al.[34,35] evaluated 18 victims of the Tokyo subway incident 6 to 8 months after exposure. All but three of these victims had plasma ChE values below normal values on the day of exposure. Sarin-exposed individuals scored significantly lower than controls on a digit symbol substitution test; they scored significantly higher than controls on a general health questionnaire (GHQ; psychiatric symptoms) and a profile of mood states (POMS; fatigue). Additionally, they had elevated scores on a post-traumatic stress disorder (PTSD) checklist; they had significantly longer P300 latencies on event-related brain-evoked potentials and longer P100 latencies on brain visual-evoked potentials; and female exposed cases had significantly greater indexes of postural sway. The elevated scores on the GHQ and POMS were positively related to the increased PTSD scores and were considered to be due to PTSD. Nakajima et al.[28,36] performed a cohort study of victims of the Matsumoto City sarin exposure 1 and 3 years following the incident. At 1 year following the exposure, they report that 20 victims still felt some symptoms (fatigue, asthenopia, blurred vision, asthenia, shoulder stiffness, and husky voice), and they had lower erythrocyte ChE activity than those who did not have symptoms and had all lived close to the sarin release site. (Note: Not all the symptoms seen at 1 year have been related to nerve agent exposure historically.) At 3 years, some victims still complained of experiencing these symptoms, although with a reduced degree and frequency. There have been two brief reports of severely poisoned nerve agent victims (one sarin, one VX) in Japan who experienced retrograde amnesia, possibly due to prolonged periods of seizures and/or hypoxia.[29,37] Additionally, one of the Matsumoto victims who experienced prolonged seizure activity was followed for at least 1 year and was found to have sporadic, sharp-wave complexes in the EEG during sleep and frequent premature ventricular contractions on Holter monitoring of the electrocardiogram.[38]

Finally, Yanno and Musiychuk published a short summary of 209 acute poisonings by sarin, soman, or VX in Russian nerve agent production facilities.[39] Twenty-eight percent of the victims required hospitalization that ranged from a few to 120 days. Long-term consequences of these exposures were described as memory loss, signs of asthenia, sleep disorders, diencephalic paroxysms, "vegetative changes in the cardiovascular system," and "microorganic disorders of the CNS" (not further defined in this paper). It was noted that CNS symptoms were most prominent and persistent following soman poisoning, confirming observations made by Sidell.[8]

In one of the major OP pesticide studies, Savage et al.[40] retrospectively (~9 years after the poisoning) examined 100 individuals with documented acute OP pesticide poisoning and compared them with matched-pair nonpoisoned controls. They reported no differences between the two groups in visually inspected EEG or a number of neurological tests. There were, however, significant differences between the two groups in their performance on a number of neuropsychological tests, as well as self- and family-assessment of functioning ratings. They stated that their results showed subtle, long-term neuropsychological sequelae to acute OP poisoning that are difficult to detect with standard neurological exams that stress sensory and motor function. Rosenstock et al.[41] performed a retrospective neuropsychological study of OP-poisoned agricultural workers and compared them with a matched control group. They found that when tested 2 years after exposure, poisoned workers self-reported significantly higher numbers of neuropsychological difficulties and had significantly lower test scores than controls on tests of verbal attention, visual memory, and visuomotor and motor functions, as well as tests of visuomotor sequencing and problem solving. Likewise, Steenland et al.[42] found deficits in vibrotactile sensitivity and sustained attention among previously intoxicated subjects vs. controls. These effects showed a rough dose-response relationship in that there were significant trends to worse performance on other neurobehavioral tests by those subjects who were more severely poisoned (longer hospitalization, took more time off from work). However, as with other studies, nerve conduction tests and neurological examinations were negative. There was no evidence of changes in postural sway in poisoned subjects as was reported for the sarin-exposed subjects from Tokyo indicating, perhaps, a difference between OP and nerve agents with regard to effects on motor activity.[34,35]

Studies in animals of long-term effects of acute, non-lethal exposures to nerve agents are numerous in the literature since 1980. Following high-dose exposure (~0.6 LD_{50} and higher), seizures are a prominent sign of nerve agent intoxication, and these prolonged seizures can produce both neural and cardiovascular lesions if not promptly treated.[43] Neurological, behavioral and cardiac deficits are predictable long-term effects following exposure to such doses. Animals exposed to convulsant doses of nerve agent can develop spontaneous seizures, display hyperreactive and aggressive behavior (rats), and display profound deficits in learning and/or performance of a variety of behavioral tasks. In fact, animal studies have demonstrated deficits in acquisition of several types of operant tasks (differential reinforcement of low rates, alternation), performance of serial probe recognition task, maze learning, and passive avoidance learning following acute intoxication with nerve agents.[44–47] Invariably, animals displaying such behavioral changes also are shown to have brain lesions in cortical and subcortical limbic structures. Exemplifying the relationships seen between such neuropathology and deficient behavior is the demonstration of EEG and performance changes following near-LD_{50} challenge with soman (GD) in a comprehensive paper by Philippens et al.[48] In this study, rats were intoxicated with an LD_{50} of soman and immediately treated with an antidotal combination of atropine and diazepam (described as a "low-dose combination"). These rats had previously trained to an over-learned criteria of 80% correct avoidance response (i.e., avoidance of a signaled foot shock). After a period of recovery of motor capacity, animals demonstrated impaired performance of the conditioned response for three test sessions before they approached the pre-challenge performance level. Similarly, electrographic correlates of lesions and ultimately, light microscopic observation of lesions, suggested the neuroanatomic basis for this deficit. By contrast, animals exposed to the same challenge dose of agent that received the "high-dose combination" of atropine and diazepam performed at or near the level of pre-challenge performance and were somewhat (but not completely) protected from electrographic and neuropathologic changes. This report demonstrates a general theme found in most of these "high-dose" exposure studies: animals exposed to nerve agents that develop seizures that are not promptly controlled, develop brain damage and consequent neurobehavioral problems; animals that do not develop seizures or those that seize and are rapidly and effectively treated with drugs that stop the seizures, suffer no brain lesions and display no long-term neurobehavioral deficits.

In the case of acute "low-level" exposures to nerve agents, exposures that produce minimal or no acute CNS signs of intoxication, an earlier study by Burchfiel et al.[30] suggested that a "clinically sign-free" dose of sarin (1 μg/kg, i.m.) given repeatedly (1/week for 10 weeks) to non-human primates (three rhesus monkeys) resulted in subtle, but persistent EEG changes (increases in the percentage of high frequency beta activity) that were virtually identical to those already described above that are seen after an acute high dose exposure (5 μg/kg, i.v.) that provoked seizures. More recently, Pearce et al.[49] followed behavioral and electrographic outcomes in nine marmoset monkeys for up to 15 months following exposure to a single, low dose of sarin (2.5–3 μg/kg, i.m.). Although the dose of sarin caused 36 to 67% inhibition of RBC AChE, there were no acute behavioral signs of intoxication (thus, the exposure was "subtoxic"), there was no significant change or decrement in performance on a series of touch screen mediated discrimination tasks either immediately or over a 12- to 15-month period following the exposure. There were also no significant long-term changes in EEG patterns in this study. Although there were changes in beta-2 amplitude that approached significance ($p = 0.07$), this was entirely due to a long-term change in the EEG of a single subject of the nine animals that were exposed. The parameters chosen in this study were employed because they had been used to demonstrate deficits caused by cholinergic lesions or ChE inhibitors in previous studies.[50,51] Marrs et al.[7] have provided an extensive overview of human studies of nerve agent exposures conducted by the United States and United Kingdom military as well as accidental exposures that occurred at production or test facilities. While this report does not come to any conclusions about long-term effects, there is no indication that asymptomatic exposures to nerve agents have produced long-term, adverse health effects. This is the same conclusion reached by the National Academy of Sciences committee that reviewed the then-available literature, to include the EEG studies of Burchfiel and Duffy.[30,32] They stated that while there may be subtle long-term EEG changes, the clinical significance and functional relevance of such changes had not been demonstrated.[10]

III. CHRONIC HEALTH EFFECTS OF REPEATED LOW-LEVEL EXPOSURE

For chronic or repeated subclinical exposures to OP compounds, be they CW nerve agents or OP pesticides, the data in regards to long-term health effects are less consistent. In regards to the nerve agents, the report of Burchfiel et al.[30] about the effects of repeated low doses of sarin to rhesus monkeys producing a long-term increase in relative power in the EEG beta frequency bands is the most-cited study in support for a long-term health effect. There are no human studies known to the authors of this review other than the National Academy of Sciences report on the volunteer program mentioned earlier, that directly address the possible adverse, long-term health effects of repeated subclinical exposures to nerve agents.[10] Workers exposed to small amounts of nerve agents that produced mild, non-threatening medical signs of exposure, reported CNS effects such as headache, insomnia, excessive dreaming, restlessness, drowsiness, and weakness.[52] Medical officers describing these patients suggested that "Mental processes used in making judgments and decisions were also affected."[52] Of 53 patients with mild exposures not requiring antidotal therapy, CNS symptoms often were fully resolved within 3 days after exposure. However, Sidell and Hurst[19] caution that psychological symptoms are probably more common than usually recognized and may persist in more subtle forms for much longer (days, weeks) than physical symptoms.[19]

Reports in the literature of animal studies show that nerve agents can be administered repeatedly with minimal overt neurobehavioral effects if care is taken in choosing the dose and the time between doses.[53,54] Blood and brain AChE levels can be reduced to >20% of normal with no observable signs of toxicity with appropriate dosing schedules. Animal studies performed at the United States Air Force School of Aerospace Medicine have demonstrated a progressive and long-lasting inhibition of ChE in the CNS following repeated administration of low doses of the nerve agent soman,[55] a finding recently corroborated by Olson et al.[56] using the nerve agent sarin. There appear to be differential sensitivities among various brain regions, with frontal and piriform cortex being most sensitive to the ChE inhibiting effects of CW nerve agents, whereas the neostriatum and the hypothalamus are relatively less sensitive. These studies and others did not demonstrate a tolerance to the CNS ChE-inhibiting effects of repeated administration of low levels of CW nerve agents.

Recovery from the ChE inhibition produced by CW nerve agents or other OP compounds is not a simple matter, however. The recovery of CNS ChE does not parallel the recovery of plasma ChE, with plasma ChE often recovering much more rapidly than RBC-AChE, which more closely parallels the recovery of brain AchE.[57,58] Thus, these and other tolerance studies suggest that behavior has often recovered to near baseline, while AChE is still significantly depressed.[59,60] This has been noted clinically and Sidell[6] cautions, "Analysis of blood for ChE is useful for occupational monitoring, but in an exposed patient, one treats the patient, not the ChE activity."

Recently, Olson and Benschop et al.[56,61] have provided reports of animal studies of effects of repeated low-level exposure to nerve CW agents. In rats, Olson determined the LOAEL and NOEL of subacute dosages of sarin, administered i.m. He found that the dose of sarin (GB) needed to produce a low but measurable blood ChE inhibition was 0.75 μg/kg once a day for 4 days. Thus, the exposure in Olson's study would be described as "subclinical." GB was paired with a variety of other chemicals to include chlorpyrifos, DEET (N,N-diethyl-m-toluamide), carbaryl, and PB. No neurobehavioral or neuropathologic effects could be attributable to dosing with GB alone or in any combination with the other chemicals. Rats were also evaluated using a Functional Observational Battery and a Figure 8 Activity Monitor with no significant behavioral effects reported. Benschop et al.[61] reported on the toxicokinetics of low-level inhalation exposure to soman in atropinized guinea pigs. Animals were exposed to 200 ppb for 5 h of a toxic steroisomer of soman, which resulted in a gradual inhibition of RBC-AChE to approximately 10% of baseline. This level of exposure resulted in an insignificant reduction of AChE activity in brain and diaphragm, although it was equivalent to a Ct value of 48 mg · min/m^3, a dose well above that sufficient to cause an incapacitating miosis. The observed lack of inhibition of AChE in brain and diaphragm at the end of the long-term, low-level exposure was interpreted to mean that systemic intoxication is unlikely despite extensive inhibition of blood AChE. Furthermore, Benschop et al. argued that the development of persistent neuropsychological disorders under these conditions would be unlikely. The authors cautioned that studies in animals without the benefit of carboxylesterase binding sites, such as primates, would most probably reflect a different outcome. This last study points out the influence of dose rate in determining whether a given exposure would be "nonlethal," subtoxic," or "subclinical," a point made as long ago as 1975 by Sim.[62] The latter wrote that a patient appearing in a clinic without measurable ChE, yet not appearing to be intoxicated, "emphasizes that the poison is cumulative and if taken into the body slowly, can be accommodated without the appearance of critical illness."

The most notable effect of repeated low doses of nerve agents seen in animal experiments is the development of tolerance to the disruptive effects of each acute exposure on certain behaviors.[58,60,63–65] This is primarily thought to be brought on by downregulation (i.e., reduction in the number) of muscarinic receptors in the brain which will remain lowered (maximal reduction 30–40%) for the duration of the exposure and then recover in parallel with the recovery in erythrocyte ChE activity following the cessation of exposure.[57,66] During the period of reduction in muscarinic receptor numbers, the animals are subsensitive to anticholinesterases or direct acting muscarinic agonists and suprasensitive to the effects of antimuscarinic drugs (e.g., atropine).[67] In this respect, nerve agents act much like other OP compounds and the possibility and mechanisms of tolerance development have been addressed in several studies (see the review by Russell and Overstreet[68] for an overview of these animal studies). The question of behavioral tolerance is seen by some authors as a masked toxicity in a vulnerable organism,[17] or as an adaptive response to a changed internal physiologic state.[69] Bignami et al.[70] have suggested that whereas some of the tolerance developed to OP may be attributable to cholinergic receptor changes, behavioral (test) variables may also play a role. These authors studied feeding and drinking responses and examined the role of practice factors in tolerance to paraoxon. Specifically, they measured the decrease in food consumption and the strength of a conditioned flavor aversion (CFA) produced by repeated doses of low levels of paraoxon. They qualified their finding of reductions in the depression of food intake or extinction of CFA by stating that treatment-behavior interactions may produce apparent attenuations of toxicity which are often not maintained when the situation is changed, leaving entirely open the nature of the purported response.

No review of subacute, subchronic, or chronic toxicity of chemical warfare nerve agents would be complete without discussion of the significant paper by Munro et al.[71] that reviewed both animal and human studies of the nerve agents tabun (GA), sarin (GB), and VX. These studies included subacute, subchronic, and chronic toxicity studies in animals. Special attention was paid to the phenomenon of Organophosphorus-Induced Delayed Neuropathy (OPIDN). Reproductive toxicity and carcinogenicity tests were reviewed as well as *in vitro* studies of mutagenicity. Munro et al.'s findings can be summarized as follows:

1. For the nerve agent GA, no evidence of subchronic toxicity was observed at any dose other than effects on ChE activity. No evidence of teratogenicity was found and GA was a weakly active mutagen.
2. For nerve agent GB, no evidence of acute or chronic toxicity was found at low intermittent exposure levels, sufficient to significantly depress AChE levels. No evidence for carcinogenicity, teratogenicity, or mutagenicity was found for GB, but a data gap in the area of reproductive toxicity was noted.
3. For nerve agent VX, no evidence for likely development of OPIDN was found. VX exposure sufficient to significantly depress RBC-AChE activity produced no subchronic toxicity, no evidence of carcinogenicity was found, and based on multiple studies, VX was not considered a potential mutagen or a teratogen.

Thus, the authors concluded that the overriding concern with regard to exposure to GA, GB, or VX was their extraordinarily high acute toxicity.[71]

For chronic or repeated subclinical exposures to OP pesticides, there are numerous studies of humans that have addressed this issue and there is no common consensus among the results. Korsak and Sato[72] reported that OP pesticide workers with relatively high occupational levels of exposure to OP pesticides, in comparison with workers with low levels of exposure, tended to have increased EEG power within the beta frequencies, primarily in frontal areas of the brain. In addition, the high exposure level group had lower performance on a Trail-Making Test and the Bender Visual Motor Gestalt test of a neuropsychological test battery. The authors suggested that these results indicated subtle frontal lobe dysfunction in the exposed subjects. Stephens et al.[73] studied a population of 146 sheep dippers with an average of 15 years of potential exposure to several OP (diazinon, propetamphos, chlorfenvinphos) and found that these individuals compared to a control group (quarry workers) has slower simple reaction time latencies, slower latencies in a symbol-digit substitution test, and slower correct reaction times in a syntactic reasoning test. Only the syntactic reasoning effect showed a significant dose-effect relationship when analyzed with an analysis of covariance. Tests of memory and learning showed no effect and there were no self-reported drops in intellectual performance. On the other hand, Ames et al.[74] surveyed 45 pesticide applicators, each of whom had at least one documented episode of asymptomatic OPP exposures. He reported no CNS or PNS effects. Rodnitsky et al.[75] evaluated 23 workers chronically exposed to a mild degree to pesticides (farmers and commercial pesticide applicators) using a battery of neurobehavioral tests. They found no differences between the exposed vs. a control group on tests of memory, signal processing, vigilance, language, and proprioceptive feedback performance even though plasma cholinesterase levels of the exposed group were depressed below the control values. Similarly, Daniell et al.[76] studied 49 pesticide applicators over a 7-month period of pesticide spraying. Performance of the cohort was compared to a control group of 40 subjects (slaughterhouse workers). Both groups were given a computerized neuropsychological test battery. The test battery consisted of visuomotor coordination tests, memory and cognition tests, and tests of motor coordination. The pesticide applicators were known to have generally well-controlled, low, intermittent exposure, as part of a program of occupational health training and monitoring. The authors found no evidence for clinically significant decrements in neuropsychological performance following one 7-month season of such exposure among pesticide applicators, the main one being Guthion (azinphosmethyl). Maizlish et al.[77] examined 99 pest control workers (46 exposed workers vs. a group of non-applicators). The 46 workers applied diazinon to residential lawn properties. Both applicators and non-applicators were monitored for the appearance of diethylthiophosphate (DETP) in their urine. Subjects were given a comprehensive neurobehavioral test battery before and after their 8-h work shift, as well as having urine samples taken. The application season lasted 39 days. The following tests were included in the neurobehavioral battery: Continuous Performance test (measures attention/vigilance), Finger Tapping (motor speed), Digit Symbol Substitution (visual/motor speed), etc. Median diazinon exposure per workday for applicators vs. non-applicators was 2.1 mg vs. 0.03 mg, respectively. No adverse DETP-related changes were found in pre- or post-shift neurobehavioral function. The authors concluded that there were no demonstrable behavioral effects of short-term, low-level diazinon exposure in a pest control program characterized by adequate personal protective equipment and direct supervision.[77]

More recently Bazylewicz-Walczak et al.[78] published their study of greenhouse workers occupationally exposed to pesticides. They gave greenhouse workers and a matched control group neuropsychological tests (simple reaction time, digit symbol, digit span, Benton visual retention test, Santa Ana test, aiming test, profile of mood states), and a subjective symptom questionnaire before and then 4 months later after the heaviest period of OP pesticide application. Overall, when compared to the controls (kitchen workers and administrative workers), the exposed subjects showed slower simple reaction times, lower hand movement efficiency on the aiming test, and reported a higher degree of anxiety, anger, depression, and fatigue-inertia. World Health Organization guidelines were used to characterize the level of exposure of these experimental groups and it was considered to be low. In addition they also reported more complaints relating to absent-mindedness and neurological symptoms. There were no differences in the exposed group over the one season. There was no change or improvement in scores on the neuropsychological tests across the season and there was an improvement in mood and general feeling scores between the pre-season test and the post-season test. The authors concluded that even low, long-term OP pesticide exposure may be associated with subtle adverse behavioral effects, characterized by increased tension and anxiety states, depression, fatigue, and a slow-down of perceptual-motor functions.

IV. THE CONTRIBUTIONS OF *IN VITRO* STUDIES TO DETERMINE HEALTH EFFECTS OF LOW-LEVEL EXPOSURE

By their nature, *in vitro* studies of the physiological effects of nerve agents tend to involve acute or short, subacute exposures. Isolated tissues and organ systems have a limited viability that limits the durations of nerve agent studies, as does the stability of an agent under physiological conditions. Even cultures of isolated cells have a limited useful lifetime *in vitro* although this can approach months under optimal conditions. However, the ability to dissect complicated phenomena into simpler processes *in vitro* often provides a unique opportunity to examine putative mechanisms of nerve agent toxicity. A critical examination of the range of nerve agent concentrations (rather than doses) that evoke these *in vitro* pathologies can then be used to assess the relative involvement of the mechanisms in nerve agent pathology.

Many of the earliest *in vitro* studies of nerve agent toxicity involved isolated smooth or striated muscle tissue and relatively high concentration of nerve CW agents. Under these conditions the actions of the nerve agents as inhibitors of AChE in the muscle tissues could be clearly discerned and related to the degree of AChE inhibition. Smooth muscle, with a predictable response to enhanced stimulation of muscarinic receptors, produced enhanced and sustained contractions in response to physiologic stimulation at relatively low concentrations and spontaneous contractions at high concentrations.[79,80] The actions of the nerve agents on skeletal muscle were more varied but clearly involved initial stimulation of neuromuscular transmission followed rapidly by desensitization of the postsynaptic nicotinic receptors during the course of physiologic tetanic stimulations, eventually leading to paralysis.[81–85] However, not all effects of organophosphorus nerve CW agents were completely consistent with the degree of AChE inhibition.[86,87] These observations led to studies of direct interactions between nerve CW agents and various cholinergic receptors.

One direct receptor effect of various organophosphate cholinesterase inhibitors noted *in vitro* was an apparent increase of the desensitization rate for nicotinic receptors at the skeletal neuromuscular junction.[88,89] Since this is a major factor in skeletal muscle failure subsequent to AChE inhibition and subsequent accumulation of endogenous ACh, such a direct action of the nerve CW agent on the nicotinic receptor would be expected to be synergistic with its actions on AChE. Separation of these phenomena was initially accomplished by comparing rates of desensitization in *in vitro* preparations in which the presence or absence of receptor agonists was controlled and where the effects of AChE inhibition were masked by prior irreversible inhibition of the enzyme. Later studies have used a variety of techniques, including direct measurement of receptor activity without AChE present,[90] or with ligand competition binding studies as indicators of receptor state.[91] Although a variety of nerve CW agents are capable of enhancing desensitization in skeletal muscle receptors, they do so with different mechanisms. VX appears to bind to the open form of the nicotinic receptor ion channel, leading to a block of ion flow and subsequent receptor desensitization,[90,92] while GB and GD appear to bind to an allosteric site on the nicotinic receptor that stabilizes the desensitized form of the receptor.[90,91] A consistent observation of this phenomenon, regardless of agent or mechanism, was that the concentrations of nerve CW agent needed to produce a significant increase in the rate of nicotinic receptor desensitization were equal to or higher than those needed for complete inhibition of AChE activity, with concentrations in the micromolar range often required to show the effect. Clearly, these direct effects on skeletal muscle do not represent low-dose effects of nerve CW agent.

Similar phenomena have been noted for effects of nerve CW agents in ganglionic nicotinic receptor systems although the origins of these effects on transmission may be different and more complicated than in the skeletal muscle endplate. In mammalian superior cervical ganglia, the nerve CW agents produce an initial enhancement of transmission followed by depression of the postsynaptic response with repetitive stimulation.[93,94] However, unlike the endplate, these effects do not appear to follow from desensitization of the postsynaptic nicotinic receptors alone since, in both mammalian and amphibian ganglia, there was no evidence of desensitization to exogenous cholinergic agonists.[94,95] The frequency-dependent decrease in ganglionic transmission has been attributed variously to presynaptic inhibition of release mediated by muscarinic autoreceptors or by cumulative and tonic depolarization of postsynaptic neurons due to endogenous ACh build-up.[94,95] The relative contributions of these phenomena may be dependent upon the source and properties of the ganglia studied. However, as with the direct effects on the skeletal neuromuscular junction, the concentrations of nerve CW agents, or other organophosphorus AChE inhibitors, responsible for depression of ganglionic transmission are at least as high, 0.1 to 100 micromolar if not higher, than those required for nearly complete AChE inhibition.

As indicated in the ganglionic effects of nerve CW agents, a number of mammalian synapses have presynaptic muscarinic autoreceptors that can modulate release of acetylcholine.[94] The concept of presynaptic receptors that can modulate evoked release of other classes of neurotransmitters suggests that organophosphorus nerve agents could have actions on cholinergic systems that would be expressed primarily in noncholinergic systems. Such interactions could occur either through inhibition of AChE and accumulation of extracellular ACh, or via direct actions of organophosphorus nerve agents on cholinergic membrane targets. Such phenomena are most readily observed in the central nervous system, where the organizational and pharmacological complexity of synaptic connections is substantially greater than in skeletal muscle. Since *in vitro* techniques to study function in such central systems have only recently become common, some of the details of such interactions are only now becoming apparent.

Studies in brain slices have indicated that presynaptic muscarinic receptors are involved in reductions of evoked transmitter release in dopaminergic, GABAergic, and glutaminergic synapses.[96,97] The effects on GABAergic neurotransmitter release were particularly potent with a small depression seen at 10 picomolar VX.[97] Although these presynaptic effects could be prevented with the general antagonist atropine, specific binding studies suggest that the effects are mediated via the M3 class of muscarinic receptors.[96,98] Studies in cloned cell lines have suggested that activation of these muscarinic receptors and the resulting activation of phospholipase C can occur with normal agonists at the orthosteric binding site and with certain organophosphorus anticholinesterases acting at an allosteric modulation site. The organophosphate anticholinesterases do not compete with the standard muscarinic ligands QNB or N-methyl scopolamine at the orthosteric site, normally occupied by agonists and competitive antagonists, except at micromolar concentrations in acute assays.[99–102] However, exposure of the muscarinic receptors for longer periods causes a noncompetitive reduction in muscarinic receptor ligand binding that reaches a maximum at 24 h.[100,103] Unlike competitive reactions, the concentrations of organophosphorus anticholinesterases causing the allosteric modulation were in the picomolar to nanomolar range.[99,104] Further, use of an unconventional ligand, *cis*-methyldiololane, to examine a subset of muscarinic binding sites also indicated competition with many organophosphates, including the nerve CW agents, at picomolar to nanomolar concentrations.[99,104] The loss of orthosteric binding sites in response to either muscarinic agonists or anticholinesterases did not result in a corresponding loss of phospholipase C activity. Rather, IP_3 production remained activated while the receptor was in a "sequestered" state.[103,105] The concentrations of organophosphorus nerve agents involved in the allosteric modulation of muscarinic receptors, and, hence, possibly involved in modulation of synaptic transmitter release, are substantially less than those required for AChE inhibition and suggest that the effects may be observed at low-doses *in vivo,* particularly with prolonged acute or subacute exposures.

Other effects of nerve CW agents on presynaptic events have been identified *in vitro* that do not involve muscarinic receptor binding. Nanomolar concentrations of VX and micromolar concentrations of GD greatly enhance spontaneous release of both GABA and glutamate from rat hippocampal brain slice synapses.[97] A similar phenomenon was observed in glutaminergic synapses in the insect, where micromolar VX caused bursts of activity,[92,106] and in amphibian sympathetic ganglia, where repetitive firing of neurons was observed subsequent to a nerve CW agent-induced reduction in a calcium-activated potassium after hypolarization.[107] These cellular effects are not mediated through muscarinic receptors. They appear to be direct actions upon ion channels in the cells. However, except for the effects of VX on spontaneous miniature postsynaptic potentials in the hippocampus, these effects are seen only at concentrations of nerve CW agent that would cause profound inhibition of AChE.

A potentially subtle role for AChE inhibition in affecting synaptic transmission in the central nervous system is suggested by the differential sensitivity of different neuronal nicotinic acetylcholine receptors. Neuronal nicotinic receptors include the $\alpha7$ and the $\alpha4\beta2$ classes.[108] The $\alpha7$ class can be found presynaptically on both glutaminergic and GABAergic neurons where it enhances spontaneous and evoked transmitter release.[109–111] Similar functions can be ascribed to the $\alpha4\beta2$ receptors, although their time course, particularly for desensitization, is slower. Significantly, the $\alpha7$ receptors are activated by both acetylcholine and choline with similar sensitivity while the $\alpha4\beta2$ receptors are activated only by acetylcholine. This suggests that a decrease in the concentration of choline present due to AChE inhibition with nerve CW agents could alter the glutaminergic or GABAergic tone in regulated synapses. The exact result of the inhibition of AChE would depend upon a number of factors, including the number of each neuronal nicotinic receptor type present, the transmitter normally released by the presynaptic terminal, and the ambient level of ACh present in the tissue. These effects would be dependent upon concentrations of the nerve CW agents that would inhibit a substantial fraction of the available AChE in the tissue. Hence, the differential activity of the neuronal nicotinic receptors would be expected to be significant only at concentrations that would provoke serious symptoms of acute anticholinergic toxicity.

Of the various mechanisms of action ascribed to nerve CW agents in various *in vitro* model systems, most appear to be active at concentrations where significant inhibition of AChE would be expected to occur. These conditions are unlikely to be seen in the absence of recognizable cholinergic toxicity and hence would not be classified as a low-dose effect. The one effect that seems to occur at a sufficiently low concentration of organophosphorus anticholinesterase to be considered a low-dose phenomenon is the slow allosteric modulation of muscarinic receptors regulating presynaptic release of other neurotransmitters. The observation that this allosteric modulation requires several hours to develop suggests that this mechanism would most likely be applicable to a subacute or subchronic exposure rather than one seen after a brief, acute dose.

V. SUMMARY AND CONCLUSIONS

We believe that studies to determine the potential long-term psychologic/neurologic sequelae following repeated low-level exposures to OP are confounded by factors such as low response rates, possible selection and follow-up biases (which is certainly the case for nerve CW agent), compensatory psychological response, possible co-exposures and the like. Although the recent national investment into additional research has emphasized animal research, we are hopeful that a more comprehensive assessment of this problem is unfolding. However, the significance of and the biochemical basis for chronic health effects resulting from high-level, acute exposure should not be overlooked with all the current emphasis on repeated low-level exposure.

ACKNOWLEDGMENTS

The authors wish to thank Mrs. Patricia Little for her skillful editorial assistance and persistence in coordinating the contributions of the four authors of this chapter.

REFERENCES

1. Gray, P.J. and Dawson, R., Kinetic constants for the inhibition of eel and rabbit brain acetylcholinesterase by some organophophates and carbamates of military significance, *Toxicol. Appl. Pharmacol.*, 91, 140, 1987.
2. Maxwell, D. and Doctor, B.P., Enzymes as pretreatment drugs for organophosphate toxicity, in *Chemical Warfare Agents,* Somani, S., Ed., Academic Press, New York, 1992.
3. Koelle, G.B., Cholinesterases and anticholinesterase agents, *Handbuch Der Experimentellen Pharmakologie XV,* Springer-Verlag, Berlin, Germany, 1963.
4. Taylor, P., Anticholinesterase agents, in *The Pharmacological Basis of Therapeutics,* 6th ed., Gilman, A.G., Goodman, L.S., Rall, T.W., and Murad, F., Eds., Macmillan, New York, 1985, 110.
5. Somani, S.M., Solana, R.P., and Dube, S.N., Toxicodynamics of nerve agents, in *Chemical Warfare Agents,* Somani, S., Ed., Academic Press, Inc., New York, 1992, 68.
6. Sidell, F.R., Clinical considerations in nerve agent intoxication, in *Chemical Warfare Agents,* Somani, S., Ed., Academic Press, New York, 155, 1992.
7. Marrs, T.C., Maynard, R.L., and Sidell, F.R., Organophosphorus nerve agents, in: *Chemical Warfare Agents—Toxicology and Treatment,* Wiley and Sons, New York, 1996, 83.
8. Sidell, F.R., Soman and sarin: Clinical manifestations and treatment of accidental poisoning by organophosphates, *Clin. Toxicol.*, 7, 1, 1974.
9. Dunn, M.A. and Sidell, F.R., Progress in medical defense against nerve agents, *JAMA,* 262, 649, 1989.
10. National Academy of Science, Committee on Toxicology, Possible long-term health effects of short-term exposure to chemical agents. Vol I: Anticholinesterases and anticholinergics, Prepared by Panel on Anticholinesterase Chemicals, Panel on Anticholinergic Chemicals, Committee on Toxicology, National Academy of Science, National Academy Press, Washington, D.C., 1982.
11. Presidential Advisory Committee on Gulf War Veterans' Illnesses: Final Report Washington, D.C., U. S. Government Printing Office, December 1996.
12. *Annual Report to Congress—Federally Sponsored Research on Gulf War Veterans' Illness for 1997, Department of Veterans Affairs,* URL http://www.va.gov/resdev/annual 97.doc
13. U.S. General Accounting Office Report to Congressional Requesters, Chemical Weapons: DoD Does Not Have a Strategy to Address Low-Level Exposures, GAO/NSIAD-98–228, September 1998.
14. Brown, M. and Brix, K., Review of health consequences from high-, intermediate- and low-level exposure to organophosphorus nerve agents, *J. Appl. Toxicol.*, 18, 393, 1998.

15. Roy, D.E., Chronic effects of low-level exposure to anticholinesterases—a mechanistic review, *Toxicol. Lett.,* 102 and 527, 1998.
16. Klaasen, C.D. and Eaton, D.L., Principles of toxicology, in *Casarett and Doull's Toxicology: The Basic Science of Poisons,* 4th Ed., Amdur, M., Doull, J., and Klaasen, C.D., eds., Pergamon Press, New York, 1991.
17. DeMenti, B., Cholinesterase literature review and comment, *Pesticides People Nature,* 1(2), 59, 1999.
18. Davis, D.R., Holland, P., and Reumens, M.J., The relationship between the chemical structures and neurotoxicity of alkyl organophosphorous compounds, *Br. J. Pharmacol.,* 15, 271, 1960.
19. Sidell, F.R. and Hurst, C.G., Long-term health effects of mustard in nerve agents, in *Textbook of Military Medicine—Medical Aspects of Chemical and Biological Warfare,* Zajtchuk, R. and Bellamy, R.F., Eds., Office of The Surgeon General, Department of the Army, Washington, D.C., 1997, 129.
20. Sidell, F.R., Nerve agents, in *Textbook of Military Medicine—Medical Aspects of Chemical and Biological Warfare,* Zajtchuk, R. and Bellamy, R.F., Eds., Office of The Surgeon General, Department of the Army, Washington, D.C., 1997, 129.
21. Grob, D., Harvey, A.M., Langworthy, O.R., and Lilenthal, J.L., The administration of di-isopropyl fluorophosphate (DFP) to man, III. Effect on the central nervous system with special reference to the electrical activity of the brain, *Bull. Johns Hopkins Hosp.,* 81, 257, 1947.
22. Holmes, J.H. and Gaon, M.D., Observations on acute and multiple exposure to anti-cholinesterase agents, *Trans. Am. Clin. Climatol.,* 68, 86, 1956.
23. Grob, D. and Harvey, A.M., Effects in man of the anticholinesterase compound sarin (isopropyl methyl phosphonofluoridate), *J. Clin. Invest.,* 37, 350, 1958.
24. Grob, D. and Harvey, A.M., The effects and treatment of nerve gas poisoning, *Am. J. Med.,* 14, 52, 1953.
25. Bowers, M.B., Goodman, E., and Sim, V.M., Some behavioral changes in man following anticholinesterase administration, *J. Nerv. Ment. Dis.,* 138, 383, 1976.
26. Lekov, D., Dimitrov, V., and Mizkow, Z., Clinical observations of individuals contaminated by a pinacolic ester of methylfluorophosphine acid (soman), *Voenno Meditsinsko Delo,* 4, 47, 1966.
27. Inoue, N., Psychiatric symptoms following accidental exposure to sarin: A case study, *Fukuokaishi Igaku Zasshi,* 86, 373, 1995.
28. Nakajima, T., Ohta, S., Morita, H., Midorikawa, Y., Mimura, S., and Yanagisawa, N., Epidemiological study of sarin poisoning in Matsumoto City, Japan, *J. Epidemiol.,* 8, 33, 1997.
29. Nozaki, H., Aikawa, N., Fujishima, S., Suzuki, M., Shinozawa, Y., Horis, S., and Nogawa, S., A case of VX poisoning and the difference from sarin, *Lancet,* 346, 698, 1995.
30. Burchfiel, J.L., Duffy, F.H., and Sim, V.M., Persistent effects of sarin and dieldrin upon the primate electroencephalogram, *Toxicol. Appl. Pharmacol.,* 35, 365, 1976.
31. Lattal, K.A., Maxey, G.C., and Wilbur, E. M., Effects of a single 1/2 LD_{50} of GB upon delayed response and conditional avoidance tests, Edgewood Arsenal Technical Report 4489, 1971.
32. Duffy, F.H., Burchfiel, J.L., Bartels, P.H., Gaon, M., and Sim, V.M., Long-term effects of an organophosphate upon the human encephalogram, *Toxicol. Appl. Pharmacol.,* 47, 161, 1979.
33. Metcalf, D.R. and Holmes, J.W., EEG, psychological and neurological alterations in humans with organophosphorus exposure, *Ann. N.Y. Acad. Sci.,* 160, 357, 1969.

34. Yokoyama, K., Araki, S., Murata, K., Nishikitani, M., Okumura, T., Ishimatsu, S., and Takasu, N., A preliminary study on delayed vestibulo-cerebellar effects of Tokyo subway sarin poisoning in relation to gender difference: Frequency analysis of postural sway, *J. Occup. Environ. Med.,* 40, 17, 1998a.
35. Yokoyama, K., Araki, S., Murata, K., Nishikitani, M., Okumura, T., Ishimatsu, S., and Takasu, N., Chronic neurobehavioral and central and autonomic nervous system effects of Tokyo subway sarin poisoning, *J. Physiol.* (Paris), 92, 317, 1998b.
36. Nakajima, T., Ohta, S., Morita, H., Midorikawa, Y., Mimura, S., and Yanagisawa, N., Epidemiological study of sarin poisoning in Matsumoto City, Japan, *J. Epidemiol.,* 8, 33, 1998.
37. Hatta, K., Miura, Y., Asukai, N., and Harnabe, Y., Amnesia from sarin poisoning, *Lancet,* 347, 1343, 1996.
38. Sekijima, Y., Morita, H., Shindo, M., Okudera, H., Shibata, T., and Yanagisawa, N., A case of severe sarin poisoning in the sarin attack at Matsumoto—one-year follow-up on the clinical findings and laboratory data, *Rinsho Shinkeijaku (Clin. Neurol.),* 35, 1241, 1995.
39. Yanno, L.V. and Musiychuk, Y.I., Acute poisoning by neuroparalytic agents and its long-term effects, *Meditsina Truda Promyshlennaia Ekologiia,* 6, 5, 1997.
40. Savage, E.P., Keefe, T.J., Mounce, L.M., Heaton, R.K., Lewis, J.A., and Burcar, P.J., Chronic neurological sequelae of acute organophosphate pesticide poisoning, *Arch. Environ. Hlth.,* 43, 38, 1988.
41. Rosenstock, L., Keifer, M., Daniell, W.E., McDonnell, R., and Claypoole, Chronic central nervous system effects of acute organophosphate pesticide, *Lancet,* 338, 223, 1991.
42. Steenland, K., Jenkins, B., Ames, R.G., O'Malley, M., Chrislop, D., and Russo, J., Chronic neurological sequelae in organophosphate pesticide poisoning, *Am. J. Public Health,* 84, 731, 1995.
43. McDonough, J.H. and Shih, T.-M., Neuropharmacological mechanisms of nerve agent-induced seizure and neuropathology, *Neurosci. Behav. Rev.,* 21, 559, 1997.
44. McDonough, J.H., Smith, R.F., and Smith, C.D., Behavioral correlates of soman-induced neuropathology: Deficits in DRL acquisition, *Neurobehav. Toxicol. Teratol.,* 8, 179, 1986.
45. Modrow, H.E. and Jaax, N.K., Effect of soman exposure on the acquisition of an operant alternation task, *Pharmacol. Biochem. Behav.,* 32, 49, 1989.
46. Castro, C.A., Larsen, T., Finger, A.V., Solana, R.P., and McMaster, S.B., Behavioral efficacy of diazepam against nerve agent exposure in rhesus monkey, *Pharmacol. Biochem. Behav.,* 41, 159, 1991.
47. Raffaele, K., Hughey, D., Wenk, G., Olton, D., Modrow, H., and McDonough, J.H., Long-term behavioral changes in rats following organophosphonate exposure, *Pharmacol. Biochem. Behav.,* 27, 407, 1987.
48. Philippens, I.H.C.H.M., Melchers, B.P.C., De Groot, D.M.G., and Wolthuis, O.L., Behavioral performance, brain histology and EEG sequelae after combined atropine/diazepam treatment of soman-intoxicated rats, *Pharmacol. Biochem. Behav.,* 42, 711, 1992.
49. Pearce, D.C., Crofts, H.W., Maggleton, N.G., Redout, D., and Scott, E.A.M., The effects of acutely administered low-dose sarin on cognitive behavior and the electroencephalogram in the common marmoset, *Psychopharmacology,* 13(2), 128, 1999.
50. Roberts, A.C., Robbins, T.W., Everitt, B.J., and Muir, J.L., A specific form of cognitive rigidity following excitotoxic lesions of the basal forebrain in marmosets, *Neuroscience,* 47, 251, 1992.
51. Sohakian, B.J. and Coull, T.J., Tetrahydroaminoacridine (THA) in Alzheimer's Disease: An assessment of attentional and mnemonic function using CANTAB, *Acta Neurol. Scand.,* 149(Suppl.), 29, 1993.

52. Craig, A.B. and Freeman, G., Clinical observations in workers accidentally exposed to "G"-agents, AD003393, Medical Laboratory Research Report 154, Edgewood Arsenal, MD, 1953.
53. Sterri, S.H., Lyngaas, S., and Fonnum, F., Toxicity of soman after repetitive injection of sublethal doses in rats, *Acta Pharmacol. Toxicol.,* 46, 1, 1980.
54. Sterri, S.H., Lyngaas, S., and Fonnum, F., Toxicity of soman after repetitive injection of sublethal doses in guinea pig and mouse, *Acta Pharmacol. Toxicol.,* 49, 8, 1981.
55. Hartgraves, S.L. and Murphy, M.R., Behavioral effects of low-dose nerve agents, in *Chemical Warfare Agents,* Somani, S., Ed., Academic Press, New York, 1992.
56. Olson, C.T., Blank, J.A., Kinney, P.H., and Singer, A.W., Neurologic assessment of rats following low doses of sarin, pyridostigmine, chlorpyrifos and DEET, *Toxicology,* 54(1), 265, 2000.
57. McDonough, J.H., Shih, T.-M., Kaminskis, S.A., Jackson, J., and Alvarez, R., Depression and recovery of rat blood and brain acetylcholinesterase activity after repeated exposure to soman, *Soc. Neurosci. Abstr.,* 9, 964, 1983.
58. Russell, R.W., Booth, R.A., Lauretz, S.D., Smith, C.A., and Jenden, D.J., Behavioral, neurochemical, and physiological effects of repeated exposure to subsymptomatic levels of the anticholinesterase, soman, *Neurobehav. Toxicol. Teratol.,* 8, 675, 1986.
59. Clement, J. G., Survivors of soman poisoning: recovery of the soman LD_{50} to control value in the presence of extensive acetylcholinesterase inhibition, *Arch. Toxicol.,* 63, 150, 1989.
60. Clement, J.G., Hypothermia: limited tolerance to repeated soman administration and cross-tolerance to oxotremorine, *Pharmacol. Biochem. Behav.,* 39, 305, 1991.
61. Benschop, H.P., Trap, H.C., Spruit, H., van der Wiel, H., Langenberg, J.P., and DeJong, L.P.A., Low-level nose-only exposure to the nerve agent soman: Atropinized guinea pigs. *Toxicol. Appl. Pharmacol.,* 153, 179, 1998.
62. Sim, V.M. Anticholinesterase poisoning, in Waser, P.G., Ed., *Cholinergic Mechanisms,* Raven Press, New York, 1975.
63. Hymowitz, N., Brezenoff, H.E., McGee, J., Campbell, K., and Kargut, V., Effects of repeated intraperitoneal injections of soman on schedule-controlled behavior in the rat, *Psychopharmacology,* 86, 404, 1985.
64. van Dongen, C.J. and Wolthuis, O.L., On the development of behavioral tolerance to organophosphates, I. Behavioral and biochemical aspects, *Pharmacol. Biochem Behav.,* 34, 473, 1989.
65. Wolthuis, O.L., Philippens, I.H.C.H.M., and Vanwersch, R., On the development of behavioral tolerance to organophosphates, III: Behavioral aspects, *Pharmacol. Biochem. Behav.,* 35, 561, 1990.
66. Churchill, L., Pazdernik, T.L., Jackson, J.L., Nelson, S.R., Samson, F.E., and McDonough, J.H., Topographical distribution of decrements and recovery in muscarinic receptors from rat brains repeatedly exposed to sublethal doses of soman, *J. Neurosci.,* 4, 2069, 1984.
67. Modrow, H.E. and McDonough, J.H., Changes in atropine dose effect curve after subacute soman administration, *Pharmacol. Biochem. Behav.,* 24, 845, 1986.
68. Russell, R.W. and Overstreet, D.H., Mechanisms underlying sensitivity to organophosphorus anticholinesterase compounds, *Prog. Neurobiol.,* 28, 97, 1987.
69. Young, R.A., Opresko, D.M., Watson, A.D., Ross, R.H., King, J., and Choudhoury, H., Deriving toxicity values for organophosphate nerve agents: A position paper in support of procedures and rationale for deriving Oral RfDs for chemical warfare agents, *Human Ecolog. Risk Assess.,* 5(3), 589, 1999.
70. Bignami, G., Giardini, V., and Scorrano, M., Behaviorally augmented versus other components in organophosphate tolerance: The role of reinforcement and response factors. *Fundam. Appl. Toxicol.,* 5, S213, 1985.

71. Munro, N.B., Ambrose, K.R., and Watson, A.P., Toxicity of the organophosphate chemical warfare agents GA, GB and VX: Implications for public protection, *Environ. Hlth. Perspect.,* 102(1), 18, 1994.
72. Korsak, R.J. and Sato, M.M., Effects of chronic organophosphate pesticide exposure on the central nervous system, *Clin. Toxicol.,* 11, 83, 1977.
73. Stephens, R., Spurgeon, A., Calvert, I.A., Beach, J., Levy, L.S., Berry, H., and Harrington, J.M., Neuropsychological effects of long-term exposure to organophosphates in sheep dip, *Lancet,* 345, 1135, 1995.
74. Ames, R., Steenland, K., Jenkins, B., Chrislop, D., and Russo, J., Chronic neurologic sequelae to cholinesterase inhibition among agriculture pesticide applicators, *Arch. Environ. Hlth.,* 50, 440, 1995.
75. Rodnitsky, R.L., Levin, H.S., and Mick, D.L., Occupational exposure to organophosphate pesticides: A neurobehavioral study, *Arch. Environ. Hlth.,* 30, 98, 1975.
76. Daniell, W., Barhart, S., Demers, P., Costa, L.G., Eaton, D., Miller, M., and Rosenstock, L., Neuropsychological performance among agricultural pesticide applicators, *Environ. Res.,* 59, 217, 1992.
77. Maizlish, N., Schenker, M., Weisskopf, C., Seiber, J., and Samuels, S., A behavioral evaluation of pest control workers with short-term, low-level exposure to the organophosphate Diazinon, *Am. J. Indust. Med.,* 12, 153, 1987.
78. Bazylewicz-Walczak, B., Majczakowa, W., and Szymczak, M., Behavioral effects of occupational exposure to organophosphate pesticides in female greenhouse planting workers, *Neurotoxicology,* 20, 819, 1999.
79. Paton, W.D.M., Vizi, E.S., and Zar, M.A., The mechanism of acetylcholine release from parasympathetic nerves, *J. Physiol.* (Lond.), 215, 819, 1971.
80. Aas, P., Veiteberg, T., and Fonnum, F., *In vitro* effects of soman on bronchial smooth muscle, *Biochem. Pharmacol.,* 35, 1793, 1986.
81. Karczmar, A.G., Neuromuscular pharmacology, *Ann. Rev. Pharmacol.,* 7, 241, 1967.
82. Faff, J., Rabsztyn, T., and Rump, S., Investigation on the correlation between abnormalities of neuromuscular transmission due to some organophosphates and activity of acetylcholinesterase in the skeletal muscle, *Arch. Toxicol.,* 31, 31, 1973.
83. French, M.C., Wetherell, J.R., and White, P.D.T., The reversal by pyridostigmine of neuromuscular block produced by soman, *J. Pharm. Pharmacol.* 31, 290, 1979.
84. Smith, A.P. and Wolthius, O.L., HI-6 as an antidote to soman poisoning in rhesus monkey respiratory muscles *in-vitro, J. Pharm. Pharmacol.,* 35, 157, 1983.
85. van Helden, H.P.M., van der Wiel, H.J., and Wolthius, O.L., Therapy of organophosphate poisoning: The marmoset as a model for man, *Br. J. Pharmacol.,* 78, 579, 1983.
86. Barnes, J.M. and Duff, J.L., The role of cholinesterase in the myoneural junction, *Br. J. Pharmacol.,* 8, 334, 1953.
87. Barstad, J.A.B., Cholinesterase inhibition and the effect of anticholinesterases on indirectly evoked single and tetanic muscle contractions in the phrenic nerve-diaphragm preparation from the rat, *Arch. Int. Pharmacodyn.,* 128, 143, 1960.
88. Kuba, K., Albuquerque, E.X., Daly, J.W., and Barnard, E.A., A study of the irreversible cholinesterase inhibitor diisopropylfluorophosphate on the time course of end plate currents in frog sartorius muscle, *J. Pharmacol. Exp. Ther.,* 189, 499, 1974.
89. Karczmar, A.G. and Ohta, Y., Neuromyopharmacology as related to anticholinesterase action, *Fund. Appl. Toxicol.,* 1, 135, 1981.
90. Tattersall, J.E.H., Effects of organophosphorus anticholinesterases on nicotinic receptor ion channels at adult mouse muscle endplates, *Br. J. Pharmacol.,* 101, 349, 1990.

91. Katz, E.J., Cortes, V.I., Eldefrawi, M.E., and Eldefrawi, A.T., Chlorpyrifos, parathion and their oxons bind to and desensitize a nicotinic acetylcholine receptor: Relevance to their toxicities, *Toxicol. Appl. Pharmacol.,* 146, 227, 1997.
92. Albuquerque, E.X., Deshpande, S.S., Kawabuchi, M., Aracava, Y., Idriss, M., Rickett, D.L., and Boyne, A.F., Multiple actions of anticholinesterase agents on chemosensitive synapses: Molecular basis for prophylaxis and treatment of organophosphate poisoning, *Fund. Appl. Toxicol.,* 5, S182, 1985.
93. Holaday, D.A., Kamijo, K., and Koelle, G.B., Facilitation of ganglionic transmission following inhibition of cholinesterase by DFP, *J. Pharmacol. Exp. Ther.,* 111, 241, 1954.
94. Yarowsky, P., Fowler, J.C., Taylor, G., and Weinreich, D., Noncholinesterase actions of an irreversible acetylcholinesterase inhibitor on synaptic transmission and membrane properties in autonomic ganglia, *Cell. Molec. Neurobiol.,* 4, 351, 1984.
95. Heppner, T.J. and Fiekers, J.F., The effects of irreversible acetylcholinesterase inhibitors on transmission through sympathetic ganglia of the bullfrog, *Neuropharmacology,* 30, 843, 1991.
96. Grillner, P., Bonci, A., Svensson, T.H., Bernardi, G., and Mercuri, N.B., Presynaptic muscarinic (M3) receptors reduce excitatory transmission in dopamine neurons of the rat mesencephalon, *Neuroscience,* 91, 1999.
97. Rocha, E.S., Santos, M.D., Chebabo, S.R., Aracava, Y., and Albuquerque, E.X., Low concentrations of the organophosphate VX affect spontaneous and evoked transmitter release from hippocampal neurons: Toxicological relevance of cholinesterase-independent actions, *Toxicol. Appl. Pharmacol.,* 159, 31, 1999.
98. Katz, L.S. and Marquis, J.K., Modulation of central muscarinic receptor binding *in vitro* by ultra low levels of the organophosphate paraoxon, *Toxicol. Appl. Pharmacol.,* 101, 114, 1989.
99. Bakry, N.M., el-Rashidy, A.H., Eldefrawi, A.T., and Eldefrawi, M.E., Direct actions of organophosphate anticholinesterases on nicotinic and muscarinic acetylcholine receptors, *J. Biochem. Toxicol.,* 3, 235, 1988.
100. Viana, G.B., Davis, L.H., and Kauffman, F.C., Effects of organophosphates and nerve growth factor on muscarinic receptor binding number in rat pheochromocytoma PC12 cells, *Toxicol. Appl. Pharmacol.,* 93, 257, 1988.
101. Abdallah, E.A., Jett, D.A., Eldefrawi, M.E., and Eldefrawi, A.T., Differential effects of paraoxon on the M3 muscarinic receptor and its effector system in rat submaxillary gland cells, *J. Biochem. Toxicol.,* 7, 125, 1992.
102. Ehrich, M., Intropido, L., and Costa, L.G., Interaction of organophosphorus compounds with muscarinic receptors in SH-SY5Y human neuroblastoma cells, *J. Toxicol. Environ. Health,* 43, 51, 1994.
103. Katz, L.S. and Marquis, J.K., Organophosphate-induced alterations in muscarinic receptor binding and phosphoinositide hydrolysis in the human SK-N-SH cell line, *Neurotoxicology,* 13, 365, 1992.
104. Silveira, C.L., Eldefrawi, A.T., and Eldefrawi, M.E., Putative M2 muscarinic receptors of rat heart have high affinity for organophosphorus anticholinesterases, *Toxicol. Appl. Pharmacol.,* 103, 474, 1990.
105. Baumgold, J., Cooperman, B.B., and White, T.M., Relationship between desensitization and sequestration of muscarinic cholinergic receptors in two neuronal cell lines, *Neuropharmacology,* 28, 1253, 1989.
106. Idriss, M.K., Aguayo, L.G., Rickett, D.L., and Albuquerque, E.X., Organophosphate and carbamate compounds have pre- and postjunctional effects at the insect glutamatergic synapse, *J. Pharmacol. Exp. Ther.,* 239, 279, 1986.

107. Heppner, T.J. and Fiekers, J.F., VX enhances neuronal excitability and alters membrane properties of *Rana catesbeiana* sympathetic ganglion neurons, *Comp. Biochem. Physiol.*, 102C, 335, 1992.
108. Léna, C. and Changeux, J.-P., Allosteric nicotinic receptors, human pathologies, *J. Physiol.* (Paris), 92, 63, 1998.
109. McGehee, D.S., Heath, M.J.S., Gelber, S., Devay, P., and Role, L.W., Nicotine enhancement of fast excitatory synaptic transmission in CNS by presynaptic receptors, *Science,* 269, 1692, 1995.
110. Alkondon, M., Pereira, E.F.R., Eisenberg, H.M., and Albuquerque, E.X., Choline and selective antagonists identify two subtypes of nicotinic acetylcholine receptors that modulate GABA release from CA1 interneurons in rat hippocampal slices, *J. Neurosci.,* 19, 2693, 1999.
111. Alkondon, M., Pereira, E.F.R., Barbosa, C.T.F., and Albuquerque, E.X., Neuronal nicotinic acetylcholine receptor activation modulates gamma-aminobutyric acid release from CA1 neurons of rat hippocampal slices, *J. Pharmacol. Exp. Ther.,* 283, 1396, 1997.

2 Toxicokinetics of Nerve Agents

Hendrik P. Benschop and Leo P. A. De Jong

CONTENTS

I. Introduction 25
II. Nerve Agent Stereoisomers: Chiral Analysis, Isolation, and Toxicology 26
III. Trace Analysis of Nerve Agents in Biological Samples 30
IV. Intravenous Toxicokinetics of Soman and Sarin in Various Species 33
V. Subcutaneous Toxicokinetics of Soman 42
VI. Inhalation Toxicokinetics of Soman and Sarin 45
VII. Inhalation Toxicokinetics of Soman upon Low-Level Exposure 52
VIII. Elimination Pathways of Phosphofluoridates 57
A. Elimination by Hydrolytic Degradation 58
B. Elimination by Covalent Binding 59
C. Renal Excretion 62
D. Elimination Products as Tools for Retrospective Detection of Exposure 63
IX. Physiologically-Based Modeling of the Toxicokinetics of Soman 64
X. The Influence of Prophylaxis and Therapy upon the Toxicokinetics of Soman 69
XI. Toxicokinetics of V Agents 72
XII. Future Directions 74
References 74

I. INTRODUCTION

Toxicokinetic studies of nerve agents deal with the *in vivo* absorption, distribution, and elimination of these agents as a function of animal species, route of administration, dose, and time after administration. Such studies are essential to provide a quantitative basis for the toxicology of nerve agents and, in combination with toxicodynamic studies, are the starting point for development of causal treatment of intoxications with these agents. Toxicodynamic studies of nerve agents have been

0-8493-0872-0/01/$0.00+$.50

the subject of a long and rich tradition of investigations since their introduction as potential agents of chemical warfare during World War II. These studies have led to (e.g., the development of prophylaxis of intoxication based on partial inhibition of cholinesterase activity with carbamates and therapy of intoxication through administration of the muscarinic cholinergic antagonist atropine) reactivation of phosphylated cholinesterases with oximes, often in combination with the administration of a central nervous depressant in order to suppress convulsions and other central effects.

Toxicokinetic studies of nerve agents were initiated in the last two decennia of the twentieth century. The reasons for this relatively late development were twofold. First, it was often assumed that nerve agents, especially at supralethal doses, act so quickly and are so rapidly degraded *in vivo* that toxicokinetic studies were not relevant for treatment of intoxications, i.e., nerve agents should be regarded as so-called "hit and run poisons." Second, it was intuitively assumed that *in vivo* concentrations of the extremely toxic nerve agents are too low for bioanalysis. However, Wolthuis et al.[1] showed in 1981 that rats initially surviving a challenge with a supralethal dose of soman by immediate treatment with atropine and the oxime HI-6 became fatally re-intoxicated 4–6 h later. Hence, soman appeared to be far more persistent than previously assumed. This suggested that toxicokinetic investigations of nerve agents are toxicologically relevant, especially in view of the refractoriness of intoxication with these agents towards treatment. Moreover, the development of analytical techniques, particularly for gas chromatographic analyses of the rather volatile nerve agents, had evolved to a level that detection limits of a few picograms (10^{-12} g) of these agents became feasible. Finally, the development of chiral gas chromatography opened up the possibility to analyze the separate stereoisomers of nerve agents, which is a *conditio sine qua non* for toxicological interpretation of toxicokinetic studies of nerve agents see (Section II).

II. NERVE AGENT STEREOISOMERS: CHIRAL ANALYSIS, ISOLATION, AND TOXICOLOGY

Interpretation and understanding of the toxicokinetics of nerve agents would not be possible without taking into consideration that these agents consist of mixtures of stereoisomers, which are often extremely different in their toxicokinetic and toxicodynamic properties. A common feature of these agents is the presence of chirality (asymmetry) around the phosphorus atom. Therefore, O-isopropyl methylphosphonofluoridate (sarin) and O-ethyl S-(2-diisopropylaminoethyl) methylphosphonothioate (VX) consist of equal amounts of stereoisomers, denoted as (+)- and (−)-sarin and (+)- and (−)-VX, respectively. In the case of O-1,2,2-trimethylpropyl methylphosphonofluoridate (soman), an additional chiral center resides in the 1,2,2-methylpropyl (pinacolyl) moiety, leading to the presence of four stereoisomers. Synthetic soman, i.e., a mixture of the four stereoisomers, is denoted as C(±)P(±)-soman, whereas the individual four stereoisomers are denoted as C(+)P(+), C(+)P(−), C(−)P(+), and C(−)P(−), in which C stands for chirality in the pinacolyl moiety and P for chirality around phosphorus. The enantiomeric pairs [C(+)P(+) + C(−)P(−)] and [C(+)P(−) + C(−)P(+)] are present in synthetic

C(±)P(±)-soman in a ratio of 45:55, with equal amounts of the two enantiomers within each pair.

Separation of the various stereoisomers of the nerve agents for analytical purposes became feasible with the advent of optically active coating materials for columns as used in capillary gas chromatography (GC) and in high performance liquid chromatography (HPLC). The complete separation of the four stereoisomers of soman and of the two stereoisomers of sarin with GC on capillary columns is described in Section III. So far, (+)- and (−)-VX could not be separated by means of capillary gas chromatography, but HPLC on a so-called Chiralcel OD-H column yields complete separation of the two stereoisomers of this agent.[2]

Using these analytical procedures to monitor progress, the four stereoisomers of soman, as well as (−)-sarin, could be isolated on a mg-scale for toxicological purposes by using judicious combinations of synthetic and enzymatic separation techniques.[3–5] In the case of C(±)P(±)-soman, synthetic resolution of the stereoisomers of (±)-pinacolyl alcohol and subsequent synthesis of soman from these stereoisomers gave C(+)P(±)-soman and C(−)P(±)-soman, i.e., two diastereoisomeric mixtures of two soman stereoisomers.[4] These pairs were separated enzymatically by incubation of C(+)P(±)- and C(−)P(±)-soman with α-chymotrypsin, which binds the P(−)-stereoisomers of soman. In this way C(+)P(+)- and C(−)P(+)-soman, respectively, could be isolated. Incubation with rabbit plasma hydrolyzes the P(+)-stereoisomers and provides therefore C(+)P(−)- and C(−)P(−)-soman. Similarly, incubation of (±)-sarin with α-chymotrypsin gave optically pure (−)-sarin. The two stereoisomers of VX are easily obtained synthetically from optically resolved precursors.[5] X-ray analysis of the (+)-enantiomer of the O-ethyl analog of VX established the absolute configuration R for this analog and for (+)-VX,[5,6] whereas the R-configuration for the P(+)-stereoisomers of sarin and soman has been determined beyond reasonable doubt by means of chemical correlation reactions with the abovementioned precursors of V agents and from interaction of the most actively inhibiting stereoisomer with the active site of acetylcholinesterase (AChE).[7–9]

With sufficient amounts of the various stereoisomers of the major nerve agents available, it became feasible to investigate the acute lethality of these stereoisomers.[10-12] *A priori,* it should be expected that the degree of lethality will correlate with the inhibitory potency towards AChE. Therefore, bimolecular rate constants of inhibition of AChE with these stereoisomers were measured, as well as their LD_{50} values in mice. A summary of the results is given in Table 2.1. Apparently, the P(−)-stereoisomers of soman and sarin inhibit AChE with rate constants which are 3–4 orders of magnitude higher than those of the corresponding P(+)-stereoisomers. At the time of these investigations, only the upper limit for the rate constants of the P(+)-stereoisomers could be determined due to the presence of trace amounts of the P(−)-stereoisomers. Concomitantly, it appeared that the P(−)-stereoisomers of soman are at least two orders of magnitude more acutely lethal than the P(+)-counterparts. For practical purposes the difference in acute lethality is such that the P(+)-stereoisomers should be regarded as a nontoxic impurity in synthetic soman, taking into consideration that the lower limit for the acute lethality of the P(+)-stereoisomers is difficult to determine in view of possible *in vivo* racemization. The same extreme differences will probably

TABLE 2.1
Stereoselectivity in Anticholinesterase Activity and Acute Lethality of Nerve Agent Stereoisomers

Nerve Agent Stereoisomer	Rate Constant for Inhibition of AChE[a] (M^{-1} min^{-1})	LD_{50} Mouse (μg/kg)	Ref.
C(+)P(−)-soman	2.8×10^8	99[b]	9
C(−)P(−)-soman	1.8×10^8	38[b]	9
C(+)P(+)-soman	$<5 \times 10^3$	>5000[b]	9
C(−)P(+)-soman	$<5 \times 10^3$	>2000[c]	9
C(±)P(±)-soman		156[b]	9
(−)-sarin	1.4×10^7	41[c]	5, 10
(+)-sarin	$<3 \times 10^3$[d]		10
(±)-sarin		83[c]	5
(−)-VX	4×10^8	12.6[c]	5, 11
(+)-VX	2×10^6	165[c]	5, 11
(±)-VX		20.1[c]	5

[a]Electric eel AChE (pH 7.5, 25°C) for soman stereoisomers; bovine erythrocyte AChE for sarin and VX stereoisomers (pH 7.7, 25°C).

[b]Subcutaneous administration.

[c]Intravenous administration.

[d]Estimated from an experiment with optically enriched sarin (64% enantiomeric excess).

Source: From Benschop, H.P. and De Jong, L.P.A., *Acc. Chem. Res.*, 21, 368, 1988. With permission.

hold for (+)- and (−)-sarin, although this cannot be made explicit since methods to isolate optically pure (+)-sarin are not yet available.

In contrast to soman and sarin, the rate of inhibition of AChE by (+)-VX is only two orders of magnitude less than that of the (−)-stereoisomer. In this case the LD_{50} of the (+)-stereoisomer could also be determined, revealing that (−)-VX is only 8-fold more acutely lethal than the (+)-stereoisomer.

Very recently, the P(+)-stereoisomers of soman could be exhaustively purified and the rate constants for inhibition of human AChE were determined.[9] As shown in Table 2.2, the rates of inhibition of the P(−)- and P(+)-stereoisomers differ by 4–5 orders of magnitude, i.e., even one order of magnitude more than estimated previously for electric eel AChE (see Table 2.1). These data in combination with the absolute configuration of the soman stereoisomers, the detailed three-dimensional structure of the active site of human AChE based on X-ray analysis, and molecular modeling were used to create a detailed model of the Michaelis complex for the inhibition of the enzyme by the soman stereoisomers, the stability of which should be regarded as a reflection of the reactivity of the stereoisomer.[9] Figure 2.1 gives these models for the C(−)P(−)- and C(−)P(+)-stereoisomers of soman. A detailed discussion of the interactions of the stereoisomers with the catalytic system of the active

TABLE 2.2
Rate Constants of Phosphonylation (ki) of Human AChE by the Stereoisomers of C(±)P(±)-soman (pH 8.0, 24°C)

k_i (10^4 $M^{-1}min^{-1}$)			
C(+)P(−)-soman	C(−)P(−≳)-soman	C(+)P(+)-soman	C(−)P(+)-soman
15,000 ± 3,000	8,000 ± 400	0.2 ± 0.1	0.2 ± 0.1

Source: Data from Ref. 9.

site is beyond the scope of this book. However, it should be noted that the P = O bond has to be polarized by interactions in the so-called oxyanion hole. These conditions determine the position of the substituents in the pinacolyl moiety of the soman stereoisomers. It appears that the extremely low reactivity of the P(+)-stereoisomers is due to steric constraints which prevent accomodation of the bulky t-butyl group in the pinacolyl moiety and practically exclude it from the acyl pocket.

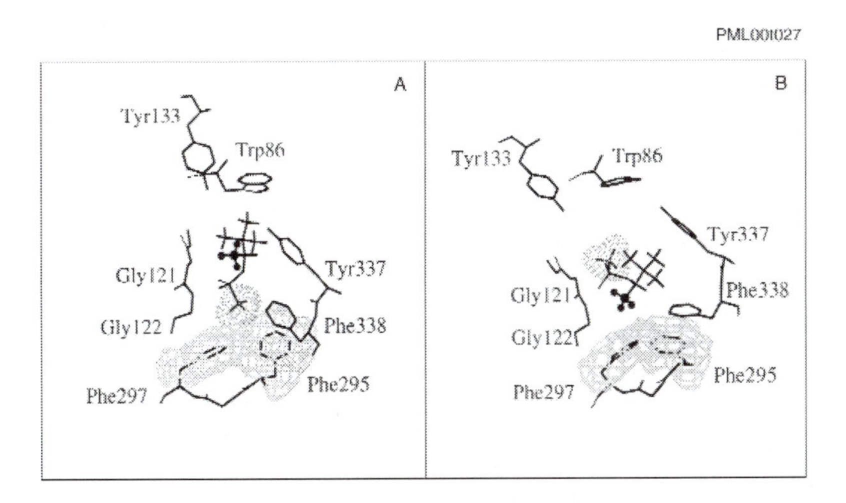

FIGURE 2.1 Michaelis complexes of human AChE with (a) C(−)-P(−) and (b) C(−)P(+) stereoisomers of soman. Only amino acids adjacent to the inhibitor are shown, while hydrogen atoms of the protein are omitted for clarity. The soman C_α-methyl substituent is displayed as balls and sticks. Molecular volumes of the phosphorus methyl substituent and of the aromatic moieties of residues Phe295 and Phe297 are shown with dots and grids, respectively. Note that in the C(−)P(+)-soman–AChE complex, the acyl pocket cannot accommodate the bulkyl tert-butyl portion of the C(−)P(+)-soman alkoxy moiety and it points away from the phenyl groups defining this acyl pocket. (From Ordentlich, A., Barak, D., Kronman, C., Benschop, H.P., De Jong, L.P.A., Ariel, N., Barak, R., Segall, Y., Velan, B., and Shafferman, A., *Biochemistry,* 38, 3055, 1999. With permission.)

III. TRACE ANALYSIS OF NERVE AGENTS IN BIOLOGICAL SAMPLES

Toxicokinetic investigations of nerve agents are only relevant if these agents can be analyzed at minimum levels in blood or tissue samples which are still toxicologically significant. Such relevance in case of anticholinesterases should be related to their capacity to inhibit the enzyme AChE. Since nerve agents inhibit this enzyme with rate constants up to 10^8–10^9 $M^{-1}min^{-1}$ (see Section II), it can be derived that blood levels down to a few picograms per milliliter blood (approximately 10^{-10} M) can still cause significant inhibition over a period of time of several hours. Obviously, the need for such extremely low minimum detectable concentrations requires analytical procedures which provide the utmost detection limits and selectivity. Moreover, as elucidated in the previous paragraph, differential analysis of the various stereoisomers of a nerve agent is also required.

Early attempts to investigate the *in vivo* disposition of the nerve agents (±)-sarin and C(±)P(±)-soman, and their metabolites were based on the use of ^{3}H-labeled agents.[13–15] Although this approach affords sufficiently low minimum detectable concentrations, the separation of intact nerve agent and metabolites was based solely on liquid extraction schemes, which provide insufficient selectivity. In principle, the approach of Harris et al. based on the measurement of inhibition of bovine AChE added to the sample to be analyzed will provide concentrations of nerve agent stereoisomers having high anticholinesterase activity.[16] This approach was mentioned in a preliminary paper dealing with the toxicokinetics of C(±)P(±)-soman, (±)-VX, and the VX-analog O-cyclopentyl S-diethylaminoethyl methylphosphonothiate in mechanically respirated rabbits and cynomolgus monkeys at very high doses of the nerve agents (10–30 LD_{50}).[16]

The relatively volatile nature of nerve agents, the extremely low detection limits of modern detectors for gas chromatography, and the recent advances in chiral separation in gas chromatography have led to the extensive use of this technique for toxicokinetic investigations of nerve agents. Primarily, the procedure was developed for analysis of the four stereoisomers of soman, based on separation of these stereoisomers on a capillary column coated with a derivative of L-valine bound to a siloxane backbone (Chirasil-L-Val).[3,4] As shown in Figure 2.2, this column separates the C(+)P(+)- and C(−)P(+)-stereoisomers of soman perdeuterated in the pinacolyl moiety from the four stereoisomers of soman. Hence, the deuterated stereoisomers are highly useful internal standards for quantitation of soman stereoisomers, without resorting to the use of expensive mass spectrometric detection systems. Instead, highly sensitive alkali flame (NPD) and pulsed flame photometric (PFPD) detectors can be used with absolute detection limits for nerve agents of 1–5 pg. This is approximately one order of magnitude higher than can be obtained with single ion detection in tandem mass spectrometric detection, for which a detection limit of 0.1 pg is reported.[17] In order to further increase the selectivity of the analytical procedure, a two-dimensional system was introduced in which a cut containing the analytes is trapped from a precolumn, e.g., a CPSil 8CB column, into a cold trap from which this cut is re-injected onto the chiral column by means of flash heating.

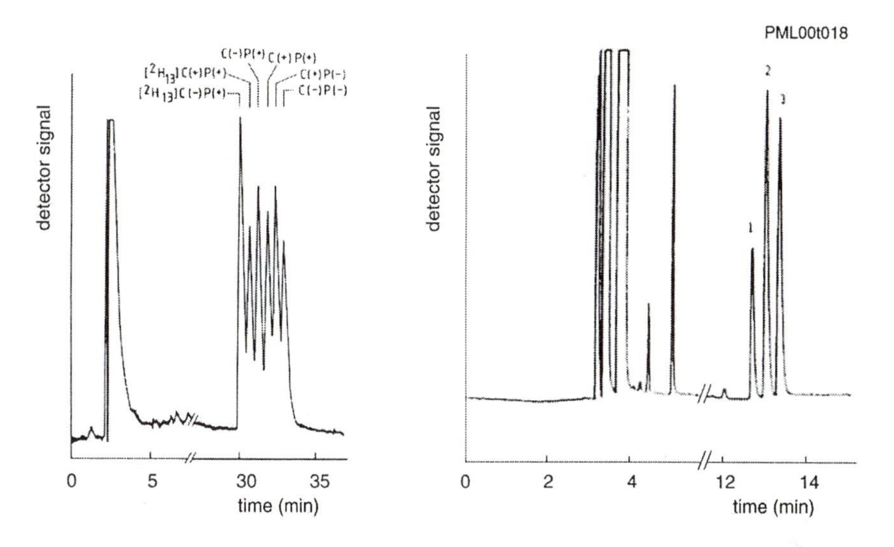

FIGURE 2.2 Gas chromatographic separation (left panel) of the four stereoisomers of soman and two deuterated stereoisomers on a Chirasil-L-Val column and (right panel) of the two stereoisomers of sarin (Peaks 2 and 3) and one deuterated stereoisomer (Peak 1) on a CycloDexB column. The deuterated stereoisomers are used as internal standards for quantitation of the stereoisomers. (Left panel from Benschop, H.P., Bijleveld, E.C., Otto, M.F., Degenhardt, C.E.A.M., Van Helden, H.P.M., and De Jong, L.P.A., *Anal. Biochem.,* 151, 242, 1985. With permission.)

The pronounced volatility of nerve agents, especially of sarin and soman, prevents their concentration into a sufficiently small sample volume of ca. 1–5 μl for injection into the two-dimensional gas chromatograph. Therefore, an on-line large volume injection system was introduced based on the application of the analytes in an organic solvent, up to a volume of 500 μl on Tenax absorption material, from which the solvent is blown off selectively. Next, the analytes are thermally desorbed from the Tenax absorbent into a cold trap for subsequent injection into the two-dimensional system by means of flash heating. A schematic drawing of the complete analytical system is given in Figure 2.3.

For analysis of the stereoisomers of sarin, the optically active Chirasil-L-Val column is replaced by an optically active CyclodexB column coated with β-cyclodextrin.[18] As shown in Figure 2.2, this column separates the two stereoisomers of sarin from the two stereoisomers of sarin which are perdeuterated in the isopropyl moiety. Therefore, these deuterated stereoisomers are convenient internal standards for analysis of sarin stereoisomers in biological samples.

When properly installed, the analytical system involving thermodesorption cold trap injection and two-dimensional chromatography can be used routinely for analysis of the stereoisomers of soman and sarin in blood and tissue samples at minimum detectable concentrations of 1–5 pg of stereoisomer per ml blood or gram tissue. In recent toxicokinetic experiments in pigs,[17] the analytical system comprised chiral gas chromatography on a Chirasil-L-Val column with splitless injection and detection

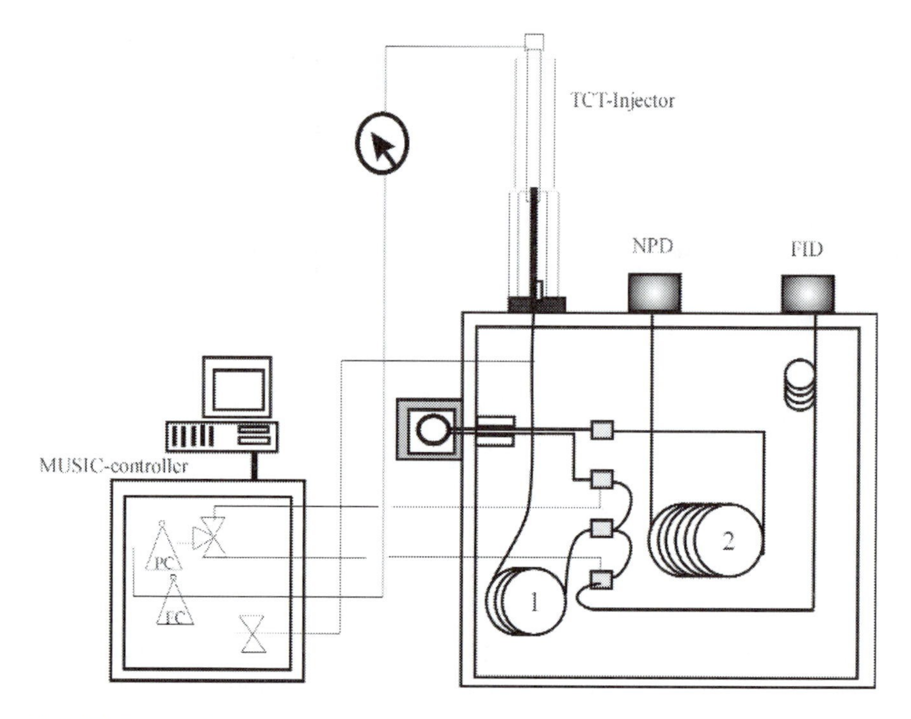

FIGURE 2.3 Scheme for analysis of nerve agent stereoisomers with a two-dimensional (MUSIC) gas chromatographic system with thermodesorption-cold trap (TCT) injection: 1, pre-column; 2, analytical column; PC and FC, constant pressure and constant flow controllers, respectively; NPD and FID, detectors.

with a hybrid tandem mass spectrometer in the single ion mode, yielding approximately the same minimum detectable concentration as the two-dimensional system.

In vivo, the stereoisomers of soman and sarin are subject to rapid processes of elimination, including spontaneous and enzymatically catalyzed hydrolysis and phosphylation of protein binding sites (see Section VIII). These processes should be "frozen" at the moment that the sample is taken for a period of time that is sufficient for further work-up. Stabilization procedures of nerve agent stereoisomers in biological samples were developed with stringent validation for all separate stereoisomers since these have widely differing rates of degradation.[4] It appeared that spontaneous and enzymatic hydrolysis of nerve agent stereoisomers can be sufficiently suppressed by immediate acidification of the sample to pH 4 with an acetate buffer. This was validated by adding known amounts of soman to rat blood samples that had been preincubated with soman in order to "saturate" irreversible binding sites and from which excess of soman had been removed. However, it then appeared that fluoride ions in the blood, present either from natural sources or from hydrolysis of soman, reactivated soman from phosphylated binding sites such as carboxylesterases (CaE) which led to substantially higher levels of soman in the samples than added for the purpose of validation. This complication was effectively suppressed by addition of aluminum sulfate which binds fluoride ions, mostly in the complex [AlF_2^+]. Finally, binding of soman stereoisomers to unoccupied phosphylation sites, which can be an overriding phenomenon when investigating the toxicokinetics at low level exposures, is blocked effectively by adding a large excess of O-neopentyl methylphosphonofluoridate (neopentyl sarin). This agent saturates the unoccupied binding sites without interfering with the gas chromatographic analysis.[4]

The combined use of acidification to pH 4, addition of aluminum sulfate and of neopentyl sarin, proved to be sufficient to stabilize the stereoisomers of sarin and soman. In subsequent work-up, the analytes and internal standard are extracted from the stabilized blood or tissue sample by means of solid-phase extraction and elution with ethyl acetate, for gas chromatographic analysis. The same work-up and analytical procedure can be used for homogenized brain and diaphragm tissue samples.

IV. INTRAVENOUS TOXICOKINETICS OF SOMAN AND SARIN IN VARIOUS SPECIES

Initial investigations on the toxicokinetics of nerve agents were performed after intravenous (i.v.) administration of doses corresponding with multiple LD_{50} values. This route of administration provides basic toxicokinetic data, which can subsequently be compared with results for more realistic routes of administration, e.g., the subcutaneous (s.c.), percutaneous (p.c.), and respiratory routes. With gradually improving methods of bioanalysis, the administered doses could be lowered. Nevertheless, data obtained at multiple LD_{50} values are highly relevant since these pertain to exposure scenarios where immediate medical treatment of casualties should be applied.

Animal species selected for initial investigations were rats, guinea pigs, and marmosets, with the latter species serving as a primate model for man. In order to perform toxicokinetic measurements at high doses, the anesthetized* animals were provided with a tracheal cannula for artificial respiration and with a carotid cannula. Shortly before administration of nerve agent in the dorsal penis vein, the animals were atropinized intraperitoneally (i.p.) and blood samples were taken from the carotid cannula at various points of time after intoxication for analysis of nerve agent stereoisomers. Blood levels of the individual soman stereoisomers in rats and guinea pigs were measured at each time point randomly in at least six animals, while complete toxicokinetic curves were measured in each individual marmoset.

The LD_{50} values of C(±)P(±)-soman are highly species-dependent, since the amount of CaE in the blood is species-dependent. These enzymes act as scavengers of nerve agents by means of irreversible binding and are present in large amounts in the blood of rats, in significantly smaller amounts in guinea pigs, and are almost absent in the blood of marmosets. Accordingly, the LD_{50} values decrease in the order rat > guinea pig > marmoset.

Blood levels in rats of the relatively nontoxic C(−)P(+)-stereoisomer at doses of 6 and 3 LD_{50} of C(±)P(±)-soman are shown in Figure 2.4.[19] It was observed that

*Sodium barbital/sodium hexobarbital (i.p.).

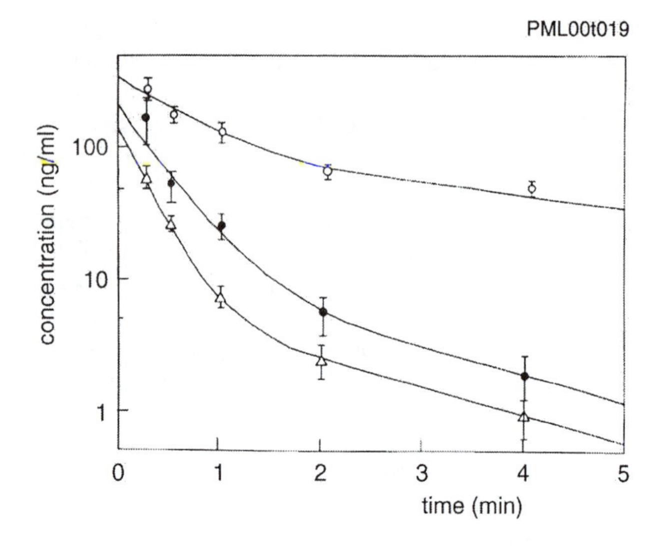

FIGURE 2.4 Semilogarithmic plot of the concentrations in blood (±s.e.m; n = 6) of C(−)P(+)-soman vs. time after i.v. administration of 3 (Δ, 248 μg/kg) and 6 (●) LD_{50} of C(±)P(±)-soman to anesthetized, atropinized, and mechanically ventilated rats. The lines represent optimal fits of bi-exponential functions to the data. The early data points from the tri-exponential curve of the C(+)P(−)-stereoisomer (○) at a dose of 6 LD_{50} are also given. (From Benschop, H.P. and De Jong, L.P.A., *Neurosci. Biobehav. Rev.*, 15, 73, 1991. With permission.)

the C(+)P(+)-stereoisomer has completely disappeared from the bloodstream at all doses and in all species at the first time point of analysis, i.e., at 0.3 min after administration. The C(−)P(+)-stereoisomer is somewhat more stable and can be observed for approximately 4 min in rats (see Figure 2.4) and guinea pigs, whereas it disappears completely from the bloodstream of marmosets within 1 min. The rapid decrease of the concentration of the C(±)P(+)-stereoisomers is largely due to rapid enzymatic hydrolysis (see Section VIII).

Figures 2.5–2.7 give a survey of the concentrations in blood of the highly toxic C(+)P(−)- and C(−)P(−)-stereoisomers of C(±)P(±)-soman at doses varying from 0.8–6 LD_{50} in rats, guinea pigs, and marmosets, respectively.[19,20] In contrast with the C(±)P(+)-stereoisomers, the highly toxic C(+)P(−)- and C(−)P(−)-stereoisomers of soman can be measured in all species for periods of almost 1 h up to several hours, depending on the species and on the dose, in spite of very steep initial decline of all blood levels (see insets in Figures 2.5–2.7) due to rapid distribution and covalent binding (see Section VIII). A summary of toxicokinetic data for the toxic stereoisomers of soman is given in Table 2.3. All toxicokinetic curves are best described with three-exponential equations, except for those at the lowest dose (1 LD_{50}) in rats and guinea pigs (0.8 LD_{50}), for which the data can be fitted to a two-exponential equation. Areas under the curve (AUC) and terminal half lives have been calculated from these equations.

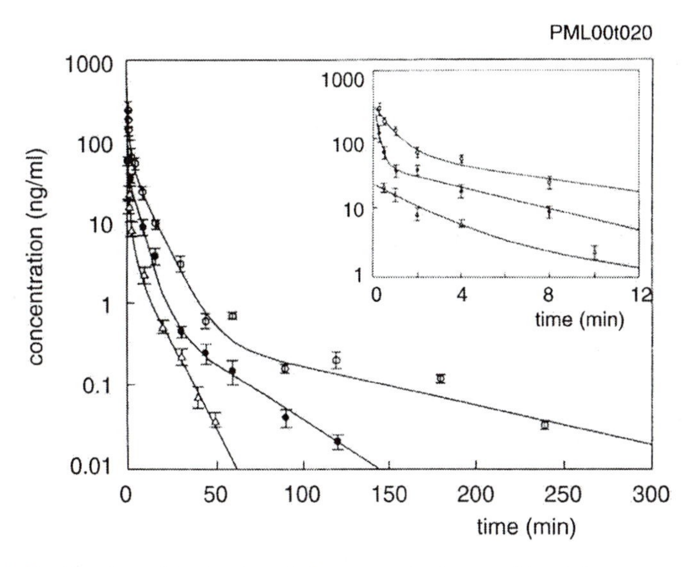

FIGURE 2.5 Semilogarithmic plot of the concentrations in blood (± s.e.m; n = 6) of C(+)P(−)-soman vs. time after i.v. administration of 1 (Δ, 83 μg/kg), 3 (●), and 6 (○) LD_{50} of C(±)P(±)-soman to anesthetized, atropinized, and mechanically ventilated rats. The lines represent optimal fits of a bi-exponential function to the data at a dose of 1 LD_{50}, and of tri-exponential functions at higher doses. The inset shows the data for the first 12 min plotted on an expanded scale. (From Benschop, H.P. and De Jong, L.P.A., *Neurosci. Biobehav. Rev.*, 15, 73, 1991. With permission.)

The derivation of the time period during which acutely toxic levels of (summed) C(±)P(−)-soman stereoisomers are present is based, somewhat arbitrarily, on a scenario of intoxication in which an animal resumes spontaneous respiration presumably due to about 5–7% reactivation by oxime (or protection by carbamate) of completely inhibited AChE in diaphragm. Since the concentration of AChE in diaphragm of guinea pigs is approximately 2–2.6 nM, this reactivated fraction corresponds with approximately 150–200 pM AChE. Based on a bimolecular rate constant for inhibition of AChE by C(±)P(−)-soman of about 10^8 $M^{-1}min^{-1}$, it is calculated that this reactivated fraction of AChE can be re-inhibited by 150 pM (30 pg · ml^{-1}) of C(±)P(−)-soman with a half life of about 1 h. An order of magnitude lower concentration of C(±)P(−)-soman can only cause toxicologically insignificant re-inhibition. Therefore it is assumed that 150 pM C(±)P(−)-soman represents approximately the lowest concentration having toxicological relevance. In a more generalized way, it may be reasoned that an area under the curve (AUC) of 30 pg · ml^{-1} × 60 min = 1.8 ng · min · ml^{-1} in the last part of the blood level curve is needed for toxicological relevance. The period of time in between intoxication and the point on the time axis at which this area starts can be regarded as the period of time in which toxicologically relevant levels of C(±)P(−)-soman are present.

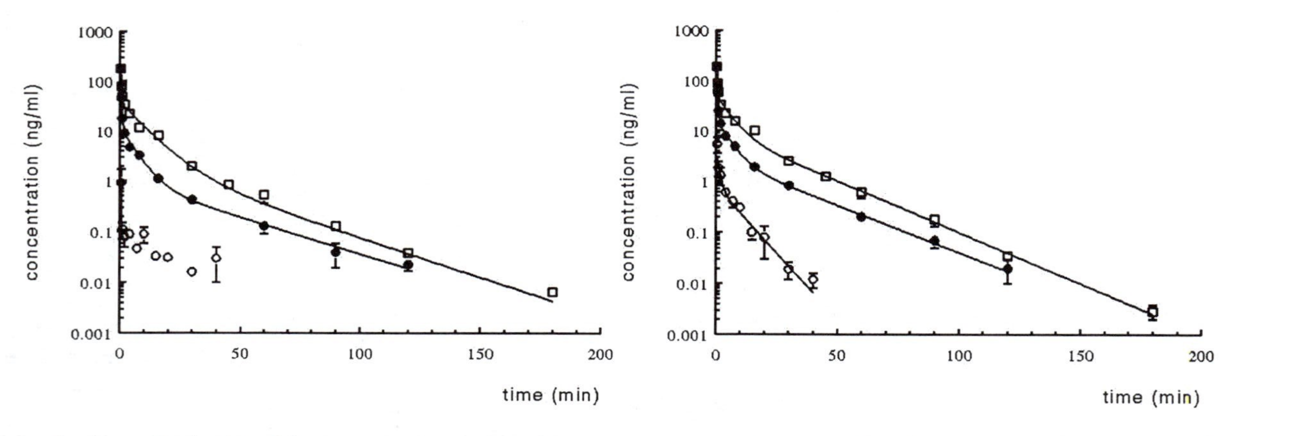

FIGURE 2.6 Semilogarithmic plot of the concentrations in blood (± s.e.m; n = 6) of (left panel) C(+)P(−)-soman and of (right panel) C(−)P(−)-soman vs. time after i.v. administration of 0.8 (○, 22 μg/kg), 2 (●), and 6 (□) LD_{50} of C(±)P(±)-soman to anesthetized, atropinized, and mechanically ventilated guinea pigs. The lines represent optimal fits of a bi-exponential function to the data at a dose of 0.8 LD_{50}, and of tri-exponential functions at higher doses.

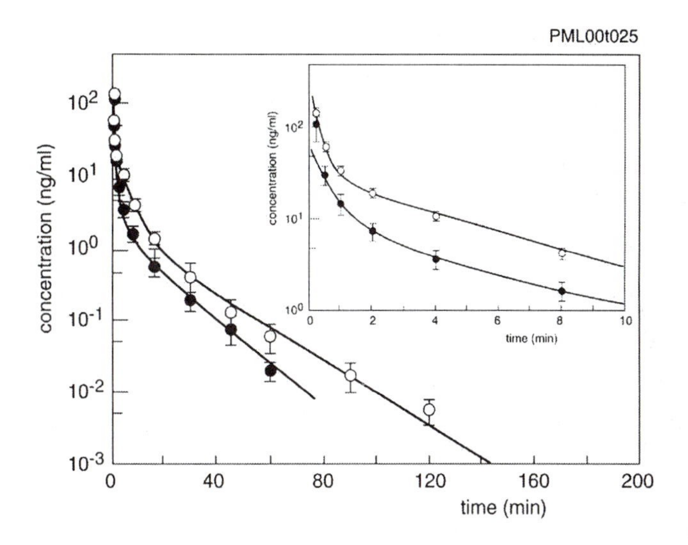

FIGURE 2.7 Semilogarithmic plot of the concentrations in blood (± s.e.m; n = 6) of C(+)P(−)-soman vs. time after i.v. administration of 2 (●, 20 μg/kg) and 6 LD_{50} (○) of C(±)P(±)-soman to anaesthetized, atropinized, and mechanically ventilated marmosets. The lines represent optimal fits of tri-exponential functions to the data. The inset shows the data for the first 10 min plotted on an expanded scale.

Based on the earlier mentioned ratio of stereoisomers in C(±)P(±)-soman and on equal bioavailability of C(+)P(−)- and C(−)P(−)-soman, it should be expected that the AUC for C(+)P(−)-soman should be approximately 20% larger than those of C(−)P(−)-soman in the same experiment. Instead, it is observed that the AUC for C(+)P(−)-soman is at most equal to but often smaller than that of the other stereoisomer. This phenomenon could be made plausible by measuring the rate constants of inhibition of CaE by the two stereoisomers. These enzymes are considered as the major covalent binding sites in blood for these two stereoisomers (see Section VIII). It appeared that the C(+)P(−)-stereoisomer reacts 30-fold faster with CaE in guinea pig blood than the C(−)P(−)-stereoisomer, which may explain the lesser bioavailability of C(+)P(−)-soman. The most striking example of this "stereospecific bioavailability" is observed at a dose of 0.8 LD_{50} in rats. In this case the C(+)P(−)-stereoisomer is almost instantly scavenged and only blood levels of the C(−)P(−)-stereoisomer can be measured (see Section VIII).

As mentioned before, the amounts of CaE in blood decrease in the order rat > guinea pig > marmoset. Accordingly, it can be seen from the AUC in Table 2.4 that the intraspecies linearity with dose for the toxicokinetics of the two toxic stereoisomers of soman is reasonable in marmosets and in guinea pigs. Similarly, a reasonable "interspecies linearity" exists between guinea pigs and marmosets as indicated, e.g., by the AUC in guinea pigs at 2 LD_{50} with that in marmosets at a dose of 6 LD_{50}. In contrast, the AUC pertaining to 6 LD_{50} in marmosets is 2.5 times larger than that at 1 LD_{50} in rats, while the absolute dose in marmosets is approximately 30% less.

TABLE 2.3
Survey of Toxicokinetic Parameters[a] of C(+)P(−)- and C(−)P(−)-soman in Anesthetized, Atropinized, and Mechanically Ventilated Rats, Guinea Pigs, and Marmosets at Intravenous Doses Corresponding with 0.8 LD_{50} of C(±)P(±)-soman

Parameter	Rat						Guinea pig						Marmoset			
	6 LD_{50}		3 LD_{50}		1 LD_{50}		6 LD_{50}		2 LD_{50}		0.8 LD_{50}		6 LD_{50}		2 LD_{50}	
	C(+)P(−)	C(−)P(−)	C(+)P(−)	C(−)P(−)	C(+)P(−)	C(−)P(−)	C(+)P(−)	C(−)P(−)	C(+)P(−)	C(−)P(−)	C(+)P(−)	C(−)P(−)	C(+)P(−)	C(−)P(−)	C(+)P(−)	C(−)P(−)
Dose[a] (μg/kg)	136	111	68	56	22.7	18.6	45.4	37.1	15.1	12.4	6.0	5.0	16.5	13.5	5.5	4.5
A (ng/ml)	253	233	301	259	18	15	339	406	318	354	–	3.8	285	172	61	52
B (ng/ml)	63	61	41	37	3.9	5.9	35	40	11	15	–	0.80	30	22	9.9	9.1
C (ng/ml)	0.55	1.0	0.9	1.1	–	–	2.8	9.9	1.0	1.7	–	–	1.9	1.6	1.8	2.1
a (min^{-1})	1.3	1.2	5.0	4.7	0.45	0.57	3.8	4.3	3.8	3.9	–	0.95	3.9	3.0	2.2	2.0
b (min^{-1})	0.11	0.10	0.19	0.15	0.096	0.12	0.12	0.19	0.19	0.21	–	0.12	0.27	0.22	0.35	0.30
c (min^{-1})	0.011	0.017	0.032	0.042	–	–	0.034	0.046	0.033	0.042	–	–	0.052	0.047	0.073	0.073
Terminal half-life (min)	64	40	22	16	7	6	20	15	21	16.5	–	5.8	13	15	9.5	9.4
Area under curve (ng · min/ml)	806	877	308	320	81	76	458	520	169	228	–	10.6	218	191	81	85
Acutely toxic levels C(±)P(−)-soman until (min)[b]	317		95		37		126		104		–		74		49	

Note: The concentration of each isomer at time t ($conc_t$) is described by: $conc_t = Ae^{-at} + Be^{-bt} + Ce^{-ct}$.

[a]Calculated on the basis of 55/45 ratio of the [C(+)P(−) + C(−)P(+)]-stereoisomers and the [C(+)P(+) + C(−)P(−)]-stereoisomers.

[b]After administration of C(±)P(±)-soman. It is assumed that the area under the curve of 1.8 ng.min/ml for C(±)P(−)-soman is the minimum area with toxicological relevance.

TABLE 2.4
Toxicokinetic Parameters of (−)-sarin and C(−)P(−)-soman after i.v. Administration of 0.8 LD_{50} (19.2 μ/kg) of (±)-sarin and 0.8 LD_{50} of C(±)P(±)-soman (22 μ/kg) to Anesthetized, Atropinized, and Mechanically Ventilated Guinea Pigs

Parameter	(−)-sarin	C(−)P(−)-soman
Dose (μg/kg)	9.6	4.95
A ($ng \cdot ml^{-1}$)	35.9	3.75
B ($ng \cdot ml^{-1}$)	0.09	0.80
a (min^{-1})	4.6	0.95
b (min^{-1})	0.012	0.12
Distribution half-life (min)	0.2	0.7
Terminal half-life (min)	58	5.8
Area under curve ($ng \cdot min \cdot ml^{-1}$)	15.3	11

Note: The concentration of each stereoisomer at time t ($conc_t$) is described by $conc_t = Ae^{-at} + Be^{-bt}$.

Toxicokinetics in rats are strikingly nonlinear. Evidently, disproportionally large amounts of soman stereoisomers are consumed by CaE at low dose in the rat. Since this is a stoichiometric process, scavenging at higher dose consumes a smaller fraction of the total dose. Of necessity, this phenomenon related to the large amount of CaE in rat blood leads to a period of time of more than 5 h during which toxicologically relevant concentrations of C(±)P(−)-soman are present (see Figure 2.5). Therefore, it is not surprising that Wolthuis et al.[1] observed fatal re-intoxication in rats challenged at a dose corresponding with 6 LD_{50} while these animals were initially saved by immediate treatment with atropine and HI-6. Neither is it surprising in view of the data in Table 2.3 that this re-intoxication phenomenon was not observed at lower dose in the rat or at any dose (≤6 LD_{50}) in guinea pigs or in marmosets. In view of the large discrepancies in toxicokinetics between rats on the one hand and guinea pigs and marmosets on the other hand, guinea pigs are considered better model animals for primates than rats in toxicological and therapeutic investigations for nerve-agent intoxication. Pretreatment of rats with the specific CaE inhibitor 2-(o-cresyl)-4*H*-1:3:2-benzodioxaphosphorin-2-oxide (CBDP) blocks binding of nerve agents to these enzymes completely in blood and lungs, and partially in kidney and liver.[21] The LD_{50} of C(±)P(±)-soman in these pretreated animals is lowered to a value in the same range as that in marmosets. Accordingly, the AUC of C(+)P(−)-soman and C(−)P(−)-soman at a dose of 6 LD_{50} in the "CBDP-rats" is lowered to the same range as the AUC in marmosets at a dose of 6 LD_{50}.[22]

Levels of soman stereoisomers in tissues rather than in blood have been measured to a limited degree in brain and diaphragm, which are considered target organs for central and peripheral toxic effects of nerve agents, respectively.[23] Figure 2.8 shows the levels of C(+)P(−)-soman and of C(−)P(−)-soman decreasing with time in blood, diaphragm, and homogenized brain samples of rats upon i.v. administration of a dose

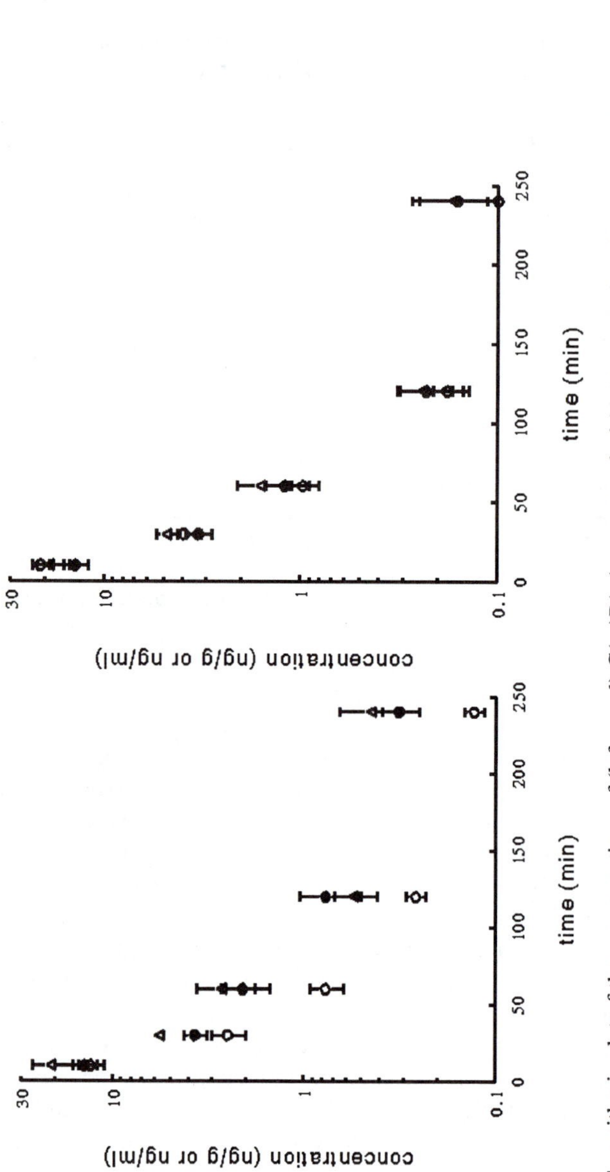

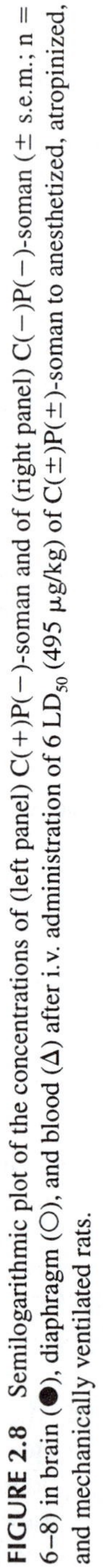
FIGURE 2.8 Semilogarithmic plot of the concentrations of (left panel) C(+)P(−)-soman and of (right panel) C(−)P(−)-soman (± s.e.m.; n = 6–8) in brain (●), diaphragm (○), and blood (Δ) after i.v. administration of 6 LD_{50} (495 μg/kg) of C(±)P(±)-soman to anesthetized, atropinized, and mechanically ventilated rats.

corresponding with 6 LD_{50} of C(±)P(±)-soman. The levels of C(+)P(−)-soman in brain appeared to be significantly (Scheffé F-tests, $p < 0.05$) lower than those in blood, which is not the case for the C(−)P(−)-stereoisomer. Due to experimental restrictions, the first time point is at 10 min after administration of C(±)P(±)-soman. Interestingly, in recent toxicokinetic experiments in pigs at i.v. doses of 0.75–3 LD_{50}, C(+)P(−)- and C(−)P(−)-soman were detected in cerebrospinal fluid (CSF) at > 1 min after administration and were observed to increase for an initial period of 3 min.[17] Evidently, the toxic stereoisomers of soman penetrate rapidly through the blood-brain barrier, which is in accordance with the pronounced central effects upon intoxication with C(±)P(±)-soman.

In comparison with C(±)P(±)-soman, little work has been done so far on the i.v. toxicokinetics of (±)-sarin. In order to obtain reference data for the inhalation toxicokinetics of (±)-sarin (see Section VI), the i.v. toxicokinetics of (±)-sarin was investigated at a sublethal dose of 0.8 LD_{50} in anesthetized, atropinized, and mechanically ventilated guinea pigs.[18,24] Blood levels of (−)-sarin vs. time after administration are shown in Figure 2.9, in which the blood concentration-time curve for C(−)P(−)-soman at equitoxic dose (0.8 LD_{50}) is also given for comparison. The toxicokinetic parameters derived from these two curves, as summarized in Table 2.4, show that a striking difference can be noted in the toxicokinetics of the two nerve agents. While the distribution phase of (−)-sarin is nearly an order of magnitude more rapid than that of C(−)P(−)-soman, its elimination phase is approximately an order of magnitude slower. The latter finding is rather surprising since (−)-sarin was not expected to be more persistent than C(−)P(−)-soman. Further studies should

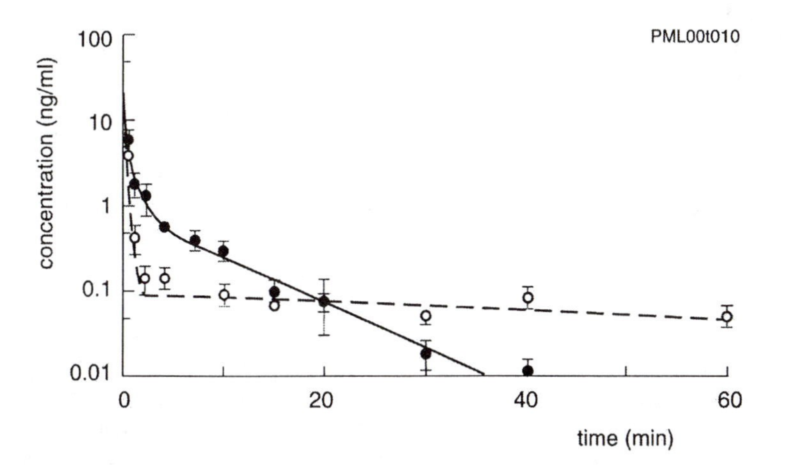

FIGURE 2.9 Semilogarithmic plot of the mean concentrations in blood (± s.e.m.; n = 6) of (−)-sarin (O) vs. time after i.v. bolus administration of 0.8 LD_{50} (19.2 μg/kg) of (±)-sarin to anaesthetized, atropinized, and mechanically ventilated guinea pigs. For comparison the concentration-time course of C(−)P(−)-soman (●) at an equitoxic dose (i.v. 22μg/kg) of C(±)P(±)-soman is also shown.

reveal whether slower rates of enzymatic hydrolysis and of binding might explain the relative *in vivo* persistence of (−)-sarin. Although (−)-sarin is rather persistent, it can only inhibit blood AChE (rate constant 10^7 $M^{-1}min^{-1}$) up to 40% during the elimination phase, due to the relatively low concentration, i.e, < 0.09 ng/ml (see Table 2.4).

V. SUBCUTANEOUS TOXICOKINETICS OF SOMAN

Animal exposure to volatile nerve agents by way of the most realistic route of exposure, i.e., respiratory exposure with concomitant absorption in the respiratory tract, is complex and difficult to analyze experimentally. Intramuscular (i.m.) and s.c. administration of the agent are often considered reasonable substitutes for respiratory exposure in efficacy studies on treatment and pretreatment against nerve agent poisoning. In order to investigate whether the toxicokinetics of nerve agents are reasonably similar upon respiratory exposure (see Section VI) and subcutaneous administration, Due et al.[25] investigated the toxicokinetics of C(±)P(±)-soman in anesthetized,* atropinized, and mechanically ventilated guinea pigs after a bolus injection of a dose corresponding to 6 LD_{50} (148 μg/kg) in the scruff of the neck. Averaged results for the C(+)P(−)- and C(−)P(−)-stereoisomers are shown in Figure 2.10. The toxicokinetics were regarded as a discontinuous process, with a mono-exponential equation for the absorption phase and a bi-exponential equation for the distribution phase. The derived toxicokinetic data are summarized in Table 2.5, in which the bioavailability upon s.c. administration is defined as the ratio of the AUC upon s.c. and i.v. administration.

As evident from Figure 2.10, the blood levels of the C(±)P(−)-stereoisomers have reached clearly measurable levels within 1 min after administration, which shows that these toxic stereoisomers of soman rapidly penetrate the walls of the capillary vessels at the site of injection. In contrast, the C(±)P(+)-stereoisomers of soman never surpassed the minimum detectable concentration (approximately 5 pg/ml) in blood. This example of almost absolute stereospecificity in the absorption phase is readily explained by the ubiquitous presence of phosphoryl phosphatases in blood, skin, and other tissues. These enzymes rapidly hydrolyze the C(±)P(+)-stereoisomers of soman and are presumably available in extra amounts due to tissue damage at the site of injection. Figure 2.10 also shows that the blood levels of the C(±)P(−)-stereoisomers of soman increase during the first 7 min after administration, from which a half-life of absorption of 3.2–3.6 min is derived (see Table 2.5). In comparison with the C(−)P(−)-stereoisomer, the absorption of the C(+)P(−)-stereoisomers is clearly lagging behind. The AUC of the latter stereoisomer is significantly (27%) less than that of the C(−)P(−)-stereoisomer, in spite of the large excess (23%) of the C(+)P(−)-stereoisomer in C(±)P(±)-soman. As in the case of i.v. administration (see Section IV), this decreased bioavailability of C(+)P(−)-soman is explained by its more rapid covalent binding (30-fold in blood) than of the C(−)P(−)-stereoisomer to CaE, which is a major route of elimination for these stereoisomers (see Section VIII). Nevertheless, the

*Hynorm® (im)/Nembutal® (ip)

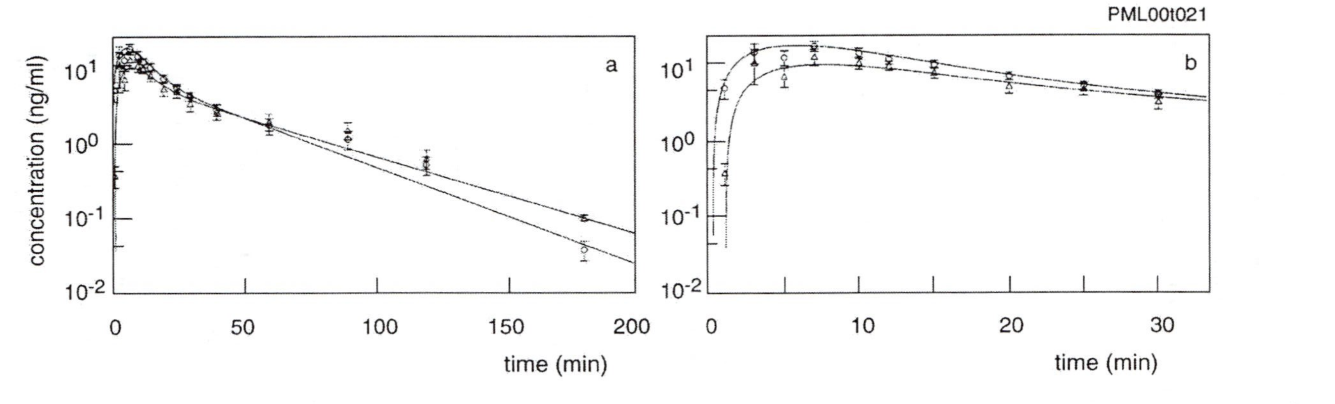

FIGURE 2.10 Semilogarithmic plots of the concentrations (± s.e.m.; n = 6–8) in blood of C(+)P(−)-soman (Δ) and of C(−)P(−)-soman (○) vs. time after s.c. bolus administration of 6 LD_{50} (148 μg/kg) C(±)P(±)-soman to anaesthetized, atropinized, and mechanically ventilated guinea pigs. Left panel: all data; right panel: data for the first 35 min plotted on an expanded time scale. (From Due, A.H., Trap, H.C., Langenberg, J.P., and Benschop, H.P., *Arch. Toxicol.*, 68, 60, 1994. With permission.)

TABLE 2.5
Toxicokinetic Parameters of C(+)P(−)-soman and C(−)P(−)-soman after s.c. Administration of 6 LD_{50} (148 μg/kg) of C(±)P(±)-soman to Anesthetized, Atropinized, and Mechanically Ventilated Guinea Pigs

Parameter	C(+)P(−)-soman	C(−)P(−)-soman
Dose[a] (μg/kg)	40	33
A ($ng \cdot ml^{-1}$)	348	328
B ($ng - ml^{-1}$)	337	318
C (ng/ml^{-1})	6.2	8.3
a (min^{-1})	0.20	0.22
b (min^{-1})	0.19	0.20
c (min^{-1})	0.023	0.029
Half-life absorption (min)	3.6	3.2
Terminal half-life (min)	30	24
Area under curve[b] ($ng \cdot min \cdot ml^{-1}$)	303 (411)	385 (466)
Biological availability[c](%)	74	83
Acutely toxic levels C(±)P(−)-soman until[d](min)	228	

Note: The concentration of each stereoisomer at time t ($conc_t$) is described by $conc_t = -Ae^{-at} + Be^{-bt} + Ce^{-ct}$, in which the first exponential function describes the absorption phase.

[a]Calculated on the basis of 55/45 ratio of the [C(+)P(−) + C(−)P(+)]-stereoisomers and the [C(−)P(−) + C(+)P(+)]-stereoisomers.

[b]The AUC after i.v. administration of the same dose of C(±)P(±)-soman is given between brackets, assuming linearity with dose upon a dose reduction from 165 to 148 μg/kg.

[c]The biological availability is calculated as the ratio between the AUC after s.c. and i.v. administration × 100%.

[d]It is assumed that an AUC of 1.8 $ng \cdot min \cdot ml^{-1}$ for C(±)P(−)-soman is the minimum area with toxicological relevance.

Source: From Due, A.H., Trap, H.C., Langenberg, J.P., and Benschop, H.P., *Arch Toxicol.*, 68, 60, 1994. With permission.

relative bioavailability of the C(+)P(−)-stereoisomer (74%) is only slightly less than that of the C(−)P(−)-stereoisomer (83%), since this parameter is calculated relative to the AUC for i.v. administration where the same stereospecific elimination of C(+)P(−)-soman is encountered.

Consistently higher levels of the (±)P(−)-stereoisomers of soman are present in the terminal elimination phase after s.c. bolus administration than after i.v. bolus administration (see Tables 2.5 and 2.3). Consequently, the time period (228 min) during which the C(±)P(−)-stereoisomers are present at toxicologically relevant concentrations is almost twice as long as after i.v. administration. Tentatively, the more pronounced persistence of C(±)P(−)-soman upon s.c. administration has been explained by gradual absorption of these stereoisomers from the site of s.c. injection.

These results show that s.c. administration can only be regarded as a toxicokinetic model for respiratory exposure in cases where the duration of the latter exposure is in the range of several minutes, which is not often realistic at (supra)lethal doses. The extended period of time in which toxicologically relevant concentrations of C(±)P(−)-soman are present after s.c. administration suggests that this route is a more rigorous challenge for prophylaxis and therapy of intoxication than other routes. However, the extremely high blood levels immediately after i.v. bolus administration may also provide a considerable challenge for that route of administration, depending on the pharmacokinetic and pharmacodynamic parameters of the specific treatment.

VI. INHALATION TOXICOKINETICS OF SOMAN AND SARIN

In the case of intoxications with nerve agents under realistic conditions, the primary route of entrance into the body of volatile nerve agents such as sarin, tabun, soman, and GF is the respiratory route. The latter route is almost as effective as parenteral administration, with approximately 70% of an inhaled dose of (±)-sarin being retained in guinea pigs, dogs, monkeys, and humans.[26,27] It was anticipated that the shapes of the toxicokinetic curves for inhalation of nerve agent stereoisomers would differ considerably from those for other routes of exposure, which may have important consequences for the efficacy of pretreatment and therapy of intoxications with these agents. Therefore, Langenberg et al. investigated the inhalation toxicokintics of C(±)P(±)-soman and (±)-sarin,[18] using an apparatus which they constructed for the continuous generation of nerve agent vapor in air, nose-only exposure of guinea pigs and monitoring of respiratory minute volume and respiratory frequency during exposure.[28] Figure 2.11 gives a schematic representation of this apparatus, as well as a short explanation of the functioning of the various elements.

The inhalation and subsequent absorption of nerve agent vapor, largely in the upper part of the respiratory tract, may involve a time period of a few seconds or minutes, up to several hours in the case of low-level exposure (see Section VII).[29] Initial investigations of inhalation toxicokinetics involved exposure periods of 4–8 min which was regarded as a compromise between the often shorter exposure time to volatile agents in the case of chemical warfare and the desire to measure in a reasonable time frame the increasing blood levels due to inhalation and absorption. Since it was considered as too involved to use mechanically ventilated animals in inhalation toxicokinetics, the investigations were restricted to sublethal doses in the range of 0.4–0.8 LCt_{50} of C(±)P(±)-soman and (±)-sarin, inhaled by anesthetized* and atropinized guinea pigs.

Concentrations of C(−)P(+)-soman < 13 pg/ml were observed in the blood samples taken during an 8-min exposure to 0.8 LCt_{50} of C(±)P(±)-soman. In all other cases, this stereoisomer was not observed in detectable concentrations, whereas

*Ketamine hydrochloride (i.m.).

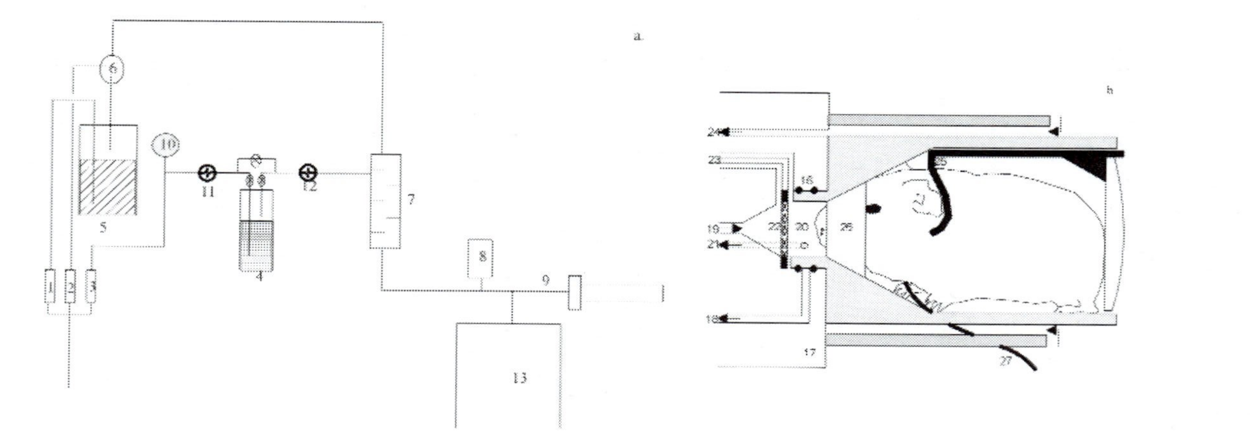

FIGURE 2.11 (a) Apparatus constructed for generation of nerve agent vapor: (1), (2), and (3), mass flow controllers; (4), vial containing the nerve agent; (5), thermostatted water bath; (6) and (7), mixing chambers; (8), temperature/relative humidity meter; (9), towards the exposure modules; (10), overpressure security; (11) and (12), splash heads; and (13), gas chromatograph with gas sampling valve. (b) Guinea pig (14) positioned in the modified Battelle tube (15); (16) O-rings for gastight connection of the tube to the body of the exposure apparatus (17); (18) tubing with a critical orifice, which is connected to an underpressure check for gastight connection of the tube; (19) tubing through which the nerve agent is transported to the exposure chamber; (20) front chamber of the modified Battelle tube, from which the animal breathes; (21) tubing with a critical orifice, which is the outlet of the front chamber; (22) wire mesh resistance; (23) differential pressure measuring device; (24) tubing with a critical orifice, which sucks air from the "underpressure chamber" surrounding the tube; (25) fork for positioning the animal; (26) rubber mask; and (27) carotid artery cannula. Arrows indicate flow directions. (From Langenberg, J.P., Spruit, H.E.T., Van Der Wiel, H.J., Trap, H.C., Helmich, R.B., Bergers, W.W.A., Van Helden, H.P.M., and Benschop, H.P., *Toxicol. Appl. Pharmacol.*, 151, 79, 1998. With permission.)

the C(+)P(+)-stereoisomer was not observed in any case. The mean concentration-time courses of C(+)P(−)- and C(−)P(−)-soman for exposures to 0.8 LCt_{50} in 8 min are presented in Figure 2.12. The effect of inhaled dose on the toxicokinetics is shown in Figure 2.13, where the time course of the concentration of C(−)P(−)-soman is given for an 8-min exposure to 0.4 and 0.8 LCt_{50}, whereas Figure 2.14 shows the effects on toxicokinetics of the time period of exposure to the same dose, i.e., 0.8 LCt_{50} in a 4 and 8 min period of time.

As in the case of s.c. toxicokinetics, the kinetics of C(+)P(−)- and C(−)P(−)-soman were described mathematically as a discontinuous process, with an equation for the exposure period and an equation for the post-exposure period. In view of the limited number of data points during exposure, the absorption phase was described with a mono-exponential function. In order to describe the exposure phase of C(+)P(−)-soman, lag times of 2 and 4 min were selected for the 8-min exposures to 0.8 and 0.4 LCt_{50}, respectively. These lag times correspond with the earliest time points at which this stereoisomer could be detected. Toxicokinetic parameters derived from the various calculated concentration-time curves are given in Table 2.6. There were no measurable effects of the exposures on the respiratory minute volume (RMV) and respiratory frequency (RF).

The data in Figures 2.12–2.14 suggest that the systemic penetration of C(−)P(−)-soman during nose-only exposure is very rapid, since this stereoisomer can be measured in blood at 30 s after starting the exposure. Moreover, the

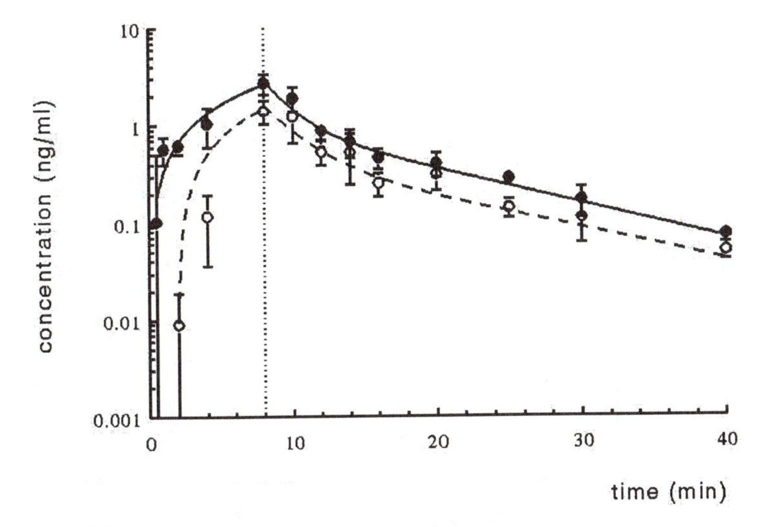

FIGURE 2.12 Semilogarithmic plot of the mean concentrations in blood (± s.e.m., n = 6) of C(+)P(−)-soman (○) and C(−)P(−)-soman (●) vs. time during and after nose-only exposure of anesthetized, atropinized, and restrained guinea pigs to 48 ± 5 $mg.m^{-3}$ of C(±)P(±)-soman vapor in air for 8 min, which corresponds with 0.8 LCt_{50}. The dotted line marks the end of the exposure period. (From Langenberg, J.P., Spruit, H.E.T., Van Der Wiel, H.J., Trap, H.C., Helmich, R.B., Bergers, W.W.A., Van Helden, H.P.M., and Benschop, H.P., *Toxicol. Appl. Pharmacol.*, 151, 79, 1998. With permission.)

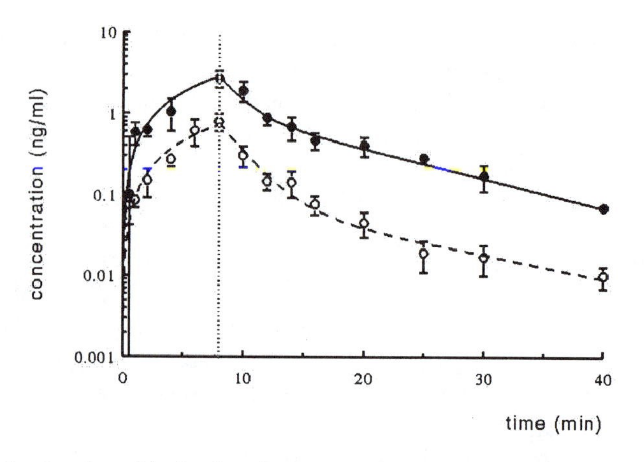

FIGURE 2.13 Semilogarithmic plot of the mean concentrations in blood (± s.e.m.; n = 6) of C(−)P(−)-soman vs. time during and after nose-only exposure of anesthetized, atropinized, and restrained guinea pigs to C(±)P(±)-soman vapor in air yielding 0.4 (○) and 0.8 (●) LCt_{50} in 8 min. The dotted line marks the end of the exposure period. (From Langenberg, J.P., Spruit, H.E.T., Van Der Wiel, H.J., Trap, H.C., Helmich, R.B., Bergers, W.W A., Van Helden, H.P.M., and Benschop, H.P., *Toxicol. Appl. Pharmacol.*, 151, 79, 1998. With permission.)

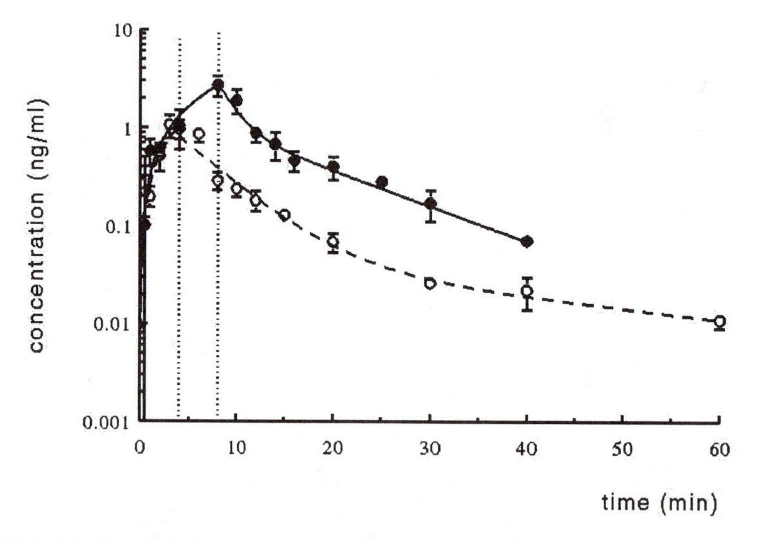

FIGURE 2.14 Semilogarithmic plot of the mean concentrations in blood (± s.e.m.; n = 6) of C(−)P(−)-soman vs. time during and after nose-only exposure of anesthetized, atropinized, and restrained guinea pigs to C(±)P(±)-soman vapor in air yielding 0.8 LCt_{50} in 8 min (●) or 4 min (○). The dotted lines mark the end of the exposure period. (From Langenberg, J.P., Spruit, H.E.T., Van Der Wiel, H.J., Trap, H.C., Helmich, R.B., Bergers, W.W.A., Van Helden, H.P.M., and Benschop, H.P., *Toxicol. Appl. Pharmacol.*, 151, 79, 1998. With permission.)

concentration of this stereoisomer does not increase further after terminating the exposure. In contrast herewith, there is a lag time of several minutes before the C(+)P(−)-stereoisomer can be detected in blood. Furthermore, as observed earlier for i.v. and s.c. administration, the concentrations in blood and the AUC of C(+)P(−)-soman are consistently lower than those of the C(−)P(−)-stereoisomer, despite the 23% excess of the C(+)P(−)-stereoisomer over the C(−)P(−)-stereoisomer in C(±)P(±)-soman. Based on the earlier mentioned 30-fold higher reaction rate of the C(+)P(−)- stereoisomer with guinea pig plasma CaE, this phenomenon can at least partly be explained by preferential binding of the C(+)P(−)-stereoisomer, e.g., to CaE at the absorption site(s) in the respiratory tract and in blood.[30,31]

The data in Table 2.6 show that the apparent elimination half-life of the C(−)P(−)-stereoisomer after respiratory exposure to 0.8 LCt_{50} in 8 min is somewhat longer than for the equitoxic i.v. dose. Moreover, the maximum concentration in blood of C(+)P(−)-soman in case of the 4-min exposure is reached not earlier than 2 min after cessation of the exposure. This suggests that, despite rapid absorption, some depot formation occurs at the absorption site, from which absorption continues after termination of the exposure. A further argument for some depot formation in the respiratory tract can be gleaned from Figure 2.15, where the concentration-time profiles for C(−)P(−)-soman are compared for the 8-min respiratory exposure and for an 8-min i.v. infusion of an equitoxic dose (0.8 LD_{50}) of C(±)P(±)-soman.[24] Evidently, the absorption phase of respiratory absorption is closely mimicked by the i.v. infusion, but blood levels subsequent to the respiratory exposure are distinctly higher than those after the i.v. infusion.

Half-life of elimination appears to increase when the exposure time was shortened from 8 to 4 min, while the exposure concentration is increased two-fold. Apparently, such a phenomenon seems hard to understand. The calculated elimination half-life would probably be appreciably longer if a data point at 60 min would be available.

Rather unexpectedly, the concentrations of C(+)P(−)- and C(−)P(−)-soman decrease over the entire time period of exposure and elimination with a concomitant fourfold lowering of the AUC when the exposure time to 0.8 LCt_{50} is lowered from 8 to 4 min. Intuitively, one would expect a higher maximum concentration but approximately the same AUC upon shortening the exposure time at equitoxic dose. Moreover, the longer terminal half-life after exposure for 4 min seems in contradiction with the lower AUC. Since the RMV and RF are almost the same during the period of exposure, this nonlinearity with exposure time might be due to decreasing retention of soman vapor at higher concentrations of soman vapor in air. Unfortunately, data on the overall retention of soman in experimental animals are not available, and are urgently needed for interpretation and physiologically based modeling of the toxicokinetics of the stereoisomers of soman.

During and after nose-only exposure to 0.4 LCt_{50} of C(±)P(±)-soman in 8 min, the maximum concentration of the C(+)P(−)- and C(−)P(−)-stereoisomers are approximately fourfold lower than the mean maximum concentrations after an 8-min exposure to a dose corresponding to 0.8 LCt_{50}. Furthermore, the AUC of the two stereoisomers for the former experiment are about fourfold lower. Assuming that retention of the agent in the respiratory tract is constant, it is likely that the

TABLE 2.6
Toxicokinetic Parameters of C(+)P(−)-soman and C(−)P(−)-soman in Anesthetized, Atropinized, and Restrained Guinea Pigs during and after Nose-Only Exposure for 4–8 min to 0.4–0.8 LCt_{50} of C(±)P(±)-soman. Comparative Parameters for C(−)P(−)-soman for i.v. Bolus Administration of a Dose Corresponding with 0.8 LD_{50} Are also Given

Parameter	0.8 LCt_{50} (8-min exposure)		0.4 LCt_{50} (8-min exposure)		0.8 LCt_{50} (4-min exposure)		0.8 LD_{50} (i.v., bolus)
	C(+)P(−)[a]	C(−)P(−)	C(+)P(−)[b]	C(−)P(−)	C(+)P(−)	C(−)P(−)	C(−)P(−)[c]
C_{max}[d] ($ng \cdot ml^{-1}$, n = 6)	1.4 ± 0.4	2.7 ± 0.6	0.2 ± 0.2	0.8 ± 0.2	0.5 ± 0.2[e]	1.1 ± 0.3	—
A ($ng \cdot ml^{-1}$)	−55.4	−35.8	−34.7	−34.7	—	−30	—
B ($ng \cdot ml^{-1}$)	32.6	141	1.74	24.8	0.7	1.7	3.8
C ($ng \cdot ml^{-1}$)	0.83	1.9	0.014	0.14	0.03	0.05	0.80
D ($ng \cdot ml^{-1}$)	55.4	35.8	34.7	34.7	—	30	—
a (min^{-1})	0.004	0.009	0.001	0.002	—	0.009	—
b (min^{-1})	0.42	0.54	0.28	0.45	0.19	0.20	0.95
c (min^{-1})	0.08	0.08	0.005	0.07	0.011	0.026	0.12
Elimination half-life (min)	9.2	8.4	—	10.1	63	27	5.8
AUC[f] ($ng \cdot min \cdot ml^{-1}$)							
0 → end exposure	7.2	8.5	1.2	2.5	0.6[h]	2.1	
end exposure → ∞	8.8	15.2	3.1	2.7	3.7	5.6	
Total	16.0	23.7	4.3	5.2	4.3	7.7	10.6
RMV[g] ($ng \cdot ml^{-1}$; n = 12)	52 ± 3		43 ± 6		37 ± 5		
RF[h] (Hz; n = 12)	0.65 ± 0.05		0.66 ± 0.05		0.80 ± 0.11		

Note: The inhalation results were fitted with discontinuous functions: [nerve agent] = $D + A^{-at}$ for the absorption phase, and [nerve agent] = $Be^{-bt} + Ce^{-c}$ for distribution and elimination.

[a] Assuming a lag time of 2 min.

[b] Assuming a lag time of 4 min.

[c] Toxicokinetic parameters could not be obtained from the low and rather erratic concentration of the C(+)P(−)-isomer.

[d] At end of exposure period, unless noted otherwise.

[e] At t = 6 min (2 min after ending of exposure).

[f] AUC measured with the trapezoidal method.

[g] RMV, respiratory minute volume. Values are means ±s.e.m.

[h] RF, respiratory frequency. Values are means ±s.e.m.

Source: From Langenberg, J.P., Spruit, H.E.T., Van Der Wiel, H.J., Trap, H.C., Helmich, R.B., Bergers, W.W.A., Van Helden, H.P.M., and Benschop, H.P., *Toxicol. Appl. Pharmacol.*, 151, 79, 1998. With permission.

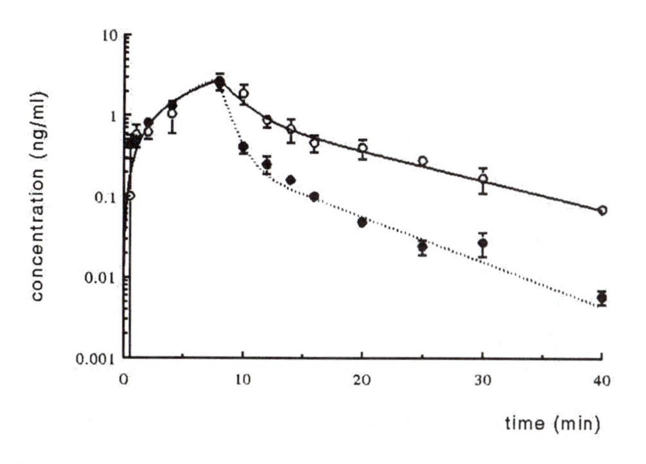

FIGURE 2.15 Semilogarithmic plot of the mean concentrations (± s.e.m.; n = 6) of C(−)P(−)-soman in blood of anesthetized and atropinized guinea pigs during and after an 8 min nose-only exposure to 0.8 LCt_{50} (○) and an 8 min i.v. infusion of 0.8 LD_{50} (●) of C(±)P(±)-soman.

toxicokinetics are nonlinear with dose, as observed for i.v. bolus administration of a sublethal dose.

Experiments on the inhalation toxicokinetics of (±)-sarin have been restricted to 8-min exposures to 0.8 and 0.4 LCt_{50} of this agent. The concentration-time profile for (−)-sarin at an exposure level of 0.8 LCt_{50} (38 mg · m^{-3}), together with that of C(+)P(−)-soman for an 8-min exposure to an equitoxic dose of C(±)P(±)-soman (exposure level 48 mg · m^{-3}) are given in Figure 2.16. The mathematical description of the concentration-time profiles for (−)-sarin was analogous to that for the inhalation experiments of C(±)P(±)-soman. Toxicokinetic parameters derived from the equations are given in Table 2.7. When comparing the measured blood levels in the absorption phase of (−)-sarin with those of C(+)P(−)-soman and C(−)P(−)-soman at equitoxic dose, it appears that the absorption of (−)-sarin resembles that of C(+)P(−)-soman, i.e., featuring a relatively slow build-up of blood levels in comparison with C(−)P(−)-soman. Tentatively, this behavior of (−)-sarin can be ascribed, as in the case of C(+)P(−)-soman, to scavenging by irreversible binding sites prior to and subsequent to entering the bloodstream. Also, the maximum concentration reached by (−)-sarin is rather similar to that of C(+)P(−)-soman. Characteristically for (−)-sarin (see Section IV), the terminal half-life of this stereoisomer at a dose of 0.8 LCt_{50} is about fourfold longer than for the two C(±)P(−)-stereoisomers of soman (8.4–9.2 min). Due to unrealistically high blood levels of (−)-sarin in the terminal elimination phase after exposure to 0.4 LCt_{50}, it is difficult to judge the (non)linearity with dose relative to exposure to 0.8 LCt_{50} of (±)-sarin. For the time period 0–120 min, the AUC are rather similar. However, when taking into account that the RMV during exposure to 0.4 LCt_{50} was about 1.5-fold

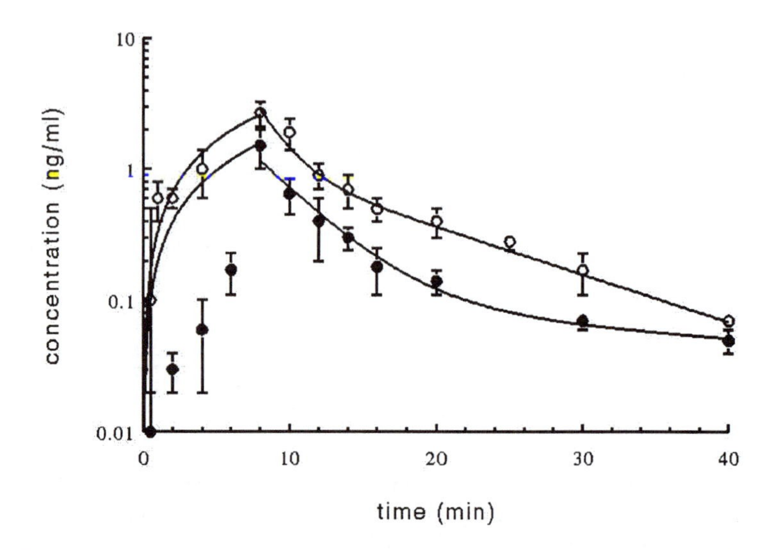

FIGURE 2.16 Mean concentrations in blood (±s.e.m.; n = 6) of (−)-sarin (●) and of C(−)P(−)-soman (○) vs. time after nose-only exposure of anesthetized and atropinized guinea pigs to 0.8 LCt_{50} of (±)-sarin and C(±)P(±)-soman, respectively, in the course of 8 min.

higher than during exposure to 0.8 LCt_{50}, nonlinearity with dose seems probable. As in the case of inhalation toxicokinetics of soman stereoisomers, a final interpretation of results will depend on additional data with regard to the retention of sarin vapor in the respiratory system under various conditions.

VII. INHALATION TOXICOKINETICS OF SOMAN UPON LOW-LEVEL EXPOSURE

As evident from the previous paragraphs of this chapter, investigations on the toxicokinetics of nerve agents have centered on lethal and supralethal doses of nerve agent. However, the controversy on the possible relationship between the so-called Gulf War Syndrome and exposure to traces of nerve agent shortly after the Gulf War has emphasized that knowledge on the acute and delayed effects of trace exposure to nerve agents is almost nonexistent.[32,33] Nevertheless, several situations can be envisaged in which trace exposures becomes realistic. In the case of chemical warfare, small amounts of agent may penetrate into gas masks and protective clothing or into a collective protective shelter. Small amounts of nerve agent may desorb from contaminated skin, clothing, or painted surfaces, posing a risk of long-term, low-level exposure. Miosis, rhinorrhea, dyspnea, and tightness of the chest were observed in rescue workers and medical personnel in hospitals due to secondary exposure to small amounts of agent subsequent to the terrorist attacks with sarin in Matsumoto and in metropolitan Tokyo.[34–38] Sarin vapor could be detected in houses up to 12 h after the attack with sarin in Matsumoto. Some victims in this city reported their first

TABLE 2.7
Toxicokinetic Parameters of (−)-sarin in Anesthetized, Atropinized, and Restrained Guinea Pigs after Nose-Only Exposure for 8 min to 0.8 and 0.4 LCt_{50} of (±)-sarin and after i.v. Bolus Administration of 0.8 LD_{50} of (±)-sarin for Comparison.

Parameter	0.8 LCt_{50} (8 min)	0.4 LCt_{50} (8 min)	0.8 LD_{50} (i.v. bolus)
C_{max}[a] ($ng.ml^{-1}$, ±s.e.m., n = 6)	5	0.3 ± 0.1	–
A ($ng.ml^{-1}$)	164	55.6	–
B ($ng.ml^{-1}$)	−164	−55.6	–
C ($ng.ml^{-1}$)	8.6	0.86	35.9
D ($ng.ml^{-1}$)	0.11	0.036	0.09
b (min^{-1})	0.0012	0.00085	–
c (min^{-1})	0.26	0.18	4.6
d (min^{-1})	0.019	0.00063	0.012
Terminal half-life (min)	36	–	58
Area under curve ($ng.min.ml^{-1}$)			
0–8 min	6.3	0.8	
8–120 min	8.5	5.5	
Total	14.8	6.3	15.3
RMV[b] ($min.ml^{-1}$, ±s.e.m., n = 12)	31 ± 4	47 ± 6	–
Rf[c] (Hz, ±s.e.m., n = 12)	0.70 ± 0.06	0.64 ± 0.04	–

Note: The inhalation results were fitted with a discontinuous function: $[(-)\text{-sarin}] = A + Be^{-bt}$ for the absorption phase, and $[(-)\text{-sarin}] = Ce^{-ct} + De^{-dt}$ for the distribution and elimination phase. The i.v. results were fitted according to $[(-)\text{-sarin}] = Ce^{-ct} + De^{-dt}$.

[a]At the end of the exposure period (t = 8 min).

[b]RMV = respiratory minute volume.

[c]RF = respiratory frequency.

symptoms as late as 20 h after the incident, presumably due to the cumulative effect of persistent low-level exposure.[39]

Systematic investigations on the effects of exposure to small amounts of nerve agents, mostly sarin, on human volunteers pertain invariably to short term (≤ 30 min) exposures, and have led to the definition of so-called "no-effect levels." In other experiments, volunteers inhaled Cts up to 15 mg · min · m^{-3} of sarin and experienced slight acute phenomena of intoxication.[40] A subsequent epidemiological study revealed no difference in health status between exposed and non-exposed individuals. Okumura et al.[41] investigated 640 victims of the terrorist attack with sarin in the Tokyo subway. After discharge from the hospital, patients in the severe and moderate exposure categories required follow-up by the hospital's outpatient system to observe for late effects, especially neurotoxic and behavioral effects, partly due to posttraumatic-stress disorder induced by exposure to sarin.[42,43] Similarly, long-term effects were observed in the victims of the sarin attack at Matsumoto, 3.5 years after the event.[44] It is concluded that exposure to nerve agents at doses leading to acute effects may lead to delayed and persistent adverse effects, mostly of a neuropsychological order. No evidence exists for such effects after single exposure to nerve agents in which acute signs of exposure were absent.

In order to initiate a quantitative basis for the toxicology of low-dose exposure to nerve agents, Benschop et al.[45] investigated the toxicokinetics of the four stereoisomers of C(±)P(±)-soman upon nose-only exposure of anesthetized, atropinized, and restrained guinea pigs to 20 ppb (160 $\mu g \cdot m^{-3}$) of C(±)P(±)-soman over a 5 h exposure period, providing blood levels of the toxic C(±)P(−)-soman stereoisomers at Ct-values accumulating from 0–48 $mg \cdot min \cdot m^{-3}$. Concomitantly, the progressive inhibition of AChE in erythrocytes was measured. The exposures were performed using the apparatus as described in Figure 2.11, with minor adaptations. A bi-exponential equation sufficed to describe the gradually increasing concentrations of the C(+)P(−)- and C(−)P(−)-soman stereoisomers (Figure 2.17), when adopting a lag time of 30 min for the C(+)P(−)-stereoisomer. Table 2.8 summarizes the toxicokinetic data derived from these equations, while Figure 2.17 illustrates the fit of the derived equations to the blood levels of the C(+)P(−)- and C(−)P(−)-soman stereoisomers as measured during exposure.

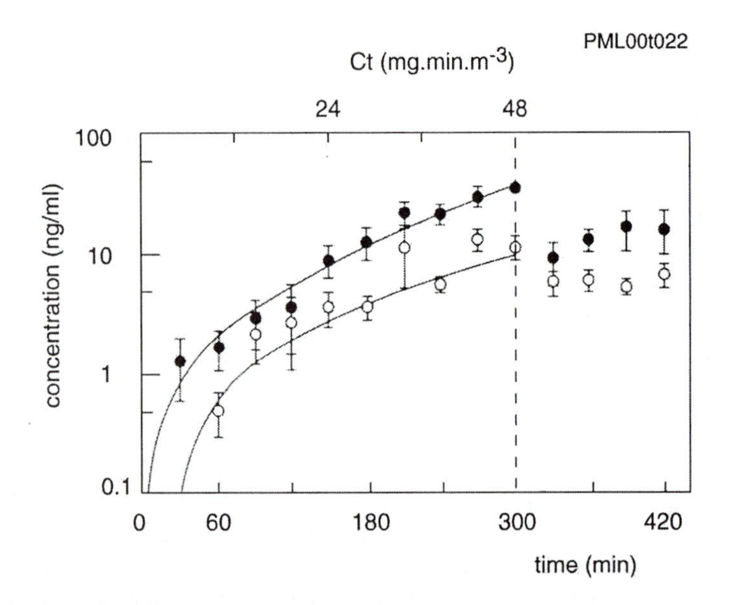

FIGURE 2.17 Semilogarithmic plot of the mean concentrations (± s.e.m., n = 6) in blood of C(−)P(−)-soman (●) and C(+)P(−)-soman (○) vs. time during nose-only exposure of anesthetized, atropinized, and restrained guinea pigs to 160 ± 16 $\mu g.m^{-3}$ of C(±)P(±)-soman for 300 min and up to 120 min after exposure. Accumulated Ct-values are also shown. The solid lines represent optimal fits of bi-exponential functions to the data. The dotted line marks the end of the exposure period. (From Benschop, H.P., Trap, H.C., Spruit, E.T., Van Der Wiel, H.J., Langenberg, J.P., and De Jong, L. P.A., *Toxicol. Appl. Pharmacol.,* 153, 179, 1998. With permission.)

TABLE 2.8
Toxicokinetic Parameters for C(+)P(−)-soman and C(−)P(−)-soman in Anesthetized, Atropinized, and Restrained Guinea Pigs during Nose-Only Exposure to C(±)P(±)-soman at a Concentration of 160 ± 16 μg.m^{-3} (20 ppb) in Air for 300 min.

Parameter	C(+)P(−)-soman	Parameter	C(−)P(−)-soman
A_+ (ng.ml^{-1})	0.00107	A^- (ng.ml^{-1})	0.0262
b_+ (min^{-1})	0.00885	b^- (min^{-1})	0.00599
c_+ (min^{-1})	-0.00947	c^- (min^{-1})	0.00509
$AUC^{a}_{0-300\ min}$ (ng.min.ml^{-1})	1.1	$AUC_{0-300\ min}$ (ng.min.ml^{-1})	3.5

Note: The data for concentrations in blood determined during exposure were fitted with bi-exponential function: $[C(+)P(-)\text{-soman}] = A_+\{e^{b+*(t-30)} - e^{c+*(t-30)}\}$ and $[C(-)P(-)\text{-soman}] = A_-\{e^{b-*t} - e^{c-*t}\}$, in which the subscript + refers to C(+)P(−)-soman and the subscript − to C(−)P(−)-soman.

[a]AUC = area under the curve.

Source: From Benschop, H.P., Trap, H.C., Spruit, E.T., Van Der Wiel, H.J., Langenberg, J.P., and De Jong, L.P.A., *Toxicol. Appl. Pharmacol.*, 153, 179, 1998. With permission.

No attempt was made to describe the time-concentration course in the 120 min post-exposure period. The blood levels decrease clearly in the first 90 min post-exposure but remain remarkably constant over the next 90 min period. One intriguing explanation, although needing validation, is the formation of a "depot" of intact soman, for example in the epithelial tissue of the respiratory tract, from which the C(+)P(−)- and C(−)P(−)-stereoisomers diffuse into the bloodstream.

In case of s.c. and short-term respiratory exposure, the C(−)P(−)-soman stereoisomer penetrated almost immediately into the bloodstream, whereas the appearance of the C(+)P(−)-stereoisomer lagged a few minutes behind. In the present case of low-level respiratory exposure, it takes approximately 30 min before even the C(−)P(−)-stereoisomer has penetrated, while it takes another 30 min before the C(+)P(−)-stereoisomer appears in measurable concentrations in the bloodstream. Thus, as should be expected, the preferential scavenging of C(+)P(−)-soman becomes more evident upon lowering the doses of C(±)P(±)-soman. In this extreme case of low-dose exposure, the AUC of the C(+)P(−)-stereoisomer is more than 3-fold lower than that of the C(−)P(−)-stereoisomer (Table 2.8), in spite of the 22% excess of the former stereoisomer in C(±)P(±)-soman. It should also be noted that enzymes such as CaE which scavenge C(+)P(−)-soman preferentially, are abundantly available in the epithelial tissue of the upper respiratory tract where, by analogy with sarin, most of the soman is presumably absorbed.[30] Not surprisingly, the C(+)P(+)- and C(−)P(+)-stereoisomers were not detected at any stage of the exposure period.

The progressive inhibition of erythrocyte AChE as well as the concentrations of the C(+)P(−)- and C(−)P(−)-soman stereoisomers in blood were measured independently. Therefore attempts were made to find out whether the sets of data are consistent, using the measured bimolecular rate constants for inhibition of AChE in erythrocyte cell walls by these two stereoisomers and kinetic equations derived to calculate the progression of AChE inhibition by the two stereoisomers in their concentration-rate profile. As shown in Figure 2.18 the calculated progression of inhibition is only slightly slower than the actually observed values.

At 120 min after ending the 300-min exposure period, AChE activities were measured in brain and diaphragm of the exposed guinea pigs, in order to estimate the relative importance of peripheral and central effects of long-term, low-level exposure to C(±)P(±)-soman. No inhibition of AChE in these two organs was observed, whereas almost complete inhibition of the activity of AChE in erythrocytes occurs during and after the exposure period. These results corroborate those of Jacobson et al., who observed negligible inhibition of brain AChE in dogs after exposure to Ct values of 10 mg · min · m^{-3} of sarin for a 6-h period daily over 6 months, which yielded 80% inhibition of erythrocyte AChE within a few days.[46] The results are also in accordance with an efficient detoxification in the blood by covalent binding to CaE and other binding sites which are first exposed to C(±)P(±)-soman (*vide infra*). The lack of inhibition of two major target organs for intoxication with C(±)P(±)-soman, i.e., brain and diaphragm, indicates that signs of systemic intoxication, either due to peripheral or central effects, are rather improbable in these long-term, low-level exposures. The same holds true for the occurrence of (delayed) neuropsychological

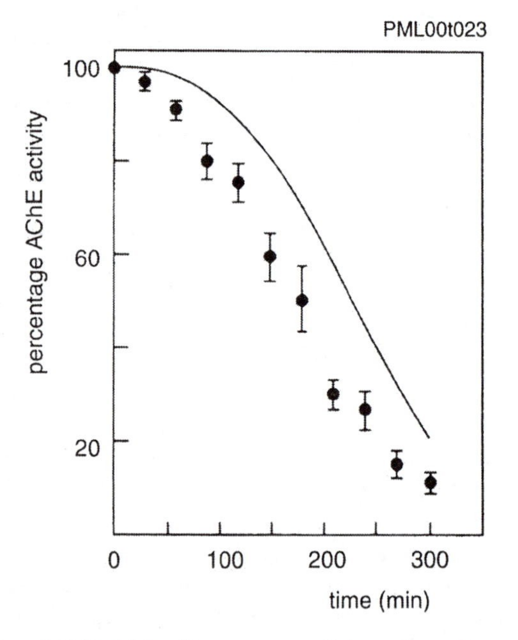

FIGURE 2.18 Mean AChE activity (± s.e.m.; n = 6) in erythrocytes vs. time during nose-only exposure of anesthetized, atropinized, and restrained guinea pigs to 160±16 μg.m^{-3} (20 ppb) of C(±)P(±)-soman for 300 min. The solid line represents AChE activities calculated from the measured concentrations of C(±)P(−)-soman in blood.

disorders. During exposure of human volunteers to sarin vapor (< 30 min) at low wind speeds, it has been observed that Ct-values > 2–3 mg · min · m^{-3} will cause incapacitating miosis due to direct penetration of the nerve agent into the eye.[47–49] This value would even decrease significantly with increasing wind speeds and presumably with replacement of sarin by soman. Therefore, the occurrence of miosis and effects thereof might be a much more probable outcome of the long-term low-level exposure to C(±)P(±)-soman at the accumulated Ct of 48 mg · min · m^{-3} than systemic anticholinergic effects, albeit that measurements of miosis upon long-term, low-level exposure have not been performed. Moreover, it should be remarked that prolonged exposure of the airways to soman vapor can also be a cause for direct systemic effects, which might be more pronounced in nonatropinized animals.

A further effect that should be taken into account when attempting to extrapolate results of the above-mentioned, low-level exposures to humans is the absence in primates of most of the scavenging effect of CaE as is available in guinea pigs. Therefore, the same challenge levels of C(±)P(±)-soman vapor might give more pronounced AChE inhibition in primates than in guinea pigs.

It is remarkable that intact C(−)P(−)-soman becomes detectable in blood after a 30-min exposure to 160 μg · m^{-3} (20 ppb), i.e., at a total dose of 4.8 μg · min · m^{-3}, which is only two orders of magnitude higher than the Ct value allowed by the U.S. Army during an 8-h occupational exposure to sarin.[50] Inhibition of AChE is still marginal under these conditions, but becomes clearly detectable at 60 min, i.e., at Ct = 9.6 μg · min · m^{-3}. The present investigations are at the limit of possibilities for measuring intact agent in blood upon low-level exposure. A further reduction of the dose can only be investigated on a basis of nerve agent accumulated by internal scavengers like butyrylcholinesterase (BuChE), CaE, and albumin. Recently, methodology for such an approach has been developed in which sarin is released from sarin-inhibited BuChE by means of fluoride ion-induced reactivation and subsequent isolation and analysis of released sarin (see Section VIII). In this way, ≥0.01% inhibition of BuChE in blood of primates can be quantified.[51] Preliminary investigations indicate that such a degree of inhibition occurs at or only slightly above the occupational exposure limit.[50,52]

VIII. ELIMINATION PATHWAYS OF PHOSPHOFLUORIDATES

The elimination routes of C(±)P(±)-^{14}C-soman were investigated by Benschop and De Jong in a series of experiments after i.v. administration of doses corresponding to 2–6 LD_{50} in anesthetized and mechanically ventilated rats, guinea pigs, and marmosets.[19] O-pinacolyl methylphosphonic acid (PMPA) and soman bound covalently to proteins accounted for more than 80% of the radioactivity 1 h after administration of the agent. Obviously, hydrolysis of the phosphorus-fluorine bond and reaction with binding sites are the major elimination pathways for C(±)P(±)-soman. Similar results were obtained for C(±)P(±)-^{3}H-soman and (±)-^{3}H-sarin after i.v. administration at sublethal doses to mice.[13,15] In the latter investigations, the highest concentrations of both hydrolyzed and bound organophosphate were already achieved 1 min after administration of the agents, indicating that both elimination processes proceed very rapidly. This result is corroborated by the observed rapid decrease of the concentrations in blood of intact C(±)P(±)-soman and (±)-sarin in various toxicokinetic studies (*vide supra*).

A. Elimination by Hydrolytic Degradation

Soman and other phosphofluoridates are degraded by enzymatic as well as by spontaneous hydrolysis. Already in 1946, Mazur[53] described that rabbit and human plasma and tissues contain enzymes that accelerate the hydrolysis of phosphorofluoridates yielding O,O-dialkyl phosphoric acid and fluoride ions. These so-called phosphoryl phosphatases were extensively studied by Mounter and co-workers and were also observed to catalyze the hydrolysis of phosphonofluoridates and of the phosphoramidocyanidate tabun.[54–57] The importance of hydrolysis as an elimination route of phosphofluoridates was illustrated by Cohen and Warringa[58] who found some protection of rats challenged (s.c.) with supralethal doses of O,O-diisopropyl phosphorofluoridate (DFP) or sarin by means of pretreatment (i.v.) with purified phosphoryl phosphatase from hog kidneys. A similar approach to protection was followed in the attempted conversion of human BuChE into a hydrolytic enzyme for phosphofluoridates by means of site-directed mutagenesis.[59,60]

Large amounts of O-isopropyl methylphosphonic acid were found in blood and urine of victims of the terrorist attacks with sarin in Matsumoto and Tokyo.[61,62] Until recently, it was assumed that hydrolysis of phosphofluoridates in plasma and tissues of mammals proceeds exclusively by cleavage of the P−F bond. For example, treatment of C(±)P(±)-^{32}P-soman with rat plasma or liver homogenate did not lead to any conversion of PMPA into the secondary hydrolysis product methylphosphonic acid (MPA).[63] Ramachandran[64] observed that the primary hydrolysis product of DF^{32}P, i.e., O,O-diisopropyl ^{32}P-phosphoric acid, is not metabolized after s.c. administration to mice. Rather, the product was excreted unchanged into urine. However, Nakajima et al.[65] reported that MPA was detected (in urine) until the third day after hospitalization of a victim of the terrorist attack with sarin in Matsumoto. This discrepancy needs further investigation.

Stereoselectivity of the enzymatic hydrolysis of the chiral phosphonofluoridates C(±)P(±)-soman and (±)-sarin has been studied extensively.[3,66–72] In various tissues of several species, (+)-sarin and C(±)P(+)-soman are much more rapidly degraded by phosphoryl phosphatases than the more toxic stereoisomers of these nerve agents. The relative order for the rate of hydrolysis of the four stereoisomers of C(±)P(±)-soman is C(+)P(+)- >> C(−)P(+)- >> C(−)P(−)- ≥ C(+)P(−)-stereoisomer (Table 2.9).

Acceleration of hydrolysis by means of enzymatic catalysis of the toxic C(±)P(−)-stereoisomers of soman and of (−)-sarin in various tissues of rats, guinea pigs, marmosets, and in human plasma is low or even absent.[67,69,72] Obviously, the overall rates of hydrolysis of C(±)P(−)-soman vary with the type of tissue within one species.[72,73] The rates of hydrolysis are consistently low in target organs of the

TABLE 2.9
First-Order Rate Constants for Hydrolysis of the Four Stereoisomers of C(±)P(±)-soman, Catalyzed by Diluted Liver Homogenate and Plasma from Rat, Guinea Pig, and Marmoset, and by Diluted Human Plasma (pH 7.5, 37°C)

Source		Rate constant (10^{-3} min^{-1})			
		C(+)P(−)	C(−)P(−)	C(−)P(+)	C(+)P(+)
Plasma:	Man	0.7	3.2	480	860
	Marmoset	1.2	5.5	310	1200
	Guinea pig	0.2	0.8	80	220
	Rat	1.5	3.2	1200	3400
Liver:	Marmoset	1.0	1.9	40	90
	Guinea pig	0.9	1.6	270	510
	Rat	2.8	4.4	420	2500

Note: Calculated for hydrolysis by 1 ml 0.6% plasma or 0.15% liver homogenate.

Source: From De Jong, L.P.A., Van Dijk, C., and Benschop, H.P., *Biochem. Pharmacol.,* 37, 2939, 1988. With permission.

three species, such as brain and muscle (see Section IX). In contrast with the phosphofluoridates, the stereoselectivity of phosphoryl phosphatases for hydrolysis of (±)-tabun is less pronounced and appears to be species dependent.[74,75]

B. Elimination by Covalent Binding

The second major elimination route for phosphofluoridates is covalent reaction with binding sites. It has been demonstrated that reaction with CaE is an important detoxification route for organophosphates *in vivo.*[76–78] In addition to CaE, albumin appears to bind covalently to organophosphates.[79,80] The active site concentrations of AChE and BuChE are only a small percentage of the total concentration of binding sites in rats and guinea pigs.[73,81] Consequently, this binding has a negligible effect on the toxicokinetics of phosphonofluoridates, albeit that binding to AChE is crucial from a toxicological point of view.

A comparison of the amounts of soman bound in various tissues of rat, guinea pig, and marmoset at 1 h after i.v. administration of a dose corresponding with 6 LD_{50} of C(±)P(±)-^{14}C-soman with the amounts bound *in vitro* in plasma and homogenized tissues after incubation with excess of C(±)P(±)-soman reveals that complete occupation of available binding sites has not occurred *in vivo,* even after administration of this relatively high dose of C(±)P(±)-^{14}C-soman, except for binding in plasma and lung (*vide infra*).[19,72,73] This outcome is in accordance with the observations of Maxwell et al.,[81] who calculated that the total concentration of CaE in the rat corresponds with more than 14 LD_{50} of C(±)P(±)-soman.

Pretreatment (s.c.) of mice, rats, and guinea pigs with tri-o-cresyl phosphate (TOCP) or with the CaE-inhibitor CBDP, which is the active metabolite of TOCP, reduces the LD_{50} of sarin and soman to approximately the same level as the LD_{50} value in nonhuman primates, e.g., marmosets.[21,82,83] Marmosets contain a lower concentration of binding sites in various tissues than rats and guinea pigs,[72] whereas the concentration of CaE in plasma is also low.[84] A further indication for the importance of CaE in the detoxification of phosphofluoridates was obtained by Sterri and coworkers.[78,85,86] They found that repeated s.c. administration of sublethal doses of C(±)P(±)-soman in mouse, rat, and guinea pig and of (±)-sarin to guinea pigs resulted in cumulative LD_{50} doses which are substantially higher than the acute value. For instance, most of the guinea pigs survived a total exposure to 5–6 times the acute LD_{50} dose when the animals were challenged every 24 h with a dose corresponding with 0.5 LD_{50} of C(±)P(±)-soman. One hour after administration of C(±)P(±)-soman, 70% of the CaE activity in plasma was inhibited, but the activity was restored to control values within 24 h. The results of these studies suggest strongly that the lower toxicity of C(±)P(±)-soman and (±)-sarin for mice, rats, and guinea pigs than for marmosets is mainly due to a higher number of covalent binding sites in the former three species. These sites serve as endogenous scavengers for detoxification of the phosphofluoridates.

It should be expected on the basis of these results that pretreatment of nonhuman primates with exogenous scavengers that bind covalently to organophosphates will offer some protection against intoxication with nerve agents. Indeed, very promising results were obtained in studies along this line in rhesus monkeys. For example, Doctor et al. and Maxwell et al. obtained protection against ≤5 LD_{50} of soman (i.v.) by pretreatment with fetal bovine AChE,[87,88] whereas Ashani and coworkers protected the animals against ≤4 LD_{50} of soman (i.v.) and guinea pigs against 4 times the median inhalatory dose of the agent by pretreatment with human plasma BuChE.[89,90]

Upon increasing the i.v. dose of C(±)P(±)-^{14}C-soman from 2–3 LD_{50} to 6 LD_{50}, the concentrations of bound soman in blood and lungs of rats and guinea pigs increased only slightly. In fact, the highest concentrations of bound soman were almost equal to the concentrations of binding sites of soman determined from *in vitro* binding experiments with C(+)P(−)-soman. It does not follow from this almost complete occupancy of binding sites that these sites in blood and lung have the highest intrinsic reactivity. Upon i.v. administration, binding sites in blood are first exposed to soman followed by those in the lungs. Only the fraction of soman surviving passage of the lungs is available for binding in other organs. This description, according to the principle of "first come, first served," is further supported by investigations in CBDP-pretreated rats and guinea pigs.[21] Administration of CBDP at a s.c. dose of 2 mg/kg, i.e., a dose which potentiates the lethality of phosphofluoridates considerably (*vide supra*), produced complete inhibition of CaE in plasma and lung of both species and of CaE in rat kidney, but inhibition of CaE in liver of both species and in the kidney of the guinea pig was only marginal. Moreover, only a small further potentiation of the lethality was achieved by petreatment of the animals with an 8-fold higher dose of CBDP. Apparently, a substantial fraction of a lethal dose of soman is eliminated by the binding sites in blood and lung of guinea pig and rat, first served after an i.v. or s.c. administration, and in kidney of the rat. Binding in other organs plays a less important role in the detoxification of soman. By analogy, a higher uptake of radioactivity in the liver and reduced incorporation into other organs was observed when the route of administration of DFP to rats was shifted from s.c. or i.v. to i.p. Moreover, the LD_{50} of i.p.-administered DFP was about 2-fold higher than the LD_{50} after i.v. or s.c. administration, stressing the major contribution of the first pass through the liver to the detoxification of DFP.[91]

Rate constants of binding for C(±)P(−)-^{14}C-soman were determined in guinea pig blood and tissue homogenates.[73] The determinations were performed in the presence of an equimolar concentration of C(±)P(+)-soman in order to account for the possible competition by the less toxic stereoisomers. The relatively high rate constants obtained for part of the binding sites, i.e., approximately 10^6–10^7 $M^{-1}min^{-1}$, suggest reaction with highly reactive enzymes such as cholinesterases and CaE. These investigations also show hetereogeneity of the binding sites in the various tissues with respect to their reactivity towards C(±)P(−)-soman. Heterogeneity may be due to different proteins, e.g., CaE and albumin (*vide supra*), serving as binding sites, but may also be related to hetereogeneity within CaE as observed by Sterri and Fonnum.[78] *Inter alia,* the interspecies nonlinearity found for the toxicokinetics of C(±)P(−)-soman in rat *vs.* guinea pigs and marmosets (see Section IV) may be explained on the basis of such heterogeneity, assuming that some of these binding sites in rats have a very high reactivity. These sites will rapidly bind a fraction of the administered dose, resulting in the sequestration of a larger fraction of C(±)P(−)-soman at a low dose than at a high dose.

In summary, the rapid initial decay of C(±)P(±)-soman in blood after i.v. or respiratory administration is due to three processes, i.e., (i) distribution to various tissues, (ii) spontaneous and enzymatic hydrolysis, and (iii) covalent binding. It has been established that the toxic C(±)P(−)-stereoisomers react rapidly with covalent binding sites. Since the less toxic C(±)P(+)-stereoisomers are hydrolyzed several orders of magnitude faster than the C(±)P(−)-stereoisomers whereas hydrolysis and covalent binding contribute almost to an equal extent (*vide supra*) to the elimination of C(±)P(±)-soman, it was expected that elimination of C(±)P(+)-stereoisomers proceeds almost exclusively by hydrolysis whereas the C(±)P(−)-stereoisomers are almost exclusively eliminated by covalent binding. However, experiments performed with an i.v. dose of 6 LD_{50} of C(±)P(±)-soman reconstituted from C(±)P(−)-^{14}C-soman and an equimolar amount of unlabeled C(±)P(+)-soman and vice versa showed that the differences in elimination of the stereoisomers are not as extreme as expected (Table 2.10).[19] Indeed, there is a clearcut preference for hydrolysis of the C(±)P(+)-stereoisomers and for covalent binding of the C(±)P(−)-stereoisomers, but binding of the C(±)P(+)-stereoisomers and hydrolysis of the C(±)P(−)-stereoisomers also take place to a considerable extent. Evidently, while C(±)P(+)-soman is also sequestrated by covalent binding, the low toxicity of these stereoisomers is primarily due to their rapid hydrolysis by phosphoryl phosphatases (see Table 2.9) and to their low intrinsic reactivity towards AChE.[9]

The high rates of the processes initially taking place indicate that pretreatment can only offer protection against nerve agent intoxication if it provides for very rapid detoxification in the organs that are first passed by the organophosphate, i.e., blood and lung after i.v., s.c., and respiratory exposure. Consequently, scavengers should have a reactivity towards the toxic stereoisomers of organophosphates comparable with that of cholinesterase and CaE. Alternatively, hydrolytic enzymes should

TABLE 2.10
Radiometrically Determined Stereospecificity of *In Vivo* Covalent Binding and Hydrolysis in Guinea Pigs 1 h after i.v. Administration of 6 LD_{50} of C(±)P(±)-soman Reconstituted from ^{14}C-C(±)P(−)-soman and Equimolar Unlabeled C(±)P(+)-soman, or *Vice Versa*. Percentages of Total Recovery.

	Hydrolysis		Covalent binding	
Tissue	^{14}C-C(±)P(−)	^{14}C-C(±)P(+)	^{14}C-C(±)P(−)	^{14}C-C(±)P(+)
Plasma	23	58	73	39
Lung	54	82	47	16
Liver	10	52	84	45
Brain	31	75	70	25

Note: Mean values from 5 measurements.

Source: From Benschop, H.P. and De Jong, L.P.A., *Neurosci. Biobehav. Rev.*, 15, 73, 1991. With permission.

degrade these stereoisomers at rates comparable with those for hydrolysis of C(±)P(+)-soman in plasma and tissues of rats and guinea pigs. In this connection it is interesting to reconsider the results of Cohen and Warringa[58] with regard to the protective effect against intoxication with (±)-sarin due to pretreatment with phosphoryl phosphatases. At first, the low capacity, if any, of these enzymes to catalyze the hydrolysis of the toxic stereoisomer of sarin seems difficult to reconcile with such a protective effect. However, it can be speculated that a more rapid decrease of the level of the less toxic (+)-stereoisomer of (±)-sarin due to hydrolysis by this enzyme will lead to less competition of the (+)-stereoisomer with the toxic (−)-stereoisomer for reaction with covalent binding sites and consequently to a more effective scavenging of the latter stereoisomer (*vide infra*).

Whereas the C(±)P(+)-stereoisomers of soman are completely eliminated in the distribution phase of the toxicokinetics, the toxic C(±)P(−)-stereoisomers are still present in the elimination phase. It cannot be derived from the presently available data as to what extent elimination of the C(±)P(−)-stereoisomers in the later phase proceeds either by binding or hydrolysis. The catalytic hydrolysis of C(±)P(−)-soman in various organs participating in central elimination is sufficiently high to account for the terminal half-life of the C(±)P(−)-stereoisomers. Alternatively, elimination may also proceed by binding since a large fraction of the binding sites is still unoccupied. The rate of binding will be much lower than in the initial phase due to the much lower residual concentration of C(±)P(−)-soman, and possibly also due to the hetereogeneity of the binding sites.

C. Renal Excretion

As a consequence of the very rapid degradation processes, only small amounts of C(±)P(±)-soman and of related organophosphates are renally excreted. Lenz et al.[92] found that more than 99% of the renally excreted labeled compounds was hydrolyzed at 1 h after s.c. administration of C(±)P(±)-^{14}C-soman (about 1 LD_{50}) to rats. No unaltered compound was detected by Heilbronn et al.[93] in the urine of rats after i.v. administration of tabun. Only after administration of the somewhat less reactive ^{3}H-DFP, 5–10% of the labeled compounds that were renally excreted during the first hour after intoxication of guinea pigs and cats were identified as intact agent.[94,95] At most, approximately 1% of C(±)P(±)-soman was renally excreted within 4 h after administration to rats, mainly as C(±)P(−)-soman.[96] Nevertheless, the levels in rats were at least two orders of magnitude higher than those for guinea pigs and marmosets after intoxication with an equitoxic dose of C(±)P(±)-soman (6 LD_{50}). The levels in rat urine surpass the blood levels which were measured 1 min after i.v. administration, suggesting that most of the C(±)P(−)-soman was excreted within the first few minutes after intoxication. This suggestion is supported by the presence of C(−)P(+)-soman in the urine, which disappears from the blood within a few minutes. These results also indicate that phosphoryl phosphatase activity is absent in the urine and in the bladder. Concomitant with the high levels in urine, levels of intact C(±)P(−)-^{14}C-soman were much higher in rat kidney at 1 h after i.v. administration of a dose equivalent to 6 LD_{50} of C(±)P(±)-^{14}C-soman than in guinea pig and marmoset kidneys after administration of an equitoxic dose, in spite of the relatively high binding capacity that is also available in rat kidneys. Presumably, the stereoisomers of soman in the kidney are protected against elimination when these are, *de facto,* present in urine that has not yet been transported to the bladder.

The amount of renally excreted C(±)P(−)-soman in rats, although very small compared with the administered dose, should be sufficient to be of toxicological relevance (*vide supra*) if the agent can be reabsorbed from the bladder. This process can indeed take place as was deduced from the lethal effect of an intravesical administration of 1.4 LD_{50} of C(±)P(±)-soman.[96] On the basis of these results it may be speculated that C(±)P(−)-soman excreted in the bladder serves as a "depot" and contributes to the "late death" of rats intoxicated with ≥6 LD_{50} of C(±)P(±)-soman several hours after initial recovery as a result of treatment with oxime. However, survival times of intoxicated rats immediately treated with oxime decreased significantly when accumulation of C(±)P(−)-soman in the bladder was prevented by rinsing and drainage of the bladder.[96] Therefore, eventual reabsorption of C(±)P(−)-soman from the bladder does not explain the "late death" in rats, whatever its relevance is for the persistence of the toxic stereoisomers.

D. Elimination Products as Tools for Retrospective Detection of Exposure

Elimination products, i.e., hydrolyzed organophosphate or covalently bound organophosphate, are the biomarkers of choice for detection of exposure, since the persistence of phosphofluoridates, although sufficiently high to interfere with therapeutic measures, is too short for this purpose. The hydrolysis product formed upon intoxication with a phosphofluoridate is rapidly excreted and seems also unsuitable for retrospective detection. However, it has been found that CaE inhibited by organophosphates reactivate spontaneously leading to gradual formation of the hydrolyzed agent which is consequently excreted over a relatively long period of time.[78] A procedure for retrospective detection of exposure to sarin, in which its hydrolysis product serves as a marker, was developed by Noort et al.[61] for analysis in blood and urine. By using this method they were able to show hydrolyzed sarin in blood samples obtained from victims of the Tokyo incident in 1995. In addition, bound organophosphate can serve as a biomarker for exposure. Treatment of organophosphate inhibited CaE and BuChE with fluoride ions can inverse the inhibition reaction yielding a phosphofluoridate and restored enzyme.[51,97] Human serum does not contain CaE but its BuChE concentration is relatively high (70–80 nM).[70,76] Based on the fluoride-induced reactivation reaction, a method for retrospective detection of exposure to nerve agents has been developed, by which exposure of victims of the Tokyo incident to an organophosphate, probably sarin, could be established from analysis of their blood samples.[51]

IX. PHYSIOLOGICALLY-BASED MODELING OF THE TOXICOKINETICS OF SOMAN

Toxicokinetic studies will be more generally applicable, and therefore more valuable, if the results can be described in a physiologically-based model.[98–101] These models represent the mammalian system in terms of compartments, i.e., specific tissues or groups of tissues, which are connected by arterial and venous blood flow pathways (see Figure 2.19). Processes taking place in the compartments are characterized by physiological parameters, e.g., tissue volumes and blood flow rates and parameters specific for the toxicant under investigation, such as tissue/blood partitioning coefficients and metabolic parameters. The model consists of differential equations describing the mass-balance in the various compartments by which time-dependent toxicokinetic data can be simulated. Physiologically-based toxicokinetic models are especially suitable for studying the effect of changes in physiological or toxicant-specific parameters induced, e.g., by treatment or pretreatment drugs, on the toxicokinetics of a toxicant. Furthermore, models provide a physiologically-based means for interspecies scaling and for extrapolation from animal results to those in human beings, which is an ultimate goal in the toxicology of nerve agents.

The description of the toxicokinetics of C(±)P(±)-soman is complicated by the high *in vivo* reactivity as well as by distinct differences in metabolic properties of the toxic C(±)P(−)-stereoisomers and the less toxic C(±)P(+)-stereoisomers. An early physiologically-based model for C(±)P(±)-soman in the rat was described by Maxwell et al.,[102] whereas Gearhart et al.[103] developed a similar model for DFP. Both models were validated indirectly, based on the time course of AChE inhibition in blood and tissues. The first model in which the chirality of C(±)P(±)-soman was taken into account has been described by Langenberg et al.[73] This model was validated on the basis of toxicokinetic data of the intact stereoisomers obtained after i.v. administration of C(±)P(±)-soman at doses corresponding with 0.8, 2, and 6 LD_{50}. As shown in Figure 2.19, the model follows the general form for modeling of

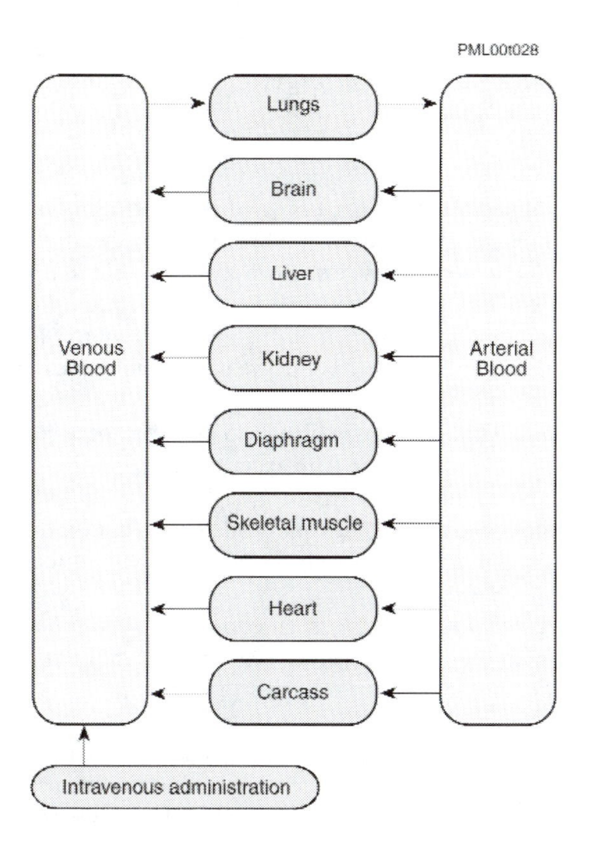

FIGURE 2.19 Schematic outline of the physiologically-based model for the toxicokinetics of C(±)P(±)-soman. (From Langenberg, J.P., Van Dijk, C., Sweeney, R.E., Maxwell, D.M., De Jong, L.P.A., and Benschop, H.P., *Arch. Toxicol.,* 71, 320, 1997. With permission.)

organophosphorus compounds.[102,103] Since exhalation as a pathway of elimination can be neglected after i.v. administration of the agent, this pathway was not taken into account.[94,104] While some of the defined tissues are chosen on the basis of their high metabolic capacity (liver, kidney, lung), other tissues are selected since these are target tissues (brain, diaphragm). The relevant processes in each compartment are (1) partitioning of C(±)P(−)- and C(±)P(+)-soman from blood, (2) elimination defined as reaction with covalent binding sites and (3) elimination by enzymatic and spontaneous hydrolysis. Covalent binding is described as a bimolecular reaction with AChE as well as with rapidly and slowly reacting binding sites, mainly CaE. Combined enzymatic and spontaneous hydrolysis are described as first-order processes. The summed effect of biochemical reaction and mass transfer processes in each compartment is described in differential equations based on mass-balance principles.

A survey of the data used for the parameters in the model is given in Table 2.11. In comparison to values used by Langenberg et al.,[73] the following improvements and refinements were recently made.[105]

- The partition coefficients were determined from partitioning of C(±)P(−)-soman over diluted blood and air, and over tissue homogenates and air after complete inhibition of the enzymatic hydrolysis of the stereoisomers by addition of EDTA, while previously hydrolysis was blocked by acidification to pH 3.3.
- Measured rate constants for inhibition of guinea pig erythrocyte AChE were introduced into the model instead of the values for inhibition of bovine erythrocyte AChE.
- Concentrations of binding sites in blood and lung were derived from results obtained of *in vivo* instead of *in vitro* experiments.
- The rate constants for hydrolysis of C(±)P(−)-soman and C(−)P(+)-soman and the concentrations of rapidly and slowly reacting binding sites in plasma were halved since these parameters are used for the blood compartment in the model.
- Finally, the liver blood flow has been corrected in the present model in accordance with literature data, i.e., 10.7 times higher than the hepatic artery flow.[106]

The model discriminates between two stereoisomeric pairs, i.e., the C(±)P(−)-stereoisomers and the C(±)P(+)-stereoisomers. Further refinements were introduced in order to achieve a proper description of i.v. toxicokinetics. Since C(+)P(+)-soman could not be detected in blood even at 0.25 min after administration, it was assumed that this stereoisomer is completely hydrolyzed immediately after administration. Therefore, only the doses of C(−)P(+)-soman and of C(±)P(−)-soman were introduced into the model and the parameters for hydrolysis of the C(±)P(+)-stereoisomers were adjusted to those of C(−)P(+)-soman. Furthermore, it was assumed that C(+)P(−)-soman is completely bound in blood immediately after administration of a dose corresponding with 0.8 LD_{50} of C(±)P(±)-soman. This assumption is based on (1) the concentrations of C(+)P(−)-soman in blood which are approximately one order of magnitude lower at this dose than those of the C(−)P(−)-stereoisomer, (2) preferential elimination of C(+)P(−)-soman due to the 30-fold faster inhibition rate of CaE in plasma of guinea pigs by this stereoisomer than by C(−)P(−)-soman, and (3) the excess of rapidly reacting binding sites in blood, being mainly CaE, over the amount of C(+)P(−)-soman at this dose. In view of these considerations, the C(+)P(−)-stereoisomer was omitted from the model at this sublethal dose, with a corresponding decrease of the amount of rapidly reacting binding sites. Furthermore, estimated rate constants for rapid binding of C(−)P(−)-soman were used instead of the overall binding constants for C(±)P(−)-soman.

With this refined model, the blood levels of C(−)P(+)-soman and of the C(±)P(−)-stereoisomers can be simulated in a satisfactory way for the three i.v.

TABLE 2.11
Values Used for Parameters in the Physiologically-Based Model for the Toxicokinetics of C(±)P(±)-soman after i.v. Administration to Anesthetized, Atropinized, and Mechanically Ventilated Guinea Pigs weighing 500 g.

Tissue	Tissue volume (ml)	Blood flow[a] (ml/min)	Partition coefficient[b] (tissue/blood)	A[c] (g tissue) (nmole/g tissue)	F[b,d] (nmole/g tissue)	S[b] (nmole/	f[b,e] ($M^{-1}.min^{-1}$)	s[b,e] ($M^{-1}.min^{-1}$)	h[f] (min^{-1})	h′[f,g] (min^{-1})
Blood	40	80	1.0	0.0097	0.495	0.184	3.0×10^7	8.2×10^4	0.05	3.33
Lung	5	80	0.9	0.0045	0.495	0.278	9.3×10^6	2.1×10^4	0.105	7.0
Brain	2.2	0.9	0.7	0.12	0.065	0.0435	2.3×10^7	2.5×10^4	0.039	2.6
Heart	1.8	4.0	1.4	0.0023	0.031	0.084	1.4×10^7	5.8×10^4	0.044	3.0
Diaphragm	1.5	0.2	1.4	0.00135	0.031	0.084	1.4×10^7	5.8×10^4	0.044	3.0
Liver	19	12.8	5.3	0.0063	20.2	37.6	7.8×10^6	1.5×10^4	0.844	56.0
Kidney	4.4	5.0	2.2	0.0044	5.29	4.69	6.5×10^5	6.5×10^3	0.265	17.6
Muscle	250	37.5	1.4	0.0016	0.031	0.084	1.4×10^7	5.8×10^4	0.044	3.0
Carcass	176.1	19.6	1.4	0.00001	0.031	0.084	1.4×10^7	5.8×10^4	0.0088	0.6

Note: A, F, and S denote the initial concentrations of AChE, rapidly and slowly reacting binding sites, respectively; a, f and s and h and h′ denote the rate constants for AChE inhibition, binding to rapidly reacting binding sites and to slowly reacting binding sites by C(±)P(−)-soman and for hydrolysis of C(±)P(−)-soman and C(−)P(+)-soman, respectively.

[a]The blood flow to lung and liver is equal to, and 16% of, the cardiac output, respectively; the blood flow to the carcass is cardiac output minus blood flow to brain, heart, diaphragm, liver, kidney, and muscle.

[b]Values for carcass, heart, and diaphragm are chosen equal to that for muscle.

[c]The value for carcass is arbitrarily set at a very low value of 0.01 pmole/g tissue; rate constants for inhibition are 8.25×10^7 and 8.25×10^3 $M^{-1}min^{-1}$ for C(±)P(−)-soman and C(±)P(+)-soman, respectively.

[d]Concentration of radioactivity bound in blood 1 h after i.v. administration of 6 LD_{50} of C(±)P(±)^{14}C-soman is taken as the values for rapidly reacting binding sites in blood and lung.

[e]Rate constants for C(±)P(+)-soman are chosen 0.01× the corresponding values for C(±)P(−)-soman.

[f]Values for heart and diaphragm are chosen equal to that for muscle; the value for carcass is 1/5 of the value for muscle, and the value for blood is half the value for plasma.

[g]Calculated as $\{(h_{tissue}/h_{plasma}) \times h'_{plasma}\}$

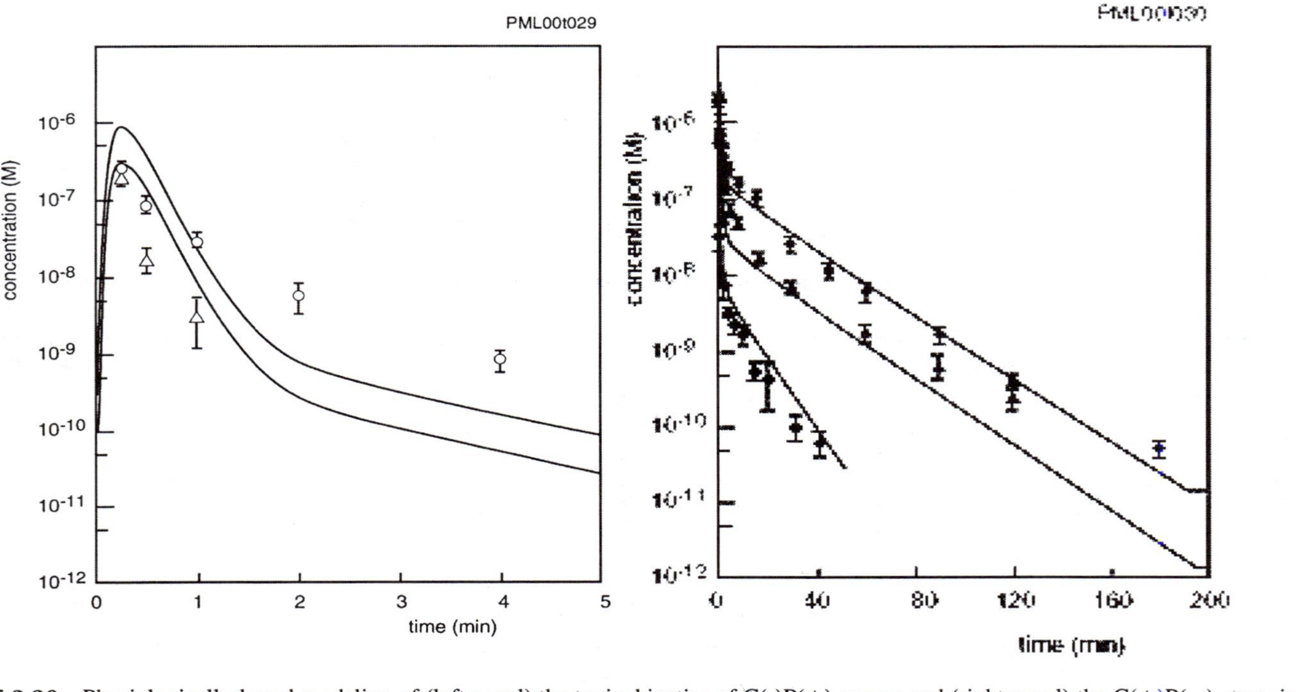

FIGURE 2.20 Physiologically-based modeling of (left panel) the toxicokinetics of C(-)P(+)-soman and (right panel) the C(±)P(−)-stereoisomers of soman for i.v. bolus administration of C(±)P(±)-soman to anesthetized, atropinized, and mechanically ventilated guinea pigs. The solid lines depict the predicted courses of the concentration in blood vs. time. The data points indicate the actually measured concentrations (± SE; n = 6) of C(−)P(+)-soman (open symbols) and of C(±)P(−)-soman (filled symbols) after a dose corresponding with 6 (○) and 2 LD_{50} (△) and of C(−)P(−)-soman (□) after a dose corresponding with 0.8 LD_{50}.

doses of C(±)P(±)-soman shown in Figure 2.20. Especially for the lowest dose (0.8 LD_{50}), the refinements in the model result in an improved simulation relative to earlier published data.[73] As should be expected, binding in blood becomes less dominant in overall toxicokinetics when the dose of C(±)P(−)-soman exceeds the capacity of the binding sites in blood, e.g., at doses corresponding to 2 LD_{50} and especially so at 6 LD_{50} of C(±)P(±)-soman. Accordingly, predictions of the blood levels of C(±)P(−)-soman at the latter doses were hardly affected by the changes introduced for modeling of C(±)P(−)-soman at a dose of 0.8 LD_{50} of C(±)P(±)-soman.

In a preliminary study, the effect of pretreatment with human plasma BuChE on the toxicokinetics of C(±)P(±)-soman was simulated with the developed model.[105] The rate constants for reaction of the soman stereoisomers with human BuChE were calculated from the overall rate constant reported by Ashani et al.,[107] and the ratio of the rate constants for inhibition of equine plasma BuChE found by Keijer and Wolring.[108] Since C(+)P(−)-soman has at least a 10-fold higher anti-BuChE activity than the other stereoisomers, it was assumed that this stereoisomer reacts preferentially and instantaneously after i.v. administration of human BuChE. Furthermore, it is assumed that any residual C(+)P(−)-soman will subsequently be eliminated by instantaneous binding to endogenous rapidly reacting binding sites in blood up to complete saturation of these binding sites. This is similar to the assumption made for modeling of the toxicokinetics of C(±)P(−)-soman after i.v. administration of a dose corresponding to 0.8 LD_{50} of C(±)P(±)-soman (*vide supra*). In order to interpret our modeling results, it is assumed that the scavenger provides sufficient protection if the predicted C(±)P(−)-soman concentrations in blood are similar to the concentrations calculated after i.v. administration of ≤ 0.7 LD_{50} of C(±)P(±)-soman in nonprotected guinea pigs, since 70% of the inhaled LD_{50} dose of C(±)P(±)-soman has been reported as a maximum sign-free dose in nonprotected guinea pigs.[90] According to this criterion, sufficient protection against intoxication with 1.5–3 LD_{50} and 5–6 LD_{50} of C(±)P(±)-soman is offered by a dose of human BuChE corresponding with 0.5 and 0.7 times the dose of C(±)P(±)-soman, respectively. This is illustrated in Figure 2.21. The predictions correspond with results obtained in protection experiments performed at similar conditions in mice, rats, and rhesus monkeys.[107]

The description of the model highlights the need for further refinement by discriminating between the four stereoismers of soman. Nevertheless, the model is a promising basis for extension to other routes of administration, for scaling to other species including man, and for making predictions on the efficacy of (pre)treatment drugs.

X. THE INFLUENCE OF PROPHYLAXIS AND THERAPY UPON THE TOXICOKINETICS OF SOMAN

In principle, the influence of prophylaxis and treatment on the toxicokinetics of C(+)P(−)- and C(−)P(−)-soman is a major item that should be addressed. Preliminary investigations have indicated that immediate treatment with HI-6 (150

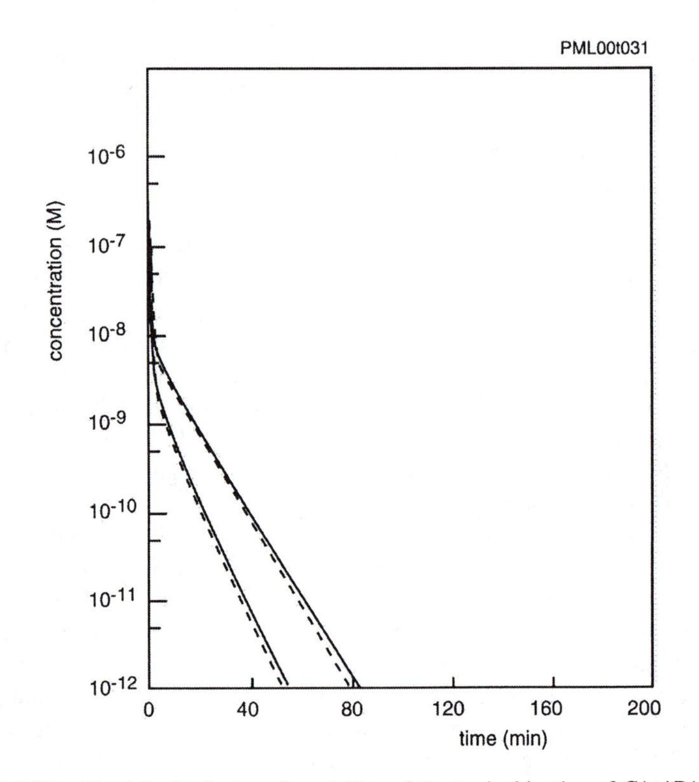

FIGURE 2.21 Physiologically-based modeling of the toxicokinetics of C(±)P(−)-soman for i.v. bolus administration of a dose corresponding with 2 LD_{50} C(±)P(±)-soman to anesthetized, atropinized, and mechanically ventilated guinea pigs pretreated with human BuChE. The solid lines depict the predicted courses of the concentration in blood vs. time of C(±)P(−)-soman in pretreated animals in which the ratio of the initial amount of scavenger in blood over the dose of C(±)P(±)-soman is 0.3 (upper curve) or 0.5 (lower curve). For comparison, the dashed lines represent predictions for the concentrations of C(±)P(−)-soman in blood subsequent to administration of doses corresponding with 0.8 (upper curve) or 0.5 (lower curve) LD_{50} of C(±)P(±)-soman (i.v.) to animals which were not pretreated with BuChE.

μmole · kg^{-1}, i.v.) of anesthetized, atropinized, and mechanically ventilated rats had hardly any influence on the toxicokinetics of C(+)P(−)- and C(−)P(−)-soman at an i.v. dose equivalent to 3 LD_{50} of C(±)P(±)-soman.[23] Pretreatment with pyridostigmine (11.8 μg · kg^{-1}, i.p.) at 1 h before i.v. administration of a dose of 3 LD_{50} of C(±)P(±)-soman to such rats caused a slight decrease of the blood levels of the abovementioned stereoisomers in the same time interval. However, omission of atropine in the latter experiments caused a significant increase of these blood levels in the same time interval.[20]

Based on the abovementioned preliminary data, the influence of atropine sulfate (50 mg · kg^{-1}, i.p.), administered 5 min before a challenge of anesthetized and

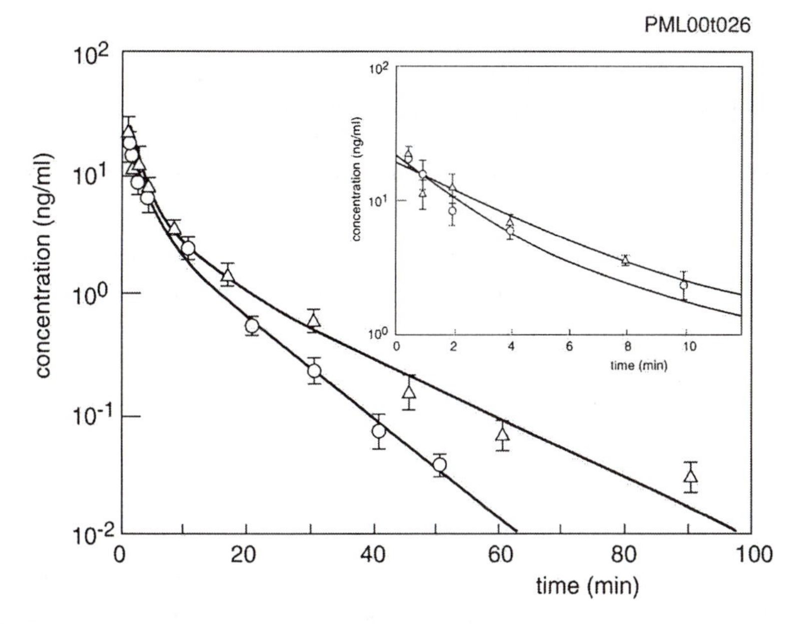

FIGURE 2.22 Semilogarithmic plot of the concentration in blood (± s.e.m.; n = 6) of C(+)P(−)-soman vs. time after administration of 1 LD_{50} (82.5 μg.kg^{-1}) of C(±)P(±)-soman to anesthetized and mechanically ventilated rats, pretreated (○) or not pretreated (△) by administration of atropine sulfate (50 mg.kg^{-1}, i.p.) 5 min before the challenge with C(±)P(±)-soman. The inset shows the data for the first 10 min plotted on an expanded scale.

mechanically ventilated rats with an i.v. dose of 1 LD_{50} of C(±)P(±)-soman was investigated in more detail.[20] The effect of pretreatment with atropine is visualized for the C(+)P(−)-stereoisomer in Figure 2.22, whereas the toxicokinetic data derived from the two-exponential equations fitting the concentration-time profiles are given in Table 2.12. Linear regression analysis of the two separate exponential terms of the equations showed that pretreatment with atropine had no significant influence in the first 10 min after intoxication. However, the subsequent elimination phase was significantly slower in the nonatropinized than in the atropinized animals, with terminal half-lives decreasing from 9.6–12.0 min to 6–7 min upon atropinization. Consequently, toxicologically relevant levels of C(±)P(−)-soman are present until 63 min after intoxication of the nonatropinized animals, but only until 37 min in the atropinized animals. In view of the antagonistic effects of atropine on the increased cardiovascular resistence caused by a dose of 1 LD_{50} of C(±)P(±)-soman in rats, the more efficacious elimination of the toxic stereoisomers of soman in atropinized animals is probably due to a better blood circulation in the latter animals, thus leading to an increased rate of transport to sites in the animal where the stereoisomers are bound, hydrolyzed, and excreted.[109,110]

TABLE 2.12
Toxicokinetic Parameters of Soman Stereoisomers in Anesthetized and Mechanically Ventilated Rats which Were Not Atropinized or Were Atropinized with Atropine Sulfate (i.p., 50 mg.kg^{-1}) 5 min Prior to a Challenge with an i.v. Dose of C(±)P(±)-soman Corresponding with 1 LD_{50} (82.5μg.kg^{-1}).

	C(+)P(−)-soman		C(−)P(−)-soman	
Parameter	Atropinized	Nonatropinized	Atropinized	Nonatropinized
Dose (μg/kg)	22.7	22.7	18.6	18.6
A (ng.ml^{-1})[a]	18.4	17.0	15.2	12.2
B (ng.ml^{-1})	3.9	2.7	5.9	4.4
a (min^{-1})	0.45	0.29	0.57	0.35
b (min^{-1})	0.096	0.058	0.12	0.072
Terminal half-life (min)	7.2	12.0	5.8	9.6
Area under curve (ng.min.ml^{-1})	81.5	105	76	96
Acutely toxic levels of	*——	——37——	——*	
C(±)P(−)soman until[b] (min)		*——	——63——	——*

[a]The concentration of each stereoisomer at time t (conc$_t$) is described by conc^t 5 Ae^{-at} + Be^{-bt}.

[b]After administration of C(±)P(±)-soman. It is assumed that an AUC of 1.8 ng.min.ml^{-1} is the minimum area with toxicological relevance (see text).

It should be concluded that the conventional treatment of C(±)P(±)-soman intoxication with atropine and oximes, such as HI-6, as well as prophylaxis with pyridostigmine have, at best, a modest beneficial effect on the elimination of the toxic stereoisomers of soman. On the other hand, the often-observed persistence of toxic levels of nerve agent for protracted periods of time after intoxication may profoundly affect the efficacy of treatment with the above-mentioned antidotes. For example, model calculations show that the efficacy of prophylaxis with pyridostigmine may be particularly limited by the persistence of C(±)P(−)-soman stereoisomers in the terminal elimination phase.[111] It follows that much can be gained in terms of efficacy of treatment if additional antidotes can be developed which aim specifically at sequestration of the toxic stereoisomers of soman. Such scavengers, e.g., phosphoryl phosphatase-type enzymes and BuChE from various sources are being developed. The investigation of the toxicokinetics of the nerve agent in the presence of scavenger will be essential to evaluate and validate the efficacy of this particular approach (see Section IX).

XI. TOXICOKINETICS OF V AGENTS

In contrast with G agents, few results of toxicokinetic investigations with V agents have been published. For various reasons it is worthwhile to compare the toxicokinetics of these two types of nerve agents. Several pathways for degradation of

G agents appear to be less effective for V agents. For example, phosphoryl phosphatases hydrolyze the P(+)-stereoisomers of G agents rapidly, but hardly degrade V agents.[112,113] Covalent scavenging by CaE is a major pathway for the degradation of the P(−)-stereoisomers of G agents, while these enzymes are much less effective in binding V agents.[114] Unless alternative *in vivo* pathways are available, these relative reactivities suggest a greater *in vivo* persistence of V agents than of G agents. What little is known about the toxicokinetics of V agents corroborates this assumption. Harris and co-workers investigated the toxicokinetics of the V agent O-cyclopentyl S-(2-diethylaminoethyl) methylphosphonothioate (45 $\mu g \cdot kg^{-1}$, i.v.) in anesthetized, atropinized, and mechanically ventilated rabbits (approximately 14.5 LD_{50}) and cynomolgus monkeys (approximately 10 LD_{50}).[16] They estimated the concentrations of V agent on the basis of the overall inhibitory potency of blood samples towards bovine AChE. Although this approach cannot account for stereospecificity in the degradation of the two stereoisomers of the V agent, their results (Figure 2.23) indicate that > 1 ng of V agent per ml blood is still circulating at 90–120 min after intoxication, albeit that the doses were extremely high. Similar results were obtained with VX at equitoxic doses. The authors suggest that the marginal effectiveness of pretreatment with the carbamate pyridostigmine against intoxication with V agents can be explained by the remarkable persistence of these agents. Recent studies by Rocha et al.[115] show that similar concentrations of VX can be sufficient to induce neurotoxic effects.

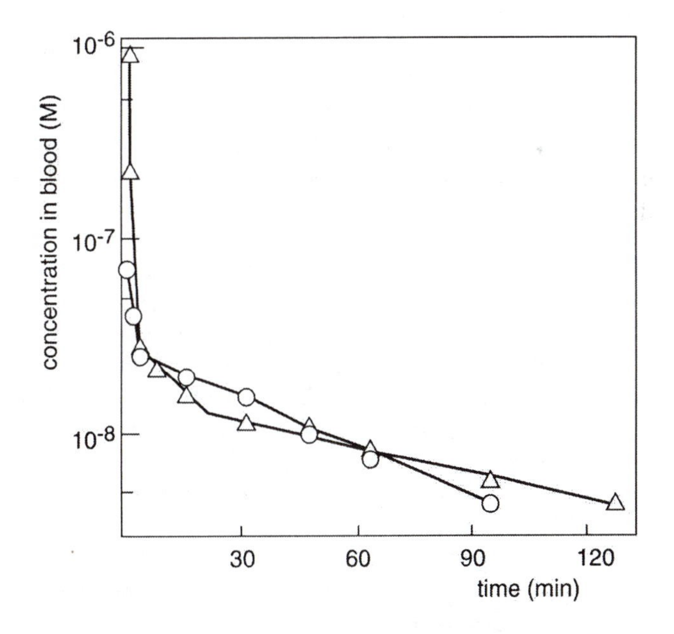

FIGURE 2.23 Semilogarithmic plot of the concentration in blood of O-cyclopentyl S-(2-diethylaminoethyl) methylphosphonothioate vs. time after administration of 45 $\mu g.kg^{-1}$ of this V agent to anesthetized, atropinized, and mechanically ventilated cynomolgus monkeys (Δ, approximately 10 LD_{50}) and white rabbits (O, approximately 14.5 LD_{50}).

Systematic investigation by Van der Schans et al.[116] on the toxicokinetics and metabolism of VX at doses corresponding with 1–2 LD_{50} in guinea pigs and marmosets via the i.v. and p.c. route are based on analysis of blood samples by means of chiral HPLC and gas chromatography. Preliminary results confirm the marked *in vivo* persistence of VX also at these more realistic doses and show that the overall stereospecificity for sequestration of (+)- and (−)-VX is much less pronounced than for the stereoisomers of G agents.

XII. FUTURE DIRECTIONS

With the methodology that has been developed for the analysis of nerve agents like soman, sarin, and VX, minimal detectable concentrations in the range of 1–50 $pg \cdot ml^{-1}$ blood have been obtained. These concentrations are at the lower limit of possibilities at the present or foreseeable state of the art in analytical chemistry. For most investigations on nerve agents, these procedures suffice to investigate the toxicokinetics at a wide range of doses and for all relevant routes of exposure. For example, many practical applications of toxicokinetic measurements can be expected in investigations dealing with the efficacy of new antidotes, e.g., in development of enzymatic scavengers for nerve agents.

The exposure studies of soman at low level (see Section VII) have shown that respiratory exposures for several hours to about 20 ppb of nerve agent are near the lower limit of what can be reached with regard to toxicokinetics based on *in vivo* measurement of intact nerve agent. If the toxicokinetics at even lower exposure levels should be investigated, e.g., at levels that are in the range of occupational exposure limits, one has to rely on measurement of nerve agent accumulated by internal scavengers such as BuChE and CaE. Methodologies for this approach have been developed for sarin and are being developed for other nerve agents, mostly based on release of the protein-bound nerve agent with fluoride ions and subsequent analysis of the generated phosphofluoridate.

It should be expected that further quantitative measurements on elimination routes of nerve agents, in combination with the wealth of available toxicokinetic data, will enable further development of physiologically-based modeling of toxicokinetics. Further model developments are needed, in particular for the respiratory and percutaneous exposure routes. Ultimately, this modeling will enable reliable interspecies extrapolation of toxicokinetic results, including extrapolation to man, which is the ultimate goal.

REFERENCES

1. Wolthuis, O.L., Benschop, H.P., and Berends, F., Persistence of the anticholinesterase soman in rats; antagonism with a non-toxic simulator of this organophosphate, *Eur. J. Pharmacol.*, 69, 379, 1981.
2. Kientz, C.E., Langenberg, J.P., and Brinkman, U.A.Th., Micocolumn liquid chromatography with thermionic detection of the enantiomers of O-ethyl S-diisopropylaminoethyl methylphosphonothioate (VX), *J. High Resolut. Chromatogr.*, 17, 95, 1994.

3. Benschop, H.P., Konings, C.A.G., and De Jong, L.P.A., Gas chromatographic separation and identification of the four stereoisomers of 1,2,2-trimethylpropyl methylphosphonofluoridate (soman). Stereospecificity of in vitro "detoxification" reactions, *J. Am. Chem. Soc.,* 103, 4260, 1981.
4. Benschop, H.P., Bijleveld, E.C., Otto, M.F., Degenhardt, C.E.A.M., Van Helden, H.P.M., and De Jong, L.P.A., Stabilization and gas chromatographic analysis of the four stereoisomers of 1,2,2-trimethylpropyl methylphosphonofluoridate (soman) in rat blood, *Anal. Biochem.,* 151, 242, 1985.
5. Benschop, H.P. and De Jong, L.P.A., Nerve agent stereoisomers: Analysis, isolation and toxicology, *Acc. Chem. Res.,* 21, 368, 1988.
6. Mehlsen-Sørensen, A., (R)-(+)-O-isopropyl-S-(trimethylammonioethyl) methylphosphonothioate iodide, *Acta Crystallogr., Sect. B,* B33, 1693, 1977.
7. Benschop, H.P., Absolute configuration of chiral organophosphorus anticholinesterases, *Pestic. Biochem. Physiol.,* 5, 348, 1975.
8. De Jong, L.P.A. and Benschop, H.P., Biochemical and toxicological implications of chirality in anticholinesterase organophosphates, in *Stereoselectivity of Pesticides. Biological and Chemical Problems,* Ariens, E.J., Van Rensen, J.J.S., and Welling, W., eds., Elsevier, Amsterdam, 1988, chap. 4.
9. Ordentlich, A., Barak, D., Kronman, C., Benschop, H.P., De Jong, L.P.A., Ariel, N., Barak, R., Segall, Y., Velan, B., and Shafferman, A., Exploring the active center of human acetylcholinesterase with stereoisomers of an organophosphorus inhibitor with two chiral centers, *Biochemistry,* 38, 3055, 1999.
10. Benschop, H.P., Konings, C.A.G., Van Genderen, J., and De Jong, L.P.A., Isolation, anticholinesterase properties and acute toxicity in mice of the four stereoisomers of soman, *Toxicol. Appl. Pharmacol.,* 90, 61, 1984.
11. Boter, H.L. and Van Dijk, C., Stereospecificity of hydrolytic enzymes on reaction with asymmetric organophosphorus compounds. III. The inhibition of acetylcholinesterase and butyrylcholinesterase by enantiomeric forms of sarin, *Biochem. Pharmacol.,* 18, 2403, 1969.
12. Hall, C.R., Inch, T.D., Inns, R.H., Muir, A.W., Sellers, D.J., and Smith, A.P., Differences between some biological properties of enantiomers of alkyl S-alkyl methylphosphonothioates, *J. Pharm. Pharmacol.,* 29, 574, 1977.
13. Little, P. J., Reynolds, M. L., Bowman, E. R., and Martin, B. R., Tissue disposition of [^{3}H]sarin and its metabolites in mice, *Toxicol. Appl. Pharmacol.,* 83, 412, 1986.
14. Little, P.J., Scimeca, J.A., and Martin, B.R., Distribution of [^{3}H]diisopropylfluorophosphate, [^{3}H]soman, [^{3}H]sarin and their metabolites in mouse brain, *Drug Metab. Dispos.,* 16, 515, 1988.
15. Reynolds, M., Little, P.J., Thomas, B.F., Bagley, R.B., and Martin, B.R., Relationship between the biodisposition of [^{3}H]soman and its pharmacological effects in mice, *Toxicol. Appl. Pharmacol.*, 80, 409, 1985.
16. Harris, L., Broomfield, C., Adams, N., and Stitcher, D., Detoxification of soman and O-cyclopentyl S-diethylaminoethyl methylphosphonothioate *in vivo, Proc. West. Pharmacol. Soc*., 27, 315, 1984.
17. Göransson-Nyberg, A., Frederiksson, S.-Å., Karlsson, B., Lundström, M., and Cassel, G., Toxicokinetics of soman in cerebrospinal fluid and blood of anaesthetized pigs, *Arch. Toxicol.,* 72, 459, 1998.
18. Benschop, H.P., De Jong, L.P.A., and Langenberg, J.P., Inhalation toxicokinetics of C(±)P(±)-soman and (±)-sarin in the guinea pig, in *Enzymes of the Cholinesterase Family,* Quinn, D.M., Balasubramanian, A.S., Doctor, B.P., and Taylor, P., eds., Plenum Press, New York, 1995, 361.

19. Benschop, H.P. and De Jong, L.P.A., Toxicokinetics of soman: Species variation and stereospecificity in elimination pathways, *Neurosci. Biobehav. Rev.*, 15, 73, 1991.
20. Benschop, H.P. and De Jong, L.P.A., Toxicokinetic investigations of C(±)P(±)-soman in the rat, guinea pig and marmoset at low doses—Quantification of elimination pathways. *Final Report Grant DAMD17-87-7015,* NTIS AD-A 226 807, 1990.
21. Maxwell, D.M., Brecht, K.M., and O'Neill, B.L., The effect of CaE's inhibition on interspecies differences in soman toxicity, *Toxicol. Lett.,* 39, 35, 1987.
22. Due, A.H., Trap, H.C., Van der Wiel, H.J., and Benschop, H.P., The effect of pretreatment with CBDP on the toxicokinetics of soman stereoisomers in rats and guinea pigs, *Arch. Toxicol.,* 67, 706, 1993.
23. Benschop, H.P. and De Jong, L.P.A., Toxicokinetics of the four stereoisomers of soman in the rat., guinea pig and marmoset, *Annual/Final Report Grant DAMD17-85-G-5004,* NTIS AD-A199 573, 1987.
24. Benschop, H.P. and Van Helden, H.P.M., Toxicokinetics of inhaled soman and sarin in guinea pigs, *Final Report Contract DAMD17-90-Z-0034,* NTIS AD-A277 585, 1993. See also: Spruit, H.E.T., Langenberg, J.P., Trap, H.C., Van Der Wiel, H.J., Helmich, R.B., Van Helden, H.P.M., and Benschop, H.P., Intravenous and inhalating toxicokinetics of savin stereoisomers in atropinized Guinea pigs, *Toxicol. Appl. Pharmacol.,* in press.
25. Due, A.H., Trap, H.C., Langenberg, J. P., and Benschop, H.P., Toxicokinetics of soman stereoisomers after subcutaneous administration to atropinized guinea pigs, *Arch. Toxicol.,* 68, 60, 1994.
26. Oberst, F.W., Factors affecting inhalation and retention of toxic vapors, in *Inhaled Particles and Vapours,* Davies, C. N., ed., Pergamon Press, Oxford, 1961, 249.
27. Oberst, F.W., Koon, W.S., Christensen, M.K., Crook, J.W., Cresthull, B.S., and Freeman, G., Retention of inhaled sarin vapor and its effect on red blood cell cholinesterase activity in man, *Clin. Pharmacol. Ther.*, 9, 421, 1968.
28. Langenberg, J.P., Spruit, H.E.T., Van Der Wiel, H.J., Trap, H.C., Helmich, R.B., Bergers, W.W.A., Van Helden, H.P.M., and Benschop, H.P., Inhalation toxicokinetics of soman stereoisomers in the atropinized guinea pig with nose-only exposure to soman vapor, *Toxicol. Appl. Pharmacol.*, 151, 79, 1998.
29. Ainsworth, M. and Shepherd, R.J., The intrabronchial distribution of soluble vapors at selected rates of gas flow, in *Inhaled Particles and Vapours,* C.N. Davies, ed., Pergamon Press, Oxford, 1961, 233.
30. Reed, C.J., Drug metabolism in the nasal cavity: Relevance to toxicology, *Drug Metab. Rev.*, 25, 173, 1993.
31. Dahl, A.R. and Gerde, P., Uptake and metabolism of toxicants in the respiratory tract. *Environm. Health Perspect.*, 102, 67, 1994; see also Miller, F.J. and Kimball, J.S., Regional dosimetry of inhaled reactive gases, in *Concepts in Inhalation Toxicology,* McClellan, R.O. and Henderson, R.F., eds., Taylor & Francis, Washington, 1995, 257.
32. Ember, L., Probe of troops' exposure to chemical arms failed, *Chem. Eng. News,* Sept. 23, 40, 1996.
33. Central Intelligence Agency, *Modeling the Chemical Warfare Agent Release at the Khamisiyah Pit,* 1997, www.gulflink.osd.mil/cia_092297
34. Croddy, E., Urban terrorism-chemical warfare in Japan,. *Jane's Intelligence Rev.,* November 1995, 520.
35. Nozaki, H., Hori, S., Shinozawa, S., Fujishima, S., Takuma, K., Sagoh, M., Kimura, H., Ohki, T., Suzuki, M., and Aikawa, N., Secondary exposure of medical staff to sarin vapor in the emergency room, *Intensive Care Med.*, 21, 1032, 1995.
36. Brackett, D.W., *Armageddon in Tokyo,* Weatherhill, New York, 1996.

37. Okudera, H., Morita, H., Iwashita, H., Shibata, T., Otagiri, T., Kobayashi, S., and Yanagisawa, N., Unexpected nerve gas exposure in the city of Matsumoto: Report of rescue activity in the first sarin gas terrorism, *Am. J. Emerg. Med.*, 15, 527, 1997.
38. Suzuki, J., Kohno, T., Tsukagosi, M., Furuhata, T., and Yamazaki, K., Eighteen cases exposed to sarin in Matsumoto, Japan, *Intern. Med.*, 36, 466, 1997.
39. Nakajima, T., Sato, S., Morita, H., and Yanagisawa, N., Sarin poisoning of a rescue team in the Matsumoto incident in Japan, *Occup. Environ. Med.*, 54, 697, 1997.
40. National Research Council (Committee on Toxicology), *Possible Long-Term Effects of Short Term Exposure to Chemical Agents*. Vol. 3. Final Report. Current health status of test subjects. NTIS AD-A163 614, 1985.
41. Okumura, T., Takasu, N., Ishimatsu, S., Miyanoka, S., Mitsuhashi, A., Kumada, K., and Hinohara, S., Report on 640 victims of the Tokyo subway attack, *Ann. Emerg. Med.*, 28, 129, 1996.
42. Murata, K., Araki, S., Yokoyama, K., Okumura, T., Ishimatsu, S., Takasu, N. and White, R.F., Asymptomatic sequelae to acute sarin poisoning in the central and autonomic nervous system 6 months after the Tokyo subway attack, *J. Neurol.*, 244, 601, 1997.
43. Yokoyama, K., Araki, S., Murata, K., Nishikitani, M., Okumura, T., Ishimatsu, S., Takasu, N., and White, R. F., Chronic neurobehavioral effects of Tokyo subway sarin poisoning in relation to posttraumatic stress disorder, *Arch. Environ. Health*, 53, 249, 1998.
44. Nakajima, T., Ohta, S., Fukushima, Y., and Yanagisawa, N., Sequelae of sarin toxicity at one and three years after exposure in Matsumoto, Japan, *J. Epidemiol.*, 9, 337, 1999.
45. Benschop, H.P., Trap, H.C., Spruit, E.T., Van Der Wiel, H.J., Langenberg, J.P., and De Jong, L.P.A., Low level-nose only exposure to the nerve agent soman: toxicokinetics of soman stereoisomers and cholinesterase inhibition in atropinized guinea pigs, *Toxicol. Appl. Pharmacol.*, 153, 179, 1998.
46. Jacobson, K.H., Christensen, M.K., DeArmon, I.A., and Oberst, F.W., Studies of chronic exposures of dogs to GB (isopropyl methylphosphonofluoridate) vapor, *Arch. Ind. Health*, 19, 5, 1959.
47. Moylan-Jones, R.J. and Price Thomas, D., Cyclopentolate in treatment of sarin miosis, *Br. J. Pharmacol.*, 48, 309, 1973.
48. Ballantyne, B. and Marrs, T.C., *Organophosphates and Carbamates*, Butterworth Heinemann, Oxford, 1992, 380.
49. Rubin, L.S. and Goldberg, M.N., Effect of sarin on dark adaptation in man: threshold changes, *J. Appl. Physiol.*, 11, 439, 1957.
50. Mioduszewski, R.J., Reutter, S.A., Miller, L.L., Olajos, E.J., and Thomson, S.A., Evaluation of airborne exposure limits for G-agents: occupational and general population exposure criteria, *Report ERDEC-TR-489*, Edgewood Research, Development & Engineering Center, April 1998. See also: Fact sheet on exposure limits for sarin (GB), July 1997, http://www.gulflink.osd.mil/dugway/low_lv_chem_fact.htm.
51. Polhuijs, M., Langenberg, J.P., and Benschop, H.P., New method for retrospective detection of exposure to organophosphate anticholinesterases: Application to alleged sarin victims of Japanese terrorists, *Toxicol. Appl. Pharmacol.*, 146, 156, 1997.
52. Van Helden, H.P.M., Langenberg, J.P., and Benschop, H.P., Low level exposure to GB vapor in air: diagnosis/dosimetry, lowest observable effect levels, performance-incapacitation and possible delayed effects, *Contract DAMD17-97-1-7360*, U.S. Army Medical Research and Materiel Command. See also: Trap, H.C., Kuijpers, W.C., Groen, B., Oostdijk, J.P., Vanwersch, R.A.P., Philippens, H.C., Langenberg, J.P., Benschop, H.P., and Van Helden, H.P.M., Low-level exposure to GB in air: diagnosis/dosimetry, lowest

observable effect level (LOEL), and lowest observable adverse effect level (LOAEL), Grant DAMD17-97-1-7360, U.S. Army Medical Command.

53. Mazur, A., An enzyme in animal tissues capable of hydrolyzing the phosphorus-fluorine bond of alkyl fluorophosphates, *J. Biol. Chem.,* 164, 271, 1946.
54. Mounter, L.A., Metabolism of organophosphorus anticholinesterase agents, in *Handbuch der Experimentellen Pharmakologie. Vol. XV. Cholinesterases and Anticholinesterase Agents,* Koelle, G.B., ed., Springer, Berlin, 1963, chap. 10.
55. Adie, P.A. and Tuba, J., The intracellular localization of liver and kidney sarinase, *Can. J. Biochem. Biophys.*, 36, 21, 1958.
56. Adie, P.A., Hoskin, F.C.G., and Trick, G.S., Kinetics of the enzymatic hydrolysis of sarin, *Can. J. Biochem. Biophys.*, 64, 80, 1958.
57. Augustinsson, K.-B. and Heimburger, G., The enzymic hydrolysis of organophosphorus compounds. I. Occurrence of enzymes hydrolyzing tabun, *Acta Chem. Scand.*, 8, 753, 1954.
58. Cohen, J.A. and Warringa, M.P.G.J., Purification and properties of dialkylfluorophosphatase, *Biochim. Biophys. Acta,* 26, 29, 1957.
59. Millard, C.B., Lockridge, O., and Broomfield, C.A., Design and expression of organophosphorus acid anhydride hydrolase activity in human butyrylcholinesterase, *Biochemistry,* 34, 15925, 1995.
60. Lockridge, O., Blong, R.M., Masson, P., Froment, M.T., Millard, C.B., and Broomfield, C.A., A single amino acid substitution, Gly117His, confers phosphotriesterase (organophosphorus acid anhydride hydrolase) activity on human butyrylcholinesterase, *Biochemistry,* 36, 786, 1997.
61. Noort, D., Hulst, A.G., Platenburg, D.H.J.M., Polhuijs, M., and Benschop, H.P., Quantitative analysis of O-isopropyl methylphosphonic acid in serum samples of Japanese citizens allegedly exposed to sarin: Estimation of internal dose, *Arch. Toxicol.,* 72, 671, 1998.
62. Minami, M., Hui, D.-M., Katsumata, M., Inagaki, H., and Boulet, C., Method for the analysis of methylphosphonic acid metabolites of sarin and its ethanol-substituted analogue in urine as applied to the victims of the Tokyo sarin disaster, *J. Chromatogr. B.*, 695, 237, 1997.
63. Harris, L.W., Braswell, L.M., Fleisher, J.H., and Cliff, W.J., Metabolites of pinacolyl methylphosphonofluoridate (soman) after enzymatic hydrolysis *in vitro, Biochem. Pharmacol.,* 13, 1129, 1964.
64. Ramachandran, B.V., The influence of DFP, atropine and pyridinium aldoximes on the rate of clearance of diisopropyl phosphate ($DI^{32}P$) from the mouse circulatory system, *Biochem. Pharmacol.*, 16, 2061, 1967.
65. Nakajima, T., Sasaki, K., Sekjima, Y., Morita, H., Fukushima, Y., and Yanagisawa, N., Urinary metabolites of sarin in a patient of the Matsumoto sarin incident, *Arch. Toxicol.*, 72, 601, 1998.
66. Christen, P.J. and Van den Muysenberg, J.A.C.M., The enzymatic isolation and fluoride catalysed racemisation of optically active sarin, *Biochim. Biophys. Acta,* 110, 217, 1965.
67. Christen, P.J., Berends, F., and Cohen, E.M., The influence of stereoisomerism of sarin on various reactions with sarinase and cholinesterase, *Acta Physiol. Pharmacol. Neerl.,* 14, 338, 1967.
68. De Bisschop, H.C., Mainil, J.G., and Willems, J.L., *In vitro* degradation of the four stereoisomers of soman in human serum, *Biochem. Pharmacol.,* 34, 1895, 1985.
69. De Bisschop, H.C., De Meerleer, W.A.P., Van Hecke, P.R.J., and Willems, J.L., Stereoselective hydrolysis of soman in human plasma and serum, *Biochem. Pharmacol.,* 36, 3579, 1987.
70. De Bisschop, H.C., De Meerleer, W.A.P., and Willems, J.L., Stereoselective phosphonylation of human serum proteins by soman, *Biochem. Pharmacol.,* 36, 3587, 1987.

71. De Jong, L.P.A., Van Dijk, C., and Benschop, H.P., Hydrolysis of the four stereoisomers of soman catalyzed by liver homogenate and plasma from rat, guinea pig and marmoset and by human plasma, *Biochem. Pharmacol.,* 37, 2939, 1988.
72. De Jong, L.P.A., Van Dijk, C., Berhitoe, D., and Benschop, H.P., Hydrolysis and binding of a toxic stereoisomer of soman in plasma and tissue homogenates from rat, guinea pig and marmoset, and in human plasma, *Biochem. Pharmacol.,* 46, 1413, 1993.
73. Langenberg, J.P., Van Dijk, C., Sweeney, R.E., Maxwell, D.M., De Jong, L.P.A., and Benschop, H.P., Development of a physiologically based model for the toxicokinetics of C(±)P(±)-soman in the atropinized guinea pig, *Arch. Toxicol.,* 71, 320, 1997.
74. Augustinsson, K.-B., Enzymatic hydrolysis of organophosphorus compounds. VII. The stereospecificity of phosphoryl phosphatases, *Acta Chem. Scand.,* 11, 1371, 1957.
75. Degenhardt, C.E.A.M., Van den Berg, G.R., De Jong, L.P.A., Benschop, H.P., Van Genderen, J., and Van de Meent, D., Enantiospecific complexation gas chromatography of nerve agents. Isolation and properties of the stereoisomers of ethyl N, N-dimethylphosphoramido-cyanidate (tabun), *J. Am. Chem. Soc.,* 108, 8290, 1986.
76. Myers, D.K., Cholinesterase. VII. Determination of the molar concentration of pseudo-cholinesterase in serum, *Biochem. J.,* 51, 303, 1952.
77. Polak, R.P. and Cohen, E.M., Influence of tri-o-cresyl phosphate on the distribution of phosphorus-32 in the body of the rat after the injection of sarin-^{32}P, *Biochem. Pharmacol.,* 18, 813, 1969.
78. Sterri, S.H. and Fonnum, F., Detoxification of organophosphorus compounds, in *Cholinesterases. Fundamental and Applied Aspects,* Brzin, M., Barnard, E.A., and Sket, D., eds., de Gruyter, Berlin, 1984, 110.
79. Christen, P.J. and Cohen, E.M., Binding of ^{32}P-sarin to esterases and other proteins in plasma from rat, man and guinea-pig, *Acta Physiol. Pharmacol. Neerl.,* 15, 36, 1969.
80. Black, R.M., Harrison, J.M., and Read, R.W., The interaction of sarin and soman with plasma proteins: the identification of a novel phosphylation site, *Arch. Toxicol.,* 73, 123, 1999.
81. Maxwell, D.M., Lenz, D.E., Groff, W.A., Kaminskis, A., and Froehlich, H.L., The effects of blood flow and detoxification on *in vivo* cholinesterase inhibition by soman in rats, *Toxicol. Appl. Pharmacol.,* 88, 66, 1987.
82. Myers, D.K., Mechanism of the prophylactic action of diacetylmonoxime against sarin poisoning, *Biochim. Biophys. Acta,* 34, 555, 1959.
83. McKay, D.H., Jardine, R.V., and Adie, P.A., The synergistic action of 2-(o-cresyl)-4*H*-1:3:2-benzodioxaphosphorin-2-oxide with soman and physostigmine, *Toxicol. Appl. Pharmacol.,* 20, 474, 1971.
84. Sterri, S.H. and Fonnum, F., CaE's—the soman scavenger in rodents; heterogeneity and hormonal influence, in *Enzymes Hydrolysing Organophosphorus Compounds,* Reiner, E., Aldridge, W.N., and Hoskin, F.C.G., eds., Ellis Horwood Ltd., Chichester, 1989, chap. 14.
85. Sterri, S.H., Lyngaas, S., and Fonnum, F., Toxicity of soman after repetitive injection of sublethal doses in rats, *Acta Pharmacol. Toxicol.,* 46, 1, 1980.
86. Sterri, S.H., Lyngaas, S., and Fonnum, F., Toxicity of soman after repetitive injection of sublethal doses in guinea-pig and mouse, *Acta Pharmacol. Toxicol.,* 49, 8, 1981.
87. Doctor, B.P., Blick, D.W., Caranto, G., Castro, C.A., Gentry, M.K., Larrison, R., Maxwell, D.M., Murphy, M.R., Schutz, M., Waibel, K., and Wolfe, A.D., Cholinesterases as scavengers for organophosphorus compounds: Protection of primate performance against soman toxicity, *Chem.-Biol. Interact.,* 87, 285, 1993.
88. Maxwell, D.M., Castro, C.A., De La Hoz, D.M., Gentry, M.K., Gold, M.B., Solana, R.P., Wolfe, A.D., and Doctor, B.P., Protection of rhesus monkeys against soman and prevention

of performance decrement by pretreatment with acetylcholinesterase, *Toxicol. Appl. Pharmacol.,* 115, 44, 1992.

89. Raveh, L., Grauer, E., Grunwald, J., Cohen, E., and Ashani, Y., The stoichiometry of protection against soman and VX toxicity in monkeys pretreated with human butyrylcholinesterase, *Toxicol. Appl. Pharmacol.,* 145, 43, 1997.
90. Allon, N., Raveh, L., Gilat, E., Cohen, E., Grunwald, J., and Ashani, Y., Prophylaxis against soman inhalation toxicity in guinea pigs by pretreatment alone with human serum butyrylcholinesterase, *Toxicol. Sci.,* 43, 121, 1998.
91. Ramachandran, B.V., Distribution of $DF^{32}P$ mouse organs. I. The effect of route of administration on incorporation and toxicity, *Biochem. Pharmacol.*, 15, 169, 1966.
92. Lenz, D.E., Maxwell, D.M., Prather, R., and Ball, L., *In vivo* distribution of 14C-soman in rats, *The Pharmacologist,* 25, 111, 1983.
93. Heilbronn, E., Appelgren, I.-E., and Sundwall, A., The fate of tabun in atropine and atropine oxime treated rats and mice, *Biochem. Pharmacol.,* 13, 1189, 1964.
94. Hansen, D., Schaum, E., and Wasserman, O., Distribution and metabolism of diisopropyl phosphorofluoridate (DFP) in the guinea pig, *Arch. Toxicol.,* 23, 73, 1968.
95. Hansen, D., Schaum, E., and Wasserman, O., Serum level and excretion of diisopropyl fluorophosphate (DFP) in cats, *Biochem. Pharmacol.,* 17, 1159, 1968.
96. De Jong, L.P.A., Benschop, H.P., Due, A., Van Dijk, C., Trap, H.C., Van der Wiel, H.J., and Van Helden, H.P.M., Soman levels in kidney and urine following administration to rat, guinea pig, and marmoset, *Life Sci.,* 50, 1057, 1992.
97. De Jong, L.P.A. and Van Dijk, C., Formation of soman (1,2,2-trimethylpropyl methylphosphonofluoridate) via fluoride-induced reactivation of soman-inhibited aliesterase in rat plasma, *Biochem. Pharmacol.,* 33, 663, 1984.
98. Dedrick, R.L., Forrester, D.D., Cannon, J.N., El Dareer, S.M., and Mellett, L.B., Pharmacokinetics of 1-β-D-arabinofuranosylcytosine (ARA-C) deamination in several species, *Biochem. Pharmacol.,* 22, 2405, 1973.
99. King, F.G., Dedrick, R.L., Collins, J.M., Matthews, H.B., and Birnbaum, L.S., Physiological model of the pharmacokinetics of 2,3,7,8-tetrachlorodibenzofuran in several species, *Toxicol. Appl. Pharmacol.,* 67, 390, 1983.
100. Lutz, R.J., Dedrick, R.L., Tuey, D., Sipes, I.G., Anderson, M.W., and Matthews, H.B., Comparison of the pharmacokinetics of several polychlorinated biphenyls in mouse, rat, dog, and monkey by means of a physiological pharmacokinetic model, *Drug Metab. Dispos.,* 12, 527, 1984.
101. Ramsey, J.C. and Andersen, M.E., A physiologically based description of the inhalation pharmacokinetics of styrene in rats and humans, *Toxicol. Appl. Pharmacol.,* 73, 159, 1984.
102. Maxwell, D.M., Vlahacos, C.P., and Lenz, D.E., A pharmacodynamic model for soman in the rat, *Toxicol. Lett.,* 43, 175, 1988.
103. Gearhart, J.M., Jepson, G.W., Clewell III, H.J., Andersen, M.E., and Conolly, R.B., Physiologically based pharmacokinetics and the pharmacodynamic model for inhibition of acetylcholinesterase by diisopropyl fluorophosphate, *Toxicol. Appl. Pharmacol.,* 106, 295, 1990.
104. McPhail, M.K. and Adie, P.A., The distribution of radioactive phosphorus in the blood and tissues of rabbits treated with tagged sarin, *Can. J. Biochem. Physiol.,* 38, 945, 1960.
105. De Jong, L.P.A., Langenberg, J.P., and Benschop, H.P., TNO Prins Maurits Laboratory, unpublished data, 1999.
106. Rakusan, K. and Blahitka, J., Cardiac output distribution in rats measured by injection of radioactive microspheres via cardiac puncture, *Can. J. Physiol. Pharmacol.,* 52, 230, 1974.

107. Ashani, Y., Grunwald, J., Raveh, L., Grauer, E., Brandeis, R., Marcus, D., Papier, Y., Kadar, T., Allon, N., Gilat, E., and Lerer, S., Cholinesterase prophylaxis against organophosphorus poisoning, *Final Report Contract DAMD17-90-C-0033,* NTIS ADA277 096, 1993.
108. Keijer, J.H. and Wolring, G.Z., Stereospecific aging of phosphonylated cholinesterases, *Biochim. Biophys. Acta,* 185, 465, 1969.
109. Vojvodic, V. and Milosevic, M., Some pharmacological actions of oximes and atropine on the cadiovascular effects produced by pinacolyl methylphosphonofluoridate (soman) in rats, *Iugosl. Physiol. Pharmacol. Acta,* 7, 439, 1971.
110. Kentera, D., Susic, D., and Stamenovic, B., The effects of HS-3 and HS-6 on cardiovascular changes in rats caused by soman, *Arh. Hig. Rada Toksikol.,* 33, 143, 1982.
111. Langenberg, J.P., De Jong, L.P.A., and Benschop, H.P., Kinetic modeling of pretreatment against soman poisoning, in *Book of Abstracts 1991 EUROTOX Congress,* Maastricht, 1991, 244.
112. Bajgar, J., Fusek, J., Patocka, J., and Hrdina, V., Detoxication of phosphonothioates and phosphonofluoridates in the rat, *Acta Biol. Med. Germ.,* 37, 1261, 1978.
113. Wang, Q., Sun, M., Zhang, H., and Huang, C., Purification and properties of soman-hydrolyzing enzyme from human liver, *J. Biochem. Mol. Toxicol.,* 12, 213, 1998.
114. Maxwell, D.M., The specificity of carboxylesterase protection against the toxicity of organophosphorus compounds, *Toxicol. Appl. Pharmacol.,* 114, 306, 1992.
115. Rocha, E.S., Santos, M.D., Chebabo, S.R., Aracava, Y., and Albuquerque, E.X., Low concentrations of the organophosphate VX affect spontaneous and evoked transmitter release from hippocampal neurons: Toxicological relevance of cholinesterase-independent actions, *Toxicol. Appl. Pharmacol.,* 159, 31, 1999.
116. Van der Schans, M.J., Langenberg, J.P., and Benschop, H.P., Toxicokinetics of O-ethyl S-(2-diisopropylaminoethyl) methylphosphonothioate [(±)-VX] in hairless guinea pigs and marmosets—Identification of metabolic pathways, *Final Report Contract DAMD17-97-2-7001,* 2000.

3 Low-Level Nerve Agent Toxicity under Normal and Stressful Conditions

Satu M. Somani and Kazim Husain

CONTENTS

I. Introduction 83
A. Nerve Agents 83
B. Delayed Neurotoxicity 85
C. Stress 89
II. Cholinergic Toxicity 92
A. Biochemical Effects 94
B. Histopathological Effects 101
C. Cholinergic Toxicity under Stressful Conditions 101
III. Non-Cholinergic Toxicity 103
A. Biochemical Effects 104
B. Histopathological Effects 107
C. Non-Cholinergic Toxicity under Stressful Conditions 108
IV. Summary 108
Acknowledgments 109
References 109

I. INTRODUCTION

A. Nerve Agents

Nerve agents were developed over six decades ago for military use and continue to be a significant threat on the battlefields of the world or as terrorist weapons. Organophosphate (OP) nerve agents (tabun, sarin, soman, and VX) are the most toxic compounds that cause biological effects by inhibiting the enzyme cholinesterase. The first OP nerve agent, tabun (O-ethyl N, N-dimethyl phosphoraminocyanidate) was synthesized by German chemist Dr. Gerhard Schrader in 1936.[1] Later sarin (isopropyl methyl phosphonofluoridate) was synthesized in 1938 followed by soman (pinacolyl methyl phosphonofluoridate) in 1944. A few years after the end of World

0-8493-0872-0/01/$0.00+$.50

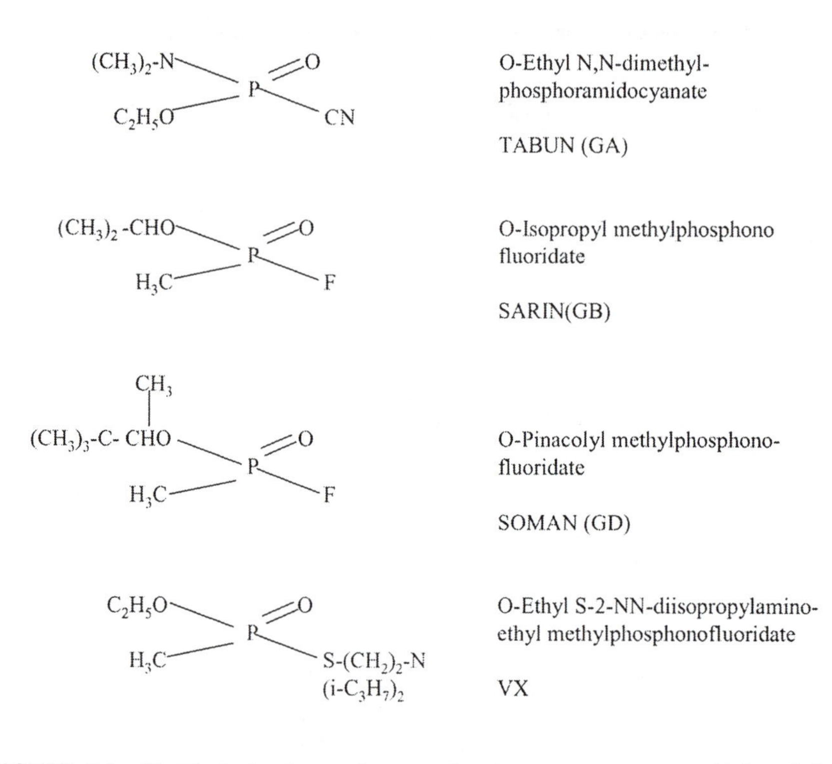

FIGURE 3.1 Chemical structures of organophosphorous nerve agents. (Adapted from Somani et al.[212])

War II, Dr. Ranajit Ghosh of England synthesized a nerve agent called VX (O-ethyl-S-2-N,N-diisopropyl amino ethyl methylphosphonothiolate) that was much more potent than sarin. The chemical structures of OP nerve agents are depicted in Figure 3.1.

Among the OP nerve agents, sarin has been used as a chemical warfare agent since its synthesis during World War II. Its use was most recently demonstrated during the Iran-Iraq conflict and during the Gulf War.[2–4] These reports indicate that Gulf War veterans might have been exposed to low doses of sarin. It has been reported that German personnel exposed to nerve agents during World War II suffered neurological problems 5 to 10 years after their last exposures.[5,6] Furthermore, long-term neurologic and psychiatric abnormalities have also been seen in personnel exposed to sarin in its manufacturing plants.[7,8] A terrorist attack with sarin gas on March 20, 1995 in Japan and an earlier killing of a Japanese terrorist member by another deadly nerve agent, VX, have attracted the world's attention about the threat to the general world population.[9,10] These episodes have added new dimensions to the dangers that humanity is facing all over the globe.

Nerve agents are inhaled as vapors or aerosols and, being lipid soluble, immediately enter systemic circulation, resulting in toxic manifestations at muscarinic, nicotinic, and CNS cholinergic sites.[11] The acute cholinergic symptoms (tremors, convulsions, salivation, lacrimation, and respiratory failure) are due to the inhibition of acetylcholinesterase at central, peripheral, and autonomic synapses, resulting in accumulation of acetylcholine at synaptic junctions.[12] The muscarinic effects include ocular (miosis, conjunctival congestion, ciliary spasm, nasal discharge), respiratory (broncho-constriction and increased bronchial secretion), gastrointestinal (anorexia, vomiting, abdominal cramps, diarrhea), percutaneous (sweating and muscular fasciculation), salivation, bradycardia, and hypotension. Nicotinic effects include muscle fasciculations and paralysis. The central nervous system effects include ataxia, confusion, loss of reflexes, slurred speech, headache, anxiety, restlessness, irritability, giddiness, insomnia, convulsions, and coma. The cause of death of persons exposed to nerve agents is generally due to peripheral and central effects leading to respiratory failure.[12] A single acute exposure of nerve gas could cause death within 5 min or in 24 h, depending on the dose, route, and type of organophosphates. If the effective therapy is not started as soon as possible, the organophosphates may cause delayed neuropathy. Convulsions, seizures, and neuropathological lesions are also a significant part of the symptoms of central nervous system toxicity caused by poisoning with nerve agents. The proposed pathway for the nerve agent-induced pathophysiological lesion is depicted in Figure 3.2.

A number of studies have shown that nerve agent-induced neuropathology can be prevented by attenuating the convulsive episodes. Diazepam, a GABA agonist and tranquilizer, was found to be an effective drug for preventing neuropathology when used in conjunction with atropine and oxime in soman-poisoned animals.[13] A review of toxicological studies of nerve agents indicates vast literature on their anticholinesterase properties[14] and on the determination of acute lethal data, especially lethal dose (LD_{50}) and lethal concentration (LCt_{50}), in various species of animals. A comparison of toxic potencies as inhibitors of acetylcholinesterase for several representative nerve agents is given in Table 3.1. The larger the value, the more potent is the agent.[14] LD_{50} and LCt_{50} of various OP nerve agents by different routes of exposure in human and different species of animals are given in Table 3.2. The LD_{50} values indicate that guinea pig, dogs, cat, monkey, and rabbit are the most susceptible, and that the rat and mouse are the most resistant species for nerve agent intoxication. It should be noted that most studies achieved nerve agent intoxication by intravenous, subcutaneous, and dermal routes of administration in experimental animals.

B. Delayed Neurotoxicity

The chronic, delayed neurotoxic effects (ataxia and paralysis) induced by nerve agents are referred to as organophosphate-induced delayed neurotoxicity (OPIDN), which are due to the inhibition of neuropathy target esterase or neurotoxic esterase (NTE) in the neuronal membrane of the nervous system.[15–19] OPIDN is a syndrome which is characterized by a delay period of 4–21 days after nerve gas exposure before clinical symptoms (ataxia and paralysis) are manifested.[17–19] The primary molecular target for the initiation of OPIDN is NTE in the nervous system.[20,21] NTE is an integral membrane-bound enzyme with a molecular weight of 155 kDa; it has no physiological substrate, but its organophosphorylation and aging in the neuronal tissue are required to trigger the pathogenesis of OPIDN. Phosphorylation (inhibition) and subsequent aging of NTE are depicted in Figure 3.3. The rapid aging of phosphorylated

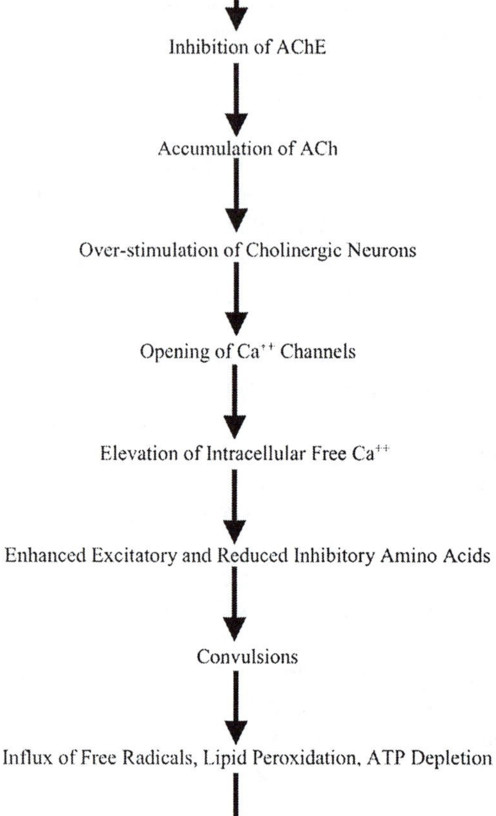

FIGURE 3.2 Proposed pathway for nerve agent-induced central and peripheral pathophysiology.

TABLE 3.1
Potency of Nerve Agents as an Inhibitor of ChE Enzyme

Nerve Agents	ChE Inhibition Potency
Tabun	8.6
Sarin	8.9
Soman	9.2
VX	8.8

TABLE 3.2
Lethal Dose (LD_{50}) and Lethal Concentration (LCt_{50}) Values of Nerve Agents in Human and Different Species of Animals by Different Routes of Administration

Exposure Route	Soman	Sarin	Tabun	VX	Ref.
Species					
Inhalation (mg · min/m³)					
Human		100	200–400	36	189, 190
Monkey		74	187	50	189
Dog		60	320	15	189
Rabbit		120	960	25	189
Guinea Pig	0.101	180		8–30	189, 191
Rat	211	220	450	17	189, 192
Mice		240–310		7–40	189
Dermal (mg/kg)					
Human		24	14–21	0.04–0.14	189, 193, 194, 195
Monkey			9.3	0.065	189
Pig		115.9		0.40	189
Dog		10.8	45	0.054	189
Cat		6.2		0.012	189
Rabbit		4.4	3	0.025	189, 196
Rat		2.5	12.6	0.10	189
Mice		1.0–9.2		0.046	189, 197
Oral (mg/kg)					
Rat		0.10	1.06		189
Intravenous (μg/kg)					
Human		14	14	8	189, 198
Goat		15		5	189
Dog		10	84	6.3	189
Cat		15–18		2.5	189, 199
Rabbit		14.7	63	8.4	189
Guinea Pig		30		4.5	200
Rat		45	70	7.9	189, 194, 201
Mice	42	70–113	311	14.1	98, 189, 202
Subcutaneous (μg/kg)					
Rabbit	29	41.7			203
Guinea Pig	28		119		204
Cat		35			88
Rat	156	158	305	21	205
Mice		190			206
Intramuscular (μg/kg)					
Monkey	3.75				207
Mice	98	179	304		204
Hen		50		30	208, 209
Intraperitoneal (μg/kg)					
Rat		450			150
Mice	440	560			210
Intracerebroventricular (μg/kg)					
Rat	66.4				211

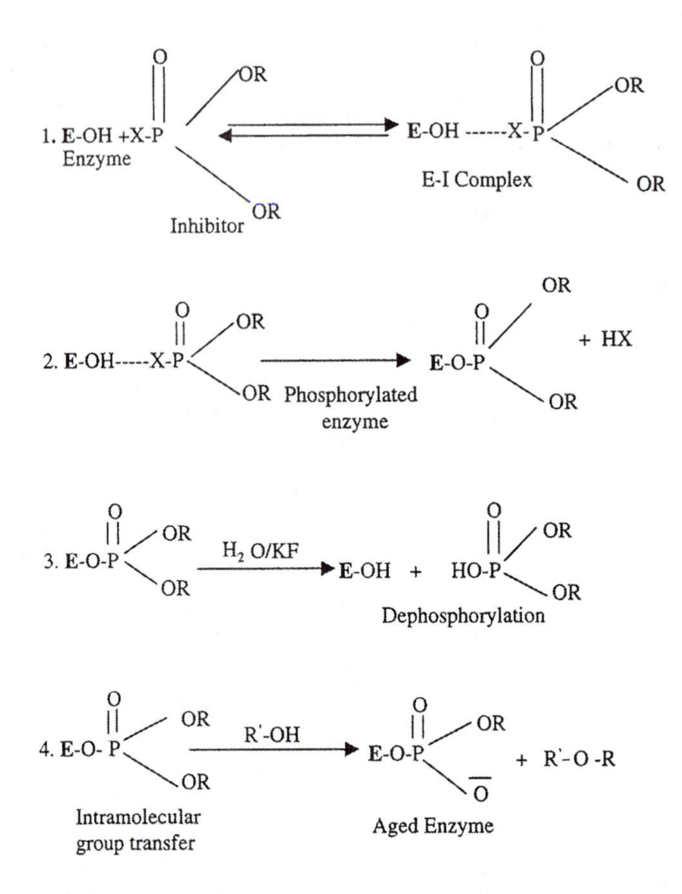

FIGURE 3.3 Interaction of OP nerve agents with enzyme, neurotoxic esterase (E-OH), and inhibitor organophosphate (I): (1) formation of Michaelis complex; (2) phosphorylation of enzyme; (3) reactivation, hydrolysis with H_2O to regenerate the enzyme; (4) aging, the proposed intramolecular transfer of an R group from the phosphate moiety to a receiving group (e.g., R′OH) within the protein structure.

NTE occurs when an alkyl group from a phosphate moiety is intramolecularly transferred to a receiving group within the protein molecule.[22] Half-lives of aging for inhibited NTE range from 1 min to 600 min.[22,23] Aging occurs rapidly with arylalkoxy and linear alkoxy groups attached to a P atom, and slowly with a highly branched alkoxy substitute.[24] NTE is defined as phenyl valerate hydrolase which is resistant to inhibition by non-neuropathic organophosphate (paraoxon) and sensitive to inhibition by neuropathic organophosphate (mipafox and DFP). The inhibitors of NTE are classified into two groups. Group A includes phosphates, phosphoramidates, and phosphonates. These are generally irreversible inhibitors and aging is possible. Group B is comprised of sulfonates, phosphonates, and carbamates. These are reversible inhibitors and aging is not possible. Trivalent phosphates such as triphenyl phosphite and trifluoromethyl ketones are also inhibitors of NTE.[25,26] Group A inhibitors are also known as agonist and Group B inhibitors as antagonist. Group B inhibitors given prior to Group A inhibitors protect animals against OPIDN.[27] When these Group B inhibitors are given after Group A inhibitors, these inhibitors potentiate the OPIDN.[28] Recent data show that NTE is comprised of two functional domains: an N-terminal regulatory domain and C-terminal effector domain. The N-terminal domain may bind a cyclic nucleotide and thereby modulate the activity of the C-terminal effector domain.[29] The non-esterase function of NTE is important for axonal maintenance and neuron-glial cells signaling. Histopathological changes in spinal cord and peripheral nerves consist of degeneration of axons followed by the demyelination.[17–19,30,31] NTE-like activity has also been reported to be present in non-nervous tissues such as lymphocytes and platelets of human, hen, and other animal species.[32,33] Various attempts have been made to use lymphocyte NTE in humans as a biomonitor of OPIDN, but many problems have been encountered.[34–36] Later platelet NTE as a peripheral biochemical marker has been proposed because platelets can be obtained from blood more easily and without contamination. Platelets are good peripheral models for central neurons and platelet NTE inhibition is well correlated with brain NTE inhibition in human, hen, and mice *in vitro* as well as *in vivo.*[17,21,32,37] Platelet NTE has now been established as a peripheral biochemical marker for OPIDN.[17,19,21,33,37] It is suggested that this is an important parameter which can be reliably measured in humans. Hens are very sensitive and have been used as a suitable model to evaluate OPIDN. However, humans are 10 to 100 times more sensitive than hens. Studies in Gulf War veterans by Jamal et al.[38] explained the mild impairment of brainstem, spinal cord, and peripheral nerve function. These studies are consistent with the spectrum of OPIDN syndrome. Sarin has been shown to produce delayed neurotoxicity at higher doses in protected hens.[39,40] However, this report describes the interactive effects of lower doses of sarin, pyridostigmine, and physical stress on biochemical and histopathological changes in tissues of animals.

C. Stress

The stressful demands of modern military duty include a broad range of activities, especially during wartime. The demanding physical tasks of a combat infantry soldier can be expected to result in significant physical and chemical changes within the body.[41] Notwithstanding this, physiological stress is still to be expected, because of the redistribution of blood flow to serve the demands of active muscle cells[42] as well as to meet the needs of temperature regulation in the body. In addition, a considerable production of metabolic acids from substrate catabolism will lead to a marked reduction of the intracellular pH.[43,44] As the time course of a drug in the body may be influenced by exercise dynamics,[45] it is important to know how physical activity interacts with low-dose nerve agent exposure under combat field conditions. Since Gulf War veterans underwent physical stress (exercise) and possibly were exposed to low doses of sarin, they make an excellent model to answer this. Therefore, the neurotoxic effects of low-dose sarin under conditions that reasonably simulate heavy military duty are discussed in this chapter.

Acute or chronic physical stress is known to influence the cholinergic system in cerebral and peripheral tissues. Acute physical exercise for 1 and 3 h increased cholinesterase (ChE) activity in the blood serum of rats.[46] In the reticulocytes and young erythrocytes of endurance-trained athletes, AChE activity was higher than that of the control group.[46] Acute exercise produces a slight elevation of ChE activity in red blood cells (RBC).[47] Husain and Somani[48] demonstrated a 55% increase in plasma cholinesterase activity after 30 min of acute exercise in rats. Acute exercise for 10 to 30 min decreased ChE activity in the heart without affecting ChE activity in the diaphragm, and muscle.[49] However, Tipton et al.[50] found no significant change in myocardial ChE activity after chronic exercise. Contrary to acute exercise, chronic exercise decreased ChE activity in RBC, heart, diaphragm, and muscle.[49] Babu et al.[51] have shown that AChE activity decreased in both EDL and soleus muscles 20 min after acute exercise; whereas, the AChE activity decreased in EDL and soleus muscles 24 h after exercise training. This is contrary to Fernandez and Donoso's findings which reported an increase in the G_4 form of AChE in fast twitch muscle due to exercise.[52] Similarly, a profound increase in the G_4 form of AChE in fast twitch muscle of rat after exercise training has been reported.[53] Different intensities of acute exercise stress produced slight decreases in brain AChE activity in rats.[47] This finding is in agreement with Ryhanen et al.[54] and is contrary to the findings of Pedzikiewiez et al.[55] who have shown a slight increase in brain ChE activity after single exercise. Acute exercise for 30 min decreased AChE activity in the striatum, medulla, and cerebral cortex without any change in the hypothalamus and cerebellum of rats.[48] Chronic exercise stress decreased AChE activity in the brain stem, in cerebral cortex, in striatum, and hippocampus.[56] These studies indicate that physical stress accelerates the nerve action in the CNS, resulting in an increased amount of acetylcholine in the nerve endings and hence increasing ChE inhibition.

Alterations in the choline acetyl transferase (ChAT) activity (biosynthetic enzyme for ACh) were differentially expressed within subregions of the brain during chronic exercise.[56] Exercise decreased ChAT activity in the adrenal gland of young rats.[57] Endurance training decreased ChAT activity in extensor digitorurn longus (EDL); whereas, in slow twitch soleus, it increased.[51] Swimming stress in rats has been shown to deplete the ACh content in various brain regions such as hippocampus and cerebral cortex.[58] Conlay et al.[59] reported a decrease in the plasma choline levels of marathon runners. Recently, Conlay et al.[60] have shown that in trained athletes, running a 26 km marathon reduced plasma choline by 40% and decreased ACh release from the neuromuscular junctions by a similar magnitude. The effect of acute exercise (swimming) in rats for 15 min resulted in a decrease in muscarinic cholinergic receptor (mACh) ligand binding in the cerebral cortex and basal ganglia; whereas, it increased in the cerebellum.[61] Chronic exercise has been reported to produce tolerance to muscarinic antagonists in rats.[62] These studies suggest that acute exercise or stress exerts rapid reversible and selective changes of cholinergic muscarinic receptors.

The cholinergic system is not only modified by physical stress but is also influenced by a variety of other stress factors. Rats exposed to repeated immobilization stress showed diminished ChAT activity in brain basal ganglia.[63] ChAT activity also decreased in the cortex, hypothalamus, hippocampus, and the mid-brain of rats after acute immobilization stress.[57] Conversely, ChAT activity increased in the rat cerebral cortex after acute and repeated electroshock.[64] After immobilization, ChAT activity has been reported unchanged in different brain regions: brain stem, striatum, hippocampus, and hypothalamus.[65,66] Chronic exposure of rats to cold conditions and other stressors has enhanced ChAT activity in basal ganglia, hypothalamus,[67] and in the medulla.[57] Acute exposure of rats to cold stress resulted in an increase in ChAT activity in the hippocampus.[58] Acute immobilization and cold stress have been shown to increase ChAT activity in the adrenal gland of rats; whereas, no change has been observed with chronic cold stress. In addition, Kita et al.[67] have reported no change in ChAT activity in rat duodenum after repeated cold stress. It is concluded from the above studies that acute and chronic stresses differentially alter ChAT activity in brain regions, thereby regulating the synthesis of the neurotransmitter and altered sensitivity of the cholinergic system. Chronic cold stress to rats has been shown to increase blood butyryl cholinesterase (BChE) activity and decrease the ChE activity in the lung;[54] whereas, ChE activity decreased in the duodenum.[67] Repeated exposure of rats to cold stress caused an enhancement of AChE activity in basal ganglia and the hypothalamus.[67] However, cholinergic parameters in various regions of the brain react differently to altered stress conditions, such as electric shock, cold, and swimming.

It has been demonstrated that the hippocampal cholinergic system is actively involved in stress response. Acute and chronic stress-induced changes in synaptic ACh release and choline uptake (parameter of cholinergic system) have been studied in rat hippocampus.[68] Acute as well as chronic intermittent immobilization stress increased ACh release; whereas, choline uptake increased after acute stress and decreased after chronic stress. Repeated cold stress has been shown to decrease the total ACh content in basal ganglia and hypothalamus, whereas its amount increased in the duodenum of rat.[67,69] Similarly, cold stress resulted in a decrease of ACh levels in the hypothalamus and hippocampus of rat.[58] It has been assumed that the stores of ACh in the hippocampus of a rat that are exposed to stress may become depleted. However, Costa et al.[70] and Mizukawa et al.[71] failed to find any change in rat ACh after stress. After electric shock stress, the ACh concentration was found to be depleted in brain regions of rat and mice.[72,73] Following 2 h of mild restrain stress, choline uptake was increased in hippocampus, septum, and frontal cortex of rat.[74] The administration of chronic electric shock to rats has increased the ACh content in the medulla.[67]

Information from animal and human studies has suggested the stress-induced hyperactivity of central muscarinic mechanisms.[75,76] Chronic immobilization stress increased muscarinic receptor binding capacity in hippocampal synaptosomes of rats.[77] Similarly, immobilization stress produced an increase in muscarinic cholinergic (mACh) binding sites in the septum, striatum, hippocampus, and pons, plus medulla oblongata of rats.[78] Immobilization stress for 30 min increased the concentration of mACh binding sites in the hippocampus of rat.[71] Restrain stress for 10 days induced hypersensitivity of the central cholinergic system in mice, whereas restrain stress for 30 days caused hyposensitivity of the central cholinergic system.[79] Similarly, shock stress also increased the hypersensitivity of the central acetylcholine receptors.[80] It has been demonstrated that hippocampal muscarinic acetylcholine receptor binding increased in rats after chronic intermittent immobilization stress.[68] It is suggested that acute stress or exercise may enhance the sensitivity of the cholinergic system, whereas chronic exercise or stress decreases the sensitivity of the cholinergic system.

II. CHOLINERGIC TOXICITY

The cholinergic system is the primary target of OP nerve agent intoxication. The cholinergic system consists primarily of synthetic (choline acetyltransferase) and degradative (acetylcholinesterase) enzymes for the neurotransmitter acetylcholine and its receptors, muscarinic and nicotinic. Nerve agents inhibit AChE activity resulting in accumulation of excess ACh at vital cholinergic sites, thereby causing toxic manifestations. Nerve gases exert their toxic effects by phosphorylating the serine hydroxyl group at the active site of the enzyme AChE, producing irreversible inhibition of the enzyme with consequent elevation of acetylcholine levels. Acetylcholine accumulates at the peripheral and central synapses, leading to cholinergic manifestations. There is depression of the respiratory center in the brain followed by peripheral neuromuscular blockade, causing respiratory paralysis and death.[81] The toxic effects of these nerve agents are dependent on their stability, rate of absorption by various routes, distribution, ability to cross the blood-brain barrier, rate of reaction with AChE and selectivity for reaction with the enzyme at specific foci, and their behavior once attached at the enzyme active site. The mechanism of action of AChE and its inhibition by a nerve agent is depicted in Figure 3.4.

The active site of AChE enzyme consists of two subsites, anionic and esteratic sites. The anionic site is represented by a glutamate ion. The esteratic site has been shown to incorporate a serine moiety and histidine as well as tyrosine residue.[82] A hydrophobic area at the active site is shown in Figure 3.4A. The normal catalytic functioning of AChE enzyme has been depicted in Figure 3.4B. After acting at the cholinergic receptor, ACh forms a reversible complex with the active site of the enzyme AChE. Next the acetyl group is transferred from the ACh molecule to the serine hydroxyl, thus forming acetylated enzyme and releasing choline. This is followed by a rate-limiting hydrolysis of the acetate ester group with a half-life of 42 s, producing acetate anion, which, in turn, provides regenerated enzyme that can be utilized to hydrolyze another molecule of ACh. The high percentage of released choline is transported back into the nerve ending for reconversion to ACh and storage. OP nerve agents bind rapidly with AChE enzyme protein. Soman reacts with AChE completely within minutes of administration to animals. The inhibition of AChE by OP nerve agent is depicted in Figure 3.4C. It is conceivable that a proton on the imidazolium ion forms a partial bond to the "onyl" oxygen (or sulphur) attached to phosphorus. At the same time, the hydrogen bonding of the tyrosine phenolic hydroxyl to the glutamate anion, and of the serine hydroxyl to the phenolic oxygen, increases electron density at the serine oxygen, which facilitates the nucleophilic attack on phosphorus and displacement of the leaving groups X (F or CN).[83] In sharp contrast to the rapid

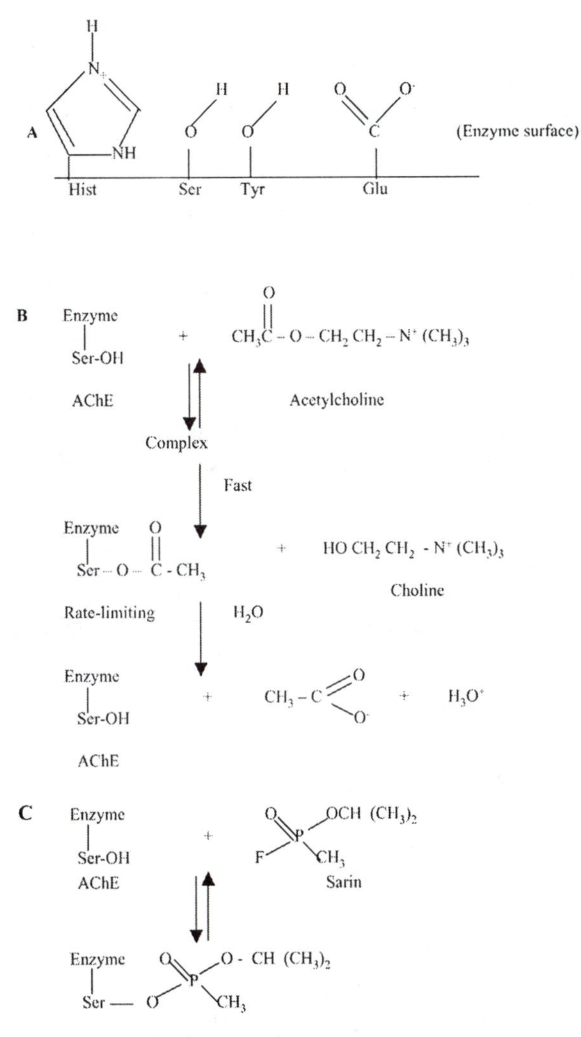

FIGURE 3.4 Mechanism of action of acetylcholinesterase inhibition: (A) Structure of AChE; (B) normal functioning of AChE; (C) inhibition of AChE by nerve agent sarin. (Adapted from Somani et al.[212])

hydrolysis of the acetylated enzyme, hydrolysis of phosphorylated enzyme is extremely slow, with a half-life of hours to days. Therefore, the enzyme is effectively inactivated and prevented from carrying out its catalytic function of hydrolyzing ACh. The enzyme molecule phosphorylated with the nerve agent undergoes the aging process (dealkylation) within a few minutes.[84] The aged enzyme is no longer reactivated with nucleophilic compounds such as oximes. The physiological, biochemical, and histopathological effects due to cholinesterase inhibition induced by low-dose nerve agents in animals and humans are described below. Behavioral effects including performance in human and animals exposed to low-level nerve agents are described in another chapter in this volume.

Rats exposed to low doses (one ninth the LD_{50}) of sarin produced alterations in motor coordination/balance.[85] Cholinesterase inhibition induced by continuous infusion of PYR (for at least 3 days before soman exposure and continuing through the exposure period) had little effect on the toxicity of repeated soman exposure in the rodents.[86] More importantly, there was no deleterious effects of PYR pretreatment and concomitant low-level exposure to soman. The cumulative effects of repeated soman exposure on serum ChE and the relatively insignificant impact of additional PYR exposure are illustrated by the convergence of ChE inhibition levels by the fifth day, regardless of treatment condition.[87] Low doses of sarin (3.5 μg/kg for 10 days or 7.0 μg/kg for 5 days, s.c.) and soman (2.5 μg/kg for 10 days or 5.0 μg/kg for 5 days) resulted in a depression of mechano receptors, conduction velocities of muscle spindle, and mechano receptor afferents in cats.[88] These authors suggested that alteration in muscle spindle function was either due to inhibition of AChE in the muscle or due to direct effects of sarin or soman on the afferents. Rats exposed to low-level sarin (0.2 and 0.4 mg/kg) through inhalation route (1 h singly or 1 h each day for 5 days or for 10 days repeatedly) caused changes in physiological parameters such as respiration rate, core body temperature, and motor activity.[89] In recent experiments, mice were exercised for 10 weeks and pyridostigmine and/or sarin administered during the fifth and sixth weeks. Exercise parameters such as respiratory exchange ratio (RER) were recorded during exercise. Respiratory exchange ratio (VCO_2/VO_2) decreased significantly at the end of the 5th week of exercise (1 week of dosing) as compared to the 4th week in both the exercise groups, treated with sarin or sarin plus pyridostigmine bromide (PB), respectively (Figure 3.5). Thereafter, a steady increase in RER values was observed with incremental exercise up to the 10th week. However, the patterns of increase in RER values were different in the sarin plus exercise group compared to the sarin plus PB plus exercise group, indicating the interactive effects of PB with low-dose sarin and physical stress.

A. Biochemical Effects

The inhibition of AChE activity in nerve tissues of animals at different times after exposure to low-dose nerve agents are depicted in Table 3.3. In most studies, subcutaneous, intravenous, inhalation, and oral routes of exposure were used. AChE activity in whole brain, spinal cord, and brain regions such as cerebral cortex, corpus striatum, medulla, and cerebellum was significantly decreased in rats,[90–94] mice,[17,95–98] and hens,[40] hours, days, and weeks after low-dose exposure to nerve agents such as soman, sarin, and tabun.

The cholinergic effects related to changes in brain AChE activity were assessed in rats repeatedly exposed to low-dose soman for 5 days. The cholinergic effects before and after each injection were examined in the brain regions such as: (1) frontal cortex, (2) piriform cortex, (3) hypothalamus, (4) hippocampus, (5) thalamus, (6) cerebellum, and (7) neostriatum.[99] Repeated administration of low-dose soman caused a significant decline in AChE activity in all regions of the brain.[99,100]

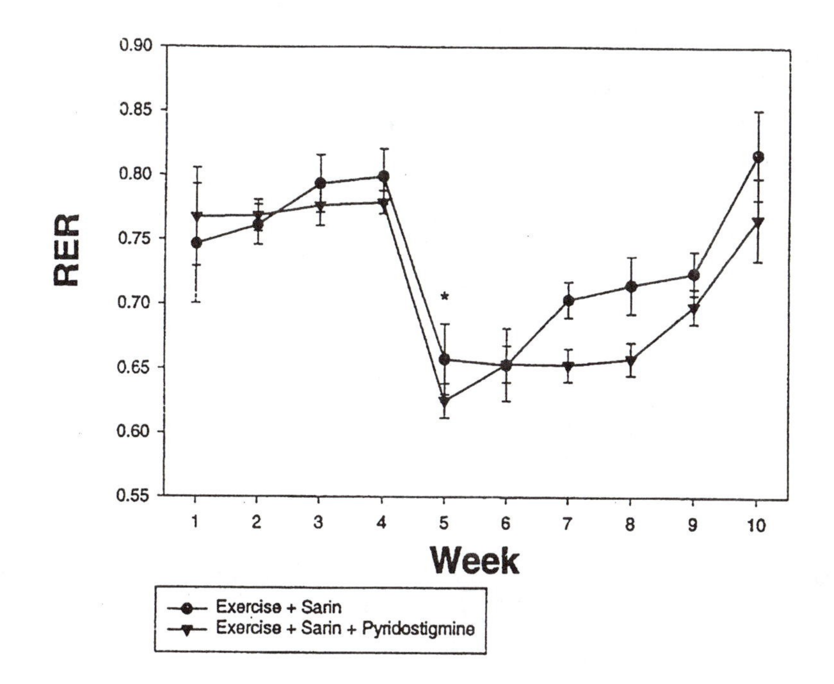

FIGURE 3.5 Effect of exercise training plus sarin (10 μg/kg, s.c.) during 5th and 6th weeks; and exercise, sarin, and pyridostigmine (1.2 μg/kg, p.o.) during 5th and 6th weeks combination on respiratory exchange ratio (R.E.R.) in mice over a period of 10 weeks. Significant decline in RER observed after the start of pyridostigmine and sarin dosing. Significant * $P < 0.05$.

Exposures to lower levels of soman (approximately 0.3 LD_{50}) in mice[98] and rats[101] caused greater inhibition of cerebellar AChE than either hippocampal or cortical AChE, and less effect on the neostriatum. Soman exposure in these two animal species showed that AChE inhibition in the hippocampus and cortex was equivalent to cerebellar AChE inhibition. The effects of sarin and tabun at 40 and 60% LD_{50} dose levels on AChE activity in different brain regions of mice were studied. The AChE activity was found to be similar to acute high-dose exposures to soman. The hippocampus and cortex were more sensitive, whereas the neostriatum was less sensitive in these studies.[98] The AChE-rich neostriatum is thus an interesting region neurochemically, being relatively resistant to nerve-agent induced AChE inhibition. It is, however, affected at the same level by soman, sarin, and tabun administered at low doses 50 to 85% of LD_{50}.[102] In another study, tolerance to low-dose soman-induced hypothermia (soman 35 μg/kg, s.c. daily for 3 days and then three times weekly for 36 days) was reported by Russell et al.[103] Body temperature was unaffected during the first 2 days of soman injections and then was reduced after the third injection. The data represent biochemical and physiological correlates of initial resistance of the hypothalamus to chronic low-levels of soman exposure. It seems likely that other processes (such as biogenic amines, other neurotransmitters and their receptors) besides ChE inhibition in the hypothalamus are responsible for tolerance

TABLE 3.3
Effects of Low-Dose Nerve Agents at Different Times on Cholinesterase (ChE) Activity in Nerve Tissues of Animals Under Normal and Stressful Conditions

			Acetylcholinesterase Activity (% of Control)					
Species	Nerve Agent Dose (μg/kg)	Time	Spinal Cord	Brain	Cortex	Striatum	Medulla	Ref.
Mice	Sarin 5 mg/m^3 20 min/day for 10 days (Inhalation)	4 days		81				17
Mice	Sarin 85 (s.c.)	1 h			47	39	38	97
Mice	Sarin 85 (s.c.)	1 h			46	41		96
Mice	Sarin 10 (s.c.) ×2 wks	4 wks	89		76	63		95
Mice	Sarin 10 (s.c.) ×2 wks	4 wks	76		81	71		95
Mice	Soman							98
	25 (i.v.)	10 min		56				
	12.5 (i.v.)	10 min		90				
Mice	Sarin							98
	60 (i.v.)	10 min		24				
	40 (i.v.)	10 min		48				
	20 (i.v.)	10 min		75				
Mice	Tabun							98
	213 (i.v.)	10 min		31				
	191 (i.v.)	10 min		52				
	170 (i.v.)	10 min		69				
	149 (i.v.)	10 min		69				
Mice	Soman							98
	37.5 (i.v.)	10 min		30				
		4 h		27				
		1 day		30				
		4 days		49				
		1 wk		55				
		2 wks		75				
Mice	Sarin							98
	80 (i.v.)	10 min		16				
		4 h		16				
		1 day		33				
		4 days		71				
		1 wk		69				
		2 wks		89				

Species	Nerve Agent Dose (μg/kg)	Time	Acetylcholinesterase Activity (% of Control)					Ref.
			Spinal Cord	Brain	Cortex	Striatum	Medulla	
Mice	Tabun							98
	213 (i.v.)	10 min		37				
		4 h		39				
		1 day		50				
		4 days		68				
		1 wk		64				
		2 wks		85				
Rat	Sarin 50 (s.c.)/day for 85 days	24 h				66		94
Rat	Sarin 50.2 mg/m^3 for 15 min	15 min	63		49			92
Rat	Soman	24 h						93
	60 (s.c.)	after		85				
	1 × wk	2 wks		85				
	for 6 wks	4 wks		85				
		6 wks						
Rat	Soman	24 h						93
	60 (s.c.)	after		60				
	3 × wk	2 wks		42				
	for 6 wks	4 wks		25				
		6 wks						
Rat	Soman 25 (s.c.) daily for 85 days	24 h after last dose					36	94
Rat	Soman 60 (s.c.)	1 h			62			90
Rat	Soman 20 (s.c.)	7 days			5	3 (midbrain)	1 (midbrain)	91
Hen	Sarin							40
	61 (p.o.)	24 h		80				
	70 (p.o.)	24 h		30				
Hen	Soman							40
	3.5 (p.o.)			150				
	7.1 (p.o.)			60				

Note: Cerebellum of mice showed 40% of control (Sikder et al., 1992) and lung of rat showed 46% of control AChE activity.

development during chronic low-level exposure. A substantial difference has been shown to exist between the recovery of serum ChE and red blood cell (RBC) AChE,[104] which more closely follows the recovery of brain AChE after soman exposure. Similar results reported that RBC AChE activity was significantly inhibited and this inhibition was matched by the responses of whole brain AChE activity.[103] McDonough et al.[104] compared regional CNS recovery patterns to those of plasma ChE and RBC AChE activity after a 4-week chronic low-dose soman administration protocol. These data indicated slow recovery patterns for RBC, cortex, hippocampus, and neostriatum; whereas, those of midbrain, brainstem, and cerebellum were faster. Plasma ChE activity showed a very fast recovery within 24 to 48 h after the last soman injection. Thus, the earlier discussion suggests that RBC AChE activity may serve as a useful clinical marker for the nerve agent-induced CNS AChE inhibition, but not that of serum ChE. Repeated administration of tabun (70 μg/kg, i.m.) for 90 days in hens significantly depressed plasma BuChE activity (44% of control), indicating cholinergic toxicity of tabun in atropine-protected hens.[105] Rats exposed to low-dose sarin (75–300 μg/kg, p.o.) for 90 days resulted in a significant inhibition of cholinesterase in plasma and RBC.[106] Low doses of tabun (28 μg/kg, p.o.) 5 days per week for 90 days in rats, significantly decreased ChE activity in plasma and RBC.[107] Rats orally exposed to VX at low-dose level (4 μg/kg) 5 days per week for 30, 60, and 90 days showed significant inhibition of plasma and RBC ChE activities and slight decrease in body weight.[108] Very low-level sarin inhalation exposure (0.1 and 1.0 $\mu g/m^3$) for 6 h per day, 5 days per week, for 4–52 weeks to beagle dogs, rats, and mice did not show any adverse toxic side effect in any species at either concentration of sarin.[109] Sarin at a very low dose 1/20 LD_{50} (10 μg/kg, s.c.) daily for 2 weeks resulted in a depression of cholinesterase activity (81, 87, 71, 88, and 68% of control) in plasma, RBC, platelets, sciatic nerve, and triceps muscle, respectively, 4 weeks after the last treatment.[95] AChE activity in cerebral tissues such as spinal cord, cerebral cortex, and corpus striatum decreased (89, 76, and 63% of control, respectively). It was suggested that the corpus striatum, which has higher basal AChE activity, is more sensitive to sarin exposure. Tabun (70 μg/kg, i.m.) for 90 days in atropine-protected hens significantly increased CPK activity (188% of control) in the plasma, indicating muscle damage.[105] The authors suggested that this may be due to increase in acetylcholine followed by mobilization of calcium ions.

The inhibition of cholinesterase in peripheral tissues of animals exposed to low dose nerve agents at different time points is summarized in Table 3.4. In most of the studies, subcutaneous, intramuscular, inhalation, as well as oral routes of exposure were employed. Cholinesterase activity in whole blood and blood constituents, such as plasma, RBC, and platelets, was significantly depressed in humans,[110] monkeys,[111,112] rats,[18,92,100,113] mice,[17,18,95–97] and hens[19,40,105,112] hours and days following low-dose exposure to nerve agents such as soman, sarin, and tabun. AChE activity in sciatic nerve and triceps muscle was decreased 4 weeks after exposure to repeated low-dose sarin in mice.[95] The inhibition of ChE in plasma, brain, and diaphragm, as well as depression of spontaneous locomotor activity and rectal temperatures, of soman-treated animals had returned to control levels within 24 h, but ChE activity was not fully recovered even after 3 days.[114] Acetycholinesterase inhibited to the

TABLE 3.4
Effects of Low-Dose Nerve Agents at Different Times on Cholinesterase (ChE) Activity in Peripheral Tissues of Various Species of Animals under Normal and Stressful Conditions

			Cholinesterase Activity (% of Control)					
Species	Nerve Agent Dose (μg/kg)	Time	Plasma/ Serum	RBC	Platelets	Sciatic Nerve	Triceps Muscle	Ref.
Mice	Sarin 5 mg/m³ for 20 min/day for 10 days (Inhalation)	4 days	73 (blood)					17
Mice	Sarin 5 mg/m³ for 20 min/day for 10 days Inhalation)	4 days		24				18
Mice	Sarin 85 (s.c.)	1 h	40 (blood)					97
Mice	Sarin 85 (s.c.)	1 h	38 (blood)					96
Mice	Sarin 10 (s.c.) × 2 wks	4 wks	81	87	71	88	68	95
Mice	Sarin 10 (s.c.) × 2 wks	4 wks	79	81	58	76	56	95
Rat	Sarin 12.5 mg/m³ 20 min/day for 10 days (Inhalation)	4 days			29			18
Rat	50.2 mg/m³ 15 min (Inhalation)	15 min	49 (Blood)					92
Rat	Soman 30 (s.c.) daily for 12 days	Day 5 Day 12		1.2 0.0				113
Rat	Soman 39 (s.c.) daily for 5 days	Day 5	14					100

TABLE 3.4 (continued)

Species	Nerve Agent Dose (μg/kg)	Time	Cholinesterase Activity (% of Control)					Ref.
			Plasma/ Serum	RBC	Platelets	Sciatic Nerve	Triceps Muscle	
Marmoset	Soman							111
Monkey	1.75 (i.m.)	1 h	15 (Blood)					
	3.5 (i.m.)	1 h	5 (Blood)					
Monkey	Sarin							112
	2.5 (i.m.)	3 h		64				
	3.0 (i.m.)	3 h		33				
Hen	50 (s.c.) daily for 10 days	4 days			30			19
Hen	Sarin							40
	61 (p.o.)	24 h	40					
	70 (p.o.)	24 h	45					
Hen	Soman							112
	3.5 (p.o.)		64					
	7.1 (p.o.)		33					
Hen	Tabun	2 h	29					105
	70 (i.m.)	6 h	34					
		24 h	54					
		30 days	50					
		60 days	45					
		90 days	44					
Human	Sarin							110
Volunteer	0.5 mg/m^3							
	for 30 min	3 h		42				
	(Inhalation)	3 days		39				

same degree in the corpus striatum and hippocampus 1 h after administration of soman and sarin.[115] It was observed that soman and sarin increased the levels of choline and ACh in both striatum and hippocampus with maximal increase at 2 h and recovery of choline levels at 4 h. Drewes and Quian[116] showed that the soman-induced increase in brain choline may be secondary to the action of ACh on muscarinic receptors coupled to phosphatidylcholine hydrolysis. Shih[117] suggested a possible relationship between elevated choline levels and soman toxicity.

The toxicity of nerve agents may include direct action on nicotinic as well as muscarinic ACh receptors[118] when their concentration in circulation rises above micromolar levels.[119] At nanomolar levels, their toxicity is the result mainly of their inhibition of AChE. However, at these low concentrations, many OP agents (e.g., soman and VX) may directly affect a small population of muscarinic ACh receptors that have a high affinity for [^{3}H]-cis-methyldioxalane binding. Aas[120] demonstrated reduction in the release of ACh from cholinergic nerves in rat bronchi after soman. Long-term inhalation exposure of soman (0.45–0.63 mg/m^3) reduced (by approximately 70%) the contraction of bronchi induced by ACh, probably as a result of the reduced number of muscarinic cholinergic receptors.

B. Histopathological Effects

Studies in animals indicate that morphologic changes in the brain may occur after low-dose nerve-agent exposure. In soman-exposed rats,[121–123] monkeys,[124,125] and baboons,[126] in sarin-exposed rats,[127] and in VX-exposed rats,[128] neuronal degeneration and necrosis were seen on necropsy as long as 45 weeks after exposure. The neuropathological effects of 5-day repeated soman at daily low doses ranging from 25 to 54 μg/kg were evaluated in rats following survival times of 7 to 35 days after soman exposure. The most sensitive area was noted to be the piriform cortex and the least sensitive, the hypothalamus and neostriatum in both the neurochemical and neuropathological studies. The neostriatum, an extremely rich area for cholinergic function, contains both intrinsic cholinergic neurons and terminals from other nuclei (nigrostriatal dopaminergic pathway) that contain AChE.[129,130] The nucleus basalis of Meynert in the ventral forebrain, which supplies up to 50% of the cholinergic innervation of frontal and parietal cortices in the rodent,[131] was not damaged even in brains showing the most massive degeneration. Petras[122,125] reported soman-induced neural degeneration in animals that showed only minor clinical symptoms (fasciculations). Limbic structures (e.g., septum, amygdala, hippocampus) involved with coordinating sensory information of the animal's external environment, as well as that of its viscera with motor function, were heavily affected.

C. Cholinergic Toxicity under Stressful Conditions

Somani and Husain[132] have reviewed the effect of physical stress on the cholinergic system indicating that it perturbs the functions of the nervous system. Earlier studies reported that physical stress enhances cholinesterase inhibition due to physostigmine, in central and peripheral tissues of the rat.[47,133,134] Physical stress is known to induce oxidative stress in the nervous system and increase lipid peroxidation.[135–137] A correlation between AChE (a membrane-bound enzyme with lipid dependence) inhibition and enhanced lipid peroxidation in specific areas of rat brain following acute and chronic physical stress have been reported.[48,138] Thus, physical stress influences the membrane lipid peroxidation and membrane-bound enzyme activity which may be related to free radicals. Forced swimming in mice has also been shown to enhance the entry of pyridostigmine (a peripheral reversible cholinesterase inhibitor) across the blood-brain barrier which resulted in inhibition of cerebral AChE activity, enhanced gene expression and cortical functions.[139] Interaction of pyridostigmine and treadmill exercise resulted in a loss of integrity of the neuromuscular system in rats.[140] Recent reports demonstrated that under physical stress, pyridostigmine enhanced AChE inhibition and increased lipid peroxidation in the triceps muscle of mice 4 weeks after the drug administration.[141]

The combined effect of physical stress and anticholinesterases on the cholinergic system has not as yet been thoroughly studied. Somani et al.[56] studied the interaction of a centrally acting anticholinesterase drug physostigmine (PHY), exercise, and the ChAT activity in brain regions of the rat. ChAT activity in the corpus striatum decreased (24, 5, and 8%) due to moderate exercise as well as PHY plus exercise training. Subacute PHY also inhibited brain stem ChAT activity (19%) after 20 min and (22%) after 24 h posttreatment. The hippocampus showed significant decreases in ChAT activity due to PHY plus exercise (28%), but not due to Phy alone. Babu et al.[51] have shown that choline acetyltransferase activity decreased in rats by trained exercise in EDL muscle (32%) and that Phy prolonged this effect even up to 24 h. Soleus muscles showed a small increase of ChAT due to exercise, but Phy plus exercise did not change the activity significantly. No recovery was observed in ChAT activity of EDL in Phy plus trained exercise group even after 24 h. Dube et al.[133] reported that the cholinesterase activity in red blood cells of exercised rats that were not exposed to physostigmine increased, while in other tissues the cholinesterase activity decreased slightly. In exercised rats exposed to physostigmine, the cholinesterase activity decreased in 10 to 30 min in red blood cells, brain, heart, diaphragm, and thigh muscles, respectively. Somani and Dube[49] reported that acute exercise, as well as endurance training, produced a slight decrease in ChE activity of the brain (3 to 9%) at various time points. Acute exercise plus physostigmine showed an increase in ChE inhibition (30% of control) as compared to physostigmine alone (40% of control) at 15 min, which recovered to control level at 60 min. Endurance training plus physostigmine showed a further decrease in ChE activity (48% of control; at 15 min it recovered to 64% of control at 60 min). Somani et al.[56] demonstrated that AChE activity decreased in the corpus striatum (18%) 20 min after subacute Phy administration and subacute Phy plus acute exercise (19%), or trained exercise (22%). Acetylcholinesterase activity remained at 89, 87, and 90% of control in Phy administered, Phy plus acute exercise, and Phy plus trained exercise, respectively, even after 24 h. The study indicated that Phy, exercise, or the combination of both, decreased AChE activity in a regionally selective pattern.

Besides physical stress, other types of stress including environmental stress have been shown to influence the cholinergic system.[61,142] The effects of different stresses have been reported in irreversible AChE inhibitors (organophosphate pesticide) intoxication.[54,143,144] Rynanen et al.[54] studied the relationship of cold stress and cholinesterase inhibiting organophosphorus compounds to cholinesterase activity in rats. They reported that cholinesterase in the liver of chronically cold-exposed rats (2 weeks) was more sensitive to diisopropyl fluorophosphate (DFP) inhibition when compared to acute cold-exposed rats. Studies have been conducted on the effects of organophosphorous pesticides and exercise on cholinesterase enzymes in rats.[143] In such studies, parathione-methyl-induced inhibition of serum cholinesterase was less marked 1 h after its termination. The activity of cholinesterase, an enzyme produced in the liver, depends upon a number of endo- and exogenous factors. It may be assumed that increased ChE activity is a secondary effect of hypoxia and the labilization of lysosomal membrane of liver cells after acute exercise. The higher values of serum ChE activity after exercise attenuates the effect of organophosphorus pesticides on this enzyme; this phenomenon is transient. However, information related to the influence of stress factors on low-dose, nerve-agent-induced cholinergic toxicity is sparse. Rats exposed to restrain stress (5 min/day for 60 days) followed by a single low dose of sarin (0.1 LD_{50} = 10 μg/kg, i.m.) resulted in a significant decrease of nicotinic acetylcholine receptor binding in cortex, brainstem, and midbrain with no change in cerebellum 24 h after sarin administration.[145] This study further demonstrated that sarin exposure caused up-regulation of M_2 muscarinic ACh receptor binding in midbrain, cortex, and brain stem, with no change in cerebellum, but stress exposure did not alter receptor binding. Plasma cholinesterase activity slightly decreased with sarin and was unaffected with stress exposure. The authors concluded that stress plays a critical role in manifestations of CNS toxicity caused by low-dose sarin exposure. Heat stress did not induce penetration of reversible cholinesterase inhibitor pyridostigmine into the brain of guinea pig.[146] However, exposure to heat stress resulted in a partial inhibition of cerebral AChE activity. Behavioral effects of soman (20–160 μg/kg, s.c.) in rats following exposure to different environmental temperatures (−1, 7, 15, 23, and 31EC) have shown that thermal stress influences soman toxicity.[147] This study concluded that interaction of thermal stress and soman influences motor activity in rats. Rats kept at high environmental temperature (40EC) showed enhanced brain AChE inhibition, hyperglycemia, lactic acidosis, depletion of glycogen in cerebral and peripheral tissues, glycogen phosphorylase and hexokinase activities, and inhibition of succinate dehydrogenase activity when exposed to organophosphorus insecticide diazinon compared to rats kept at normal room temperature.[144] The differences in the toxicity of DFP in inhibiting tissue ChE were observed in experimental animals subjected to a cold environment.[148] Wheeler[149] showed the effect of temperature on soman toxicity in rats. The toxicity of soman increased during exposure to either cold or hot environments and after removal from cold environment. The increased toxicity of soman while in or after removal from a cold environment is believed to be the result of a generalized adrenal cortical stress response. A recent study showed that physical stress enhanced sarin-induced depression of cholinesterase activity in plasma platelets, triceps muscle, sciatic nerve, and corpus striatum of mice.[95] The effect of cold environmental stress (+5 and −5EC) on the toxicity of sarin in mice and rats has been studied.[150] The authors showed that cold temperature sensitized the animals to the inhibition of brain AChE activity by sarin.

III. NON-CHOLINERGIC TOXICITY

In addition to cholinergic toxicity, certain nerve agents at low doses have been reported to induce a long-term neurotoxicity, which is not related to cholinesterase inhibition as mentioned earlier. This noncholinergic toxicity is known as organophosphate-induced delayed neurotoxicity (OPIDN). OPIDN is related to phosphorylation (inhibition) of neuropathy target esterase or neurotoxic esterase (NTE) and subsequent aging of this enzyme. A minimum of 70% NTE inhibition after single exposure and 45% after multiple exposure to organophosphorus nerve agents, and subsequent aging of NTE, is the biochemical prerequisite for the development of

OPIDN.[18,19,151,152] The main nerve agents (sarin, soman, tabun, and VX) have been shown to inhibit NTE *in vitro* as well as *in vivo.*[39,153,154] The physiological, biochemical, and histopathological effects of low-dose, nerve-agent-induced delayed neurotoxicity in various species of experimental animals and humans are described in this chapter.

Published studies dealing with delayed neurotoxicity caused by low-level exposure to nerve agents are scanty. However, it has been known since the late 1950s that exposure of normal individuals to low doses of sarin induces abnormalities in central nervous system (CNS) functions.[155] It has also been reported that workers engaged in German chemical warfare production plants showed persistent neurological abnormalities even 10 years after low-level exposure to nerve agents.[5] Delayed neurotoxic effects have also been reported 6–8 months after sarin exposure to humans in Japan.[156] However, the first reported case of delayed neurotoxicity in animals (sheep) exposed to nerve agents occurred in Skull Valley, Utah.[157] Chickens treated with low-dose tabun (70 μg/kg, i.m.) 5 days per week for 90 days did not induce OPIDN behaviorally or histopathologically.[105] Rats treated with low doses of tabun (28 μg/kg, i.p.) 5 days per week for 90 days revealed no delayed neurotoxic effects.[158,159] Antidote-protected chickens treated with VX (40 μg/kg, i.m.) for 90–100 days did not show OPIDN behaviorally and histopathologically.[160] Rats exposed to low doses of sarin (0–300 μg/kg, p.o.) 5 days per week for 90 days did not induce OPIDN.[107] Repeated intramuscular treatments of tabun (70 μg/kg) for 90 days in atropine-protected hens did not cause any delayed neurotoxic symptoms.[105] However, repeated s.c. administration of sarin (50 μg/kg) daily for 10 days to hens caused delayed neurotoxic symptoms such as ataxia 4 days after the last dose of sarin.[19] Studies in experimental animals have also shown that low-level sarin (5 mg/m^3, inhalation) for 20 min daily for 10 days to mice resulted in expression of delayed clinical symptoms such as muscular weakness of the hind limbs, muscle twitching, and mild ataxia 4 days after the last exposure.[17,18] Studies using very low doses of sarin (10 μg/kg, s.c.) daily for 2 weeks in 11 out of 15 mice showed slight to mild muscular weakness in the hind limb and motor incoordination 4 weeks after last sarin administration.[95] However, there is insufficient information about OPIDN effects at different times after exposure to nerve agents, with different routes of administration.

A. Biochemical Effects

Reports on the non-cholinergic biochemical effects such as inhibition of NTE due to low-dose exposure to nerve agents in certain species of animals are scanty. The inhibition of NTE activity in central and peripheral tissues of animals exposed to low-dose nerve agents at different time points are shown in Table 3.5. In most of the studies, subcutaneous, inhalation, and oral routes of exposure were used. NTE activity in platelets, lymphocytes, spinal cord, whole brain, and brain regions, such as cerebral cortex and corpus striatum, was significantly decreased in mice,[17,95] rats,[18] and hens[19,40] days and weeks after low-dose exposure to nerve-agent sarin. However, studies have been carried out in protected hens, which are a suitable model for OPIDN evaluation with relatively high doses of nerve agents (at lethal

TABLE 3.5
Effects of Low-Dose Nerve Agents at Different Times on Neurotoxic Esterase (NTE) Activity in Tissues of Animals under Normal and Stressful Conditions

Species	Nerve Agent Dose (μg/kg)	Stress as a Factor	Time	NTE Activity (% of Control)					Ref.
				Platelets	Lymphocytes	Sciatic Nerve	Spinal Cord	Brain	
Mice	Sarin 5 mg/m³ 20 min/day for 10 days (Inhalation)	Normal	4 days	45			53	41	17
Mice	Sarin 10 (s.c.) × 2 wks	Normal	4 wks	55		82	75	63 (Cortex) 75 (Striatum)	95
Rat	Sarin 12.5 mg/m³ 20 min/day for 10 days (Inhalation)	Normal	4 days	67			81	65	18
Rat	Sarin 300 (p.o.)	Normal	90 days					85	107
Hen	Sarin								40
	61 (p.o.)	Normal	24 h	67					
	70 (p.o.)	Normal	24 h	60	80	92			
Hen	Soman								40
	3.5 (p.o.)	Normal	24 h		90	90	86		
	7.1 (p.o.)	Normal	24 h		70				
Hen	Sarin 50 (s.c.) for 10 days	Normal	4 days	46			62	47	19
Mice	Sarin 10 (s.c.) × 2 wks	Exercise Training	4 wks	42		79	69	58 (Cortex) 72 (Striatum)	95

doses).[39,40,153,161] Soman at doses of 1.0 and 1.2 mg/kg inhibited spinal cord NTE (67 and 37% of control, respectively) in hens protected with the antidote, atropine.[161] Tabun at a dose of 12 mg/kg decreased NTE activity (67% of control) in the spinal cord of protected hens.[161] The inhibition of NTE activity in these studies were below threshold levels and suggested the inability of soman and tabun to cause OPIDN. Rats exposed to low-dose sarin (300 μg/kg, p.o.) for 90 days significantly inhibited brain NTE activity (85% of control).[107] Crowell et al.[40] studied the effects of oral graded doses of sarin (61, 70, 140, 200, 280, and 400 μg/kg) and soman (3.5, 7.1, and 14.2 μg/kg) on NTE inhibition in brain, spinal cord, and lymphocytes of hens and showed that sarin decreased lymphocyte NTE activity (33% of control) and brain NTE activity (80–60% of control) whereas, soman did not alter NTE activity in either of the tissues of hens at 24 h after dosing. The authors concluded that sarin might cause a cumulative neurotoxicity but soman appeared to be non-neuropathic.

Female mice exposed to atmospheric sarin (5 $mg \cdot m^{-3}$ for 20 min) daily for 10 days showed significant inhibition of NTE activity in the brain, spinal cord, and platelets (41, 53, and 45% of control, respectively) 4 days after sarin exposure. Results of this study indicate that sarin may induce delayed neurotoxic effects in mice following repeated inhalation exposure.[17] Rats exposed to sarin aerosols (12.5 mg/m^3 for 20 min) daily for 10 days showed significant inhibition of NTE activity in the brain, spinal cord, and platelets (65, 81, and 67% of control, respectively) 4 days after sarin exposure, but the inhibition was below the threshold level.[18] This study concluded that mice are sensitive to delayed neurotoxicity induced by repeated exposure to sarin, whereas rats were insensitive. The delayed neurotoxic effects of the known neurotoxic compound, mipafox, a chemical warfare nerve gas, sarin, and an insecticide, parathion, at low equitoxic doses (0.1 LD_{50}) were compared in hens (more susceptible to OPIDN) after repeated s.c. exposure.[19] Hens treated with mipafox (10 mg/kg, s.c.), sarin (50 μg/kg, s.c.), or parathion (1 mg/kg, s.c.) daily for 10 days resulted in significant inhibition of NTE activity in the brain, spinal cord, and platelets 4 days following sarin or mipafox exposures. This study concluded that repeated administration of equitoxic doses of mipafox, sarin, and parathion resulted in severe, moderate, and non-delayed neurotoxic effects, respectively, in hens. In a recent study using a low dose of sarin (1/20 LD_{50} = 10 μg/kg, s.c.) daily for 2 weeks, it was demonstrated that NTE activity decreased (55, 82, 75, 63, and 75% of control) in platelets, sciatic nerve, spinal cord, cerebral cortex, and corpus striatum, respectively, 4 weeks after the last dose of sarin administration in mice.[95] Although the inhibition of NTE activity in nervous tissues was below the threshold level, the platelet NTE inhibition was within the threshold limit (45%). This study suggested that platelet NTE inhibition is a more sensitive parameter for assessing OPIDN in experimental animals and humans. However, along with NTE inhibition, clinical symptoms should also be considered.

Organophosphate nerve agents interact with a variety of non-cholinergic enzymes. *In vivo* experiments have shown that nerve agents produce inhibition of succinate dehydrogenase, Na^+K^+-ATPase, aldolase,[162] Ca^{2+}-ATPase,[163] tyrosine hydroxylase,[164] and aliesterase.[165] There are reports on the effects of OP nerve agents on non-cholinergic neurotransmitters GABA (gamma amino butyric acid) in the brain,[166] catecholamines,[167] and second messengers, such as cyclic nucleotides.[168,169] The OP nerve agents produce changes in several neurotransmitters (e.g., dopamine, noradrenaline, and serotonin) in addition to ACh.[115,166,170–172] These changes may represent a compensatory mechanism in response to overstimulation of the cholinergic system or, in some instances, could result from a direct action of the OP on enzymes relevant to noncholinergic aspects of neurotransmission.[168] Soman, sarin, and tabun inhibited the adenosine receptors' binding of [$^{-3}$H]L-phenylisopropyl adenosine

([^{3}H]L-PIA) to the brain membranes in a dose-dependent manner.[173] Soman was found to be five and nine times more effective than tabun and sarin, respectively, in inhibiting [^{3}H]L-PIA binding. They suggested that nerve agents could interact directly at the A1 adenosine receptors, which could subsequently mediate changes in K^+ permeability of the synaptic membrane, with possible effects on Na^+ and/or Ca^{2+} conductance.

Effects of nerve agents reported to be mediated by hormones include hyperglycemia,[174] hyperlipidemia,[175] increase in cyclic-AMP level,[169,173] stimulation of protein synthesis,[176,177] and decrease in brain RNA levels.[178] Kokka et al.[179] studied the time course of the change in temperature and plasma levels of corticosterone, growth hormone, and prolactin following administration of soman. There was an initial rise in corticosterone level after soman administration. The time course of hypothermia after soman did not correlate with the rise in corticosterone.

Spinal cord reflexes were studied by recording the monosynaptic reflexes (MSR), dorsal root reflexes (DRR), polysynaptic reflexes (PSR), and primary afferent depolarization (PAD) due to the effects of nerve agent intoxication. Several investigators[180,181] reported that the MSR and DRR are especially depressed following OP nerve agent exposure. The OP nerve agents are reported to facilitate the MSR in cats, or depress or abolish it,[183,184] although the mechanism was found to be unrelated to changes in AChE activity or ACh content of the spinal cord.[12] It has been reported that tabun facilitates PSR and depresses MSR in cat spinal cord. Karczmar[11] reported that various mono- and polysynaptic flexor and extensor reflexes vary in their responses to OP compounds owing to variation in the circulatory responses in the spinal cord. Goldstein[185] evaluated subchronic administration of soman and sarin on the spinal MSR and DRR in spinal cord-transected cats. He showed that both agents significantly reduced the area under the MSR and DRR with only minimal changes in the excitability of the potentials. However, none of the nerve agents produced behavioral signs of delayed neurotoxicity. Pretreatment studies of carbamates (physostigmine, pyridostigmine) show that the protective carbamylation of ChE is ineffective against sarin-induced MSR depression.[186] Das Gupta et al.[187] reported that sarin-induced depression of MSR is reversed by thyrotropin-releasing hormone (TRH). They suggested that the beneficial effect of TRH in this situation may involve a noncholinergic mechanism.

B. Histopathological Effects

Studies in experimental animals have shown that repeated low-dose exposure to nerve agents cause histopathological lesions in the nervous system. The delayed neurotoxic lesions in the spinal cord due to low dose repeated inhalation exposure of sarin in two species of rodents, rats and mice, have been reported by Husain et al.[18] Rats exposed to sarin aerosols (12.5 mg/m^3 for 20 min) daily for 10 days showed swollen axons without fragmentation and loss of myelin in a few places of the spinal cord. The axonal degeneration in the spinal cord was not observed in rats exposed to sarin. Mice exposed to sarin aerosols (5 mg/m^3 for 20 min) daily for 10 days showed focal axonal degeneration in the spinal cord.[17] This study concluded that mice are sensitive to delayed neurotoxicity induced by repeated exposure to sarin, whereas rats are less sensitive.[18] Low doses of sarin (75–300 μg/kg, p.o.) for 90 days in rats caused cerebral necrosis.[106] The delayed neurotoxic effects of sarin at low dose (0.1 LD_{50} = 50 μg/kg, s.c. daily for 10 days) were studied in the hen.[19] The spinal cords of hens treated with sarin showed moderate axonal degeneration. It is suggested that the repeated exposure of nerve agent, specifically sarin, at low dose may produce OPIDN.

C. Non-Cholinergic Toxicity under Stressful Conditions

The effects of various types of stress on cholinergic toxicity due to anticholinesterase agents are known. However, the effects of various stress factors on nerve-agent-induced non-cholinergic toxicity are not well documented. The effect of low and high social stress on triorthotolyl phosphate (TOTP)-induced delayed neurotoxicity in hens has been reported by Ehrich and Gross.[188] Low social stress chickens had no competition for food or water and were housed individually in single cages with two automatic water sources and a single feeder. They were exposed to continuous soothing background music and daily socialization with animal caretakers. High social stress chickens had to compete for food, water, and group dominance, and were housed in a group of seven to eight birds per cage with a single automatic water source and a single feeder. These authors showed that clinical signs of OPIDN were less in the low social stress group, unless exposed to a high stress environment 24 h before TOTP administration. NTE activity was less than 20% of control value in all treatment groups. The authors suggested that protection of birds from OPIDN was due to reduction of conversion of TOTP to its active metabolite. A recent report showed that physical stress enhanced the clinical symptoms such as muscle weakness of the hind limb of mice treated with low-dose sarin (1/20th LD_{50} = 10 μg/kg) 4 weeks after treatment.[95] Physical stress enhanced sarin-induced inhibition of NTE in platelets, spinal cord, and cerebral cortex, and increased lipid peroxidation in triceps muscle and spinal cord in mice. Plasma creatine phosphokinase (CPK) activity was also enhanced in mice treated with sarin and exercised on treadmill, indicating neuromuscular effects of the combination. It is suggested that physical stress seems to potentiate the delayed neuro toxicity in subjects exposed to low dose sarin.

IV. SUMMARY

This chapter is a review of low-dose organophosphorus (OP) nerve agents (tabun, sarin, soman, and VX) induced toxicity under normal as well as stressful conditions. This chapter also deals with the interaction of environmental and physical stress on cholinergic as well as non-cholinergic effects induced by low-dose exposure to nerve agents and their potential for additive or synergistic neuropathologic sequelae. These agents exert their major acute toxic effects on the central and peripheral nervous system via acetylcholinesterase (AChE) inhibition. This is a high-affinity, covalent, and irreversible phosphorylation with slow reactivation or dephosphorylation. Aging occurs within a few minutes to an hour of phosphorylated AChE due to dealkylation. The physiological, biochemical, and histopathological changes due to low-dose exposure to these agents are described in human and experimental animals. These agents induce delayed neurotoxicity in human, hen, and other susceptible animals with a single high dose or repeated low dose which is characterized by a delay period of 4–21 days before clinical symptoms such as muscular weakness of the hind limb and ataxia. The molecular target for delayed neurotoxicity is a membrane-bound enzyme called neuropathy target esterase or neurotoxic esterase (NTE). Phosphorylation of NTE and subsequent aging is required to initiate axonal degeneration followed by demyelination in peripheral nerve and spinal cord. NTE is also distributed in non-nerve tissues, and platelet NTE can be used as a molecular marker for assessing delayed neurotoxicity in humans or animals exposed to neuropathic nerve agents. Delayed neurotoxicity due to low-dose exposure to these agents, specifically sarin, in terms of behavioral, biochemical, and histological changes are described. It is suggested that physical stress seems to potentiate the delayed neurotoxicity caused by low-dose exposure to sarin.

ACKNOWLEDGMENTS

The authors sincerely thank Judith M. Bryan for technical support in preparation of this manuscript. The authors are grateful to Colonel James Romano and Dr. Radharaman Ray of the U.S. Army Medical Research Institute of Chemical Defense, Aberdeen Proving Ground, MD, and Dr. Amy Arai, Assistant Professor of Pharmacology, Southern Illinois University, Springfield, IL, for a thorough review of this chapter.

REFERENCES

1. Harris, R. and Paxman, J., *A Higher Form of Killing,* Hill and Wang, New York, 1982.
2. Ivarsson, U., Nilsson, H., and Santesson, J., Eds., *Chemical Warfare Agents Today,* Vol. 16: *Chemical Weapons—Threat, Effects and Protection,* Ljungforetagen Press, Orebro, Sweden, 1992, 20.
3. Defense Science Board, Report of the Defense Science Board Task Force on Persian Gulf War Health Effects, Washington, D.C., Office of the Under Secretary of Defense for Acquisition and Technology, 1994.
4. Committee on Banking, Housing and Urban Affairs, U.S. Senate, U.S. chemical and biological warfare-related dual use export to Iraq and their possible impact on the health consequences of the Persian Gulf War, Washington D.C., U.S. Senate, 1994.
5. Spiegelberg, U., Psychopathologisch-neurologische Schaden nach Einwirkung synthetischer Gifte, in *Wehrdienst und Gesundheit,* Vol III, Wehrund Wissen Verlagsgesellschaft mbH, Darmstadt, Germany, 1961.
6. Stockholm International Peace Research Institute, Delayed toxic effects of chemical warfare agents: A Stockholm International Peace Research Institute Monograph, Alrnquist and Wiskell International, New York, 1975.
7. Sidell, F.R., Soman and sarin: Clinical manifestations and treatment of accidental poisoning by organophosphates, *Clin. Toxicol.,* 7, 1, 1974.

8. Duffy, F.H., Burchfiel, J.L., Bartels, P.H., Gaon, M., and Sim, V.M., Long term effects of an organophosphate upon the human electroencephalogram, *Toxicol. Appl. Pharmacol.*, 47, 161, 1979.
9. Nagao, M., Takatori, T., Matsuda, Y., Nakajima, M., Iwase, H., and Iwadate, K., Definitive evidence for the acute sarin poisoning diagnosis in the Tokyo subway, *Toxicol. Appl. Pharmacol.*, 144, 198, 1997.
10. Zurer, P., Japanese cult used VX to slay member, *Chem. Eng. News,* 7, 1998.
11. Karczmar, A.G., Pharmacologic, toxicologic and therapeutic properties of anticholinesterase agents, in *Physiological Pharmacology,* Root, W.S. and Hoffman, E.Z.G., eds., Academic Press, New York, 1967, 3, 163.
12. Holmstedt, B., Pharmacology of organophosphorus cholinesterase inhibitors, *Pharmacol. Rev.,* 11, 567, 1959.
13. Martin, L.J., Doebler, J.A., Shih, T.M., and Anthony, A., Protective effect of diazepam pretreatment on soman-induced brain lesion formation, *Brain Res.,* 325, 287, 1985.
14. Dacre, J.C., Toxicology of some anticholinesterases used as chemical warfare agents—a review, *Cholinesterases,* Walter de Gruyter and Co,. Berlin, New York, 1984, 415.
15. Johnson, M.K., The delayed neuropathy caused by some organophosphorus esters: Mechanism and challenge, *Crit. Rev. Toxicol.,* 3, 289, 1975.
16. Abou-Donia, M.B., Organophosphorus ester-induced delayed neurotoxicity, *Ann. Rev. Pharmacol. Toxicol.,* 21, 511, 1981.
17. Husain, K., Vijayaraghavan, R., Pant, S.C., Raza, S.K., and Pandy, K.S., Delayed neurotoxic effect of sarin in mice after repeated inhalation exposure, *J. Appl. Toxicol.,* 13, 143, 1993.
18. Husain, K., Pant, S.C., Vijayaraghavan, R., and Singh, R.. Assessing delayed neurotoxicity in rodents after nerve gas exposure, *Def. Sci.,* 44, 161, 1994.
19. Husain, K., Pant, S.C., Raza, S.K., Singh, R., and Das Gupta, S.A., comparative study of delayed neurotoxicity in hens following repeated administration of organophosphorus compounds, *Indian J. Physiol. Pharmacol.,* 39(l), 47, 1995.
20. Johnson, M.K., Organophosphorus and other inhibitors of brain 'neurotoxic esterase' and the development of delayed neurotoxicity in hens, *Biochem. J.,* 120, 523, 1970.
21. Husain K., Neurotoxicesterase, *Asia Pac. J. Pharmacol.,* 9, 119, 1994.
22. Williams, D.G., Intramolecular group transfer is a characteristic of NTE and is independent of tissue source of the enzyme, *Biochem. J.,* 209, 817, 1983.
23. Clothier, B. and Johnson, M.K., Reactivation and aging of neuro-toxic esterase inhibited by a variety of OP esters, *Biochem. J.,* 185, 739, 1980.
24. Clothier, B. and Johnson, M.K., Rapid aging of neurotoxic esterase after inhibition by diisopropyl phosphofluoridate, *Biochem. J.,* 177, 549, 1979.
25. Carrington, C.D. and Abou-Donia, M.B., Triphenyl phosphite neurotoxicity in the hen: Inhibition of NTE and a lack of prophylaxis by phenylmethylsulfonyl fluoride, *Arch. Toxicol.,* 62, 375, 1988.
26. Thomas, T.C., Szekacs, A., Rojas, S., Hammock, B.D., Wilson, B.W., and McNamee, M.G., Characterization of neuropathy target esterase using trifluoromethyl ketones, *Biochem. Pharmacol.,* 40, 2587, 1990.
27. Johnson, M.K., Sensitivity and selectivity of compounds interacting with neuropathy target esterase: Further structure/activity studies, *Biochem. Pharmacol.,* 37, 4095, 1988.
28. Pope, C.N. and Padilla, S., Potentiation of organophosphorus-induced delayed neurotoxicity phenylmethylsulfonyl fluoride, *J. Toxicol. Environ. Health,* 31, 261, 1990.
29. Glynn, P., Neuropathy target esterase, *Biochem. J.,* 3, 625, 1999.
30. Abou-Donia, M.B. and Lapadula, D.M., Mechanisms of organophosphorus ester-induced delayed neurotoxicity: type I and type II, *Ann. Rev. Pharmacol. Toxicol.,* 30, 405, 1990.

31. Bouldin, T.W. and Cavanagh, J.B., Organophosphorus neuropathy, *I. Am. J. Pathol.,* 94, 241, 1979.
32. Husain, K., Mirza, M.A., and Matin, M.A., Neurotoxic esterase activity in brain, spinal cord and platelets of certain birds and mammals, *J. Appl. Toxicol.,* 11, 61, 1991.
33. Husain K., Phenyl valerate and cholinester hydrolases in the platelets of human, hen, rat and mouse, *Hum. Exp. Toxicol.,* 13, 157, 1994.
34. Lotti, M., Biological monitoring for organophosphate-induced delayed polyneuropathy, *Toxicol. Letters,* 33, 167, 1986.
35. Lotti, M., Organophosphate-induced delayed polyneuropathy on humans: Perspectives for biomonitoring, *Trends Pharmacol. Sci.,* 8, 175, 1987.
36. Lotti, M., Moretto, A., Zoppellari, R., Dainese, R., Rizzuto, N., and Barusco, G.A., Inhibition of lymphocytic neuropathy target esterase predicts the development of organophosphate-induced delayed polyneuropathy, *Arch. Toxicol.,* 59, 176, 1990.
37. Husain, K., Peripheral biochemical marker for organophosphate induced delayed neurotoxicity, *Biochem. In.,* 24, 1051, 1991.
38. Jamal, G.A., Hansen, S., Apartopoulos, F., and Peden, A., Is there neurological dysfunction in Gulf War syndrome? *J. Neurol. Neurosurg. Psychiatry,* 60, 449, 1996.
39. Gordon, J.J., Inns, R.H., Johnson, M.K., Leadbeater, L., Maidment, M.P., Upshall, D.G., Cooper, G.H., and Rickard, R.L., The delayed neuropathic effects of nerve agents and some other organophosphorus compounds, *Arch. Toxicol.,* 52, 71, 1983.
40. Crowell, J.A., Parker, R.M., Bocci, R.J., and Dacre, J.C., Neuropathy target esterase in hen after soman and sarin, *J. Biochem. Toxicol.,* 4, 14, 1989.
41. Brooks, G.A. and Fahey, T.N., *Exercise Physiology,* John Wiley & Sons, New York, 1984, 726.
42. Connolly, R.J., Flow patterns in the capillary bed of rat skeletal muscle at rest and after repetitive tetanic contraction, in *Microcirculation,* Grayson, J., Zingg, W., Eds., Plenum Press, New York, 1976.
43. Hughson, R.L. and Green, H.J., Blood acid-base and lactate relationships studies by ramp work tests, *Med. Sci. Sports Exer.,* 14, 297, 1982.
44. Sahlin, K., Intracellular pH and energy metabolism in skeletal muscle of man with special reference to exercise, *Acta Physiol. Scand. Suppl.,* 455, 1, 1978.
45. Day, R.E., Effects of exercise performance on drugs used in musculoskeletal disorders, *Med. Sci. Sports Exer.,* 13, 272, 1981.
46. Spodaryk, K.K., Berger, L., and Hanke, S., Infusion of physical training on the functional changes of young and old red blood cells, *Mech. Agency Dev.,* 55, 199, 1990.
47. Dube, S.N., Somani, S.M., and Babu, S.R., Concurrent acute exercise alters central and peripheral responses to physostigmine, *Pharmacol. Biochem. Behav.,* 46, 827, 1993.
48. Husain, K. and Somani, S.M., Influence of exercise and ethanol on cholinesterase activity and lipid peroxidation in blood and brain regions of rat, *Prog. Neuro-Psychopharm. Biol. Psychiat.,* 21, 659, 1997.
49. Somani, S.M. and Dube, S.N., Endurance training changes central and peripheral responses to physostigmine, *Pharmacol. Biochem. Behav.,* 41, 773, 1992.
50. Tipton, C.M., Barnard, R.J., and Tharp, G.D., Cholinesterase activity in trained and non-trained rats, *Int. Z. Angew. Physiol. Einschl. Arbeitsphysiol.,* 23, 34, 1966.
51. Babu, S., Somani, S.M., and Dube, S.N., Effects of physostigmine on choline acetyltransferase and acetylcholinesterase activities in fast and slow muscles of rat, *Pharmacol. Biochem. Behav.,* 45, 713, 1993.
52. Fernandez, H.L. and Donoso, A., Exercise selectively increases G4 AChE activity in fast twitch muscle, *J. Appl. Physiol.,* 65, 2245, 1988.

53. Gisiger, V., Belisle, M., and Gardiner, P.F., Acetylcholinesterase adaptation to voluntary wheel running is proportional to the volume of activity in fast, but not slow, rat hind limb muscles, *Eur. J. Neurosci.,* 6, 673, 1994.
54. Ryhanen, R., Kajovaara, M., Harri, M., Kaliste-Korhonen, E., and Hanninen, O., Physical exercise affects cholinesterase and organophosphate response, *Gen. Pharmacol.,* 19, 815, 1988.
55. Pedzikiewicz, J., Piaskowska, E., and Pytas, M., Acetylcholinesterase (E.C.3.1.1.7) in the skeletal muscle and brain of rats after exercise and long term training, *Acta Physiol. Pol.,* 35, 469, 1994.
56. Somani, S.M., Babu, S.R., Arneric, S.P., and Dube, S.N., Effect of cholinesterase inhibitor and exercise on choline acetyltransferase and acetylcholinesterase activities in rat brain regions, *Pharmacol. Biochem. Behav.,* 39, 337, 1991.
57. Wahba, Z.Z. and Soliman, K.F.A., Effect of stress on choline acetyltransferase activity of the brain and adrenal of the rat, *Experientia,* 48, 265, 1992.
58. Fatranska, M., Budai, D., Oprsalova, L., and Kyetnansky, R., Acetylcholine and its enzymes in some brain areas of the rat under stress, *Brain Res.,* 424, 109, 1987.
59. Conlay, L.A., Wurtman, R.J., Blusztajn, J.K., Lopoz, G.I., Maher, T.J., and Evoniuk, G.B., Marathon running decreases plasma choline concentration, *N. Eng. J. Med.,* 315, 892, 1986.
60. Conlay, L.A., Sabounjian, L.A., and Wurtman, R.J., Exercise and neuromodulators: Choline and acetylcholine in marathon runners, *Int. J. Sports Med.,* 13, S141, 1992.
61. Estevez, E.E., Jerusalinsky, D., Medina, J.H., and Robertis, E.D., Cholinergic muscarinic receptors in rat cerebral cortex, basal ganglia and cerebellum undergo rapid and reversible changes after acute stress, *Neuroscience,* 13, 1353, 1984.
62. McMaster, S.B. and Carney, J.M., Chronic exercise produces tolerance to muscarinic antagonists in rats, *Pharmacol. Biochem. Behav.,* 24, 865, 1986.
63. Gottesfeld, Z., Kvetnansky, R., Kopin, I.J., and Jacobowitz, D.M., Effects of repeated immobilization stress on glutamate decarboxylase and choline acetyltransferase in discrete brain regions, *Brain Res.,* 152, 374, 1978.
64. Longoni, R., Mulas, A., Oderfeld-Novak, B., Pepeu, I.M., and Pepeu, G., Effect of single and repeated electroshock applications on brain acetylcholine levels and choline acetyltransferase activity in the rat, *Neuropharmacology,* 15, 283, 1976.
65. Gilad, G.M. and McCarty, R., Differences in chounacetyltransferase but similarities in catecholamine biosynthetic enzymes in brains of two rat strains differing in their response to stress, *Brain Res.,* 206, 239, 1981.
66. Tucek, S., Zelena, J., Ge, I., and Vyskocil, F., Choline acetyltransferase in transected nerves, denervated muscles and Schwann cells of the frog: Correlation of biochemical, electron microscopical and electrophysiological observations, *Neuroscience,* 3, 709, 1978.
67. Kita, T., Hata, T., Higashiguchi, T., Itah, E., and Kawabata, A., Changes of total acetylcholine content and the activity of related enzymes in SART (Repeated cold)-stress rat brain and duodenum, *Jpn. J. Pharmacol.,* 40, 174, 1986.
68. Gilad, G.M., Rabey, J.M., and Shenkman, L., Strain-dependent and stress-induced changes in rat hippocampal cholinergic system, *Brain Res.,* 267, 171, 1983.
69. Hata, T., Kita, T., Higash, T., and Ichide, S., Total acetylcholine content and activities of choline acetyltransferase and acetylcholinesterase in brain and duodenum of SART stressed (repeated cold stressed) rat, *Jpn. J. Pharmacol.,* 41, 475, 1986.
70. Costa, E., Tagliamonte, N., Brunello, N., and Cheney, D.L., Effects of stress on the metabolism of acetylcholine in the cholinergic pathways of extra pyramidal and limbic systems, in *Catecholamines and Stress: Recent Advances,* Usdin, E., Kvetnansky, R., and Kopin, I. J., Eds., Elsevier, New York, 1980, 59.

71. Mizukawa, K., Takayama, H., Sato, H., Ota, Z., Haba, K., and Ogawa, N., Alteration of muscarinic cholinergic receptors in the hippocampal formation of stressed rat: *In vitro* quantitative autoradiographic analysis, *Brain Res.,* 478, 187, 1989.
72. Singh, H.C., Singh, R.H., and Udupa, K.N., Electric shock-induced changes in free, bound and total acetylcholine and acetylcholinesterase in different brain regions of rats, *Indian J. Exp. Biol.,* 17, 304, 1979.
73. Cosgrove, K.A., Scudder, C.L., and Karczmar, A.C., Some aspects of acute quantitative shock on mouse whole brain levels of acetylcholine and choline, *Pharmacologist,* 15, 255A, 1973.
74. Gilad, G.M., Gilad, V.H., and Tizabi, Y., Aging and stress-induced changes in choline and glutamate uptake in hippocampus and septum of two rat strains differing in longevity and reactivity to stressors, *Int. J. Dev. Neurosci.,* 8, 709, 1990.
75. Dilsaver, S.C., Effects of stress on muscarinic mechanisms, *Neurosci. Biobehav. Rev.,* 12, 23, 1988.
76. Janowsky, D.S. and Risch, S.C., Cholinomimetic and anticholinergic drugs used to investigate an acetylcholine hypothesis of affective disorders and stress, *Drug. Dev. Res.,* 4, 125, 1984.
77. Finkelstein, Y., KaMer, B., Rabey, J.M., and Gilad, G.M., Dynamics of cholinergic synaptic mechanisms in rat hippocampus after stress, *Brain Res.,* 343, 314, 1985.
78. Takayama, H., Mizukawa, K., Ota, Z., and Ogawa, N., Regional responses of rat brain muscarinic cholinergic receptors to immobilization stress, *Brain Res.,* 436, 291 1987.
79. Zerbib, R. and Laborit, H., Chronic stress and memory: implication of the central cholinergic system, *Pharmacol. Biochem. Behav.,* 36, 897, 1990.
80. Cherek, D.R., Lane, J.D., Freeman, M.E., and Smith, J.E., Receptor changes following shock avoidance, *Soc. Neurosci. Abstr.,* 6, 543-A, 1980.
81. Koelle, G.B., *Cholinesterases and Anticholinesterase Agents,* Springer-Verlag, Berlin, 1976, 15, 123.
82. Ashani, Y., Radic, Z., Tsigelny, I., Vellon, D.C., Pickering, N.A., Quinn, D.M., Doctor, B.P., and Taylor, P., Amino acid residues that control mono and bisquaternary oxime-induced reactivation of O-ethyl methylphosphonylated cholinesterase, in *Enzymes of the Cholinesterase Family,* Quinn, D.M., Balasubramanian, A.S., Doctor, B.P., and Taylor, P., Eds., Plenum Press, New York, 1975, 133.
83. Gray, A.P., Design and structure-activity relationships of antidotes to organophosphorus anticholinesterase agents, *Drug Metabol. Rev.,* 15, 557, 1984.
84. Talbot, B.G., Anderson, D.R., Harris, L.W., Yarbrough, L.W., and Lennox, W.J., A comparison of *in vivo* and *in vitro* rates of aging of soman-inhibited erythrocyte acetylcholinesterase in different animal species, *Drug Chem. Toxicol.,* 11, 289, 1988.
85. Sirkka, U., Nieminen, S.A., and Ylitalo, P., Neurobehavioral toxicity with low doses of sarin and soman, *Methods Find. Exp. Clin. Pharmacol.,* 12, 245, 1990.
86. Kerenyi, S.Z., Murphy, M.R., and Hartgraves, S.L., Toxic interactions between repeated soman and chronic pyridostigmine in rodents, *Pharmacol. Biochem. Behav.,* 37, 267, 1990.
87. Blick, E.W., Kerenyi, S.Z., Miller, S., Murphy, M.R., Brown, G.C., and Hartgraves, S.L., Behavioral toxicity of anticholinesterase in primates: Chronic pyridostigmine and soman interaction, *Pharmacol. Biochem. Behav.,* 38, 527, 1991.
88. Goldstein, B.D., Fincher, D.R., and Searlet, J.R., Electrophysiological changes in the primary sensory neuron following subchronic soman and sarin: Alterations in sensory receptor function, *Toxicol. Appl. Pharmacol.,* 91, 55, 1987.
89. Henderson, R.F., Conn, C.A., Barr, E.B., March, T.H., Krone, J.R., Sopori, M.L., Tesfaigzi, Y., Wachulec, M., and Mash, D.B., Effect of low level sarin exposure on physiological parameters in rats, *Toxicologist,* 54(1), 266A, 2000.

90. Tonduli, L.S., Testylier, G., Pernot Marino, I., and Lallement, G., Triggering of soman-induced seizures in rats: Multiparametric analysis with special correlation between enzymatic, neurochemical and electrophysiological data, *J. Neuro. Res.,* 58, 464, 1999.
91. Yang, X.-H., Li, W., Erwin, L., and Buccafusco, J., Regulation of central muscarinic receptors after cholinesterase inhibition: Effect of clonidine, *Brain Res. Bull.,* 32, 681, 1993.
92. Vijayaraghavan, R., Husain, K., Kumar, P., Paydey, K.S., and Das Gupta, S., Time dependent protection of carbamates against inhaled sarin aerosols in rats, *Asia Pac. J. Pharmacol.,* 7, 89, 1992.
93. Shih, T.M. and McDonough, J.H., Jr., Organophosphorus nerve agents-induced seizures and efficacy of atropine sulfate as anticonvulsant treatment, *Pharmacol. Biochem. Behav.,* 64, 147, 1999.
94. Dulaney, M.D., Jr., Hoskins B., and Ho, I.K., Studies on low dose sub-acute administration of soman, sarin, and tabun in the rat, *Acta Pharmacol. Toxicol.* 57, 234, 1985.
95. Asha, T., Husain, K., Helfert, R., Verhulst, S., and Somani, S.M., Delayed interactive effects of sarin, pyridostigmine and exercise on the biochemical and histopathological changes in mice, *Toxicologist,* 54(1), 266A, 2000.
96. Katrolia, S.P., Sikder, A.K., Achrya, J., Sikder, N., Jaiswal, D.K., Bhattachrya, R., Husain, K., Dube, S.N., Kumar, D., and Das Gupta, S., Antidotal efficacy of 1-alkyl pyridinium oximes in sarin intoxication, *Pharmacol. Commun.,* 4, 317, 1994.
97. Sikder, A.K., Pandey, K.S., Jaiswal, K.K., Dube, S.N., Kumar, D., Husain, K., Bhattachrya, R., and Gas Gupta, S., The 3, 3-Bis pyridinium mono oximes as antidote against organophosphorus intoxication, *J. Pharm. Pharmacol.,* 44, 1057, 1992.
98. Tripathi, H.L. and Dewey, W.L., Comparison of the effect of diisopropylfluorophosphate, sarin, soman and tabun on toxicity and brain acetylcholinesterase activity in mice, *J. Toxicol. Environ. Hlth.,* 26, 437, 1989.
99. Murphy, M.R., Kerenyi, S.Z., Miller, S.A., Chambers, J.P., Noonan, R.F., and Hartgraves, S.L., The use of carboxylesterase inhibitors to develop an improved rodent model of soman toxicity, *Soc. Neurosci. Abstr.,* 16, 1114A, 1990.
100. Howerton, T.C., Murphy, M.R., Miller, S.A., and Hartgraves, S.L., Differential sensitivity of CNS regions to acetylcholinesterase inhibition following chronic low-dose soman treatment, *Psychopharmacology,* 105(3), 400, 1991.
101. Jimmerson, V.R., Shih, T.-M., and Mailman, R.B., Variability in soman toxicity in the rat: Correlation with biochemical and behavioral measures, *Toxicology,* 57, 241, 1989.
102. Hoskins, B.D.K., Fernando, J.C.R., Dulaney, M.D., Lim, D.K., Liu, D.D., Watanabe, H.K., and Ho, I.K., Relationship between the neurotoxicities of soman, sarin, and tabun and acetylcholinesterase inhibition, *Toxicol. Lett.,* 30, 121, 1986.
103. Russell, R.W., Booth, R.A., Lauretz, S.D., Smith, C.A., and Jenden, D.J., Behavioral neurochemical and physiological effects of repeated exposure to subsymptomatic levels of the anticholinesterase, soman, *Neurobehav. Toxicol. Teratol.,* 8, 675, 1986.
104. McDonough, J.H., Jr., Shih, T.-M., Kaminiskis, S.A., Jackson, J., and Alvarez, R., Depression and recovery of rat blood and brain acetylcholinesterase activity after repeated exposure to soman, *Soc. Neurosci. Abstr.,* 9, 964A, 1983.
105. Henderson, J.D., Higgins, R.J., Dacre, J.C., and Wilson, B.W., Neurotoxicity of acute and repeated treatments of tabun, paraoxon, diisopropyl fluorophosphate and isofenphos to the hen, *Toxicology,* 72, 117, 1992.
106. Bucci, T.J. and Parker, R.M., Toxicity studies on agents GB and BD (phase II): 90-day subchronic study of GB (sarin, type II) in CD-rats, AD A 248628, Jefferson, AR: National Center for Toxicological Research, 1992.

107. Bucci, T.J. and Parker, R.M., Toxicity studies on agents GB and GD (phase II): 90-day subchronic study of GB (sarin, type I) in CD-rats, AD A 248618, Jefferson, AR: National Center for Toxicological Research, 1991.
108. Goldman, M., Rosenblatt, L.A., Wilson, B.W., Kawakami, T.G., Culbertson, M.R., Schreider, J.P., Remsen, J.F., and Shifrine, M., Toxicity studies on agent VX, Final report, AD A201397, Ft. Detrick, Frederick, MD, U.S. Army Medical Research and Development Command, 1988.
109. Weimer, J.T., McNamara, B.P., Owens, E.J., Cooper, J.G., and Van de Wal, A., Proposed revision of limits for human exposure to GB vapor in nonmilitary operations based on one-year exposures of laboratory animals to low airborne concentrations, ARCSL-TR-78056, Aberdeen Proving Ground, MD, U.S. Army Armament Research and Development Command, 1979.
110. Baker, D.J. and Sedgwick, E.M., Single fibre electromyographic changes in man after organophosphate exposure, *Human Exp. Toxic.,* 15, 369, 1996.
111. Wolthius, O.L., Groen, B., Busker, R.W., and Van Helden, H.P.M., Effects of low doses of cholinesterase inhibitors on behavioral performance of robot-tested marmosets, *Pharm. Biochem. Behav.,* 51, 443, 1995.
112. Pearce, P.C., Crofts, H.S., Muggleton, N.G., Ridout, D., and Scott, E.A.M., The effects of acutely administered low dose sarin on cognitive behaviour and the electroencephalogram in the common marmoset, *J. Psychopharmacol.,* 13, 128, 1999.
113. Peoples, R.W., Spratto, G.R., Akbar, W.J., and Fletcher, H.P., Effect of repeated administration of soman on selected endocrine parameters and blood glucose in rats, 11, 587, 1988.
114. Reynolds, M.L., Little, P.J., Thomas, B.F., Bagley, R.B., and Martin, B.R., Relationship between the biodisposition of [^{3}H] soman and its pharmacological effects in mice, *Toxicol. Appl. Pharmacol.,* 80, 409, 1985.
115. Flynn, C.J. and Wecker, L., Elevated choline levels in brain. A non-cholinergic component of organophosphate toxicity, *Biochem. Pharmacol.,* 35, 3115, 1986.
116. Drewes, L.R. and Quian, Z., Mechanisms of elevated brain choline after organophosphate exposure, *USAMRDC Med. Def. Bio. Sci. Rev.,* 189, 1989.
117. Shih, T.M., Time course effects of soman on acetylcholine and choline levels in six discrete areas of the rat brain, *Psychopharmacology,* 78, 170, 1982.
118. Bakry, N.M., el-Rashidy, A.H., Eldefrawi, A.T., and Eldefrawi, M.E., Direct actions of organophosphate anticholinesterases on nicotinic and muscarinic acetylcholine receptors, *J. Biochem. Toxicol.,* 3, 235, 1988.
119. Churchill, L., Pazdernik, T.L., Jackson, J.L., Nelson, S.R., Samson, F., McDonough, J.H., and McLeod, C.G., Soman-induced brain lesions demonstrated by muscarinic receptor autoradiography. *Neurotoxicology,* 6, 81, 1985.
120. Aas, P., The toxic effect of an AChE-inhibitor on the cholinergic nervous system in airway smooth muscle, *Toxicology,* 49, 91, 1988.
121. Lemercier, G., Carpentier, P., Sentenac-Roumanou, H., and Morelis, P., Histological and histochemical changes in the central nervous system of the rat poisoned by an irreversible anticholinesterase organophosphorus compound, *Acta Neuropathol.* (Berlin), 61, 123, 1983.
122. Petras, J.M., Soman neurotoxicity, *Fundam. Appl. Toxicol.,* 1, 242, 1981.
123. McLeod, C.G., Jr., Singer, A.W., and Harrington, D.G., Acute neuropathology in soman-poisoned rats, *Neurotoxicology,* 5, 53, 1984.
124. Wall, H.G., Brain lesions in rhesus monkeys after acute soman intoxication, *Proceedings of the Sixth Medical Chemical Defense Bioscience Review,* AD B121516, U.S. Army Medical Research Institute of Chemical Defense, Aberdeen Proving Ground, MD, 1987.
125. Petras, J.M., Brain pathology induced by organophosphate poisoning with the nerve agent soman, *Proceedings of the 4th Annual Chemical Defense Bioscience Review,* 1984, 4.

126. Anzueto, A., Berdine, G.G., Moore, G.T., Gleiser, C., Johnson, D., White, C.D., and Johanson, W.G., Jr., Pathophysiology of soman intoxication in primates, *Toxicol. Appl. Pharmacol.,* 86, 56, 1986.
127. Singer, A.W., Jaax, N.K., Graham, J.S., and McLeod, C.G., Jr., Cardiomyopathy in soman- and sarin-intoxicated rats, *Toxicol. Lett.,* 36, 243, 1987.
128. McDonough, J.H., Jr., McLeod, C.G., Jr., and Nipwoda, M.T., Direct microinjection of soman or VX into the amygdala produces repetitive limbic convulsions and neuropathology, *Brain Res.,* 435, 123, 1987.
129. Butcher, L.L., Talbot, K., and Bilezikjian, L., Localization of acetylcholinesterase within the dopamine-containing neurons in the zona compacta of the substantia nigra, *Proc. West. Pharmacol. Sci.,* 18, 256, 1975.
130. Jacobwitz, D.M. and Palkovits, M., Topographic atlas of catecholamine and acetylcholinesterase-containing neurons in the rat brain, *J. Comp. Neurol.,* 157, 13, 1974.
131. Hartgraves, S.L., Mensah, P.L., and Kelly, P.H., Regional decreases of cortical choline acetyltransferase after lesions of the septal area and in the area of the nucleus basalis magnocellularis, *Neuroscience,* 7, 2369, 1982.
132. Somani, S.M. and Husain, K., Exercise, drugs and cholinergic system, in *Pharmacology in Exercise and Sports,* Somani, S.M., Ed., CRC Press, Inc., Boca Raton, Florida, 1996, 263.
133. Dube, S.N., Somani, S.M., and Colliver, J.A., Interactive effects of physostigmine and exercise on cholinesterase activity in RBC tissues of rat, *Part-I. Arch. Int. Pharmacodyn. Ther.,* 307, 71, 1990.
134. Somani, S.M., Giacobini, E., Boyer, A., Hallak, M., Khalique, A., Unni, L., Hannant, M., and Hurley, E., Mechanisms of action and pharmacokinetics of physostigmine in relation to acute intoxication by organofluorophosphates, *Reports submitted to U.S. Army Medical Research and Development Command,* Fort Detrick, Frederick, MD, 1986, 102 pp.
135. Somani, S.M., Rybak, L.P., and Ravi, R.P., Effect of trained exercise on antioxidant system in rat brain regions, *Pharmacol. Biochem. Behav.,* 50, 635, 1995.
136. Somani, S.M., Husain, K., Diaz-Phillips, L., Lanzotti, D.J., Kareti, K.R., and Trammell, G.Z., Interaction of exercise and ethanol on antioxidant enzymes in brain regions of rat, *Alcohol,* 13, 603, 1996.
137. Somani, S.M. and Husain, K., Interaction of exercise training and chronic ethanol ingestion on antioxidant system of rat brain regions, *J. Appl. Toxicol.,* 17, 329, 1997.
138. Husain, K. and Somani, S.M., Effect of exercise training and chronic ethanol ingestion on cholinesterase activity and lipid peroxidation in blood and brain regions of rat, *Prog. Neuro-Psychopharmacol. Biol. Psychiat.,* 22, 411, 1998.
139. Friedman, A., Kaufer, D., Shemer, J., Hendler, I., Soreq, H., and Tur-Kaspa, I., Pyridostigmine brain penetration under stress enhances neuronal excitability and induces early immediate transcriptional response, *Nature Med.,* 2, 1382, 1996.
140. Hubert, M. and Lison D., Study of muscular effects of short-term pyridostigmine treatment in resting and exercising rats, *Hum. Exp. Toxicol.,* 14, 49, 1995.
141. Somani, S.M., Husain, K., Asha, T., and Helfert, R., Interactive and delayed effects of pyridostigmine and physical stress on biochemical and histopathological changes in peripheral tissues of mice, *J. Appl. Toxicol.,* 20, 327, 2000.
142. Tsakiris, S. and Kontopoulos, A.N., Time changes in Na^+, K^+-ATPase, Mg^{++}-ATPase, and acetylcholinesterase activities in the rat cerebrum and cerebellum caused by stress, *Pharmacol. Biochem. Behav.,* 44, 339, 1993.
143. Pawlowska, D., Moniuszko-Jankoniuk, and Soltys, M., Parathion methyl effect on the activity of hydrolytic enzymes after single physical exercise in rats, *Pol. J. Pharmacol. Pharm.,* 37, 629, 1985.

144. Husain, K. and Mirza, M.A., Effect of environmental factor (temperature) on the toxicity of an organophosphorous insecticide (diazinon) in rats, *Biol. Memrs.*, 15, 84, 1989.
145. Jones, K.H., Dech Kovskaia, A.M., Khan, W.A., and Abou-Donia, M.B, Sarin and stress modulation of nicotinic and muscarinic acetylcholine receptors in sarin exposed rats. *Toxicologist*, 54(1), 324A, 2000.
146. Lallement, G., Foquin, A., Baubichon, D., Burckhart, M.-F., Carpentier, P., and Canini, F., Heat stress, even extreme, does not induce penetration of pyridostigmine into the brain of guinea pigs, *Neurotoxicology*, 19, 759, 1998a.
147. Wheeler, T.G., The behavioral effects of anticholinesterase insult following exposure to different environmental temperatures, *Aviat. Space Environ. Med.*, 58, 54, 1987.
148. Honkakoski, P., Ryhanen, R., Harri, M., Ylitahe, P., and Hanninen, O., Spontaneous recovery of cholinesterases after organophosphate intoxication: Effect of environmental temperature, *Bull. Environ. Contam. Toxicol.*, 40, 358, 1988.
149. Wheeler, T.G., Soman toxicity during and after exposure to different environmental temperatures, *J. Toxicol. Environ. Hlth.*, 26, 349,1989.
150. Kaliste-Korhonen, E., Ryhanen, R., Ylitalo, P., and Hanninen, O., Cold exposure decreases the effectiveness of atropine-oxime treatment in organophosphate intoxication in rats and mice, *Gen. Pharmac.*, 20, 805, 1989.
151. Johnson, M.K., The target for initiation of delayed neurotoxicity by organophosphorus esters: Biochemical studies and neurotoxicological applications, *Rev. Biochem. Toxicol.*, 4, 141, 1982.
152. Lotti, M. and Johnson, M.K., Repeated small doses of a neurotoxic organophosphate: Monitoring of neurotoxic esterase in brain and spinal cord, *Arch. Toxicol.*, 45, 263, 1980.
153. Willems, J.L., Nicaise, M., and DeBisschop, H.C., Delayed neuropathy by the organophosphorus nerve agents soman and tabun, *Arch. Toxicol.*, 55, 76, 1984.
154. Vranken, M.A., DeBisschop, H.C., and Willems, J.L., *In vitro* inhibition of neurotoxic esterase by organophosphorus nerve agents, *Arch. Pharmacodyn.*, 260, 316, 1982.
155. Grob, D. and Harvey, J.C., Effects in man of the anticholinesterase compound sarin (isopropyl methyl phosphonofluoridate), *J. Clin. Invest.*, 37, 350, 1958.
156. Yokoyama, K., Araki, S., Murata, K., Nishikitani, M., Okumura, T., Ishimatsu, S., and Takasu, N., A preliminary study on delayed vestibulocerebellar effects of Tokyo subway sarin poisoning in relation to gender difference: Frequency analysis of postural sway, *JOEM*, 40, 17, 1998.
157. VanKampen, K.R., James, L.F., Rasmussen, J., Huffacker, R.H., and Fawcett, M.O., Organic phosphate poisoning of sheep in Skull Valley, Utah, *JAVMA*, 154, 623, 1969.
158. Parker, R.M., Crowell, J.A., Bucci, T.J., Thurman, J.D., and Dacre, J.C., Thirteen-week oral toxicity studies of tabun (GA) using CD rats, *Toxicologist*, 10, 343, 1990.
159. Bucci, T.J., Parker, R.M., Crowell, J.A., Thurman, J.D., and Gosnell, P.A., Toxicity studies on agent GA (phase II): 90 day subchronic study of GA (tabun) in CD rats. AD A25 8042, Jefferson, AR: National Center for Toxicological Research, 1992.
160. Wilson, B.W., Henderson, J.D., Kellner, T.P., Goldman, M., Higgins, R.J., and Dacre, J.C., Toxicity of repeated doses or organophosphorus esters in the chicken, *J. Toxicol. Environ. Hlth.*, 23, 115, 1988.
161. Johnson, M.K., Willems, J.L., DeBisschop, H.C., Read, D.J., and Benschop, H.P., High doses of soman protect against organophosphorus-induced delayed polyneuropathy but tabun does not, *Toxic Applied Pharm.*, 92, 34, 1988.
162. Jovic, R., Bachelard, H.S., Clark, A.G., and Nicholas, P.C., Effects of soman and DFP *in vivo* and *in vitro* on cerebral metabolism in rat, *Biochem. Pharmacol.*, 20, 519, 1971.

163. Dierkes-Tizek, V.U., Glaser, U., Oldiges, H., and Hettwer, H., Effect of organophosphates on rat heart ATPases, *Arzneim-Forsch/Drug Res.,* 34, 671, 1984.
164. Richardson, J.S., Lamprecht, F., Kazic, T., and Kopin, I., Reduction of brain tyrosine hydroxylase activity following acetylcholinesterase blockade in rats, *Can. J. Physiol. Pharmacol.,* 54, 774, 1976.
165. Clement, J.G., Importance of aliesterase as a detoxification mechanism for soman (pinacolyl methylphosphonofluoridate) in mice, *Biochem. Pharmacol.,* 33, 3807, 1984.
166. Sivam, S.P., Nabeshima, T., Lim, D.K., Hoskins, B., and Ho, I.K., Diisopropylfluorophosphate and GABA synaptic function: Effect on levels, enzymes, release and uptake in the rat striatum, *Res. Commun. Chem. Pathol. Pharmacol.,* 42, 51, 1983.
167. Coudray-Lucas, C., Leuguen, A., Prioux-Guyonneau, M., Cohen, Y., and Wepierre, J., Changes in brain monamine content and metabolism induced by paraoxon and soman intoxication. Effect of atropine, *Xenobiotica,* 17, 1131, 1987.
168. Sellaijevic, L., Kriolica, K., and Boskovic, B., The effect of soman poisoning on phosphorylating capability and adenylate cyclase activity of isolated synaptosomal membranes. *Biochem. Pharmacol.,* 33, 3714, 1984.
169. Lundy, P.M. and Shaw, R.K., Modification of cholinergically induced convulsive activity and cyclic GMP levels in the CNS, *Neuropharmacology,* 22, 55, 1983.
170. Coudray-Lucas, C., Prioux-Guyonneau, M., Sentenac, H., Cohen, Y., and Wepierre, J., Brain catecholamine metabolism changes and hypothermia in intoxication by anticholinesterase agents, *Acta Pharmacol. Toxicol.,* 52, 224, 1983.
171. Fernando, J.E.R., Lim, D.K., Hoskins, B., and Ho, I.K., Effect on striatal dopamine metabolism and different motor behavioral tolerance following chronic cholinesterase inhibition with diisopropylfluorophosphate, *Pharmacol. Biochem. Behav.,* 20, 951, 1984.
172. Fernando, J.C.R., Hoskins, B., and Ho, I.K., A striatal serotonergic involvement in the behavioral effects of anticholinesterase organophosphates, *Eur. J. Pharmacol.,* 98, 129, 1984.
173. Liu, W.M., Freeman, S.E., and Szilagyi, M., Binding of some organophosphorus compounds at adenosine receptors in guinea pig brain membranes, *Neurosci. Lett.,* 94, 125, 1988.
174. Meller, D., Fraser, I., and Kryger, M., Hyperglycemia in anticholinesterase poisoning, *Can. Med. Assoc. J.,* 124, 745, 1981.
175. Fukuyama, G.S., Adie, P.A., and Hughes, E.R., Increased concentrations of blood lipids in guinea pigs poisoned by sarin, *Nature* (London), 200, 897, 1963.
176. Sikora-Van Meter, K.C., Wierwille, R.C., Willetts, E.J., and Van Meter, W.G., Morphological evidence for increased protein synthesis in CNS neurons after soman exposure, *Fundam. Appl. Toxicol.,* 8, 23, 1987.
177. Swisher, J., Doebler, J., and Anthony, A., Cytophotometric analysis of liver DNA and RNA changes in acute soman-toxicated rabbits, *Fed. Proc.,* 43, 1640A, 1984.
178. Anthony, A., Doebler, J.A., Bocan, T.M.A., Zerweck, C., and Shih, T.-M., Scanning-integrating cytophotometric analysis of brain neuronal RNA and acetylcholinesterase in acute soman toxicated rats, *Cell Biochem. Funct.,* 1, 30, 1983.
179. Kokka, N., Clemons, G.K., and Lomax, P., Relationship between the temperature and endocrine changes induced by cholinesterase inhibitors, *Pharmacology,* 34, 74, 1987.
180. Yang, Q.J. and Warnick, J.E., Effect of sarin and soman on spinal reflexes in the cat. *Neurosci. Abstr.,* 9, 230A, 1983.
181. Yang, Q.Z. and Warnick, J.E., Antagonism of organophosphate-induced depression of reflex activity in the neonatal rat spinal cord, *Neurosci. Abstr.,* 10, 417A, 1984.
182. Swanson, K.L. and Warnick, J.E., Tabun facilitates and depresses spinal reflexes in cat and neonatal rat spinal cords, *Neurosci. Abstr.,* 10, 817A, 1984.

183. The mechanism of action of DFP and TEPP on the patellar reflex, *J. Pharmacol. Exp. Ther.,* 110, 232, 1954.
184. Robinson, G.M., Beck, R., McNamara, B.P., Edberg, L.J., and Wills, J.H., The mechanism of action of anticholinesterase compounds on the patellar reflex. *J. Pharmacol. Exp. Ther.,* 110, 385, 1954.
185. Goldstein, B.D., Changes in spinal cord reflexes following subchronic exposure to soman and sarin, *Toxicol. Lett.,* 47, 1, 1989.
186. Das Gupta, S., Bass, K.N., and Warnick, J.E., Interaction of reversible and irreversible, cholinesterase inhibitors on the monosynaptic reflex in neonatal rats, *Toxicol. Appl. Pharmacol.,* 99, 28, 1989.
187. Das Gupta, S., Bass, K.N., and Warnick, J.E., Segmental synaptic depression caused by diisopropyl-phosphonofluoridate and sarin is reversed by thyrotropin-releasing hormone in the neonatal rat spinal cord, *Toxicol. Appl. Pharmacol.,* 95, 499, 1988.
188. Ehrich, M. and Gross, B., Modification of triorthotolyl phosphate toxicity in chickens by stress, *Toxicol. Appl. Pharmacol.,* 70, 249, 1983.
189. U.S. DOA, Chemical agent data sheets, Vol. 1, Technical Report. Edgewood Arsenal Special Report (EOSR-74001, AD B028222, Aberdeen Proving Ground, MD, U.S. Department of the Army Headquarters, 1974.
190. Dick, C.J., Soviet chemical warfare capabilities. International defense review, in Selected readings in nuclear, biological and chemical operations, 1984 (reprint). Ft. Leavenworth, KS, U.S. Army Command and General Staff College, 1981, 1.
191. Allon, N., Raveh, L., Gilat, E., Cohen, E., Grunwald, J., and Ashani, Y., Prophylaxis against soman inhalation toxicity in guinea pigs by pretreatment alone with human serum butyrylcholinesterase, *Toxicol. Sci.,* 43, 121, 1998.
192. Schoene, K., Hochrainer, D., Oldiges, H., Krugel, M., Franzes, N., and Bruckert, H.-J., The protective effect of oxime pretreatment upon the inhalative toxicity of sarin and soman in rats., *Fund. Appl. Toxicol.,* 5, S84, 1985.
193. U.S. Department of the Army and U.S. Department of the Air Force, Military chemistry and chemical compounds. Field Manual, Army FM 3–9, Air Force AFR355–7, Washington, D.C., Department of the Army, 1975.
194. Fielding, G.H., V agent information summary. NRL 5421, Washington, D.C., U.S. Naval Research Laboratory, 1960.
195. Sidell, F.R., Clinical notes on chemical casualty care. USAMRICD Technical memorandum 90–1, Aberdeen Proving Ground, MD, U.S. Army Medical Research Institute of Chemical Defense, 1990.
196. Wiles, J.S. and Alexander, T.B., Comparative toxicity of VX applied to the unclipped and clipped skin of bare and clothed rabbits, AD839329, Aberdeen Proving Ground, MD, U.S. Army Chemical Research and Development Laboratories, 1960.
197. Loomis, T.A. and Salafsky, B., Antidotal action of pyridinium oximes in anticholinesterase poisoning: Comparative effects of soman, sarin and neostigmine on neuromuscular function, *Toxicol. Appl. Pharmacol.,* 5, 685, 1963.
198. Robinson, J.P. Chemical warfare, *Science J.,* 4, 33, 1967.
199. Murtha, E.F. and Harris, L.W., Effects of 2-pyridine aldoxime methochloride on cerebral acetylcholinesterase activity and respiration in cats poisoned with sarin, *Life Sci.,* 27, 1869, 1980.
200. Worek, F. and Szinicz, L., Cardiorespiratory function in nerve agent poisoned and oxime + atropine treated guinea pigs: Effect of pyridostigmine pretreatment, *Arch. Toxicol.,* 69, 322, 1995.

201. O'Leary, J.F., Kunkel, A.M., and Jones, A.H., Efficacy and limitations of oxime-atropine treatment of organophosphorus anticholinesterase poisoning, *J. Pharmacol. Exp. Ther.,* 132, 50, 1961.
202. Schoene, K. and Oldiges, H., Die wirkungen von pyridiniumsalzen gegenuber Tabun and sarinvergiftunger, *in vivo* and *in vitro, Arch. Int. Pharmacodyn.,* 204, 110, 1973.
203. Leadbeater, L., Inns, R.H., and Rylands, J.M., Treatment of poisoning by soman, *Fundam. Appl. Toxicol.,* 5, 225, 1985.
204. Jones, D.E., Koplovitz, I., Harrington, D.G., Hilmas, D.E., and Canfield, C.J., Models for assessing efficacy of therapy compounds against organophosphates (OP). *Proc. Fourth Annual Chemical Defense Bioscience Review,* U.S. Army Medical Research and Development Command, 1, 1984.
205. Harris, L.W., Lennox, W.J., and Talbot, B.G., Toxicity of anticholinesterase: Interactions of pyridostigmine and physostigmine with soman, *Drug Chem. Toxicol.,* 7, 507, 1984.
206. Clement, J.G., Variability of sarin-induced hypothermia in mice: Investigation into incidence and mechanism, *Biochem. Pharmacol.,* 42, 1316, 1991.
207. Lallement, G., Clarencon, D., Masqueliez, C., Baubichon, D., Galonnier, M., Burckhart, M.F., Peoc'h, M., and Mestries, J.C., Nerve agent poisoning in primates: Antilethal, antiepileptic and neuroprotective effects of GK-11, *Arch. Toxicol.,* 72, 84, 1998.
208. Davies, D.R., Holland, P., and Rumens, M.J., The relationship between the chemical structure and neurotoxicity of alkyl organophosphorus compounds, *Br. J. Pharmacol.,* 15, 271, 1960.
209. Wilson, B.W., Henderson, J.D., Chow, E., Schreider, J., Goldman, M., Culbertson, R., and Dacre, J., Toxicity of an acute dose of agent VX and other organophosphorus esters in the chicken, *J. Toxicol. Environ. Health,* 23, 103, 1988.
210. Tuovinen, K. and Hanninen, O., Protection of mice against soman by pretreatment with eptastigmi and physostigmine, *Toxicology,* 139, 233, 1999.
211. Ballough, G.P., Cann, F.J., Smith, C.D., Forster, J.S., Kling, C.E., and Filbert, M.G., GM1 monosialoganglioside pretreatment protects against soman-induced seizure-related brain damage, *Mol. Chem. Neuropath.,* 34, 1, 1998.
212. Somani, S.M., Solana R.P., and Dube, S.N., Toxicodynamics of nerve agents, in *Chemical Warfare Agents,* Somani, S.M., Ed. Academic Press, Ltd., San Diego, CA, 1992, 67.

4 Blood-Brain Barrier Modulations and Low-Level Exposure to Xenobiotics

Hermona Soreq, Daniela Kaufer, Alon Friedman, and David Glick

CONTENTS

I. Introduction 122
II. The Physical Basis of Blood-Brain Barrier Properties 122
  A. Endothelial Cells in Brain Vasculature 123
  B. Adherens and Tight Junctions 124
  C. Potential Involvement of Acetylcholinesterase 124
  D. Signal-Transducing Elements 126
  E. Astrocyte Contributions to Blood-Brain Barrier Properties 128
III. Functional Characteristics of the Blood-Brain Barrier 128
  A. Inward and Outward Movement across the Blood-Brain Barrier: Physiological Considerations 128
  B. Cholinergic Involvement in Blood-Brain Barrier Functioning 129
  C. Pericellular Passage across Blood-Brain Barrier Structures 129
  D. Cell Culture, Organ Systems, and Imaging Approaches in Blood-Brain Barrier Research 130
  E. Transgenic Engineering Models for Blood-Brain Barrier Studies 131
IV. Modulators of Blood-Brain Barrier Functions and Their Interrelationships 132
  A. Nitric Oxide and Vasoactive Agents Involvement 133
  B. Immunomodulators and Multi-Drug Transporters 133
V. Conditions Inducing Blood-Brain Barrier Distruption 134
  A. Pathophysiological Induction of Blood-Brain Barrier Penetrance 134
  B. Blood-Brain Barrier Disruption Following Acute Insults 135
  C. Psychological and Physical Stressors Impair Blood-Brain Barrier Functioning 135

0-8493-0872-0/01/$0.00+$.50

D. Blood-Brain Barrier as a Complex Trait with Genetic and Physiological Components: Prospects 136
VI. Summary 137
Acknowledgments 138
References 138

I. INTRODUCTION

Separation of the brain from the peripheral blood is crucial for protecting this most delicate and important organ from various insidious agents that circulate in the blood. Conversely, the separation must allow for the nutrition of the brain and the removal from it of waste products. The existence of a physical barrier that separates the brain tissue from the general circulation was first proposed 100 years ago, by Ehrlich, who discovered that injection of a series of dyes into laboratory animals resulted in uncolored brains, as opposed to highly stained visceral organs.[1] The blood-brain barrier (BBB) is formed during the late embryonic and early postnatal period. It is an endothelial barrier present in the capillaries throughout the brain, contact-influenced by neighboring astrocytes.[2] Electron microscopic studies reveal two major factors that distinguish brain endothelial cells from their peripheral relatives: first, they contain lower amounts of endocytic vesicles, and second, the space between adjacent cells is sealed by tight junctions; both factors restrict intercellular flux. These features enable the formation of a barrier that hinders the entry of most xenobiotics into the brain, and is actively involved in exporting such substances from the brain when they do enter it. Small lipophilic molecules enter the brain fairly freely, but hydrophilic molecules enter via active transport, and specific transporters exist for required nutrients such as glucose, L-DOPA, and certain amino acids.[3]

The physical and functional complexity of the BBB has hampered research efforts to delineate its components and fully understand its mode of action. Numerous experimental approaches were developed for evaluating BBB integrity; these include *in vitro* and *in vivo* systems as well as transgenic engineering approaches. The use of these methods has revealed several modulators of BBB functioning and has demonstrated intricate relationships between these modulators in their effects on BBB integrity. Impairments of any element of these chains of factors can disrupt BBB functioning, but the extent and duration of such disruptions apparently depend on the genetics, health, and wellbeing of the involved organism. In the following, we discuss these considerations as they relate to the issue of low-level exposure to xenobiotics.

II. THE PHYSICAL BASIS OF BLOOD-BRAIN BARRIER PROPERTIES

Low-level exposure to xenobiotics would first affect the circulation; to affect the brain, the xenobiotic must traverse the BBB. In certain cases, e.g., under exposure to anticholinesterases, these agents interact with and inhibit the catalytic activity of their target enzymes, cholinesterases, in peripheral and brain systems alike. The cellular

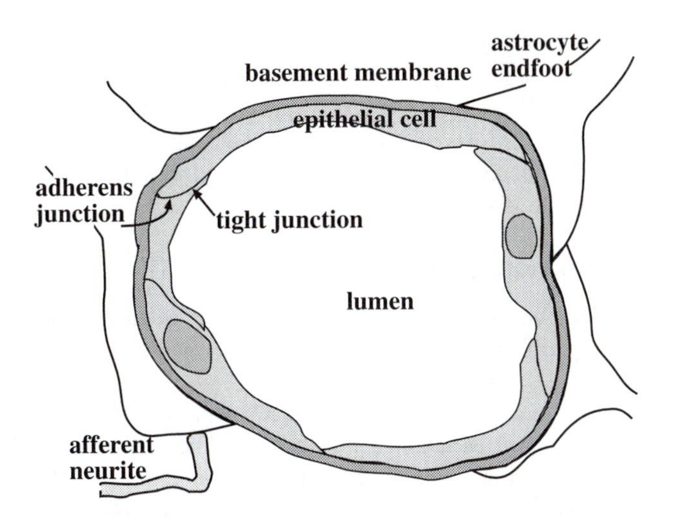

FIGURE 4.1 The physical components of the blood-brain barrier (BBB): Within the mammalian brain, blood vessels and microvessels transverse the brain tissue, bringing in essential compounds and removing metabolic end products. The three layers surrounding the microvessel lumen comprise the BBB, including endothelial cells lining the blood vessels, a basement membrane surrounding them, and astrocyte endfeet separating these structures from adjacent neurons, some of which interact with these astrocytes through contacting neurites. Two types of junctions connect endothelial cells to each other, tight and adherens junctions.

components of the BBB include the endothelial cells that line the inside of brain capillaries, the basement membrane surrounding them, and brain astrocytes, which constitute a third layer separating these blood vessels from the surrounding brain tissue. Intercellular BBB structures include adherens junctions, which attach endothelial cells to each other, as well as the tight junctions that seal them. Within these cells, surface membrane proteins transduce signals by activation of specific kinases, and the flow of information among the several cell types is affected by neuronal and astrocytic activities in the brain as well as by peripheral metabolic changes and external forces. Figure 4.1 presents a schematic view of these elements.

A. Endothelial Cells in Brain Vasculature

Many have reported the special properties of endothelial cells in brain vasculature.[4] Small hydrophobic molecules diffuse across the BBB; large and/or hydrophilic molecules may be transported only if a specific receptor or transporter exists. Thus, small hydrophobic molecules penetrate the brain by diffusion; nutrients such as glucose and certain amino acids are transported into the brain by specific transporters; and several large proteins like transferrin are transcytosed into the brain via specific receptors.

To enable these special properties, tight junctions connect brain endothelial cells, so that intercellular transport is extremely limited.[5] The expression of P glycoprotein (the multi-drug resistance mdr protein) on their surface membrane controls both penetrance of small molecule drugs into the brain and their export from the brain. Genomic disruption of the mdrla (a.k.a. mdr3) gene causes extreme drug sensitivity, for example, to ivermectin.[6] This finding highlights the importance of active transport mechanisms for the integrity of BBB functioning and may imply that cumulative exposure can modulate BBB properties.

B. Adherens and Tight Junctions

A primary difference between endothelial cells of brain vasculature and the very similar cells that line peripheral blood cells relates to the composition and properties of the tight junctions between these cells.[7–10]

Adherens junctions are similar to the attachment structures of other cells in which their functions may be more easily studied on the molecular level. Yeast, for example, can express an analogue of the adherens junction, and its assembly was shown to depend upon the Ca^{++}-dependent protein kinase pathway.[11] Genetic studies in yeast are easy to perform, and since yeast has a well-described genome, the discovery of the genes that regulate junction formation is possible, and once the yeast gene is known, it is a relatively simple matter to discover its homologues in a mammalian genome.

Unlike adherens junctions, which form homophilic intercellular adhesion sites, tight junctions are complex structures recognized as being the molecular site of pericellular transport and its regulation.[12] In addition to adherens and tight junctions, brain capillary endothelial cells have transmembrane receptors for matrix proteins (e.g., integrins).[9] Impairment of either cell-cell or cell-matrix interactions can disrupt the BBB, in processes that parallel those of the peripheral endothelium. However, such impairments occur much less frequently in the brain.

Several proteins, including cingulin and occludin, were shown to be essential for the function of tight junctions.[13,14] More recently, junction-associated proteins such as Rho were reported to regulate tight junctions and perijunctional actin organization in polarized epithelia.[15] Junction proteins are physically linked to cytotskeletal elements such as actin or linking proteins like β-catenin in a manner subject to modulation by phosphorylation or dephosphorylation of specific kinases and phosphatases. This suggests a potential opening of tight junctions by kinase regulation, however, no experimental evidence is yet available to demonstrate such opening *in vivo*. Table 4.1 catalogs key proteins assumed to be associated with BBB junctions.

C. Potential Involvement of Acetylcholinesterase

Like yeast, for nearly a century the fruit fly *Drosophila melanogaster* has served as a model for genetics studies. With the introduction of genetic engineering and genomic databases, this has also become a powerful tool for the discovery of genes and gene products that participate in physiological functions. The identification of a physiological defect in the insect can be quickly traced to a specific gene, and the homologous sequence in the mammalian genome can then be identified, where it then serves as a candidate gene for the similar function in the mammal. For instance, a defect in

TABLE 4.1
Protein Components and Candidate Components of the Blood-Brain Barrier

Component	Intra/Extra Cellular[a]	Interaction Partners[b]	Reviewed in
7H6 antigen	Intra and extra	Tight junction	10
Acetylcholinesterase	Extra	Neurexin	16
Actin	Intra	Catenin	10
Band 4.1 protein	Intra	Cytoskeleton	16
Cadherin	Extra	Catenin, p120	10; 17
CASK[c]	Intra	Neurexin II β	16
β-catenin	Intra	Cadherin, actin	10
Cingulin	Intra	Tight junction	10
Gliotactin	Intra and extra	Neurexin IV	16
ICAM-1	Extra	ICAM	18
Neuoligin	Intra and extra	Neurexin II β	16
Neurexin	Intra and extra	Neuroligin 1, CASK	16
Nitzin	Intra	Cytoskeleton	16
Occludin	Extra	ZO-1/ZO-2/p130	10
p100	Intra	Cadhedrin	10; 19
p120	Intra	Cadherin/catenin	10; 17; 19
p130	Intra	Tight junction, ZO-1	10
PSD-95[d]	Intra	Neuroligin 1, NMDA receptor	20
RPTP[d]	Intra	Cadherin/catenin	10
Selectin	Extra		21
src	Intra	Adherens junction	10
Tyrosine kinase	Intra	ZO-1, β-catenin	10
ZO-1	Intra	Tight junction, p130, tyrosine kinase	10
ZO-2	Intra	Tight junction	10

[a]BBB components may function in extracellular locations (extra), convey signals within intracellular locations (intra), or do both.

[b]Independent factors to which attachment has been shown are separated by commas; aggregates of factors to which attachment has been shown are indicated by slashes.

[c]Post-synaptic density.

[d]Calmodulin-dependent protein kinase.

[d]Receptor-type protein tyrosine phosphatase.

the hemolymph-neuron barrier, which serves a function analogous to the BBB in mammals, was shown to depend on the structural and functional integrity of the special septate junctions, which seal this barrier in insect larva.[22,23] Disruption of these structures by genomic destruction of either of two different genes, neurexin IV and gliotactin, causes severe neuronal sensitivity to the high concentrations of K^+ in the hemolymph. This leads to paralysis and death of the developing insect larva. Such genomic disruption also impaired the subcellular targeting of coracle, a band 4.1 homologue that transduces signals from the cell membrane to the cytoskeleton.

Gliotactin is one of several structural homologues of the acetylcholine-hydrolyzing enzyme acetylcholinesterase (AChE) that were discovered in the past decade. Gliotactin, however, like the other AChE homologues, has no capacity for acetylcholine (ACh) hydrolysis. Intriguingly, AChE may compete with its structural homologues for their cell-cell interactions.[16,24] This potential involvement of AChE has raised the question of which of the three variants, formed by alternative splicing of the human AChE pre-mRNA, may be involved in these interactions. These variants are: AChE-S, the synaptic form, AChE-E, the erythrocyte form, and AChE-R, a soluble monomeric form which, perhaps significantly for BBB physiology, has been shown to be over-expressed under stress.[25]

Gliotactin, like several other AChE homologues, is equipped with an extracellular domain, a transmembrane peptide and C-terminal peptide that protrudes into the cytoplasm and can transduce signals into cells. In particular, it interacts with proteins, which modulate the cytoskeleton. Therefore, these discoveries present the entire series of AChE homologues and their yet unidentified binding partners as promising candidates to participate in control of the integrity of the BBB and transduction of signals that regulate its functioning. The impressive conservation of these inter- and intracellular factors, and the chain of interactions by which they may affect cytoskeletal properties, suggest at a mechanism by which AChE levels, and/or the specific chemical properties of its variants, affect the integrity of the BBB. Kaufer et al.[26] have recently discovered a feedback process that leads to AChE-R accumulation under exposure to anticholinesterases. This points to the AChE protein as a modulator that may be intimately involved in BBB disruption under exposure to such agents. That no embryonic impairment in BBB functioning is known in mammals most likely attests to the essential role played by the BBB in mammalian embryonic development, as early lethality of such a mutant would preclude its discovery. Figure 4.2 summarizes the evolutionary conservation of the structural properties of AChE as these may be involved in BBB integrity.

D. Signal-Transducing Elements

Appropriate functioning of the BBB and its capacity to respond to environmental insults evidently depend on fast, accurate, and sensitive transduction of appropriate signals from the periphery into the brain and vice versa. Over the past few years, several molecular components were discovered which ascertain such a flow of information and ensure its reliability. The role of guanine nucleotides in regulating BBB properties is of special interest. Endothelial capillary cells are polarized, being long and flat structures linked by tight junctions. GTP-Binding Rho proteins are responsible, in these cells, for the particular organization of the filamentous actin fibers that ensure their polarization.[15] Further, transduction of intracellular signals is based, in most polarized epithelial cells, primarily on PDZ domain proteins. Named for three members of this family, PDZ proteins include the post-synaptic density protein, PSD-95, the *Drosophila* tumor-suppressor protein, discs-large (DlgA), and the tight-junction protein, ZO-1; they are often found at the plasma membrane and transduce signals into the cell, affecting cytoskeletal organization.[27,28] That this is also the case

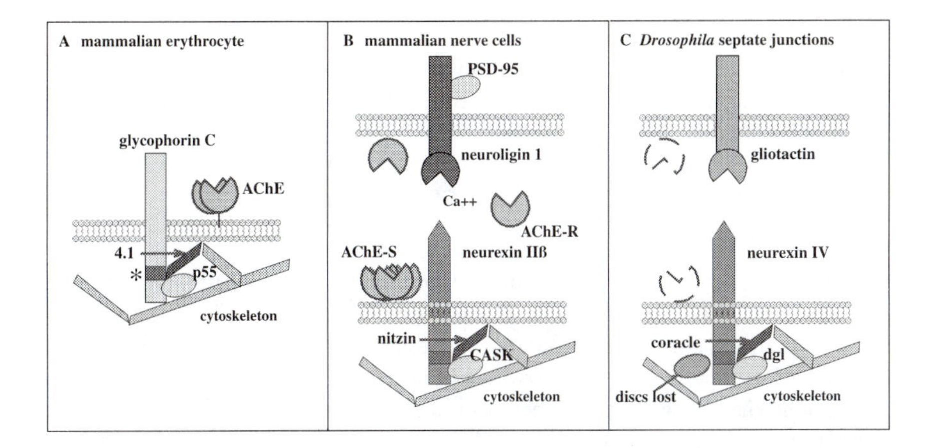

FIGURE 4.2 AChE and its structural homologues are potentially involved in BBB functioning. Shown are schematic drawings of membrane signaling and cytoskeletal components which involve AChE and its structural homologues. (A) In the mammalian erythrocyte, glycophorin C is located on the surface membrane, with one domain protruding into the cytoplasm. A conserved element within this domain (starred) interacts with the band 4.1 protein that serves as an anchor to the cytoskeleton. Another region in the cytoplasmic domain of glycophorin C binds p55, a PDZ protein. It is not yet known whether the AChE-E dimers, which are anchored by a glycophosphoinositol moiety to the outer surface of the erythrocyte membrane, are involved in glycophorin's role of modulating the erythrocyte structure. (B) In mammalian nerve cells, the major AChE isoform is AChE-S tetramers. The AChE-homologous neuronal membrane proteins, neuroligins, are expressed in the developing brain and in excitatory synapses. At least one of the neuroligins, neuroligin 1, interacts with at least one of the neurexins, neurexin II β, in a Ca^{++}-dependent manner. Both neuroligin and neurexin protrude into the cytoplasm, and both interact with PDZ proteins such as PSD-95 and CASK. Neurexins further associate with neuronal band 4.1 homologues, like nitzin,[16] creating a link with the cytoskeleton. Modulation of AChE properties under low-level exposure to an anticholinesterase (e.g., accumulation of AChE-R monomers) may therefore alter neuroligin-neurexin interactions, transducing signals to the neuronal cytoskeleton. (C) In *Drosophila* septate junctions, the AChE-homologous protein, gliotactin, protrudes into the cytoplasm. Another transmembrane protein, neurexin IV, shares with other neurexins an extracellular domain that may interact with the core domain module that is common to AChE and neuroligins. Neurexin IV also includes cytoplasmic glycophorin C elements; these regions intact with the insect band 4.1 homologue, coracle, as well as with the PDZ protein dg1 and the multi-PDZ domain protein, disc lost. Therefore, neurexin IV interactions with either gliotactin or the cytoskeleton are essential for maintaining septate junction integrity. It is not yet known whether AChE itself (broken-line structure) is expressed in these junctions.

for BBB components has recently been demonstrated in *Drosophila* embryos by Bellen and co-workers, who found a third protein that is essential for the integrity of septate junctions formation.[29] This protein, discs-lost, uses multiple PDZ domains to interact with intracellular components in a manner dependent on septate junction interactions. In general, PDZ domains interact with the carboxy terminal end of their target proteins.[27] Therefore a multi-PDZ protein can aggregate a series of target proteins within the cell, simultaneously transducing multiple signals.[29] This enables an extremely sensitive biosensor activity, as is expected from a system designed to protect the brain from low-level exposures.

E. Astroycyte Contributions to Blood-Brain Barrier Properties

Janzer and Raff recognized the key function of astrocytes that surround brain capillaries in the dynamic properties of the BBB.[30] Specific interactions between astrocyte endfeet, which surround brain capillaries, are essential to ensure BBB integrity. The discovery of astrocytic responses to altered ion (e.g., Ca^{++}) concentrations in their environment sheds new light on the specificity of astrocyte interactions and their importance for ensuring BBB integrity.[31] In a tissue co-culture model, astrocytes were shown to affect the integrity of the tight junctions between adjacent endothelial cells.[32] More recently, astrocytes were demonstrated to enhance the defense of capillary endothelial cells against reactive oxygen species.[33] Thus, astrocytes both signal higher centers of the brain that the BBB has been disrupted and themselves receive signals from higher centers that cause them to modulate the BBB.

III. FUNCTIONAL CHARACTERISTICS OF THE BLOOD-BRAIN BARRIER

Designed to protect the brain from penetrance and accumulation of unwanted molecules and cells, the BBB has distinct properties at central and peripheral structures. To properly control the inward and outward flow of constituents to and from the brain, it must sense the needs of both the brain and the peripheral system. Therefore, both the electrophysiological activity in the brain and peripheral properties such as blood pressure must affect BBB properties. Similarly, changes in BBB integrity inevitably affect brain functioning; for example, BBB disruption will allow the passage into the brain of serum constituents, which are known to affect neuronal electric activity (e.g., amino acids). The relative contribution of such agents to brain function under BBB disruption awaits further investigation.

A. Inward and Outward Movement across the Blood-Brain Barrier: Physiological Considerations

Blood-brain barrier properties largely depend on the surrounding brain tissue. This is evident from a recent study that demonstrated the development of an intact BBB in brain tissue transplants in a manner dependent on the site of transplantation.[34] The integrity of BBB functioning is affected by cellular glutathione and is sensitive to oxidative stress.[35] Neuronal activity is another important factor in BBB functioning, perhaps through the activation of afferent axons innervating brain microvasculature. Indeed, both psychotropic drugs and nicotine impair dopamine transport across the BBB,[36] suggesting that BBB functioning is affected by the state of cholinergic or dopaminergic neuronal activity, and that BBB integrity is imperative for preventing neurotoxicity under exposure to dopamine analogs (e.g., MPTP).[37] Thus, human cerebromicrovascular endothelium was shown to possess dopaminergic receptors linked to adenylyl cyclase, suggesting signal transduction activities. Adrenergic influences on BBB control were reported by Sarmento et al.,[38] who also demonstrated the influence of electrical stimulation of locus coeruleus on the rat BBB permeability to sodium fluorescein. Similar effects in a cell culture model were shown by Borges.[39]

B. Cholinergic Involvement in Blood-Brain Barrier Functioning

The importance of ACh innervation to cortical capillaries has been suggested on the basis of a body of biochemical and morphological data, and indicates the underlying mechanism. Purified capillaries are capable of releasing ACh in a Ca^{++}-dependent mechanism, in response to K^+ depolarization or electrical stimulation.[40] Moreover, specific cholinergic machinery was identified in isolated microvessels from goat cerebral cortex, as demonstrated by measuring AChE and choline acetyl transferase (ChAT) activities.[41] ChAT activity in bovine cerebral cortex capillaries does not originate from the endothelial cells, nor do they release ACh in response to electrical stimulation. Rather, cerebrovascular ACh apparently has a neuronal origin.[40] The origin of the perivascular cholinergic terminals was examined in rat brains in which the nucleus basalis of Mynert, accounting for 70% of cortical ChAT activity, was lesioned. No change was observed in the microvessel-associated ChAT activity in the lesioned animals, ruling out the basal forebrain as the origin of this pathway.[40] The existence of released ACh hints at the presence of receptors that respond to the signal, and, indeed, muscarinic ACh receptors were identified in rat brain cortical capillaries.[42,43] Taken together, all this points to the involvement of cholinergic innervation of cerebral microvessels in cerebral blood flow and BBB permeability, which is an essential requirement under various physiological and pathological insults. The cholinergic involvement in BBB functioning is particularly important under exposure to cholinesterase inhibitors such as organophosphates or carbamates, since these induce a feedback response of AChE accumulation, which would lead to cholinergic hypo-functioning.[26,44] Therefore, AChE may affect BBB disruption through two interrelated mechanisms, which involve its catalytic capacity for ACh hydrolysis or its structural resemblance to gliotactin, neuroligins, or related proteins.[20,29]

C. Pericellular Cell Passage across Blood-Brain Barrier Structures

One of the roles of the BBB likely involves protection of the brain from invasive bacteria, viruses, and fungi. However, when under BBB disruption any of these parasites invades the brain, the immune system must respond. This implies that under certain conditions, lymphocytes cross the BBB and reach those sites in the brain where their protective functions are needed. The existence of tight junctions between endothelial cells in brain vasculature complicates this process and requires specific signaling to ensure the specificity of the pericellular transport. Also, the endothelial monolayer needs to be re-sealed once this transport has been completed. A recent study demonstrates that lymphocyte migration through brain endothelial cell monolayers involves signaling through endothelial I-CAM-1 via a Rho-dependent pathway, thus expanding the list of BBB-involved molecules.[18]

Mast cells represent another cell type that might penetrate the brain through BBB structures. This is of considerable importance because of the key role of mast cells in autoimmune demyelinating diseases.[45–47] When interacting with myelin basic protein, mast cells degranulate to induce by exocytosis immediate demyelination.[48,49] Under normal circumstances mast cells are located in leptomeninges and are concentrated along blood vessels, especially in dorsal thalamic nuclei.[50] Exposure to steroid hormones induces in mast cells massive secretion of, for example, histamine.[51] Several of the other neuromodulators, neurotransmitters, and growth factors secreted by mast cells can alter BBB properties.[52] Using confocal microscopy and vital dyes, Silverman et al.[53] very recently demonstrated rapid penetrance of mast cells through BBB structures into nests of glial processes. This transport may account for the rapid increases in mast cell populations after physiological manipulations.

D. Cell Culture, Organ Systems, and Imaging Approaches in Blood-Brain Barrier Research

The complexity and plasticity of BBB properties called for experimental dissection of the disruption process in both *in vitro* and *in vivo* conditions. Multiple cell and organ cultures, animal models, and measurement techniques have been developed, each of which addresses some of the issues involved. The development of research into BBB characteristics was initially approached in avian embryos, where transplanted endothelial quail cells invaded a developing chick chimera.[54] A simpler cell culture model of the BBB was developed by Rubin and co-workers.[55] More recently, an immortalized cell line created from vascular endothelial cells was used to develop another model of the BBB in co-cultures with glioma cells and was used to demonstrate nitric oxide-induced perturbations of these cells.[56] In another cell culture model, hypoxia was shown to increase the susceptibility to oxidative stress and intercellular permeability.[57]

Measurements of Evans blue penetrance proved useful in analyzing BBB properties in animal models.[58] Recent technological breakthroughs in brain imaging now offer a previously impossible view into the integrity of human BBB under various conditions. Imaging the human brain is widely used in the clinical and research settings by two major methods: (1) computerized tomography (CT) and (2) magnetic resonance imaging (MRI). In both methods, standard techniques use contrast agents to enhance signals and unmask brain pathologies. Both approaches are therefore aimed at delineation of the site, duration, and extent of potential BBB disruption in CNS pathologies.

Iodine is the only heavy atom that possesses the chemical properties suitable for intravascular use in CT analyses. The currently available iodinated contrast agents are nonionic, are highly hydrophilic, and have low osmolarity and minimal toxicity. The paramagnetic atom gadolinium (Gd), with seven unpaired electrons, forms a stable complex with diethylenetriamine pentaacetic acid (DTPA) and is the contrast material used in MRI. The DTPA complex is well tolerated and even minimal concentrations lead to marked shortening of its observed relaxation times and increase in signal intensity.

Both contrast agents normally do not cross the BBB. When injected intravenously, neither Gd nor DTPA affects BBB integrity. In contrast, intra-arterial injection of iodinated contrast agents was shown to disrupt normal BBB functioning due to both osmotic and chemotoxic effects (reviewed by Sage et al.).[59] This study, performed to determine the safety of currently used contrast agents, is of considerable significance for predicting the risks involved in low-level exposure to xenobiotics, as it indicates that even minor arterial elevation of the concentration of potentially harmful agents may by itself disrupt BBB functions.

In healthy individuals with normally functioning BBB, CT and MRI contrast agents cannot accumulate in the extracellular fluid of the brain parenchyma. Therefore, brain structures are not enhanced and remain relatively transparent in the imaging scans. In cases when the permeability of the BBB is increased because of a pathological process, the passage of iodinated agents (in CT) or paramagnetic DTPA complex (in MRI) leads to enhancement of signals. This occurs through X-ray attenuation, creating enhanced brain images in CT scans, or shortening of relaxation times, which results in sharper images in MRI. Such alterations in BBB penetrance may be local or massive, reflecting brain tumors, infectious disease, or cerebrovascular impairments. Figure 4.3 demonstrates examples for such imaging analyses of human BBB disruption.

E. Transgenic Engineering Models for Blood-Brain Barrier Studies

Genetic manipulations of the molecular mechanisms controlling BBB functioning yield new insights into the corresponding physiological or pathological circumstances and the dissection of their effects on BBB integrity. Several transgenic and knockout models have unraveled key elements involved in BBB functioning. These included several intentional as well as serendipitous studies. Mice with a genetic disruption in the mdr1a gene (multiple drug resistance), encoding the drug-transporting P-glycoprotein, which resides in the BBB, display up to 10-fold increases in their dexamethasone uptake into the brain.[60] The effect of cytokine overproduction on BBB functioning was checked in transgenic mice that overexpress interleukin 3 (IL3) or interleukin 6 (IL6). The IL3 transgenics develop progressive demyelination and infiltrated CNS lesions associated with BBB defects.[61] The effects observed in IL6 transgenics were even more dramatic: extensive breakdown of the BBB was evident in the cerebellum of IL6-overproducing mice, followed by subsequent inflammation, reactive gliosis, axonal degeneration, and macrophage accumulation.[62] CuZn-superoxide dismutase (SOD) was discovered to have protective effects against trauma-induced BBB disruption, in a model of mice that overexpress human SOD.[63] These

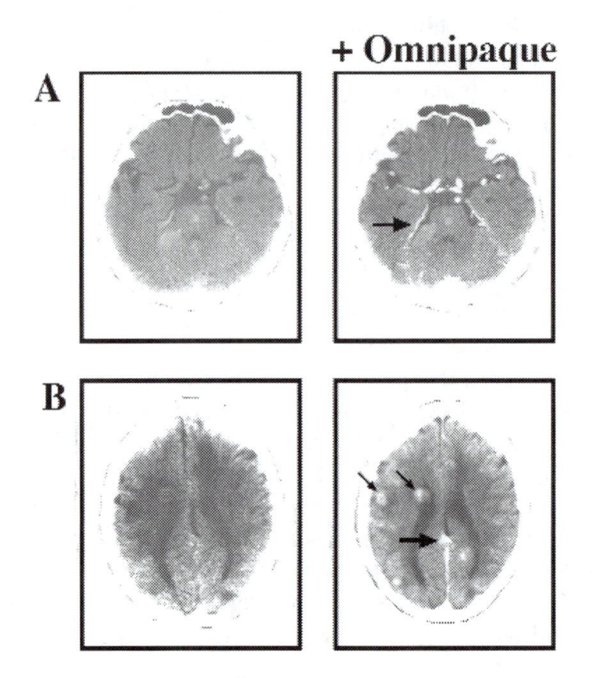

FIGURE 4.3 Blood-brain barrier disruption as revealed by computerized tomography: A. Normal CT scan, before and after injection of the enhancement material, Omnipaque™. Note the enhancement (hyperdense white) in brain arteries and vein sinuses (wide arrows) without penetration into the brain parenchyma. B. A case of multiple brain metastases in a 41-year-old female with a history of breast cancer. Following injection of Omnipaque™, several round white regions appear, reflecting focal BBB disruption in the regions harboring tumors (thin arrows).

mice were further found to display improved neurological recovery following traumatic brain injury,[64] which emphasizes the importance of oxidative stress in BBB disruption.

IV. MODULATORS OF BLOOD-BRAIN BARRIER FUNCTIONS AND THEIR INTERRELATIONSHIPS

Several considerations point to specific natural compounds and therapeutic agents as potential modulators of BBB functions. The rapid kinetics of BBB transport implicates post-translational control mechanisms in this process, simply because there is insufficient time to allow the slow transcription and translation processes to take place. The intimate relationship with vasculature properties points to vasoactive agents as potential modulators, and the necessity for penetrance of cells from the immune system suggests the involvement of immunomodulators. These, and drug transporters, should all communicate with the complex array of molecules and cellular structures that together compose the BBB.

Kinase cascades, universal pathways for rapid signal transduction in numerous biological processes, were naturally investigated for their potential relevance to BBB functioning. To regulate BBB properties, such kinase cascades should be induced in the various cell types that comprise the BBB. This prediction is verified by the finding that a pituitary adenylyl cyclase-activating polypeptide is successfully transported across the BBB, preventing the ischemia-induced death of hippocampal neurons.[65] The next logical step in this kinase cascade is tyrosine phosphorylation, which may increase tight junction permeability.[66] That such phosphorylation is actively involved in regulating BBB transport processes is evident from findings of phosphorylation of endothelial Na-K-Cl co-transport protein under changes in tonicity and hormones.[67]

A. Nitric Oxide and Vasoactive Agents Involvement

A primary mediator that was demonstrated to participate in meningitis-induced BBB disruption is nitric oxide (NO). NO is produced in response to exposure to bacterial endotoxins by the host endothelial cells. In an animal model of lipopolysaccharide-induced meningitis, BBB disruption and NO production sites in the brain co-localized,[68] and NO-synthase inhibitors reduced the meningeal-associated alterations in BBB permeability.[69] NO is likely produced by astrocytes, and it decreases endothelin-1 secretion by brain microvessel endothelial cells.[70,71]

Agents that regulate vasoactive processes, such as bradykinin and angiotensin, were shown to effect biochemical opening of the BBB.[72] In tissue culture experiments, such agents were further demonstrated to modulate tight junction structures in BBB endothelial cells co-cultured with astrocytes.[32] In cultured A431 cells, the signaling cascade induced by these agents was shown to involve tyrosine phosphorylation and reorganization of the tight junction protein Z0–1, processes that are also mediated by epidermal growth factor-1 (EGF1).[73]

B. Immunomodulators and Multi-Drug Transporters

The passage of immunomodulators across the BBB has been the subject of much research activity, especially because of the known impairment in BBB functioning in autoimmune diseases such as multiple sclerosis (MS).[74] It is generally considered that basic mechanisms of brain inflammation involve massive, yet transient, disruption of BBB functioning that plays an important role in the acute episodes of several autoimmune diseases.[75] This may indicate that individuals with an inherited susceptibility to autoimmune responses are a high-risk group for low-level exposure to xenobiotics.

The mdra1 genomic disruption studies noted above have pointed to this multidrug transporter protein as a rate-limiting factor in the bidirectional transport of drugs across the BBB. This finding explained the drug-induced neurotoxicity under chemotherapy and opens interesting options for developing BBB-regulating drugs.

Interestingly, the mouse mdr1a gene is also the earliest known endothelial cell differentiation marker during BBB development.[76]

V. CONDITIONS INDUCING BLOOD-BRAIN BARRIER DISRUPTION

The pathophysiological origin of BBB impairments is of major clinical interest for several reasons. Impairment is dangerous, as it may cause extreme susceptibility to adverse drug responses, which would necessitate individualized drug dosage; however, breaching the barrier is sometimes useful to enable delivery of needed drugs to the brain (for example, under bacterial, fungal, or viral brain infection, or in cases of malignant brain tumors).

A. Pathophysiological Induction of Blood-Brain Barrier Penetrance

Several diseases are known which are associated with BBB disruption. These include brain tumor metastases; epilepsy and the more severe condition of status epilepticus; cerebrovascular disorders; autoimmune diseases such as multiple sclerosis; acute cerebral infarcts; meningeal carcinomatosis; and ischemic white matter lesions.[77–84] Several genetic polymorphisms are known which increase the susceptibility to BBB disruption. These include polymorphisms in glutathione transferase, important for protection against oxidative stress,[85] and malfunctioning variants of serum BChE (e.g., "atypical" *BCHE*).[86] In particular, such mutations increase the risk of BBB disruption that is involved with exposure to anticholinesterases or to lead sulfate batteries, with subsequent increased risk for Parkinson's disease.[87,88]

Disruption of the BBB was reported in MS patients examined by contrast-enhanced MRI.[89] Blood-brain barrier disruption in MS patients was suggested to be the initial event in the development of the brain lesions that are characteristic of the advanced stages of this disease.[90] It correlates with the severity of symptoms, and an earlier age of the disease onset.[91] Another pathological condition in which BBB breakdown was demonstrated is epilepsy. Disruption was demonstrated by computerized tomography (CT) in a patient following generalized seizure and by Evans blue penetration in a rat model of pentylentetrazol-induced seizures.[92,93]

Cerebrovascular pathologies are abundant in Alzheimer's disease (AD) and are demonstrated by changes in the endothelium, amyloid depositions in the cerebral blood vessels, and disruption of the BBB.[94] A possible mechanism that underlies this phenomenon may be drawn from *in vitro* studies using a BBB model of a monolayer of vascular endothelial cells. Amyloid β-peptide, which deposits in plaques of AD patients, induced in these cells permeability to albumin and apoptotic cell death.[95] The potential clinical relevance of this finding was emphasized by intracarotid infusion of amyloid β-peptide, which resulted in BBB damage.[96]

BBB disruption has also been reported for CNS infections, primarily in meningitis, where it is used as a differential diagnostic tool. In an acute cytokine-induced mouse model of meningitis, endothelial selectins (glycoproteins involved in cell adhesion) were demonstrated to contribute toward the disruption of the BBB.[21]

HIV-1 infection of the CNS was also suggested to involve a component of chronic brain tissue inflammation and BBB disruption, resulting in neuronal injury and death, which lead to cognitive, motor, and behavioral impairments.[97]

B. Blood-Brain Barrier Disruption Following Acute Insults

Blood-brain barrier disruption following ischemia is well documented. Carotid artery occlusion, followed by reperfusion resulted in transendothelial leakage of a marker horseradish peroxidase in the hippocampus.[98] Unilateral BBB permeabilization in the cortex and striatum subregions was demonstrated in a rabbit model of ischemic hemisphere using contrast-enhanced MRI.[99] Extreme temperature changes appear to be an additional factor influencing BBB integrity, as both cold and heat stress impair it. Cold injury in mice induced the penetrance of Evans blue, immediately following the injury, with reversal to the normal situation of intact BBB only 24 h post-injury.[100] In a milder model, infusion of hypothermic saline into the left carotid artery of rats resulted in disruption of the BBB in the left hemisphere, which did not occur with a normothermic solution.[93] The effects of hyperthermia, on the other hand, were checked in a model of local heating of a rat's head. BBB opening was observed from 6 h to 3 days post-injury.[101] Similar results were noted in rats that were exposed to general heat stress (38°C) for 4 hours.[102] The involvement of the NO pathway in this phenomenon[103] was indicated by the up-regulation of neuronal NO synthase activity, which coincided with BBB breakdown in distinct brain regions.[104]

Traumatic brain injury, simulated by a model of closed head injury to mice, had also been shown to result in disruption of the BBB.[105] The temporal resolution of this disruption was monitored by MRI in rats subjected to closed head injury. Blood-brain barrier disruption appeared immediately after the impact, and declined gradually, until full reversal to control levels 30 min post-injury.[106] Opening of the BBB was similarly demonstrated in response to acute anticholinesterase exposure, however, low-level exposure has not yet been tested. BBB disruption under anticholinesterase exposure was proven to be seizure-dependent, as it could be blocked by the use of anticonvulsant agents.[107] The anticholinesterase effect on BBB ultrastructure did not impair endothelial tight junctions. Yet, an increased number of endothelial vesicles were observed, suggesting increased transcytosis as the mechanism involved.[108]

C. Psychological and Physical Stressors Impair Blood-Brain Barrier Functioning

Friedman et al.[58] have demonstrated enhanced brain penetrance under psychological stress of relatively small molecules such as anticholinesterases, as well as larger dye-protein complexes and DNA plasmids. This stress-induced process putatively explains some of the nervous system-associated sequelae reported by Gulf War veterans, who were exposed to unknown doses and combinations of potentially harmful xenobiotics, particularly anticholinesterases. The anticipated chemical warfare agents would have irreversibly blocked AChE. For prophylactic protection from these agents, Gulf War soldiers were administered pyridostigmine, a reversible carbamate cholinesterase inhibitor which has a quaternary ammonium group that under normal circumstances prevents its transport across the BBB. Pyridostigmine is routinely used to treat peripheral neuromuscular junction deficiencies in myasthenia gravis patients,[109] and was shown to cause mild, primarily peripheral side effects during peacetime clinical tests in healthy volunteers.[110] However, pyridostigmine use during the Gulf War caused a significant increase in reported CNS symptoms. Similarly, in animal experiments the dose of pyridostigmine required to block 50% of brain AChE in stressed mice was found to be 100-fold lower than that required in non-stressed mice, indicating a breakdown of the BBB.[58] More recently, heat stress, even extreme, reportedly failed to induce penetration of pyridostigmine into the brain of guinea pigs.[111] That BBB disruption depends on the status of neuronal activity in a brain-region specific manner was demonstrated in a study that compared stress-induced increase in BBB permeability in control and monosodium glutamate-treated rats, which reported increased BBB disruption in the hypothalamus and decreased in the brain stem, as compared with control animals.[112] Other reports demonstrate AChE overproduction in response to anticholinesterase exposure and to increases in interleukin 1.[26,113,114] Therefore, a long-term outcome of low-level exposure to an anticholinesterase may be a hypocholinergic state, due to entry of the agent into the brain, and the induction of AChE expression and excessive AChE-R accumulation.

D. Blood-Brain Barrier as a Complex Trait with Genetic and Physiological Components: Prospects

Blood-brain barrier properties are probably a complex genetic trait, in which the correlation of genotype to phenotype is difficult to dissect. Such a trait is termed complex or, if the phenotype is measured through a continuous variable, a quantitative trait. In certain cases, complex traits induce a susceptibility to a disease that depends upon environmental conditions. One example is the extreme adverse response to pyridostigmine treatment during the Gulf War that was found in a homozygous carrier of the "atypical" *BCHE* variant. The *BCHE,* with minor expression in the brain, gene encodes the butyrylcholinesterase (BChE, a.k.a. serum cholinesterase), that sequesters anticholinesterases such as pyridostigmine and prevents their reaction with AChE.[86] However, "atypical" BChE is incapable of binding pyridostigmine. Hence, homozygous carriers of this variant are at risk for extremely adverse responses, especially under the stress associated with war, to pyridostigmine doses in the circulation that would not affect individuals with normal *BCHE*. Therefore, the indirectly related *BCHE* gene becomes an important consideration for BBB disruption under the combination of stress and anticholinesterase exposure.

Loci in the genome that affect traits that may be quantified are called quantitative trait loci (QTL). Since many complex traits can be measured through a continuous variable (e.g., anxiety through cortisol measurements, Alzheimer's disease through cognition tests), QTL may serve as a general term for complex traits. Although the identification of QTL in humans and in model organisms is in its infancy, the QTL paradigm fits BBB properties from many points of view. Thanks to the human genome project and the development of related technologies, the detection of BBB genes will soon be aided by high-density single nucleotide polymorphism

(SNP) marker maps, which will allow population-based studies of greater significance than the current family-based studies. The human genome project, soon to be completed, will further provide important information on the sequence of all of the relevant genes and the homologies of their protein products. However, even with all the genome sequenced, significant additional work to assess functionality will be required. Expression analysis of endothelial cell genes through high-density microchip arrays will provide an independent dimension to increase the efficiency and efficacy of the QTL aspects of BBB research. Comparative genetics will also bring essential insights for functional determination. Finally bioinformatics and theoretical developments emerging from it will allow integration of all the various aspects of this multidisciplinary field to achieve the appropriate results. All of these efforts together will be needed to shed more light on the issue of BBB properties.

VI. SUMMARY

A comprehensive survey of the recent literature reveals an increasingly complex collection of BBB constituents and functions. For example, under low-level exposure to anticholinesterases, BBB integrity may be compromised because of four interrelated processes:

1. Anticholinesterase blockade of the ACh hydrolytic capacity of AChE induces a short-term hypercholinergic activation in the brain, leading by a rapid yet long-term feedback process to accumulation of an excess of AChE-R and modification of the cholinergic status in a manner affecting BBB properties at a later phase.
2. Anticholinesterase-AChE interactions may modify the flexible 3-dimensional structure of the AChE protein and change its capacity to compete in protein-protein interactions with its non-neuronal signal transducing homologues (e.g., gliotactin), or its neuronal homologues, like neuroligin. This could alter astrocyte or neuron properties that control BBB functioning.[115]
3. Anticholinesterase-induced AChE-R may differ from the normally present AChE-S in its ability to affect BBB integrity. Therefore, the combination of AChE's catalytic and structural properties with the anticholinesterase-induced feedback response would have a more dramatic effect on BBB properties than would any of these processes alone.
4. In individuals prone to adverse responses to stress stimuli, all of the above processes may be exacerbated in a complex manner, combining genetic and physiological mechanisms.

Blood-brain barrier disruption would affect brain functioning because of penetrance of the brain by peripheral compounds that may modulate the properties of glia and neurons. Therefore, the consequences of breaching the BBB, even for a short duration and in a limited area, may persist for long periods and involve larger brain areas. In an era when breakthroughs in molecular genetics that allow a previously unimagined dissection of biological processes, and with technological developments that provide a dynamic real-time view of brain functions, the BBB represents a medical and scientific frontier awaiting exploration.

ACKNOWLEDGMENTS

The authors acknowledge with thanks Profs. E. Reichenthal (Beersheva), A. Miller (Haifa), and C. Minini (Paris) for reviewing a draft of this manuscript. Additionally, HS and AF thank the U.S. Army Medical Research and Materiel Command (DAMD 17-99-1975) and The Israel Science Foundation and Ester Neuroscience for research support.

REFERENCES

1. Pardridge, W.M., Connor, J.D., and Crawford, I.L., Permeability changes in the blood-brain barrier: Causes and consequences, *CRC Crit. Rev. Toxicol.,* 3(2), 159, 1975.
2. Rubin, L.L. and Staddon, J.M., The cell biology of the blood-brain barrier, *Ann. Rev. Neurosci.,* 22, 11, 1999.
3. Pardridge, W.M., CNS drug design based on principles of blood-brain barrier transport, *J. Neurochem.,* 70(5), 1781, 1998.
4. Gross, P.M., Circumventricular organ capillaries, *Prog. Brain. Res.,* 91, 219, 1992.
5. Risau, W., Differentiation of endothelium, *FASEB J.,* 9(10), 926, 1995.
6. Schinkel, A.H., Smit, J.J., van Tellingen, O., Beijnen, J.H., Wagenaar, E., van Deemter, L., Mol, C.A., van der Valk, M.A., Robanus-Maandag, E.C., te Riele, H.P., et al., Disruption of the mouse mdr1a P-glycoprotein gene leads to a deficiency in the blood-brain barrier and to increased sensitivity to drugs, *Cell,* 77(4), 491, 1994.
7. Dejana, E., Corada, M., and Lampugnani, M.G., Endothelial cell-to-cell junctions, *FASEB J.,* 9(10), 910, 1995.
8. Gumbiner, B.M., Cell adhesion: The molecular basis of tissue architecture and morphogenesis, *Cell,* 84(3), 345, 1996.
9. Lum, H. and Malik, A.B., Regulation of vascular endothelial barrier function, *Am. J. Physiol.,* 267(3 Pt 1), L223, 1994.
10. Staddon, J.M. and Rubin, L.L., Cell adhesion, cell junctions and the blood-brain barrier, *Curr. Opin. Neurobiol.,* 6(5), 622, 1996.
11. Balda, M.S., Gonzalez-Mariscal, L., Matter, K., Cereijido, M., and Anderson, J.M., Assembly of the tight junction: The role of diacylglycerol, *J. Cell Biol.,* 123(2), 293, 1993.
12. Anderson, J.M. and Van Itallie, C.M., Tight junctions and the molecular basis for regulation of paracellular permeability, *Am. J. Physiol.,* 269(4 Pt 1), G467, 1995.
13. Citi, S., Sabanay, H., Jakes, R., Geiger, B., and Kendrick-Jones, J., Cingulin, a new peripheral component of tight junctions, *Nature,* 333(6170), 272, 1988.
14. Furuse, M., Hirase, T., Itoh, M., Nagafuchi, A., Yonemura, S., and Tsukita, S., Occludin: A novel integral membrane protein localizing at tight junctions, *J. Cell Biol.,* 123(6 Pt 2), 1777, 1993.
15. Nusrat, A., Giry, M., Turner, J.R., Colgan, S.P., Parkos, C.A., Carnes, D., Lemichez, E., Boquet, P., and Madara, J.L., Rho protein regulates tight junctions and perijunctional actin organization in polarized epithelia, *Proc. Natl. Acad. Sci. U.S.A.,* 92(23), 10629, 1995.

16. Grifman, M., Galyam, N., Seidman, S., and Soreq, H., Functional redundancy of acetylcholinesterase and neuroligin in mammalian neuritogenesis, *Proc. Natl. Acad. Sci. U.S.A.*, 95(23), 13935, 1998.
17. Thoreson, M.A., Anastasiadis, P.Z., Daniel, J.M., Ireton, R.C., Whellock, M.J., Johnson, K.R., Hummingbird, D.K., and Reynolds, A.B., Selective uncoupling of p120ctn from E-cadhedrin disrupts strong adhesion, *J. Cell Biol.*, 148(1), 189, 2000.
18. Adamson, P., Etienne, S., Couraud, P.O., Calder, V., and Greenwood, J., Lymphocyte migration through brain endothelial cell monolayers involves signaling through endothelial ICAM-1 via a rho-dependent pathway, *J. Immunol.*, 162(5), 2964, 1999.
19. Ratcliffe, M.J., Rubin, L.L., and Staddon, J.M., Dephosphorylation of the cadherin-associated p100/p120 proteins in response to activation of protein kinase C in epithelial cells, *J. Biol. Chem.*, 272(50), 31894, 1997.
20. Irie, M., Hata, Y., Takeuchi, M., Ichtchenko, K., Toyoda, A., Hirao, K., Takai, Y., Rosahl, T.W., and Sudhof, T.C., Binding of neuroligins to PSD-95, *Science*, 277(5331), 1511, 1997.
21. Tang, T., Frenette, P.S., Hynes, R.O., Wagner, D.D., and Mayadas, T.N., Cytokine-induced meningitis is dramatically attenuated in mice deficient in endothelial selectins, *J. Clin. Invest.*, 97(11), 2485, 1996.
22. Auld, V.J., Fetter, R.D., Broadie, K., and Goodman, C.S., Gliotactin, a novel transmembrane protein on peripheral glia, is required to form the blood-nerve barrier in Drosophila, *Cell*, 81(5), 757, 1995.
23. Baumgartner, S., Littleton, J.T., Broadie, K., Bhat, M.A., Harbecke, R., Lengyel, J.A., Chiquet-Ehrismann, R., Prokop, A., and Bellen, H.J., A Drosophila neurexin is required for septate junction and blood-nerve barrier formation and function, *Cell*, 87(6), 1059, 1996.
24. Darboux, I., Barthalay, Y., Piovant, M., and Hipeau-Jacquotte, R., The structure-function relationships in Drosophila neurotactin show that cholinesterasic domains may have adhesive properties, *EMBO J.*, 15(18), 4835, 1996.
25. Grisaru, D., Sternfeld, M., Eldor, A., Glick, D., and Soreq, H., Structural roles of acetylcholinesterase variants in biology and pathology, *Eur. J. Biochem.*, 264(3), 672, 1999.
26. Kaufer, D., Friedman, A., Seidman, S., and Soreq, H., Acute stress facilitates long-lasting changes in cholinergic gene expression, *Nature*, 393(6683), 373, 1998.
27. Fanning, A.S. and Anderson, J.M., Protein-protein interactions: PDZ domain networks, *Curr. Biol.*, 6(11), 1385, 1996.
28. Saras, J. and Heldin, C.H., PDZ domains bind carboxy-terminal sequences of target proteins, *Trends Biochem. Sci.*, 21(12), 455, 1996.
29. Bhat, M.A., Izaddoost, S., Lu, Y., Cho, K.O., Choi, K.W., and Bellen, H.J., Discs lost, a novel multi-PDZ domain protein, establishes and maintains epithelial polarity, *Cell*, 96(6), 833, 1999.
30. Janzer, R.C. and Raff, M.C., Astrocytes induce blood-brain barrier properties in endothelial cells, *Nature*, 325(6101), 253, 1987.
31. Bellen, H.J., Lu, Y., Beckstead, R., and Bhat, M.A., Neurexin IV, caspr and paranodin—novel members of the neurexin family: Encounters of axons and glia, *Trends Neurosci*, 21(10), 444, 1998.
32. Wolburg, H., Neuhaus, J., Kniesel, U., Krauss, B., Schmid, E.M., Ocalan, M., Farrell, C., and Risau, W., Modulation of tight junction structure in blood-brain barrier endothelial cells. Effects of tissue culture, second messengers and cocultured astrocytes, *J. Cell Sci.*, 107(Pt 5), 1347, 1994.

33. Schroeter, M.L., Mertsch, K., Giese, H., Muller, S., Sporbert, A., Hickel, B., and Blasig, I.E., Astrocytes enhance radical defence in capillary endothelial cells constituting the blood-brain barrier, *FEBS Lett.*, 449(2–3), 241, 1999.
34. Granholm, A.C., Curtis, M., Diamond, D.M., Branch, B.J., Heman, K.L., and Rose, G.M., Development of an intact blood-brain barrier in brain tissue transplants is dependent on the site of transplantation, *Cell Transplant,* 5(2), 305, 1996.
35. Hurst, R.D., Heales, S.J., Dobbie, M.S., Barker, J.E., and Clark, J.B., Decreased endothelial cell glutathione and increased sensitivity to oxidative stress in an *in vitro* blood-brain barrier model system, *Brain Res.,* 802(1–2), 232, 1998.
36. Martel, C.L., Mackic, J.B., Adams, J.D., Jr., McComb, J.G., Weiss, M.H., and Zlokovic, B.V., Transport of dopamine at the blood-brain barrier of the guinea pig: Inhibition by psychotropic drugs and nicotine, *Pharm. Res.,* 13(2), 290, 1996.
37. Harik, S.I., MPTP toxicity and the "biochemical" blood-brain barrier, *NIDA Res. Monogr.,* 120, 43, 1992.
38. Sarmento, A., Borges, N., and Lima, D., Influence of electrical stimulation of locus coeruleus on the rat blood-brain barrier permeability to sodium fluorescein, *Acta Neurochir.,* 127(3–4), 215, 1994.
39. Borges, N., Shi, F., Azevedo, I., and Audus, K.L., Changes in brain microvessel endothelial cell monolayer permeability induced by adrenergic drugs, *Eur. J. Pharmacol.,* 269(2), 243, 1994.
40. Galea, E. and Estrada, C., Periendothelial acetylcholine synthesis and release in bovine cerebral cortex capillaries, *J. Cereb. Blood Flow Metab.,* 11(5), 868, 1991.
41. Estrada, C., Triguero, D., Munoz, J., and Sureda, A., Acetylcholinesterase-containing fibers and choline acetyltransferase activity in isolated cerebral microvessels from goats, *Brain Res.,* 453(1–2), 275, 1988.
42. Luiten, P.G., de Jong, G.I., Van der Zee, E.A., and van Dijken, H., Ultrastructural localization of cholinergic muscarinic receptors in rat brain cortical capillaries, *Brain Res.,* 720(1–2), 225, 1996.
43. Mohr, E., Subcellular RNA compartmentalization, *Prog. Neurobiol.,* 57(5), 507, 1999.
44. Kaufer, D., Friedman, A., Seidman, S., and Soreq, H., Anticholinesterases induce multigenic transcriptional feedback response suppressing cholinergic neurotransmission, *Chem. Biol. Interact.,* 119–120, 1999.
45. Powell, H.C., Braheny, S.L., Myers, R.R., Rodriguez, M., and Lampert, P.W., Early changes in experimental allergic neuritis, *Lab. Invest.,* 48(3), 332, 1983.
46. Seeldrayers, P.A., Yasui, D., Weiner, H.L., and Johnson, D., Treatment of experimental allergic neuritis with nedocromil sodium, *J. Neuroimmunol.,* 25(2–3), 221, 1989.
47. Brosnan, C.F., Claudio, L., Tansey, F.A., and Martiney, J., Mechanisms of autoimmune neuropathies, *Ann. Neurol.,* 27(Suppl), S75, 1990.
48. Theoharides, T.C., Spanos, C., Pang, X., Alferes, L., Ligris, K., Letourneau, R., Rozniecki, J.J., Webster, E., and Chrousos, G.P., Stress-induced intracranial mast cell degranulation: A corticotropin-releasing hormone-mediated effect, *Endocrinology,* 136(12), 5745, 1995.
49. Guo, Z., Turner, C., and Castle, D., Relocation of the t-SNARE SNAP-23 from lamellipodia-like cell surface projections regulates compound exocytosis in mast cells, *Cell,* 94(4), 537, 1998.
50. Goldschmidt, R.C., Hough, L.B., and Glick, S.D., Rat brain mast cells: Contribution to brain histamine levels, *J. Neurochem.,* 44(6), 1943, 1985.
51. Silver, R., Silverman, A.J., Vitkovic, L., and Lederhendler, I., Mast cells in the brain: Evidence and functional significance, *Trends Neurosci.,* 19(1), 25, 1996.

52. Zhuang, X., Silverman, A.J., and Silver, R., Distribution and local differentiation of mast cells in the parenchyma of the forebrain, *J. Comp. Neurol.,* 408(4), 477, 1999.
53. Silverman, A.-J., Sutherland, A.K., Wilhelm, M., and Silver, R., Mast cells migrate from blood to brain, *J. Neurosci.,* 20(1), 401, 2000.
54. Stewart, P.A. and Wiley, M.J., Developing nervous tissue induces formation of blood-brain barrier characteristics in invading endothelial cells: A study using quail—chick transplantation chimeras, *Dev. Biol.,* 84(1), 183, 1981.
55. Rubin, L.L., Hall, D.E., Porter, S., Barbu, K., Cannon, C., Horner, H.C., Janatpour, M., Liaw, C.W., Manning, K., Morales, J., et al., A cell culture model of the blood-brain barrier, *J. Cell. Biol.,* 115(6), 1725, 1991.
56. Hurst, R.D. and Fritz, I.B., Properties of an immortalised vascular endothelial/glioma cell co-culture model of the blood-brain barrier, *J. Cell Physiol.,* 167(1), 81, 1996.
57. Plateel, M., Dehouck, M.P., Torpier, G., Cecchelli, R., and Teissier, E., Hypoxia increases the susceptibility to oxidant stress and the permeability of the blood-brain barrier endothelial cell monolayer, *J. Neurochem.,* 65(5), 2138, 1995.
58. Friedman, A., Kaufer, D., Shemer, J., Hendler, I., Soreq, H., and Tur-Kaspa, I., Pyridostigmine brain penetration under stress enhances neuronal excitability and induces early immediate transcriptional response, *Nat. Med.,* 2(12), 1382, 1996.
59. Sage, M.R., Wilson, A.J., and Scroop, R., Contrast media and the brain. The basis of CT and MR imaging enhancement, *Neuroimaging Clin. N. Am.,* 8(3), 695, 1998.
60. Meijer, O.C., de Lange, E.C., Breimer, D.D., de Boer, A.G., Workel, J.O., and de Kloet, E.R., Penetration of dexamethasone into brain glucocorticoid targets is enhanced in mdr1A P-glycoprotein knockout mice, *Endocrinology,* 139(4), 1789, 1998.
61. Powell, H.C., Garrett, R.S., Brett, F.M., Chiang, C.S., Chen, E., Masliah, E., and Campbell, I.L., Response of glia, mast cells and the blood-brain barrier, in transgenic mice expressing interleukin-3 in astrocytes, an experimental model for CNS demyelination, *Brain Pathol.,* 9(2), 219, 1999.
62. Brett, F.M., Mizisin, A.P., Powell, H.C., and Campbell, I.L., Evolution of neuropathologic abnormalities associated with blood-brain barrier breakdown in transgenic mice expressing interleukin-6 in astrocytes, *J. Neuropathol. Exp. Neurol.,* 54(6), 766, 1995.
63. Chan, P.H., Epstein, C.J., Li, Y., Huang, T.T., Carlson, E., Kinouchi, H., Yang, G., Kamii, H., Mikawa, S., Kondo, T., et al., Transgenic mice and knockout mutants in the study of oxidative stress in brain injury, *J. Neurotrauma,* 12(5), 815, 1995.
64. Mikawa, S., Kinouchi, H., Kamii, H., Gobbel, G.T., Chen, S.F., Carlson, E., Epstein, C.J., and Chan, P. H., Attenuation of acute and chronic damage following traumatic brain injury in copper, zinc-superoxide dismutase transgenic mice, *J. Neurosurg.,* 85(5), 885, 1996.
65. Banks, W.A., Uchida, D., Arimura, A., Somogyvari-Vigh, A., and Shioda, S., Transport of pituitary adenylate cyclase-activating polypeptide across the blood-brain barrier and the prevention of ischemia-induced death of hippocampal neurons, *Ann. NY Acad. Sci.,* 805, 270, 1996.
66. Staddon, J.M., Smales, C., Schulze, C., Esch, F.S., and Rubin, L.L., p120, a p120-related protein (p100), and the cadherin/catenin complex, *J. Cell Biol.,* 130(2), 369, 1995.
67. O'Donnell, M.E., Martinez, A., and Sun, D., Endothelial Na-K-Cl cotransport regulation by tonicity and hormones: Phosphorylation of cotransport protein, *Am. J. Physiol.,* 269(6 Pt 1), C1513, 1995.
68. Jaworowicz, D.J., Jr., Korytko, P.J., Singh Lakhman, S., and Boje, K.M., Nitric oxide and prostaglandin E2 formation parallels blood-brain barrier disruption in an experimental rat model of bacterial meningitis, *Brain Res. Bull.,* 46(6), 541, 1998.

69. Boje, K.M., Inhibition of nitric oxide synthase attenuates blood-brain barrier disruption during experimental meningitis, *Brain Res.*, 720(1–2), 75, 1996.
70. Federici, C., Camoin, L., Creminon, C., Chaverot, N., Strosberg, A.D, and Couraud, P.O., Cultured astrocytes release a factor that decreases endothelin-1 secretion by brain microvessel endothelial cells, *J. Neurochem.*, 64(3), 1008, 1995.
71. O'Donnell, M.E., Martinez, A., and Sun, D., Cerebral microvascular endothelial cell Na-K-Cl cotransport: Regulation by astrocyte-conditioned medium, *Am. J. Physiol.*, 268(3 Pt 1), C747, 1995.
72. Black, K.L., Biochemical opening of the blood-brain barrier, *Adv. Drug Deliv. Rev.*, 15, 37, 1995.
73. Van Itallie, C.M., Balda, M.S., and Anderson, J.M., Epidermal growth factor induces tyrosine phosphorylation and reorganization of the tight junction protein ZO-1 in A431 cells, *J. Cell Sci.*, 108(Pt 4), 1735, 1995.
74. Stitt, J.T., Passage of immunomodulators across the blood-brain barrier, *Yale J. Biol. Med.*, 63(2), 121, 1990.
75. Lassmann, H., Basic mechanisms of brain inflammation, *J. Neural Transm. Suppl.*, 50, 183, 1997.
76. Qin, Y. and Sato, T.N., Mouse multidrug resistance 1a/3 gene is the earliest known endothelial cell differentiation marker during blood-brain barrier development, *Dev. Dyn.*, 202(2), 172, 1995.
77. Akeson, P., Larsson, E.M., Kristoffersen, D.T., Jonsson, E., and Holtas, S., Brain metastases—comparison of gadodiamide injection-enhanced MR imaging at standard and high dose, contrast-enhanced CT and non-contrast-enhanced MR imaging, *Acta Radiol.*, 36(3), 300, 1995.
78. Cornford, E.M. and Oldendorf, W.H., Epilepsy and the blood-brain barrier, *Adv. Neurol.*, 44, 787, 1986.
79. Correale, J., Rabinowicz, A.L., Heck, C.N., Smith, T.D., Loskota, W.J., and DeGiorgio, C.M., Status epilepticus increases CSF levels of neuron-specific enolase and alters the blood-brain barrier, *Neurology*, 50(5), 1388, 1998.
80. Klatzo, I., Disturbances of the blood-brain barrier in cerebrovascular disorders, *Acta Neuropathol. Suppl.*, 8, 81, 1983.
81. Larsson, H.B., Stubgaard, M., Frederiksen, J.L., Jensen, M., Henriksen, O., and Paulson, O.B., Quantitation of blood-brain barrier defect by magnetic resonance imaging and gadolinium-DTPA in patients with multiple sclerosis and brain tumors, *Magn. Reson. Med.*, 16(1), 117, 1990.
82. Merten, C.L., Knitelius, H.O., Assheuer, J., Bergmann-Kurz, B., Hedde, J.P., and Bewermeyer, H., MRI of acute cerebral infarcts, increased contrast enhancement with continuous infusion of gadolinium, *Neuroradiology*, 41(4), 242, 1999.
83. Siegal, T., Sandbank, U., Gabizon, A., Mizrachi, R., Ben-David, E., and Catane, R., Alteration of blood-brain-CSF barrier in experimental meningeal carcinomatosis. A morphologic and adriamycin-penetration study, *J. Neurooncol.*, 4(3), 233, 1987.
84. Skoog, I., A review on blood pressure and ischaemic white matter lesions, *Dement. Geriatr. Cogn. Disord.*, 9, Suppl 1, 13, 1998.
85. Menegon, A., Board, P.G., Blackburn, A.C., Mellick, G.D., and Le Couteur, D.G., Parkinson's disease, pesticides, and glutathione transferase polymorphisms [see Comments], *Lancet*, 352(9137), 1344, 1998.
86. Loewenstein-Lichtenstein, Y., Schwarz, M., Glick, D., Norgaard-Pedersen, B., Zakut, H., and Soreq, H., Genetic predisposition to adverse consequences of anti-cholinesterases in 'atypical' BCHE carriers, *Nat. Med.*, 1(10), 1082, 1995.

87. Soreq, H. and Glick, D., Novel roles for cholinesterases in stress and inhibitor responses, in *Cholinesterases and Cholinesterase Inhibitors: Basic, Preclinical and Clinical Aspects,* Giacobini, E., ed., Martin Dunitz, Ltd., London, 47–61, 2000.
88. Kuhn, W., Winkel, R., Woitalla, D., Meves, S., Przuntek, H., and Muller, T., High prevalence of Parkinsonism after occupational exposure to lead-sulfate batteries, *Neurology,* 50(6), 1885, 1998.
89. Rosenberg, G.A., Dencoff, J.E., Correa, N., Jr., Reiners, M., and Ford, C.C., Effect of steroids on CSF matrix metalloproteinases in multiple sclerosis: Relation to blood-brain barrier injury, *Neurology,* 46(6), 1626, 1996.
90. McFarland, H.F., The lesion in multiple sclerosis: Clinical, pathological, and magnetic resonance imaging considerations, *J. Neurol. Neurosurg. Psychiatry,* 64, Suppl 1, S26, 1998.
91. Stone, L.A., Smith, M.E., Albert, P.S., Bash, C.N., Maloni, H., Frank, J.A., and McFarland, H.F., Blood-brain barrier disruption on contrast-enhanced MRI in patients with mild relapsing-remitting multiple sclerosis: Relationship to course, gender, and age, *Neurology,* 45(6), 1122, 1995.
92. Clarke, H.B. and Gabrielsen, T.O., Seizure induced disruption of blood-brain barrier demonstrated by CT, *J. Comput. Assist. Tomogr.,* 13(5), 889, 1989.
93. Oztas, B. and Kucuk, M., Intracarotid hypothermic saline infusion: A new method for reversible blood-brain barrier disruption in anesthetized rats, *Neurosci. Lett.,* 190(3), 203, 1995.
94. Hachinski, V. and Munoz, D.G., Cerebrovascular pathology in Alzheimer's disease: Cause, effect or epiphenomenon?, *Ann. N.Y. Acad. Sci.,* 826, 1, 1997.
95. Blanc, E.M., Toborek, M., Mark, R.J., Hennig, B., and Mattson, M.P., Amyloid beta-peptide induces cell monolayer albumin permeability, impairs glucose transport, and induces apoptosis in vascular endothelial cells, *J. Neurochem.,* 68(5), 1870, 1997.
96. Jancso, G., Domoki, F., Santha, P., Varga, J., Fischer, J., Orosz, K., Penke, B., Becskei, A., Dux, M., and Toth, L., Beta-amyloid (1-42) peptide impairs blood-brain barrier function after intracarotid infusion in rats, *Neurosci. Lett.,* 253, 139, 1998.
97. Epstein, L.G. and Gelbard, H.A., HIV-1-induced neuronal injury in the developing brain, *J. Leukoc. Biol.,* 65(4), 453, 1999.
98. Shinnou, M., Ueno, M., Sakamoto, H., and Ide, M., Blood-brain barrier damage in reperfusion following ischemia in the hippocampus of the Mongolian gerbil brain, *Acta Neurol. Scand.,* 98(6), 406, 1998.
99. Lo, E.H., Pan, Y., Matsumoto, K., and Kowall, N.W., Blood-brain barrier disruption in experimental focal ischemia: Comparison between in vivo MRI and immunocytochemistry, *Magn. Reson. Imaging,* 12(3), 403, 1994.
100. Murakami, K., Kondo, T., Yang, G., Chen, S.F., Morita-Fujimura, Y., and Chan, P.H., Cold injury in mice: A model to study mechanisms of brain edema and neuronal apoptosis, *Prog. Neurobiol.,* 57(3), 289, 1999.
101. Urakawa, M., Yamaguchi, K., Tsuchida, E., Kashiwagi, S., Ito, H., and Matsuda, T., Blood-brain barrier disturbance following localized hyperthermia in rats, *Int. J. Hypertherm.,* 11(5), 709, 1995.
102. Sharma, H.S., Westman, J., Cervos-Navarro, J., and Nyberg, F., Role of neurochemicals in brain edema and cell changes following hyperthermic brain injury in the rat, *Acta Neurochir. Suppl.,* 70, 269, 1997.
103. Carpentier, P., Delamanche, I.S., Le Bert, M., Blanchet, G., and Bouchaud, C., Seizure-related opening of the blood-brain barrier induced by soman: possible correlation with the acute neuropathology observed in poisoned rats, *Neurotoxicology,* 11(3), 493, 1990.

104. Alm, P., Sharma, H.S., Hedlund, S., Sjoquist, P.O., and Westman, J., Nitric oxide in the pathophysiology of hyperthermic brain injury. Influence of a new anti-oxidant compound H-290/51. A pharmacological study using immunohistochemistry in the rat, *Amino Acids,* 14(1–3), 95, 1998.
105. Chen, Y., Constantini, S., Trembovler, V., Weinstock, M., and Shohami, E., An experimental model of closed head injury in mice: Pathophysiology, histopathology, and cognitive deficits, *J. Neurotrauma,* 13(10), 557, 1996.
106. Barzo, P., Marmarou, A., Fatouros, P., Corwin, F., and Dunbar, J., Magnetic resonance imaging-monitored acute blood-brain barrier changes in experimental traumatic brain injury, *J. Neurosurg.,* 85(6), 1113, 1996.
107. Petrali, J.P., Maxwell, D.M., Lenz, D.E., and Mills, K.R., Effect of an anticholinesterase compound on the ultrastructure and function of the rat blood-brain barrier: A review and experiment, *J. Submicrosc. Cytol. Pathol.,* 23(2), 331, 1991.
108. Grange-Messent, V., Bouchaud, C., Jamme, M., Lallement, G., Foquin, A., and Carpentier, P., Seizure-related opening of the blood-brain barrier produced by the anticholinesterase compound, soman: New ultrastructural observations, *Cell. Mol. Biol. (Noisy-le-grand),* 45(1), 1, 1999.
109. Soreq, H. and Zakut, H., *Human Cholinesterases and Anticholinesterases,* Academic Press, San Diego, 1993.
110. Glikson, M., Achiron, A., Ram, Z., Ayalon, A., Karni, A., Sarova-Pinchas, I., Glovinski, J., and Revah, M., The influence of pyridostigmine administration on human neuromuscular functions—studies in healthy human subjects, *Fundam. Appl. Toxicol.,* 16(2), 288, 1991.
111. Lallement, G., Foquin, A., Baubichon, D., Burckhart, M.F., Carpentier, P., and Canini, F., Heat stress, even extreme, does not induce penetration of pyridostigmine into the brain of guinea pigs, *Neurotoxicology,* 19(6), 759, 1998.
112. Skultetyova, I., Tokarev, D., and Jezova, D., Stress-induced increase in blood-brain barrier permeability in control and monosodium glutamate-treated rats, *Brain Res. Bull.,* 45(2), 175, 1998.
113. Kaufer, D., Friedman, A., and Soreq, H., The vicious circle: Long-lasting transcriptional modulation of cholinergic neurotransmission following stress and anticholinesterase exposure, *The Neuroscientist,* 5, 173, 1999.
114. Yuekui, L., Li, Y., Liu, L., Kang, J., Sheng, J.G., Barger, S.W., Mrak, R.E., and Griffin, W.S.T., Neuronal-glial interactions mediated by interleukin-1 enhance neuronal acetylcholinesterase activity and mRNA expression, *J. Neurosci.,* 20(1), 149, 2000.
115. Bourne, Y., Grassi, J., Bougis, P.E., and Marchot, P., Conformational flexibility of the acetylcholinesterase tetramer suggested by x-ray crystallography, *J. Biol. Chem.,* 274(43), 30370, 1999.

5 Pharmacokinetics and Pharmacodynamics of Carbamates under Physical Stress

Satu M. Somani, Kazim Husain, and Ramesh Jagannathan

CONTENTS

I. Introduction 146
II. Pyridostigmine Bromide 147
A. General Aspects 147
B. Absorption, Distribution, Metabolism, and Excretion 148
C. Pharmacokinetics of Pyridostigmine 153
D. Pharmacodynamics of Pyridostigmine Bromide: Use as a Pretreatment Drug 154
E. Factors Influencing Pharmacokinetics and Pharmacodynamics of Pyridostigmine Bromide 160
1. Stress 160
2. Environmental Exposures 165
3. Gender and Age 165
III. Physostigmine 166
A. General Aspects 166
B. Pharmacokinetics of Physostigmine 166
C. Pharmacodynamics of Physostigmine 170
D. Influence of Physical Stress on Pharmacokinetics and Pharmacodynamics 173
E. Effect of Soman on Pharmacokinetics and Pharmacodynamics 176
IV. Neostigmine 177
V. Summary 180
Acknowledgments 181
References 181

0-8493-0872-0/01/$0.00+$.50

I. INTRODUCTION

Carbamates (reversible cholinesterase inhibitors) are potential pretreatment agents against nerve gas poisoning. This chapter discusses the pharmacokinetics and pharmacodynamics of pyridostigmine, physostigmine, and neostigmine in human beings and various animal species, under normal, disease states, and stressful conditions. The chemical structures of these carbamates, pyridostigmine, physostigmine, and neostigmine, are given in Figure 5.1.

Pyridostigmine (Mestinon), the N,N-dimethyl-carbamate of 3-hydroxy-N-Methyl pyridinium, was first synthesized in 1946 by R. Urban.[1] It is a quaternary ammonium compound and dispensed as a bromide salt, pyridostigmine bromide (PB). PB is a reversible anticholinesterase drug most frequently used in the treatment of patients with myasthenia gravis. It has been proposed as a pretreatment drug against nerve agent poisoning.[2] The hypothesis is well established that short-lasting AChE inhibitor drugs (PB and physostigmine) protect the ChE enzyme against inactivation by nerve agents.[2–4]

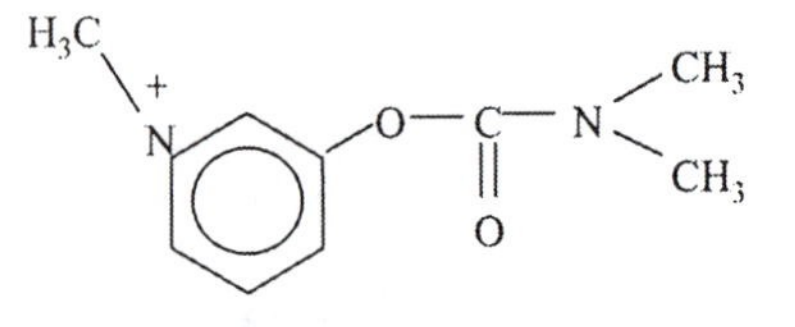

PYRIDOSTIGMINE

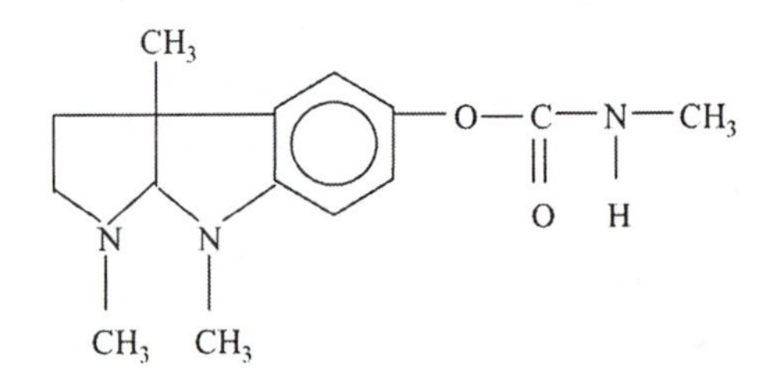

PHYSOSTIGMINE

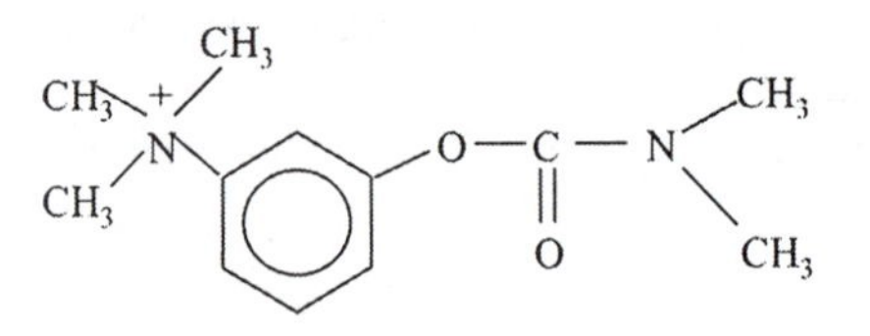

NEOSTIGMINE

FIGURE 5.1 Chemical structures of carbamates: pyridostigmine, physostigmine, and neostigmine.

The carbamate anticholinesterases such as PB bind reversibly with ChE enzyme, yet spontaneously reactivate relatively rapidly. However, nerve agents (organophosphate compounds) bind with the ChE irreversibly and form a much more stable phosphorylated enzyme (ChE-OP) complex. PB binds to peripheral ChE at anionic and esteratic sites and thus carbamylates the enzyme. The carbamylated enzyme sites cannot bind with nerve agents. In the meantime, some of the nerve agents are hydrolyzed to inactive metabolites by nonspecific hydrolases. The decarbamylation of the ChE takes place at the alcohol moiety on the esteratic site, regenerating the ChE enzyme to sustain life.

Physostigmine (also called eserine) is an alkaloid obtained from the leguminous plant Calabar or ordeal bean—the dried, ripe seed of Physostigma Venenosum Balfour, a perennial plant in tropical West Africa. The main alkaloid was first isolated from the seeds of the Calabar bean in a pure form in 1864 by Jobst and Hesse, who called it physostigmine.[5] One year later, it was obtained in a crystalline form by Vee and LeVen, who called it eserine.[6] Physostigmine (PHY) is the first anticholinesterase agent known to man and is used in the treatment of atropine-induced intoxication.

Neostigmine (also called prostigmine) was first synthesized by Aeschlimann and Reinert in 1931.[7] Among the quaternary nitrogen compounds synthesized, they found that the dimethyl carbamate ester of 3-oxy-phenyl-trimethyl ammonium (neostigmine) was one of the most active compounds, and had actions similar to physostigmine. This agent has been employed in the treatment of myasthenia gravis. It is also used to reverse the neuromuscular blockade caused by anesthetic agents.

II. PYRIDOSTIGMINE BROMIDE

A. General Aspects

Pyridostigmine is a close congener of neostigmine, with a longer duration of action and fewer muscarinic effects. Pyridostigmine acts by competing with ACh for its binding site on the enzyme acetylcholinesterase (AChE). Thus, PB interferes with the enzymatic destruction of ACh, potentiating the action of ACh on both the skeletal muscle (nicotinic effect) and gastrointestinal tract (muscarinic effect). The symptoms associated with PB intoxication are tremors, diarrhea, hypersalivation, abdominal cramps, muscle weakness, fatigue, blurred vision, fasciculations, and urinary incontinence.[8–10] PB was used as a pretreatment drug to protect soldiers in the event of nerve gas exposure during the Persian Gulf War. Many war veterans complained of various side effects ascribed to PB use and about half of the total veterans during the Gulf War complained of PB side effects.[11–13] These veterans received a 30 mg oral dose of PB three times a day for 2 weeks. This dose was used because it was suggested to be a symptom-free dose from experimental studies and is intended to produce 30–40% AChE inhalation. The most common symptoms reported by the veterans included effects on the gastrointestinal and genitourinary systems.[11] These symptoms are likely to be enhanced by other stress factors including physical stress. Subchronic oral doses of PB (0.5–60 mg/kg/day) administered for 13 weeks in rats showed toxicity (tremors and ChE inhibition) above the 5 mg/kg/day dose.[14] Pyridostigmine administered i.v. to dogs showed initial cardiopulmonary toxicity and

ChE inhibition at 2 and 5 mg/kg/day doses.[15] These and other side effects are related to dose, route of administration, and the disposition of drug from the body.

B. Absorption, Distribution, Metabolism, and Excretion

ABSORPTION: Absorption of PB, a quaternary ammonium compound, appears to be poor and erratic from the gastrointestinal tract after oral administration. This is because, being a polar compound, PB passes poorly across biological membranes. Hence much larger doses are required for the pharmacological effect by the oral route as compared to the parenteral routes. A study reported that the oral dose required to produce an effect with PB is 30 times that of the i.v. dose producing the same effect.[16] After oral administration of PB, the onset of action occurs after 30 to 45 min, and the duration of action is approximately 3 to 6 h. Oral bioavailability of PB is affected by inter-individual differences in gastrointestinal absorption, peristalsis, and possible differences in metabolism in the gastrointestinal tract and liver. These factors are likely to be of importance in relation to the symptoms experienced by the Gulf War veterans who received PB.

The absorption and bioavailability of PB have been evaluated in studies, both in healthy subjects and in patients of myasthenia gravis.[16–19] In healthy subjects, who received 60 mg oral PB, the oral bioavailability ranged from 11.5 to 18.9%, and maximum plasma levels were attained between 1.5 and 5 h after dosing.[17] The results suggested that PB absorption occurs at a slower rate than its elimination, and this may be affected further by prior food consumption. Although ingestion of food with PB delays the time to reach peak plasma concentration by about 90 min, the extent of absorption of PB though is not affected.[20] It has been suggested that the considerable variations in daily dosage requirements in myasthenic patients may be due to inter-individual differences in disease severity and in absorption or metabolism of PB.[16] In patients, oral PB (dose range of 30 to 240 mg) administered over a period of 7 to 22 h, resulted in an oral bioavailability of 3.6%. In another study, the bioavailability of PB was found to be 7%.[21] Malabsorption with orally administered PB was also found to occur in patients with myasthenia gravis.[19] This may be responsible for inadequate disease control in these patients. The authors suggest that this malabsorption may be the result of PB-induced alterations in the gastrointestinal epithelium. The pharmacokinetics of absorption are given in the following section and also in Table 5.1.

DISTRIBUTION: The distribution of 14C-pyridostigmine was studied by Birtley et al.[22] Ten per cent of the i.m. dose was present in the alimentary tract within 1 h after injection and 0.3% of the dose is secreted in the bile. A high concentration of radioactivity occurred in the kidney when excretion in the urine was at its maximum level. Lower concentrations were present in the liver, intestinal contents, heart, blood, and muscle. Radioactivity was also detected in the lungs, spleen, and skin, but not in the brain, thymus gland, intestinal wall, or body fat. The detection of radioactive respiratory CO2 suggested that to a small extent pyridostigmine may be metabolized by another route. The serum concentration was dose-dependent and was correlated with the clinical response.[23] A radio-immunoassay (RIA) method was developed to determine the plasma concentration time profiles and tissue distribution of PB in rat following its i.m. administration.[24] This study found that PB had a half-life (t1/2) of 25 min and was not detectable in plasma 6 h following its administration. Studies with ^{14}C-pyridostigmine have shown that PB gets trapped in various tissue compartments.[20, 25] It is suggested that the uptake of PB into the liver and kidney is concentration dependent and is responsible for its metabolism and elimination. The steady-state volume of distribution (Vd) of PB is relatively small (0.3–0.7 l/kg), suggesting its limited distribution to the muscle and other organs/tissues.[25] The mean plasma $t_{1/2}$ after oral PB was 200 min, while after i.v. infusion of 4 mg, the $t_{1/2}$ was 97 min.[17] It has been reported that neither PB nor its metabolite 3-hydroxy-N-methyl-pyridinium is bound to plasma proteins.[20,26] In order to find the distribution and retention of radioactivity of PB in the body, ^{14}C-pyridostigmine (463 μg, 1.78 μmol/2.2 μCi/kg) was administered s.c. twice a day for up to 16 days. Four rats were sacrificed on days 1, 4, 8, 12, and 16. Tissues such as ear, eye, heart, kidney, liver, lung, muscle, skin, sternum, and tail were analyzed. The results indicated the consequent increase in radioactivity per g of tissues from day 1 to day 16; cartilaginous tissues particularly accumulated, increasing concentrations with subsequent doses of pyridostigmine. This increase in radioactivity in the body tissues after chronic dosage was indicative of its binding to macromolecules, such as negatively charged chondroitin sulfate (unpublished data, Somani 1983). However, the study did not determine the radioactivity in brain tissue. Recently, ^{11}C labeled PB was administered to mice and its accumulation was measured in brain tissues.[27] The study documented no difference in the brain radioactivity between swim exercise-stressed mice and controls. Obviously, there was no change in blood brain barrier permeability and this may be related to variables such as age, strain, or dose of PB.[28] It is quite possible that stress may alter the expression of CNS AChE. However, it would be important to re-examine the effect of physical stress (exercise training/swim exercise) on permeability of quaternary ammonium compound through blood/brain compartments. It would also be necessary to study the possible distribution of radioactivity in brain regions after single and chronic dosages of ^{14}C-labeled PB under normal and stressful conditions.

METABOLISM: Metabolism of ^{14}C-pyridostigmine in myasthenic patients was studied by Kornfeld et al. and Somani et al.[29,30] Liver seems to be the main site of metabolism of pyridostigmine.[22] We have shown that PB is metabolized to 3-hydroxy-N-methyl pyridinium (3-HNMP) in man, (Figure 5.2),[30] and both the parent drug and metabolite seem to accumulate in the muscles after chronic administration of the drug to rats.[31] The pharmacological properties of 3-HNMP are largely unknown other than that it is less toxic than 3-hydroxy-phenyl-trimethyl ammonium, a metabolite of neostigmine.[32] The authors reported that the LD_{50} of these metabolites in mice after s.c. injection were 1350 mg/kg and 100 mg/kg, respectively.

Metabolism and excretion of PB after multiple dosing was studied in albino Sprague-Dawley rats by Somani.[31] ^{14}C-pyridostigmine (463 μg, 1.78 μmol/2.2 μ Ci/kg) was administered s.c. twice a day for 16 days. At the end of the sixteenth day, rats were found to have gained weight from 30 to 45%. Urine and feces were collected every 24 h and the radioactivity was measured in the excreta. The proportion of unchanged PB, 3-HNMP, and other metabolites were determined by paper chromatography in urine samples. The daily excretion of PB in urine ranged from 75–81% as an unchanged drug and that of 3-HNMP from 15–20% and unidentified

TABLE 5.1
Pharmacokinetics of Pyridostigmine in Humans and Animals

Species	Dose	Route of Admin.	Peak Plasma Conc. or AUC	$t_{1/2}$	Vd	Cl	F	Ref
Healthy Human	4 mg	i.v.	6.7±2.1 mg/ml × min (AUC)	97 min	1.03±0.35 l/kg	0.63±0.20 l/kg		17
Healthy Human	60 mg	Oral	20–100 mg/l	200 min			14.3% (11.5–18.9)	17
Myasthenic Patients	60 mg	Oral	40–60 mg/l	30–90 min	0.5–1.7		10%	20
Myasthenic Patients	180–1440 mg/day	Oral	180 mg/l	1.5 h	0.65 l/kg	0.29±0.30 l/h/kg	3.6% (2.9–4.2)	16
Myasthenic Patients	60–540 mg/day	Oral	12.4–64.5 mg/ml					
Myasthenic Patients	240–1080 mg/day		15.3–144.0 mg/ml					
Healthy Male	2.5 mg	i.v.		1.52 h	1.43 l/kg	0.65 l/kg × h		21
	120 mg	Oral	40–60 mg/l	1.78±0.24 hr	1.64±.29 l/kg	0.66±0.22 l/kg × h	7.6 ± 2.4%	
Human (normal patient)	0.35 mg/kg	i.v.	700–1800 mg/ml and after 3–4 h 40–60 mg/ml	112±12 min	1.1±0.3 l/kg	9±2 ml/kg/min		43
Anephric patient	0.35 mg/kg	i.v.		379±162 min		2±0.6 ml/kg/min .		43
Myasthenic Patients						349–481 ml/min		160
Elderly Patient	0.25 mg/kg	i.v.		157±56 min	1.4±0.4 l/kg	6.7±2.2 ml/kg/min		105

Species	Dose	Route of Admin.	Peak Plasma Conc. or AUC	$t_{1/2}$	Vd	Cl	F	Ref.
Young Patient	0.25 mg/kg	i.v.		140±60 min	1.8±0.7 l/kg	9.5±2.7 ml/kg/min		105
Human Volunteers	60 mg	Oral	0.28±0.18 mg/h/mo Absorption rate constant 0.23/h	Elimination rate constant 2/h		6.65±2.4 ml/min/kg (renal)		26
Myasthenic Patient	60–120 mg	Oral	588–4560 mg/ml					39
Healthy Human Male	30 mg/daily every 8 h for 21 days	Oral	311±120 l/h			221±13.4 l/hr	12%	44
Healthy Human Female		Oral	Absorption rate constant 0.32±0.02/h			172±11.4 l/h	12%	44
Human Male	3.65 mg/70 kg	i.v.		22.79 min	247–834 ml/kg	9.3–26.5 ml/min/kg		40
Myasthenic Patient	420–2100 mg daily	Oral	15–60 mg/l	Time to peak conc. 2 h				19
Myasthenic Patient	5–6 mg	i.v.	61–80 h.ng/ml	1.05±0.32 h	1.76±.54 l/kg	1.0±0.01 l/h		18
Dog	0.6 mg/kg	i.v.		8.3 hr±2.1 h	Lambda 2 8.7±1.9 l/kg Steady state 3.9±0.9 l/kg	13.0±1.0 ml/min/kg		42
Dog		Oral					33.6±9.5	
Rat	0.056 mg/kg	i.m.	1010 mg × min per ml	24.8 min	1.97 l/kg			41
Rat	0.5–2.0 μmol/kg	i.v.		24.2±4.2 min	0.35±0.05 l/kg	15.0±0.2 ml/min/kg		25

Note: Bioavailability (F) value is the fraction of the dose that reaches the systemic circulation. $t_{1/2}$ is the plasma half-life; Vd is the steady-state volume of distribution. AUC refers to the area under the plasma concentration vs. time curve, and Cl refers to the clearance.

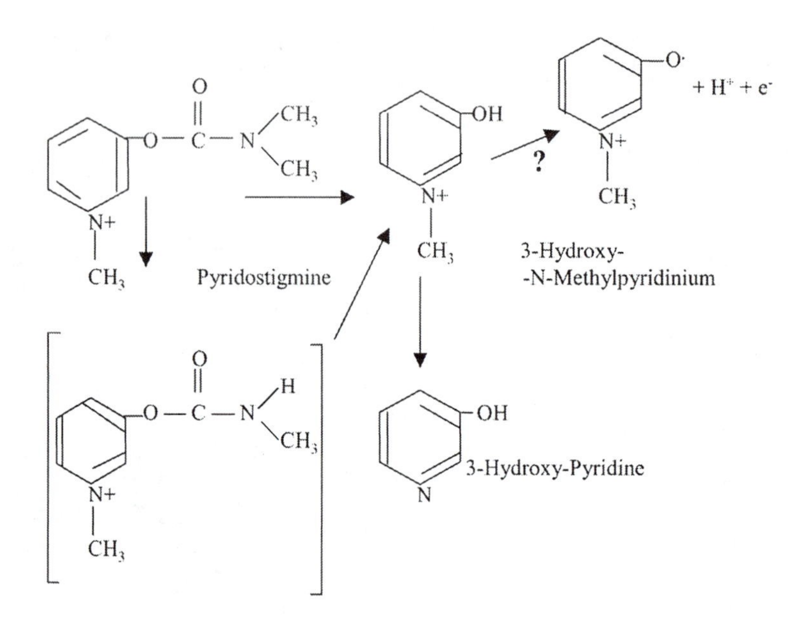

FIGURE 5.2 Metabolic pathway of pyridostigmine showing possible formation of reactive metabolite under physical stress.

metabolite (1–4%) as percent of daily dose. There was no consistent increase or decrease in the excretion of PB and its metabolites during the whole study time, suggesting that there was no stimulation or inhibition of metabolism. The elimination of radioactivity in feces ranged from 3–10% as a percent of daily dose. The average amount of PB in the body at steady state ranged from 4.83–8.77:g, corresponding to 7–12% of the administered dose. These studies suggest that PB may accumulate in the body after multiple dosing. The metabolite 3-HNMP did not form glucuronide or sulfate conjugates *in vivo,* in isolated liver perfusion, and in isolated liver microsomal studies.[31] A recent report on pyridostigmine by Dr. Beatrice A. Golomb of the Rand Corporation incorrectly mentioned the formation of glucuronide of 3-hydroxy-N-methyl pyridinium.[33] However, 3-hydroxy-N-trimethylphenyl ammonium, a metabolite of neostigmine formed the glucuronide conjugate in *in vivo* and *in vitro* studies.[34,35] This is the first study to show that the quaternary ammonium compound, despite being highly polar, formed glucuronide conjugate.

EXCRETION: Earlier work on disposition of PB reported that 50% of the dose rapidly appeared in the urine after a s.c. injection of pyridostigmine (3 mg/kg) in the dog, and a total of 67% of the dose in 24 h.[36] There was no further urinary excretion of the drug by this route of administration. However, after oral administration of pyridostigmine (7.5 mg/kg) to the dog, the unchanged drug was excreted in the urine up to 36 h, at which time 57% of the dose had been recovered. The rapid excretion of ^{14}C-pyridostigmine radioactivity in hen was blocked by prior administration of a dye, Cyanine 863, indicating PB was secreted by the renal tubules in the hen and probably also in the rat and man.[22] The maximum excretion of radioactivity occurred between 1 and 3 h after oral administration of single doses of ^{14}C-pyridostigmine to the rat.[37] In 24 h, 42% of the dose was excreted in the urine, and 38.4% was present in the feces and intestinal contents. The peak concentration of radioactivity in liver and blood occurred about 2 h after administration and about 75% of the radioactivity in the urine was present as unchanged pyridostigmine, the remainder as metabolites. The main route of excretion of PB is via the kidneys; renal clearance occurs both by glomerular filtration and tubular secretion.[20]

C. Pharmacokinetics of Pyridostigmine

The pharmacokinetics of pyridostigmine in healthy human volunteers, myasthenic patients, and in animal species such as dogs and rats are enumerated in Table 5.1. The numerical data on pharmacokinetic parameters, such as $t_{1/2}$ bioavailability (F), area under the plasma concentration vs. time curve (AUC), volume of distribution (Vd), and clearance (Cl) after i.v. and oral administration to humans and animals, are summarized in this table.

Pyridostigmine bromide is eliminated from plasma in a biexponential manner.[38] There is a direct linear relationship between the area under plasma concentration time curve and total daily dose of pyridostigmine in myasthenic patients which indicates linear pharmacokinetic modeling.[39] Calvey et al.[40] studied the pharmacokinetics of ^{14}C-pyridostigmine based on a two-compartment model after i.v. administration to myasthenic patients. The fast disposition half-life of pyridostigmine ranged from 0.61–1.78 min and the terminal half-life from 14.81–37.01 min (mean half-life = 22.79 min).[40] Pyridostigmine clearance (9.3–26.5 ml/min/kg) was invariably greater than the presumptive value for glomerular filtration rate, and the volume of distribution of the drug ranged from 0.25–0.83 l/kg. The steady state kinetics of PB in myasthenic patients indicated that the routine measurement of plasma pyridostigmine concentration had little to offer in the management of myasthenic patients.[16] These investigators reported a relatively stable kinetic parameter of PB in and between individual patients. The maximal effect of pyridostigmine with a "bell shaped" dose response curve occurred at a concentration of 30–60 ng/ml in plasma of 20 myasthenic patients (both male and female).[18] The pharmacokinetics of PB were determined in rats after intramuscular administration in the dose of 0.056 μg/kg.[41] The maximum plasma concentration (C_{max}) was found to be 21.3 μg/ml and the time to reach C_{max} was approximately 9 min. PB pharmacokinetics were determined in beagle dogs after i.v. infusion and oral doses of syrup and tablet, all in the dose of 0.6 μg/kg,[42] PB had a relatively long terminal $t_{1/2}$ of 8.3 h and a high Vd of 8.7 l/kg. The drug showed affinity for the peripheral tissues, as evidenced by a 4-fold higher residence time in tissues, as compared to plasma. The renal clearance of PB in volunteers under the influence of ranidine and pirenzepine was studied.[26] Its renal clearance average was about 74% due to tubular secretion. In an earlier similar study, it was shown that the renal route accounts for 75% of the pyridostigmine clearance in anesthetized patients.[43] PB has a short $t_{1/2}$, 1.0 h after i.v. and 3.7 h after oral administration.[17] Recently Marino et al.[44] carried out the studies to determine the time course of plasma PB concentration and RBC AChE activity after administration of 30 mg PB three times a day in healthy human subjects. They correlated the plasma concentration of pyridostigmine with RBC AChE activity and determined pharmacokinetic parameters using NONMEN-IV version 2.1. This study was population-based and carried out in 90 male and female volunteers. This study showed that pharmacokinetics of PB are both gender and weight dependent and the pharmacodynamic effects did not lag significantly from plasma concentration and returned to near normal within 8 h.

Table 5.2 depicts the differences in the pharmacokinetics of PB within and among different species. Some of the human data indicated that the Vd of pyridostigmine being a quaternary amine is higher. Although these pharmacokinetic data are under normal conditions, there is no literature on pharmacokinetic data under stressful
conditions such as under heavy military duty, mild or strenuous exercise, or after swim exercise. It is imperative to assume that exercise or stressful conditions might alter pharmacokinetics of pyridostigmine which in turn will alter the pharmacodynamics of this drug.[45] This is an area which needs further exploration especially in light of the use of PB as a pretreatment drug against possible threat from exposure to nerve gases.

D. Pharmacodynamics of Pyridostigmine Bromide: Use as a Pretreatment Drug

Pharmacodynamic actions of PB were studied as early as 1946 by Koster[3] and Koelle.[2] Anticholinesterase activity of PB was about one-fifth of the activity of neostigmine;[46] whereas, the comparative activity reported by Blaschko et al.[47] was about one-tenth. One of the most noticeable differences between neostigmine and pyridostigmine is the inability of the latter compound to produce a direct stimulant action on smooth muscle either *in vitro*,[46] or *in vivo*.[48] It has been suggested that this may account for the occurrence of fewer unpleasant side effects when pyridostigmine is used clinically.[49–51] Foldes and Smith[52] reported maximum inhibition of butylcholinesterase with 7×10^{-9}M PB at 1 h.

Pyridostigmine bromide was employed as a pretreatment drug against possible threat of the nerve agent sarin by soldiers during the Persian Gulf War. It has been suggested that the effectiveness of pyridostigmine pretreatment is due to the carbamylation of a portion of tissue AChE that protects it against irreversible inhibition by sarin. Spontaneous decarbamylation of PB produces sufficient free AChE to restore normal function.[53] The pharmacodynamics (cholinesterase inhibition) in RBC and plasma of humans and animal species following pyridostigmine administration are depicted in Table 5.2. Pyridostigmine pretreatment reversed the neuromuscular blockade produced by sarin. The rate of recovery was similar in rhesus monkey, cats, and rabbits, suggesting a common mechanism of action. The effectiveness of pyridostigmine pretreatment in nonhuman primates (marmosets and rhesus monkeys) on sarin poisoning was assessed.[54] PB produced a dose-related blood AChE inhibition in these animal species. The time to peak carbamylation occurred within 10–20 min of i.v. dosing of PB. PB pretreatment supported by therapy with atropine protected the primates against sarin poisoning. Hence, these authors suggested that pyridostigmine pretreatment could be effective in humans. The relationship between reversible

TABLE 5.2
Time Course of Pyridostigmine Bromide (PB) Action (Cholinesterase [ChE] Activity: Percent of Control) in Plasma and RBC in Various Animal Species and Humans, under Normal Conditions

		ChE Activity—% of Control			
Species	Dose (mg/kg)	Plasma	RBC	Time after PB	Ref.
Mouse	0.20 (p.o.)		70	60 min	64
	0.82 (p.o.)		40	60 min	
Rat	39.2 mg/ml		72	Day 1	69
	0.50:l/h		74	Day 4	
	(s.c.)		70	Day 8	
	for 14 days		80	Day 14	
			(blood)		
Rat	5		102	18 h	14
Male	15		57		
	30		66		
	60		51		
	(p.o.)				
	for 13 weeks				
Rat	5		95	18 h	14
Female	15		81		
	30		58		
	60		71		
	(p.o.)				
	for 13 weeks				
Rat	0.075		94	5 min	63
	(i.m.)		54	20 min	
			70	35 min	
			88	50 min	
			(blood)		
Rat	20	25	20	3 h	57
	40	50	30	3 h	
	80	20	20	2 h	
	(p.o.)	50	30	4 h	
Rat	90/day		13	Day 1	58
	(p.o.)		17	Day 2	
	for 15 days		25	Day 4	
			26	Day 7	
			10	Day 15	
Rat	12 ml/h (low		68	Day 6	161
	dose) (s.c.		52	Day 7	
	infusion) for		60	Day 14	
	14 days		95	7 days post-Λ	
	60 ml/h (high		25	Day 6	
	dose) (s.c.		30	Day 8	
	infusion) for		32	Day 14	
	14 days		100	7 days post-Λ	

continued

TABLE 5.2 (continued)

		ChE Activity—% of Control			
Species	**Dose (mg/kg)**	**Plasma**	**RBC**	**Time after PB**	**Ref.**
Rat	2.0 (i.p.)		95	Day 1	56
			115	Day 4	
			125	Day 10	
			100	Day 20	
	5.0 (infusion)		68	Day 1	
			72	Day 4	
			100	Day 10	
			90	Day 20	
	25.0 (infusion)		67	Day 1	
			50	Day 4	
			57	Day 10	
			47	Day 20	
Rat	0.2		70	30 min	55
	0.025		80		
	0.010		90		
	(i.m.)				
Guinea Pig	0.10 (i.v.)		51 (blood)	15 min	53
Guinea Pig	0.47		70	60 min	64
	1.9		40	60 min	
	(p.o.)				
Guinea Pig	0.94 (p.o.)		92	10 min	162
			74	30 min	
			53	60 min	
			45	120 min	
			50	180 min	
			57	240 min	
Guinea Pig	0.2		30	30 min	55
	0.025		55		
	0.010		70		
	(i.m.)				
Guinea Pig	0.05 μmol/kg (i.v.)		28	30 min	66
			25	45 min	
			29	90 min	
Marmoset	0.20 (i.v.)		39 (blood)		54
Dog	0.5		80	35 min	15
			80	75 min	
			80	115 min	
	2.0		55	35 min	
			58	75 min	
			58	115 min	
	5.0		50	35 min	
			50	75 min	
	i.v. infusion over 15 min		56 (blood)	115 min	

TABLE 5.2 (continued)

Species	Dose (mg/kg)	ChE Activity—% of Control Plasma	RBC	Time after PB	Ref.
Monkey	0.2 (i.v.)		46 (blood)	15 min	54
Monkey	0.25	80		Mean readings over 5 days	62
	0.50	70			
	1.0 s.c. infusion over 5 days	40 (serum)			
Monkey	0.12	66.6		30 min	59
	0.24	46.5			
	0.48	29.7			
	0.96 (s.c.)	16.9 (serum)			
Monkey	40 mg/day, 5 days of wk for 10 wk		54	2 h	61
			49	4 h	
			69	8 h	
			80	16 h	
			92	24 h	
Hen	5/day for 2 months		17		103
Human	30 mg every 8 h (p.o.) for 3 weeks		90	1 h	44
			77	2 h	
			78	4 h	
			91	8 h	
Human	30 mg every 8 h (p.o.) for 10 days		77	6th day	60
			77	9th day	
			101 (blood)	5th day post Rx	
Human	630 mg (p.o.)	46 (serum)		50 min	8
Human	30 mg every 8 hours (p.o.) for 7 days		64	Day 1	88
			59	Day 3	
			61	Day 5	
			59	Day 7	
Human (Healthy Males)	90 mg/day Wk. 3 & 4 on M, T, W only (p.o.)		72	Day 1	86
			69	Day 2	
			68	Day 3	

ChE inhibition by pyridostigmine at different doses and its efficacy against soman toxicity was studied in rats and guinea pigs. The study concluded that ChE inhibition may provide some protection against soman poisoning by as low as 10%.[55]

The effects of pyridostigmine (2, 5, and 25 mg/kg, i.p.) on RBC cholinesterase and skeletal muscle contraction were studied in rats.[56] The high dose (25 mg/kg) significantly inhibited ChE activity at different times; however, the study did not show a correlation between the ChE inhibition and decrease in muscle contraction. The acute toxicity of pyridostigmine at three oral doses (20, 40, and 80 mg/kg) produced acute focal necrosis, leukocytic infiltration, and marked changes in the motor end-plates of skeletal muscle of rat. The changes were more prominent in the diaphragm than the quadriceps muscle. The whole blood and RBC cholinesterase (ChE) activity was reduced to considerably less than one-half the normal value.[57]

Oral administration of pyridostigmine for 13 weeks in male and female Sprague-Dawley rats caused tremors and significant RBC ChE inhibition (51 to 81% of control) at 15 mg/kg dosage and higher, indicating toxicity.[14] However, no toxic effects were observed with 5 mg/kg pyridostigmine dose. This study also noted differences in ChE inhibition between male and female rats. At different doses of PB, the effects of PB (90 mg/kg/day) in diet for 15 days on blood ChE activity and myofiber morphology in diaphragm of rats was evaluated.[58] Electron microscopy showed maximal changes in the post-synaptic areas of the neuro-muscular junction of diaphragm. It was observed that subchronic PB induced primarily myopathic changes in rat diaphragm; however, some mechanism of adaptation seems to be activated that minimizes skeletal muscle injury 1 week after stoppage of pyridostigmine.

Different s.c. doses of pyridostigmine in primates indicated that serum cholinesterase activity dose-dependently decreased 30 min after pyridostigmine administration but did not significantly alter performance.[59] These investigators concluded that pyridostigmine is a very safe pretreatment drug for nerve agent poisoning. The increasing doses of pyridostigmine promptly and progressively lowered the AChE activity of blood to a minimum of 40% of control at a 5 mg/kg dose in beagle dogs.[15] With higher doses (5 mg/kg), the cardiac output was unchanged; however, airway resistance increased significantly. The lowest dose (0.5 mg/kg) produced minimal effects on the cardiovascular and respiratory systems. These authors hypothesized that PB in low doses would cause little or no adverse effects in normal humans when used as a protective agent. Acute oral pyridostigmine overdose (390–900 mg) in nine patients during the Persian Gulf War showed serum cholinesterase activity inhibition 21 to 75% of control, 20–90 min after pyridostigmine ingestion. The data indicated that serum ChE inhibition was a reliable diagnostic tool in pyridostigmine poisoning; however, no correlation between the extent of ChE inhibition and the incidence or severity of the cholinergic signs and symptoms was found.[8] The possible detrimental effects of oral pyridostigmine 30 mg three times a day on human neuromuscular function was assessed.[60] Muscle strength and endurance were tested before and after treatment; electro-diagnostic tests such as EMS nerve conduction and response to repetitive stimulation were also carried out before and after treatment (8th day). It was shown that pyridostigmine causing 20–30% ChE inhibition in healthy young men produced no significant neuromuscular adverse effects in humans.

The efficacy of orally administered pyridostigmine syrup, when used as a pretreatment for rhesus monkeys exposed to sarin and treated with antidotes such as atropine sulfate, the oxime TMB-4 and the anticholinergic agent benactyzine hydrochloride, was evaluated.[61] The authors showed that oral PB treatment followed by antidotal therapy was effective in protecting rhesus monkeys against repeated exposure to lethal concentrations of sarin. The protective period of oral pyridostigmine supported by the antidotal therapy was between 1/2 and 8 h. In another study, the continuous infusion of PB producing 30 and 60% of normal serum ChE inhibition provided protection against the behavioral toxicity induced by five daily repeated low doses of soman in monkeys.[62]

The efficacy of pyridostigmine pretreatment at symptom-free doses was studied at various times (5, 20, 35, and 50 min) prior to exposure to sarin in rats. Significant inhibition of whole blood and lung cholinesterase activity occurred 20 min after pyridostigmine administration, suggesting this to be the optimal time for protection against sarin inhalation toxicity.[63] The effects of PB pretreatment on antidotal efficacy of atropine and 2-PAM in sarin, tabun, and VX poisoning in mice and guinea pigs has been evaluated. Further the oxime-induced reactivation of VX-inhibited whole blood AChE of guinea pigs was studied.[64] This study showed that 1 h prior to organophosphate exposure, pyridostigmine induced 30 and 60% inhibition of RBC cholinesterase activity at 0.47 and 1.9 mg/kg oral doses in guinea pigs and 0.20 and 0.82 mg/kg in mice. The data also showed that PB significantly enhanced the efficacy of antidotes atropine and 2-PAM against tabun in both animal species, whereas it reduced or did not increase the efficacy of these antidotes against sarin or VX in both species. Pretreatment with PB also reduced significantly the recovery of VX-inhibited AChE activity by 2-PAM. In a previous report in male rhesus monkeys, it was found that the combination of PB pretreatment and prompt post-treatment with atropine and 2-PAM chloride resulted in greatly improved protection (protective ratio >40, compared to control) against soman intoxication.[65] The effect of pyridostigmine pretreatment on cardiorespiratory function in tabun poisoning was evaluated in guinea pigs.[66] The study found significant inhibition of RBC cholinesterase at different times after intravenous administration of pyridostigmine. The investigators concluded that pyridostigmine enhanced circulatory depression, decreased survival time and rate in tabun poisoned animals. Lintern et al.[67] have demonstrated that repeated administration of pyridostigmine (0.4 μmoles/kg, s.c.) twice a day for 3 weeks altered the molecular forms of AChE activity in the soleus, extensor digitorum longus (EDL), and diaphragm muscle of mice.

Pyridostigmine pretreatment was used by the Gulf War veterans to obtain 20–30% cholinesterase inhibition in order to enhance the efficacy of the standard therapeutic regimen for possible nerve gas intoxication.[11,12] Pyridostigmine reversibly inhibits cholinesterase (60–70% of control) in the peripheral tissues and blood when given in symptom-free doses in primates and rodents 15–30 min after dosing.[62,68] A symptom-free dose of pyridostigmine for 14 days caused inhibition of whole blood cholinesterase (70–80% of control) in rats.[69] Somani et al.[70] have demonstrated that, under resting conditions, pyridostigmine in a symptom-free dose (1.2 mg/kg, p.o.) for 14 days inhibited plasma BChE activity (87% of control) in mice 4 weeks after the last dose of pyridostigmine indicating the delayed effect of the drug. Thus, subchronic exposure to pyridostigmine might have influenced the *de novo* synthesis of enzyme. BChE is a nonspecific choline-ester hydrolase which is synthesized in the liver.[71] Moreover, data of Somani et al. further show that pyridostigmine did not significantly alter AChE activity in blood cells (RBC and platelets) indicating that most of the pyridostigmine or its metabolites interacted with plasma cholinesterase which is the scavenger of the anticholinesterase agents.[70] Pyridostigmine 30 mg every 8 h produced RBC cholinesterase inhibition greater than 10% at the time of trough in approximately 70% of individuals.[44] These investigators based their estimations using the pharmacodynamic model that best fit RBC AChE activity using an E_{max} value compared to baseline. Pharmacodynamic effects and the toxicity of pyridostigmine have clearly shown the importance of dose, time of administration, and the time course of disposition (pharmacokinetics), which could determine the protective efficacy of PB against nerve gas poisoning.

E. Factors Influencing Pharmacokinetics and Pharmacodynamics of Pyridostigmine Bromide

1. Stress

The effects of physical stress on the disposition and pharmacokinetics of drug is not well recognized. Recently, Somani and Kamemori[45] have reviewed the effect of exercise on absorption, distribution, metabolism, and excretion of drugs and chemicals. During exercise, cardiac output increases with the intensity of workload and concomitant changes in regional blood flow distribution occurs. Thus, blood flow to skeletal muscles and skin is greatly increased, while, on the other hand, hepatic blood flow decreases during exercise.[72] Therefore, the clearance of drugs that are primarily eliminated by liver metabolism may be impaired due to decrease in hepatic blood flow.[73,74] Hepatic clearance is the product of hepatic blood flow and the hepatic extraction ratio. The decrease in hepatic blood flow could theoretically result in a diminished clearance of drug or chemical, thereby resulting in the body's accumulation of the drug and metabolites during chronic administration. This increased concentration could cause potentially detrimental effects of the drug or chemical. Several studies with PB under stressful conditions have been carried out by various investigators and are summarized in Table 5.3.

The effect of ChE inhibition induced by pyridostigmine pretreatment on endurance, thermoregulation, and pathophysiology during exercise in a hot environment was studied.[75] As a result of exercise, after PB administration, no change in ChE inhibition was observed due to the hot environment and exercise, after PB administration. However, the endurance time for pyridostigmine treated animals was only 23 min compared to 35 min for the control animals. It was concluded that intense ChE inhibition induced by pyridostigmine administration severely limited the endurance capacity of rats working in the heat.[75] The effects of chronic oral PB treatment and a single i.m. atropine injection on thermoregulatory effector responses of patas monkeys was evaluated.[76] The study reported a 20–25% decrease in serum ChE activity with daily oral PB treatment; however, thermoregulatory or cardiovascular functions were not affected. The potential muscle damage produced by pyridostigmine pretreatment (14 days) when given alone or when combined with physical exercise was investigated in a mouse model.[69] It was found that only the combination of pyridostigmine plus physical exercise contributed to a loss of integrity in skeletal muscle,

TABLE 5.3
Time Course of Pyridostigmine Bromide (PB) Action (Cholinesterase [ChE] Activity: Percent of Control) in Plasma and RBC in Various Animal Species and Humans, under Stressful Conditions

		ChE Activity—% of Control				
Species	**Dose (mg/kg)**	**Stress as a Factor**	**Plasma**	**RBC**	**Time after PB**	**Ref.**
Mouse	1.2 (p.o.)	None	87	96	4 weeks after	70
	2 weeks	Exercise training	79	103	PYR treatment	
Rat	0.6 (i.p.)	None	36		105 min	75
		Heat (35°C) + Exercise	38		120 min	
Guinea Pig Male	0.20 (s.c.)	Low Stress		35	2 h	80
		Median Stress		35		
		High Stress		31		
Monkey	0.4 every 8 h (p.o.) for 7 days	None	70		Day 1	76
			75		Day 7	
		Heat Stress	NC (serum)			
Human	30 mg (p.o.)	None		61	150 min	82
		Exercise (58% VO_2 max) for 30 min		NC	150 min	
Human	30 mg (p.o.)	None		33.4	110 min	87
		Cold Stress		30.2	127 min	
Human	60 mg	Healthy		28.4	150 min	83
	30 mg (p.o.)	Mildly Asthmatic		76.7 (blood)	150 min	
Human	30 mg every 8 h (p.o.)	Healthy		80.7	9.6 h	89
		Symptomatic		81.2 (blood)	7.1 h	
Human	30 mg (p.o.)	Heat (35° C) and Hypohydration		73.1	90 min	84
				66.9	160 min	
				69	220 min	
Human (Male Soldiers)	30 mg every 8 h (p.o.) for 7 days	Heat Stress plus Exercise		76.3	Day 1	85
				72.9	Day 3	
				58.8	Day 5	
				65.7	Day 7 (2 h after PB)	
Human Volunteers	30 mg × 4 at 8 h intervals (p.o.)	Normal		75.8	60 min	81
				64.4	120 min	
				67.2	240 min	
		Heat-Exercise Stress			NC	

as evidenced by increased CPK activity in serum and enhanced urinary creatinine excretion rate. A recent study that used an acute, much higher PB dose has demonstrated that forced swim exercise enhances brain AChE inhibition, AChE mRNA, and neuronal c-fos oncogene by pyridostigmine in mice.[77]

The delayed and interactive effects of PB and exercise training on BChE, AChE, creatine phosphokinase (CPK), and malondialdehyde (MDA) in peripheral and cerebral tissues of mice were reported.[70] The mice were sacrificed 4 weeks after the last dose of PB or saline and 24 h after the last exercise. Blood, muscle, and nerve tissues were isolated and analyzed. BChE and AChE activity significantly decreased to 79% of control in plasma and 78% of control in triceps muscle of mice, respectively, in PB + exercise group. Creatine phosphokinase activity increased 122% of control in plasma in PB + exercise group indicating enhanced neuromuscular effect of combination. Malondialdehyde concentration (lipid peroxidation end product) significantly increased to 124% of control in triceps muscle in PB + exercise group indicating the oxidative stress of the combination. This study showed that the interactive and delayed effects of PB (even after 4 weeks of stoppage of dosage) and exercise training occurred primarily in peripheral tissues.

The delayed effects of PB and exercise training on muscle tension in the mouse lower extremity were studied.[78] This study reported the interactive effects of PB and exercise training on muscle tension elicited in mouse lower extremity anterior muscular compartment by dorsiflexion of the foot with stimulation of the peroneal nerve. Experiments on muscle tension were conducted 4 weeks after the last dose of PB or saline and 24 h after exercise training. The muscle tension was measured in right and left legs using a tension transduction device connected to a polygraph. There was a significant increase in the muscle tension of combined legs ($p < 0.05$) in the group treated with PB plus exercise as compared to control and exercise groups (Figure 5.3). A significant reduction in acetylcholinesterase activity ($p < 0.01$) was also observed in the triceps muscle in mice treated with PB plus exercise when compared to control and exercise groups. These results suggest that delayed effects of PB (even after 4 weeks of stoppage of dosage) and interaction with exercise training leads to a reduction in muscle AChE activity. This may be due to an increase in acetylcholine leading to increased muscle tension. Exercise in combination with pyridostigmine may lead to a loss of some muscle fibers, which may be compensated for by hypertrophy of the remaining fibers involving contractile proteins. These changes in triceps muscle proteins and muscle fiber profile can alter the muscle tension in mice. A recent study examined the relationship between intake of PB and handgrip strength in Gulf War veterans, in comparison to control humans.[79] The report showed that handgrip strength was negatively associated with age and female gender. Further, the data suggested no relationship between PB intake and postwar handgrip strength.

The AChE activity in different brain areas of heat-stressed guinea pigs was measured after administration of pyridostigmine for 2 h at temperatures of 41.5°C, 42.6°C, and 44.3°C. The study found no entry of pyridostigmine into cortex, striatum, and hippocampus following tritiated pyridostigmine administration and autoradiographic evaluation. The results indicated that heat stress did not induce pyridostigmine penetration into brain areas of guinea pigs.[80] Pyridostigmine inges-

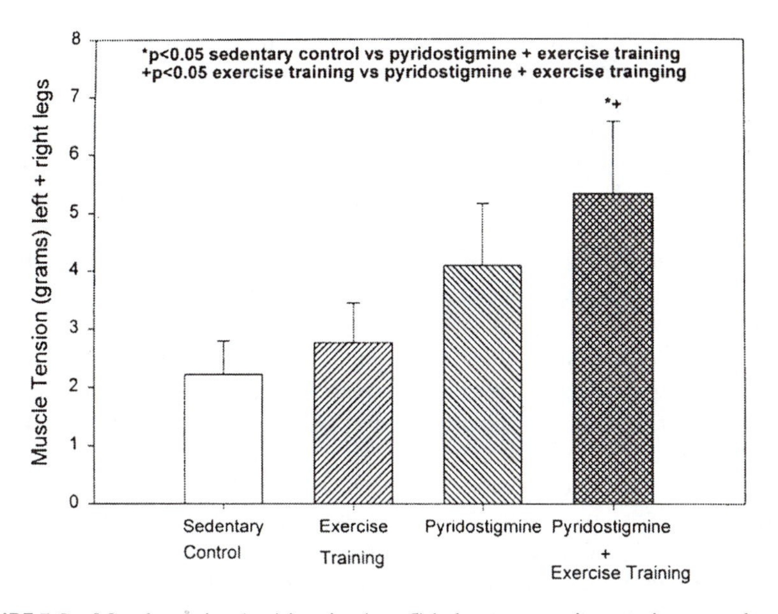

FIGURE 5.3 Muscle tension (gm) in mice (n = 5) in lower extremity anterior muscular compartment by dorsiflexion of left plus right legs with stimulation of peroneal nerve.

tion at a dose of 30 mg 8 hourly for 4 days resulted in significant inhibition of whole blood ChE activity; however, no significant differences were found between treatments on physiological responses and heat-balance parameters.[81] The study concluded that pyridostigmine ingestion did not increase the physiological strain resulting from exercise stress in hot conditions.

Oral PB ingestion in human subjects who underwent exercise at high environmental temperature (36°C), reduced skin blood flow, which may limit exercise thermoregulation.[82] The respiratory function in healthy and asthmatic volunteers following a single oral dose of pyridostigmine was evaluated.[83] The study found a correlation between ChE inhibition and forced expiratory volume in 1 s (FEV_1) at 60 mg oral dose in healthy subjects. Based on this result, 30 mg PB was suggested to be an appropriate dose for mild asthma patients producing similar level of ChE inhibition. The authors concluded that a certain threshold of ChE inhibition must be reached before the effect of pyridostigmine on respiratory function could be observed.[83] A few measurable physiological effects (responses to heat, exercise, and hypohydration) were studied after 30 mg oral pyridostigmine ingestion. Pyridostigmine had little significant and practical effect on human physical responses to moderate exercise-heat stress (35% VO_2 maximum at 35°C temperature and up to 75% relative humidity.[84]

A double-blind study investigated the effects of multiple-dose oral PB on physiological responses to heat stress tests in a hot, dry environment (42°C and 20% relative humidity) in seven male soldiers.[85] The authors found that a 7 day oral PB course caused very mild effects on physiological responses in men undergoing moderate treadmill exercise in a hot environment.

The deleterious effects, if any, of pyridostigmine on acceleration tolerance or performance was evaluated in healthy male human volunteers.[86] These authors found that any significant decline in the optimum performance of normal and healthy aircrew receiving prophylactic doses of pyridostigmine during aerial combat was unlikely. The effects of oral pyridostigmine bromide (PB) on human thermoregulation during cold water immersions (20°C) was also investigated.[87] These authors concluded that a 30 mg oral dose of pyridostigmine did not increase individual susceptibility to hypothermia during cold water immersion. However, in combination with cold stress, pyridostigmine may result in marked abdominal cramping and limit cold tolerance. Acute and chronic oral ingestion of pyridostigmine (30 mg three times a day) by healthy human volunteers did not alter thermoregulatory or metabolic effects during moderate activity in cold climates.[88] The levels of cholinesterase inhibition did not correlate with the type or severity of symptoms in Gulf War veterans after oral ingestion of pyridostigmine.[89]

Since Gulf War veterans underwent physical stress (exercise) and were exposed to pyridostigmine, it is possible that exercise could cause significant effects on pharmacokinetics and pharmacodynamics of PB under conditions that reasonably simulate heavy military duty. Based on the data from previous studies, we believe that physical exercise will increase the inhibition of ChE activity by PB after its administration. The stressful demands of modern military duty include a broad range of activities, especially during war time. The demanding physical tasks of a combat infantry soldier can be expected to result in significant physical and chemical changes within the body.[90] Notwithstanding this, physiological stress is still to be expected, because of the redistribution of blood flow to serve the demands of active muscle cells[91] as well as to meet the needs of temperature regulation in the body. In addition, a considerable production of metabolic acids from substrate catabolism will lead to a marked reduction of the intracellular pH.[92, 93] The time course of a drug in the body may be influenced by exercise dynamics.[94, 95] Hence, it is important to investigate further how physical activity interacts with a drug that may potentially be administered under combat field conditions.

Physical stress increases the oxygen consumption in the body, generates reactive oxygen species in tissues, and exerts oxidative stress response.[96, 97] The interaction of exercise and ethanol resulted in an inhibition of AChE in certain brain regions of rats and the enzyme inhibition was well correlated with enhanced lipid peroxidation in corresponding brain regions.[98,99] AChE is a membrane-bound enzyme with lipid dependence.[100] The present data show that interaction of pyridostigmine and exercise enhanced AChE inhibition and lipid peroxidation in triceps muscle of mice indicating an enhanced oxidative stress response of the combination. Earlier studies from our laboratory have also reported that physical stress enhanced the inhibition of cholinergic enzymes elicited by a centrally acting reversible cholinesterase inhibitor (physostigmine) in skeletal muscle of rats.[101] The accumulation of pyridostigmine and metabolite 3-HNMP may likely be due to its existence in zwitter ionic form at the acidic pH of the muscle of mice that have undergone physical stress which can cause detrimental effects on skeletal muscle. Further studies are needed to confirm this phenomenon. A recent preliminary report on the neuro-endocrine-immune effects of either PB and/or exercise training on treadmill for 14 days showed a significant decrease in plaque-forming cell response and altered splenic and thymic CD4/CD8 sub-populations following 40 and 60 min exercise and/or PB treatment to adult female mice. However, no effect was observed in lymphoproliferation or natural killer cell activity after either treatment. Administration of PB did not show any effect on thymic and splenic weights, but the physical stress resulted in a significant decrease in both spleen and thymus weight at 40 and 60 min exercise training protocols on treadmill.[102]

Stress is an important factor that can alter the pharmacokinetic and pharmacodynamic effects of PB resulting in increased toxicity, and this area of research needs further investigation.

2. Environmental Exposures

Single and combined effects of pyridostigmine, DEET, and permethrin on plasma butyrylcholinesterase activity and lethality in hens was assessed.[103] It was found that pyridostigmine (5 mg/kg/day) for 2 months resulted in significant inhibition of plasma ChE activity in hens. The authors concluded that carbamylation of peripheral esterases by pyridostigmine reduces the hydrolysis of DEET and permethrin and increases their availability to the nervous system. The doses of pyridostigmine and chlorpyrifos and the duration of exposure used by Abou-Donia et al.[103] were approximately four times higher than what veterans used during the Persian Gulf War. These investigators had used pyridostigmine (5 mg/kg, p.o.) for 2 months, along with other chemicals, to show neurotoxic effects in hens.

3. Gender and Age

The effect of oral doses of pyridostigmine in different dosages (0.5, 15, 30, and 60 mg/kg/day) on RBC cholinesterase inhibition was evaluated in male and female Sprague-Dawley rats.[14] The study showed differences in RBC cholinesterase inhibition between male and female rats at different dosages. The study suggested that male rats may be more sensitive to RBC cholinesterase inhibition, especially at doses of 15 mg/kg/day and higher.

We have recently studied the influence of physical stress (acute exercise) and PB on ChE activity in blood and brain regions of male (NIH Swiss) and female (C_3H-He/N-ve) mice.[104] This study examined the interaction of acute exercise and a single PB dose (2 mg/kg, p.o.) on ChE activity in blood and brain regions of these two different strains of male and female mice. PB significantly decreased BChE activity (61% of control) in male and (31% of control) in female mice. PB significantly decreased AChE activity in RBC (72% of control) in male and (61% of control) in female mice. The interaction of PB and exercise resulted in a significant inhibition of plasma BChE activity (58% of control), RBC AChE activity (72% of control), and cortical AChE activity (84% of control) in male mice. However, there was a significant inhibition of plasma BChE activity (37% of control), RBC AChE activity (51% of control) and cortical AChE activity (80% of control) in female mice. These results showed the differences in ChE activity in male and female, which are differentially influenced by exercise. Physical stress seems to increase the permeability of PB in brain, thereby inhibiting cerebral AChE activity in both species.

The influence of age on the pharmacokinetics of PB has been evaluated in subjects under anesthesia and paralyzed with neuromuscular blockers.[105] The plasma concentrations of PB were determined by radioimmuno-assay and after 1 h were found to be greater in the elderly (71–85 years of age) compared to the younger patients (21–51 years of age). It was found that the plasma clearance of PB was significantly reduced in the elderly (6.7 ml/kg/min). The study concluded that the prolonged duration of action of pyridostigmine in the elderly is due to the reduced plasma clearance of the drug. The relationship between the pharmacokinetic parameters and variables such as gender and weight in response to PB was evaluated in healthy subjects.[44] The relation between oral clearance and gender was found to be significant. The clearance in men was found to be 221 l/h while that in women was found to be 172 l/h. On the other hand, the relationship between the pharmacodynamic parameters elimination rate constant from the effect compartment (K_{eo}) and concentration at steady state giving half maximal effect (EC_{50}) was not significant. There have been very few studies of pyridostigmine evaluating its pharmacokinetics and pharmacodynamics with respect to gender and age. These could be very significant especially with the use of pyridostigmine as a pretreatment under stressful conditions.

III. PHYSOSTIGMINE

A. General Aspects

Physostigmine (PHY) is one of the oldest drugs, isolated from Calabar beans and successfully used for the treatment of glaucoma in 1864. It gained further prominence due to its use in the clinical trials of Alzheimer's disease. Physostigmine is also a potent prophylactic antidote for organophosphate poisoning. It is a reversible cholinesterase inhibitor and has a short duration of action. Being a tertiary amine structurally, it is lipid soluble and hence crosses the blood-brain barrier readily to produce central actions.

B. Pharmacokinetics of Physostigmine

The pharmacokinetics of PHY in rat showed a biexponential disappearance after i.v. dosage, suggesting a two-compartment model.[106] The half-life of distribution phase ($t_{1/2}\ \beta = 1.31$ min) of PHY suggests rapid equilibration of the drug with tissues. The half-life of elimination phase was found to be 16 min following i.v. administration and 17 min after i.m. administration in rat.[107] These half-lives are different from the half-lives of PHY in dog (30.7 ± 17.1 min),[108] in man (21.7 min),[109] and in guinea pig (40–50 min).[110] The Vd of 1.36 l/kg is higher than the total body volume which is indicative of a sequestration of this drug in tissues. The Vd in the dog and man was also higher than body water. Studies with radioactive 3H-PHY have shown that about

30–40% of the radioactivity (RA) was irreversibly bound to liver after i.v. or i.m. administration.[111] The clearance of 62 ml min^{-1} kg^{-1} in rat was higher than the dog (41 ml min^{-1} kg^{-1}) and man (22 ml min^{-1} kg^{-1}). The high clearance in rat may be related to increased metabolism of PHY in this species. Hepatic clearance plays a major role in the elimination of PHY. Physostigmine excretion in urine is less than 4% in 24 h after a single dose of i.m. administration.[112] In comparison to renal excretion, biliary excretion plays a greater role in that about 27% of the dose is excreted in bile of the rat.[112] The pharmacokinetics of physostigmine in Alzheimer patients showed an elimination half-life of 20–30 min.[113] Clearance and volume of distribution were 7.7 ± 0.9 (SE) l/min and 2.4 ± 0.6 (SE) l/kg, respectively. Butyrylcholinesterase inhibition half-life was 83.7 ± 5.2 (SE) minutes. During sustained steady-state infusion, plasma physostigmine concentration (r = 0.95) and butyrylcholinesterase inhibition (r = 0.99) were linearly correlated with the dose.[114]

The absorption rate constant (Ka) and the elimination rate constant (Ke) of PHY after oral administration are 0.1 ± 0.07 and 0.036 ± 0.024, respectively. Cp_{max} and t_{max} are 3.3 ng/ml and 16 min. The clearance of PHY is 80.95 ml min^{-1} kg^{-1}. The bioavailability (F) value, the fraction of the dose that reaches the systemic circulation, is 0.02. The lack of parent drug in systemic circulation and low AUC (160.57 ng min^{-1} kg^{-1}) and high extraction ratio (0.98) after oral administration strongly suggest "first pass" effect.[115]

The half-life of PHY in the liver and the muscle was found to be 24 and 20 min, respectively, after i.v. administration, whereas the studies after i.m. administration gave a half-life of 26 min in the liver. The elimination rate constant for liver (Kl) and muscle (Km) after i.v. administration was found to be 0.0288 and 0.0351 min^{-1}, respectively, and after i.m., Kl was 0.027 min^{-1} for the liver.[111]

The half-life of PHY in rat brain was 11 min. PHY is rapidly concentrated in the rat's brain. The drug, or metabolite, appears to be concentrated in mitochondria in greater amounts by a mechanism other than simple diffusion. The effect of the drug on mitochondrial function are not known, nor is it known how these effects are related to the toxicity of PHY in humans.[116]

Plasma protein binding studies are important in determining drug distribution and excretion. The Scatchard plot for the binding of PHY to rat plasma and rat serum albumin resulted in negative slope.[117] Somani et al.[118] reported that the binding of PHY to plasma proteins decreases in the presence of quinidine, furosemide, acetaminophen, theophylline, and verapamil. The binding of [3H]physostigmine to crystallized human serum albumin was investigated using equilibrium dialysis. The percentage bound to 1% (w/v) human serum albumin decreased from 18 to 4% as the total concentration of physostigmine increased from 3.3 nM to 2.7 μM (0.9 to 750 ng ml-1). A single class of specific binding sites with a large affinity constant, K = 8 × 10(7) l mol-1, was identified.[119]

The distribution and metabolism of PHY in the body determines its duration of action. After i.m. administration, the time course of PHY indicates that PHY is metabolized rapidly in plasma and comparatively slowly in the brain. PHY is distributed in all tissues and sequestered in the liver. Distribution studies showed that the radioactivity per gram of tissue was highest in kidney and liver, whereas the percentage of the administered dose in terms of radioactivity was maximum in muscle, followed by liver.[106,107] Lukey et al.[110] studied the pharmacokinetics of PHY in guinea pigs following i.m. doses of 5, 27, and 146 μg//kg. Plasma PHY concentrations were analyzed by HPLC and it was found that the peak concentration of the drug was reached at 30 min for all doses. There was a linear relationship between the PHY dose and AUC and C_{max}. Therefore, i.m. PHY administration to guinea pigs resulted in rapid absorption, distribution, and elimination and showed linear pharmacokinetics.

A physiological model for physostigmine disposition was developed in the rat which incorporated anatomical, physiological, and biochemical parameters, i.e., tissue volume, plasma flow rates, drug metabolism, and tissue-to-plasma partition coefficients.[120] Predicted concentrations of physostigmine in different tissue compartments were consistent with the experimental observations in the rat following an i.v. dose. Part of this study also compared the time course changes in measured effect, as percentage change in cholinesterase activity in brain and related these changes to the plasma or brain drug level in either a combined pharmacokinetic-pharmacodynamic (plasma physostigmine-effect relationship) or a dynamic model (brain physostigmine-effect relationship).

One of the major factors that determines the duration and intensity of the pharmacological activity of a drug is the rate and pattern of its metabolism. Little is known about the fate of PHY in the body. Although PHY has been in use for more than a century, to date there is still no data available on its metabolism and mechanism of toxicity other than from excessive cholinergic activity. PHY is metabolized to eseroline and three other metabolites (M_1 M_2, and M_3) that have not been identified.[111] This reference showed that about 90% of the drug reaching the liver is metabolized within 5 min indicating the importance of hepatic elimination of this drug. The nature of the metabolites and their possible toxic effects have not been studied. The liver sequestered radioactivity after i.m. or i.v. administration of ^{3}H-PHY.[106,107] When PHY is administered to humans by the oral route, it is anticipated that most of the drug will be metabolized as soon as it reaches the liver. PHY is hydrolyzed to eseroline, which has a phenyl hydroxyl group and is capable of forming conjugates with endogenous substrates. However, eseroline could be further oxidized to catechol and then to quinone (rubreserine type).[121] Such metabolites are highly reactive and can act as strong electrophiles (Figure 5.4). These electrophiles can form covalent conjugates with nucleophilic amine and sulfhydryl groups of cysteine and glutathione. The metabolism of physostigmine was studied by its incubation with the microsomal fraction of mouse liver. The metabolites formed were separated by reversed-phase ion-pair liquid chromatography and detected amperometrically by dual electrodes. Two major and six minor metabolites were found. Retention times and electrochemical characteristics were studied for these and compared with the hydrolyzed products of physostigmine: eseroline and rubreserine. None of the major metabolites was identical with these standards.[122]

The time course of subcellular distribution of radioactivity in rat brain after i.v. administration of ^{3}H-PHY was studied.[116] The concentration of radioactivity was

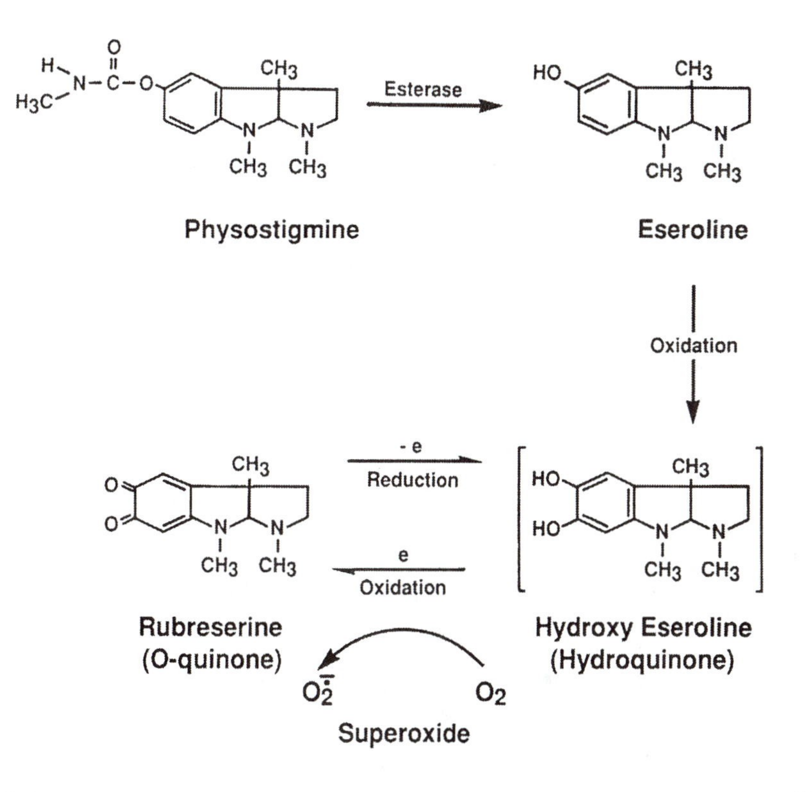

FIGURE 5.4 Possible metabolic pathway of physostigmine. (Adapted from Somani et al.[123])

higher in mitochondrial fraction and continuously increased from 5–60 min. The amount of radioactivity in synaptosome and microsomes increased up to 30 min and then declined at 60 min. Irreversible binding of xenobiotics to proteins can result in toxicity. The "covalent binding" is an experimental parameter that serves as an index of the formation of highly reactive metabolites that are difficult to measure by other means. "Reactive" metabolites can possibly reduce molecular oxygen to superoxide anion, which can in turn produce highly reactive singlet oxygen, and then to hydrogen peroxide and hydroxyl radical.

PHY is metabolized to eseroline which is further hydroxylated to form catechol and its oxidative product rubreserine (o-quinone). Eseroline causes damage to neuronal cells.[123] In another investigation, the changes in antioxidant enzymes were studied in brain regions in response to chronic infusion of PHY (34.5 μg/kg/hr) in rats that were sacrificed at the end of days 1, 7, and 12 of infusion. PHY infusion increased superoxide dismutase (SOD) activity in brain stem (122 and 123% of control) and in striatum (119 and 117% of control) on days 7 and 12, respectively. PHY infusion depressed catalase activity in the brain stem, while glutathione peroxidase activity increased in the brain stem (153 and 151% of control) and in cortex (114 and 138% of control) on days 7 and 12 of PHY infusion, respectively. This study suggests

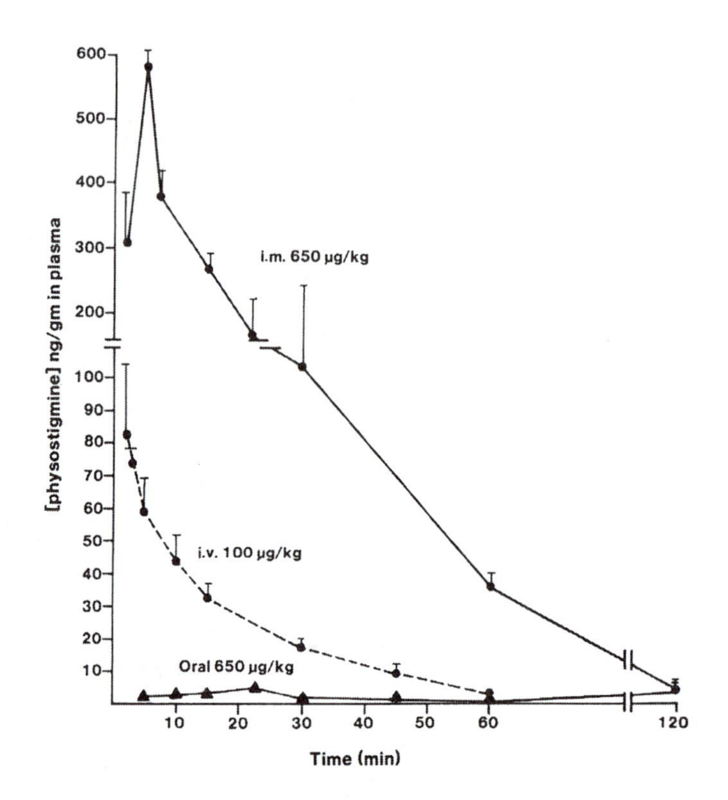

FIGURE 5.5A Time course of plasma and brain physostigmine concentration and corresponding plasma BChE and brain AChE inhibition (% of control) after different routes of administration: (A) time course plasma physostigmine concentration.

that PHY and its metabolites influenced the antioxidant enzyme activity selectively in different brain regions possibly as a compensatory mechanism of electrophilic stress of PHY metabolites.

Time course of PHY concentration in plasma and brain was compared after 650 μg/kg dose i.m. and oral and 100 μg/kg after i.v. administration as shown in Figure 5.5. BuChE activity in plasma and AChE activity in brain was also compared after these doses. The figure shows the pharmacokinetic and pharmacodynamic effects of PHY. PHY does not reach an effective concentration in the brain after oral administration because of its first-pass effect. However, it is an effective pretreatment drug after i.v. and i.m. routes of administration.

C. Pharmacodynamics of Physostigmine

PHY is a short-acting anticholinesterase agent and could potentially be utilized as a prophylactic agent against OP intoxication as it reversibly inhibits a portion of tissue cholinesterase, thereby preventing phosphorylation and aging of this enzyme by organophosphates.[2, 124–126] ChE is an important parameter to monitor the efficacy of PHY in OP intoxication.

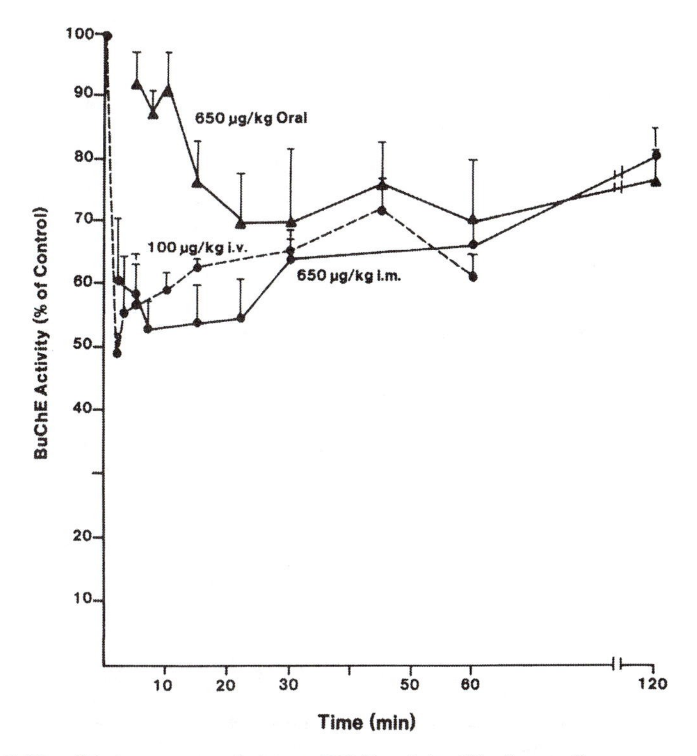

FIGURE 5.5B (B) time course of plasma BChE activity (% of control).

Earlier studies showed that PHY afforded protection against several lethal doses of DFP. The increased toxicity of PHY following the previous administration of DFP is to be expected due to the fact that a large portion of the tissue cholinesterase would be inactivated at the time when PHY was given. As a result, only a small amount of cholinesterase would have to be inactivated by PHY to produce death.[127] It is conceivable that the protective action exerted by PHY when injected prior to DFP probably results from the reversible combination of PHY with the active groups of the cholinesterase molecules, thereby blocking access to DFP and the subsequent formation of an irreversible ChE-OP complex. During the time necessary for the dissociation of the PHY-cholinesterase complex, part of the uncombined DFP would be excreted or hydrolyzed and the decarbamylated cholinesterase would then resume its physiological function. Therefore, the hypothesis was well established that short-lasting anticholinesterase drugs (carbamates) may protect ChE enzymes against inactivation by irreversible anticholinesterase (organophosphates).[2,4] This type of protective action is of interest in many respects, but particularly with a view to developing drugs which are potentially effective against poisoning by nerve agents.[3, 54,125, 128–130]

Physostigmine has been used as a prophylactic treatment regimen to antagonize the toxic effects of soman in animals.[126, 131–136] Studies done on the protection afforded by the carbamate AChE inhibitors generally used survival as an end point. Solana et al.[137]

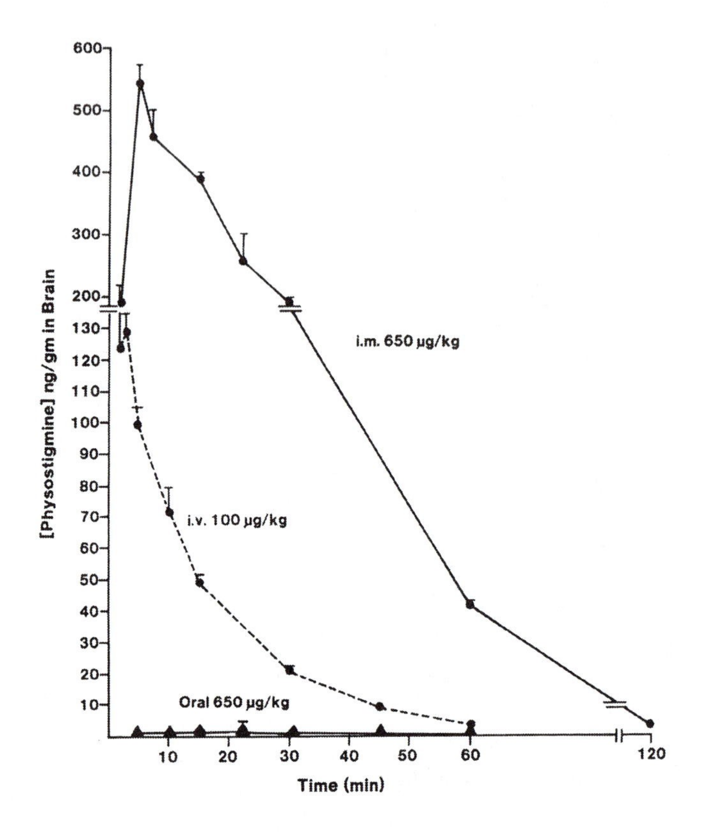

FIGURE 5.5C (C) Time course of brain physostigmine concentration.

evaluated the combination pretreatments PHY/pyridostigmine in guinea pigs challenged with 2 LD_{50} of soman. Both carbamates contributed to blood AChE inhibition. However, PHY alone seems to protect the ChE inhibition by soman as much as the protection by optimal dose of combination. The pretreatment regimen against soman, sarin, and VX intoxication in guinea pigs was studied by Lennox et al.[138] These regimens include PHY and an adjunct: aprophen, atropine, azaprophen, benactyzine, bentropine, scopolamine, or trihexyphenidyl. These investigators reported that several regimens were effective against several organophosphates. PHY subacute in conjunction with acute adjunct (scopolamine or trihexyphenidyl) is effective as pretreatment against $5LD_{50}$ of soman and 2 LD_{50} of VX in guinea pigs.[139]

Harris et al.[140] administered sustained-release PHY (0.4, 10, or 50 mg/ml) to rat at a rate of 2.5 μl/h for 28 days. The blood AChE was inhibited about 11, 42, and 66% corresponding to the above rates, respectively. These PHY dosages did not decrease the performance of rat on an accelerating rotarod. Harris et al.[140] suggested that "in a pretreatment mode, 42–66% inhibition of AChE by sustained exposure to PHY, with an acute dose of cholinolytic, would suffice to protect against lethality and motor performance decrement by a toxic level of soman.

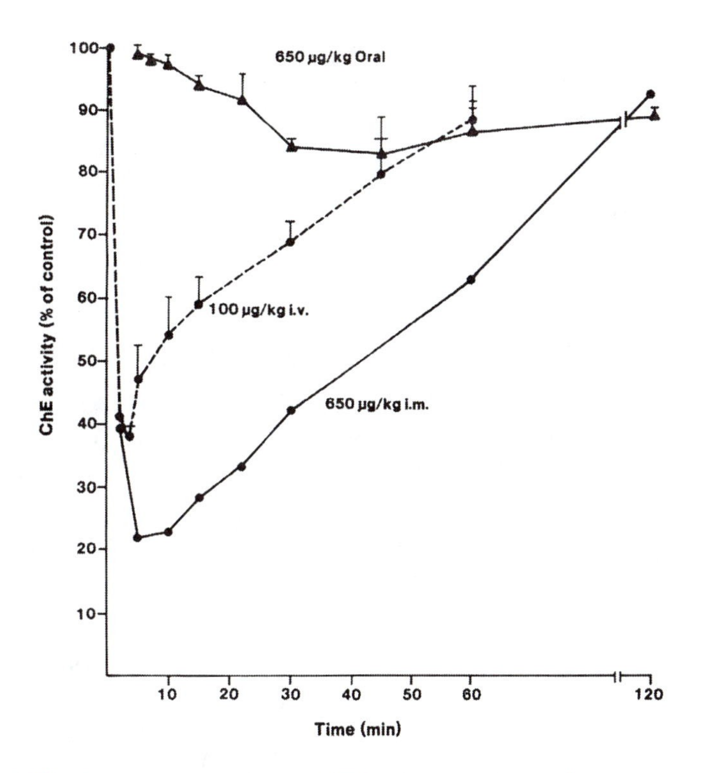

FIGURE 5.5D (D) time course of brain AChE activity (% of control).

D. Influence of Physical Stress on Pharmacokinetics and Pharmacodynamics

This section discusses the observation that exercise alters the pharmacokinetic and pharmacodynamic parameters of PHY, a flow-limited drug. Exercise increases cardiac output but diverts blood flow away from the liver[73,141] and could decrease the clearance of drugs, particularly those flow-limited drugs which are hepatic extractable, such as propranolol.[142,143] PHY is highly extracted by liver and its clearance may be dependent on hepatic blood flow.[144] A decrease in liver blood flow due to exercise will decrease the amount of PHY reaching the parenchymal cells, which in turn reduces the metabolism of PHY, thereby decreasing the clearance of PHY and increasing the area under the curve and $t_{1/2}$. The pharmacokinetics and disposition of flow-limited drugs are more likely to be affected by exercise, whereas the pharmacokinetics and disposition of capacity-limited drugs which are strongly bound and poorly extracted are less likely to be influenced by exercise.[144] Exercise also causes a plasma shift[145] which results in a decrease in plasma volume and a change in the volume of distribution.

The time course of PHY distribution is different in tissues in trained exercise rats compared to control rats.[106] Training exercise altered the time taken by different tissues to reach peak concentration of the drug plus metabolites. These results showed that the radioactivity of PHY + metabolites was higher as percent of control in brain (133%), liver (126%), heart (191%), kidney (385%), lung (106%), and muscle (180%) at 2 min post exercise in endurance-trained rats. However, radioactivity in trained rats declined below control at 5 min post exercise in kidney, muscle, brain, heart, and lung; whereas, in the liver, radioactivity declined below control at 15 min post exercise. The amount of total radioactivity in the different tissues reveals the distribution of PHY and its metabolite's affinity to different tissues. The highest amount of radioactivity accumulates in liver when compared to other tissues. Peak concentrations of RA were observed in 2 min time in heart and lung—the organs of very high blood flow. It seems the distribution of RA was dependent on blood flow. Peak concentration of RA was observed in brain at the 5-min time point and decreased within 30 min. Muscle showed peak concentration of RA only after 15 min. The blood flow to different organs changes with intensity of exercise. During exercise the arterioles in muscle will inflate, and during cessation of exercise these arterioles will return to normal condition, which will not allow higher blood flow from arterioles, thereby helping to sequester the drug in muscle mass. In plasma, RA showed a decreasing trend from the beginning. The half lives of RA in trained vs. control rats were for brain, 18 vs. 20 min; liver, 25 vs. 35 min; heart, 31 vs. 26 min; kidney, 30 vs. 28 min; lung, 26.5 vs. 30 min; and muscle, 45 vs. 31 min. There was no significant difference in $t_{1/2}$ except muscle and liver. Exercise influences the profile of distribution of RA in all tissues and pharmacokinetics of PHY. It appears that these influences may be due to the flux of blood flow after the cessation of exercise, severity of exercise, pH changes due to lactic acid production, ionization of the drug, lipid solubility, and other undetermined factors.

Acute exercise increases behavioral sensitivity of PHY.[146] Carbamate-induced decrease in performance has been shown to be restored with diazepam and atropine.[147] The combined effect of physical exercise and physostigmine on AChE activity in different tissues of rat has been extensively studied by Somani and co-workers.[148] Matthew and co-workers[149] have studied the acute and chronic administration of physostigmine on ChE inhibition and performance (endurance) of exercising rats. Acute physostigmine administration in exercising rats resulted in an inhibition of blood ChE and a reduction in endurance (performance); whereas, chronic administration attenuated the decrease in ChE activity and the endurance of exercising rats. These studies suggest that decreases in performance, caused by acute drug administration, may be attenuated through accommodation with chronic administration. Dube et al.[150] reported the interactive effects of physostigmine and exercise on cholinesterase activity in RBC and tissues of rat. The results indicate that in control rats not given physostigmine, different intensities of acute exercise affect the cholinesterase enzyme to a moderate degree in red blood cells and heart without affecting brain, diaphragm, and thigh muscles. Acute exercise modifies the effect of physostigmine by increasing the cholinesterase inhibition in red blood cells and brain without affecting other tissues.

The central and peripheral responses of rats were altered due to the interactive effect of acute exercise and endurance training in the presence of physostigmine. The data indicated that endurance training delayed ChE recovery; however, there was almost complete recovery in rats given acute exercise plus physostigmine and slower recovery in endurance training plus physostigmine as compared to physostigmine alone. Physostigmine's rate of decarbamylation of cholinesterase enzyme (K_d) due to acute and/or trained exercise in brain, heart, diaphragm, and muscle of rat have been studied.[151] Acute exercise + PHY increased, whereas endurance training + PHY decreased ChE activity in brain, red blood cells, and various tissues as compared to PHY alone. The results shown in Table 5.4 suggested that acute exercise and endurance training have opposite effects on K_d after PHY administration.

PHY and exercise have significant effects on the synthetic (ChAT) and degradative (AChE) enzymes of acetylcholine in active EDL muscle. Exercise has prolonged the inhibitory effect of PHY on ChAT and AChE activities both in active EDL and passive soleus muscles.[101] The interactive effect of PHY and concurrent acute exercise resulted in a slight decrease in ChE activity in the brain.[150] The interaction of exercise and subacute PHY decreased AChE activity in both corpus striatum and hippocampus after PHY, as well as PHY plus acute or trained exercise. AChE activity in cerebral cortex was inhibited by PHY plus exercise, (acute or trained). AChE activity decreased in the brain stem in all groups except in PHY plus acute or trained exercise rats.[148] The study indicated that PHY, exercise, or the combination of both decreased AChE activity in a regionally selective pattern. The data are consistent with the hypothesis that elevation in ACh levels down-regulates the ongoing cholinergic neurotransmission through a negative feedback mechanism.

TABLE 5.4
Effect of Acute or Trained Exercise on Rate of Decarbamylation (K_d)—in min^{-1} of ChE in RBC and Tissues of Rat

GROUP		IV	V	VI
Treatment		PHY	AE + PHY	ET + PHY
RBC	K_d min^{-1}	0.021	0.024	–
	r	0.93	0.95	–
	$t_{1/2}$ min	33.5	29.0	–
Brain	K_d min^{-1}	0.014	0.0252	0.009
	r	0.97	0.90	0.91
	$t_{1/2}$ min	50.0	27.5	75.0
Heart	K_d min^{-1}	0.019	–	0.008
	r	0.95	–	0.98
	$t_{1/2}$ min	37.5	–	85.0
Diaphragm	K_d min^{-1}	0.01	0.039	.008
	r	0.97	0.99	0.98
	$t_{1/2}$ min	67.5	17.5	84.0
Muscle	K_d min^{-1}	0.012	0.008	0.012
	r	0.91	0.89	0.79
	$t_{1/2}$ min	55.0	83.5	60.0

Note: r is the correlation coefficient for % ChE inhibition vs. time for the declining curve; $t_{1/2}$ is the half-life in min for recovery of ChE enzyme.

E. Effect of Soman on Pharmacokinetics and Pharmacodynamics

Soman is a potential irreversible cholinesterase (ChE) inhibitor. Its extreme toxicity and rapid irreversible inhibition of ChE have been studied by several authors. Poisoning by soman does not respond to treatment with a combination of atropine and an oxime. Pretreatment with a carbamate offered the possibility of devising a drug treatment that would be effective against poisoning by an organophosphate (OP) anticholinesterase, including soman. Recently, emphasis has been placed on two carbamates, physostigmine and pyridostigmine, because they are effective as pretreatment compounds in mammals for treatment of the reversal of central anticholinergic syndrome.

Pharmacokinetics of physostigmine (PHY) was compared after pretreatment with different routs of administration and then soman challenge.[152] Rats were dosed with ^{3}H-PHY (i.v., 100; i.m., 500; oral, 650 μg/kg), 5 or 15 min prior to soman (105 μg/kg s.c.; 1.5 LD_{50} or 35 μg/kg s.c. 0.5 LD_{50}) treatment and were sacrificed at various times. Pharmacokinetic parameters were determined for PHY using JANA and PC-NONLIN programs. AUC decreased from 1372 to 603 and from 4502 to 1610, indicating 56% and 64% reduction in systemic availability of PHY after i.v. and i.m. dose, respectively, in the presence of soman. Cl increased from 73 to 165 and from 111 to 310 ml min^{-1} kg^{-1} in the pretreated rats with i.v. and i.m. PHY, respectively. On the other hand, systemic availability of PHY increased by about 100% (an increase in AUC from 152 to 312), and total Cl decreased from 4254 to 2080 ml min^{-1} kg^{-1} after oral pretreatment with PHY. In the presence of soman, hepatic Cl decreased from 31.85 to 29.9 ml min^{-1} kg^{-1} and intrinsic Cl from 1592.5 to 373.7 ml $min^{-1}kg^{-1}$. PHY was slightly less metabolized in soman-challenged rats.

Time course of ^{3}H-PHY concentration and ChE activity in plasma, muscle, and brain were studied in rats pretreated with PHY and then soman challenge.[152] BuChE activity in plasma was 5% of control from 7–30 min after PHY (100 μg/kg, i.v.) pretreatment and then soman challenge (105 μg/kg, s.c.), or treatment with soman alone. Plasma PHY concentration steadily declined from 32.6 ng/ml at 7 min to 15.0 ng/ml at 30 min. ChE activity in muscle was 60–50% of control for PHY pretreatment, but soman alone gave 85–72% of control activity from 2–30 min. Brain ChE activity was about 5% of control within 2 min after soman challenge; however, with PHY pretreatment and soman challenge, the activity was about 40% at 10 min, 28% at 15 min, which recovered to 45% of control at 30 min, indicating that PHY protected brain ChE. Brain PHY concentration steadily declined from 58.6 ng/gm at 7 min to 11.7 ng/gm at 30 min. However, pretreatment of rat with a higher dose of PHY (500 μg/kg, i.m.) and then soman (105 μg/kg) challenge showed BChE in plasma and ChE activity in brain and muscle to be about 25, 30, and 62% of control in comparison to about 5% of control in plasma and brain with soman alone, indicating the protection of ChE enzyme with higher PHY pretreatment dose.[152] The protective role of PHY seen in total brain was not consistent for all brain regions. Soman alone produced a 95% ChE inhibition and there were no differences in its effect between total brain or brain areas.[152] Pretreatment of the rat with PHY produced a protective effect upon ChE activity up to 30 min.

However, after pretreatment with oral administration of PHY (650 μg/kg), the BChE activity in plasma was lowest (12.4% of control) at 20 min whereas PHY concentration was maximum (5.5 ng/ml) at 15 min. BuChE activity remained the same up to 90 min and recovered to 30% at 120 min. Brain ChE did not show any protection after oral administration.

ChE activity in total brain was 12, 30, and 24% at 5, 15, and 30 min after PHY (100 μg/kg) pretreatment with a higher dose than soman challenge (105 μg/kg, s.c.). After pretreatment with a higher dose of PHY (500 μg/kg), ChE activity was found to be 4, 13, and 19% at 5, 15, and 30 minutes. The non-significant difference in ChE activity from 100–500 μg PHY/kg might indicate that higher doses of PHY do not necessarily provide more protection of the enzyme from soman than lower doses. However, protective role of PHY seen in total brain was not consistent for all brain regions. Soman alone produced a 95% ChE inhibition and there were no differences in its effect between total brain or brain areas. Pretreatment of the rat with PHY produced a protective effect upon ChE activity up to 30 min. However, no protective effect on survival was observed.

The effects of soman challenge on ChE activity in diaphragms of rats pretreated with PHY were studied by Somani et al.[152] After an i.m. PHY dose (500 μg/kg), ChE activity was 49, 76, and 74% of control in diaphragm at 5, 20, and 45 min, respectively, which rapidly recovered to 88% at 60 min. A semilog plot of % ChE inhibition vs. time gave a 0.02 min^{-1} rate of recovery of the enzyme. ChE activity in soman (105 μg/kg, s.c.) challenged rats was about 16–22% of control from 2 to 30 min; however, with i.m. PHY (500 μg/kg) pretreatment, the activity recovered to about 35% of control at 20 min and 43% of control at 30 and 45 min. ChE activity after oral administration of PHY (650 μg/kg) was found to be 57% of control at 5 min, which rapidly recovered to about 83% at 22 min, then slowly recovered to 98% of control in 120 min. The rate of recovery of fast phase was 0.053 min^{-1} and slow phase was 0.017 min^{-1}. ChE activity in the soman (105 μg/kg, s.c.) challenged rat was found to be 16–22% of control from 2 to 30 min. However, with oral pretreatment and then soman challenge, the ChE activity was found to be 67, 76, and 87% of control at 15, 45, and 120 min, respectively. These results indicate that PHY pretreatment by oral route of administration gave some protection to diaphragm ChE.

In conclusion, pretreatment of PHY and then soman challenge decreased systemic availability of PHY after i.v. and i.m. administration. However, systemic availability of PHY after oral administration was increased in the presence of soman. Pretreatment with PHY produced protective effect upon ChE enzyme in CNS and peripheral tissues.

IV. NEOSTIGMINE

Neostigmine (NEO) is a reversible cholinesterase inhibitor that was introduced into therapeutics in 1931 due to its stimulant action on the intestinal tract. This quaternary ammonium compound has greater stability and potency compared to physostigmine and pyridostigmine. Neostigmine has been used in the treatment of myasthenia gravis for more than 60 years.[153] This drug is widely used in anesthesia to antagonize the effects of muscle relaxants after operative surgery. Early work had shown that the anticholinesterase activity of neostigmine was five times greater than that of pyridostigmine.[46] Further, an important difference between neostigmine and pyridostigmine is the inability of the latter compound to produce a direct action on smooth muscle either *in vitro*[46] or *in vivo*.[48] This may account for the occurrence of fewer unpleasant side-effects when pyridostigmine is used clinically.[49–51]

The pharmacokinetics and metabolism of neostigmine has been studied in rats after intramuscular and oral administration.[153–156] The metabolic pathway of neostigmine was elucidated as shown in Figure 5.6. 3-Hydroxyphenyltrimethylammonium (3-HPTMA) and the glucuronide of 3-HPTMA were isolated and characterized. Small amounts of 3-hydroxyphenyldimethylamine (3-HPDMA) and two unidentified metabolites were also detected. Liver is the organ of metabolism for neostigmine. This drug is rapidly taken up by the hepatic cells as shown by the liver perfusion technique.[157] The drug and its charged metabolites were retained for prolonged time within the rat liver. The formation of a glucuronide of a quaternary ammonium compound (3-HPTMA) was shown *in vitro* in rat liver microsomes supplemented with UDP-glucuronic acid.[35] However, 3HNMP did not form the glucuronide in *in vivo* as well as *in vitro* studies utilizing rat liver microsomes.

The distribution and metabolism of neostigmine in different tissues of rat were studied after acute and chronic s.c. administration of ^{14}C-neostigmine.[34] The $t_{1/2}$ of

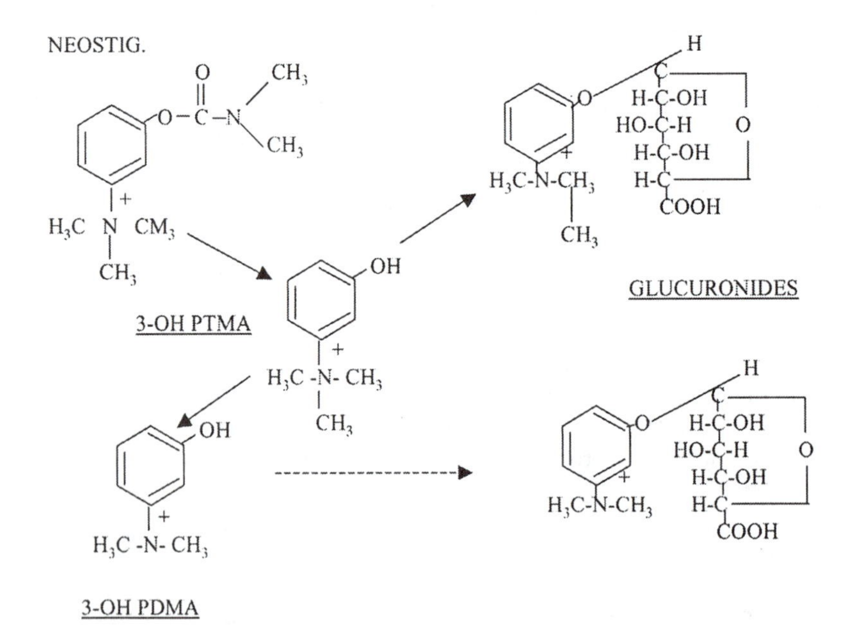

FIGURE 5.6 Probable metabolic pathways of neostigmine. (Adapted from Somani et al.[153])

neostigmine and its metabolites on average is 10 min in plasma, 33 min in liver, and 1.5 h in muscle. Neostigmine is metabolized in the liver at the rate of 2.24 H10^{-2} μmol/min/g and 3-HPTMA is metabolized at the rate of 2.89 H10^{-2} μmol/min/g. Metabolic degradation of neostigmine proceeds in muscle at the rate of 2.1 H10^{-2} μmol/g. During chronic administration, the concentration of neostigmine and its metabolites rose in the liver between days 1 to 8 from 0.63 to 5.89 μmol/g of tissue and in muscle from 0.085 to 0.39 μmol/g of tissue. Liver contained the highest concentration of glucuronide of 3-HPTMA (G-3-HPTMA) followed by neostigmine and 3-HPTMA. G-3-HPTMA concentration was 0.321 μmol on day 1 and increased to 2.89 μmol/g liver on day 8. The neostigmine concentration increased from 0.14 to 1.62 μmol/g of liver during the same period. Similar increases in the concentration of neostigmine and its metabolites were also observed in the muscle under the same conditions. 3-HPDMA, a probable metabolite and an unknown metabolite, also consistently increased in the liver and muscle.[34] Thus, the liver is a significant reservoir for neostigmine and G-3-HPTMA. Neostigmine and its major metabolites also accumulate in the muscle. Muscle and cartilage tissues contain chondritin sulfate, a negatively charged macromolecule, and a constituent of mucopolysaccharide. The binding of quaternary amines to chondroitin sulfate was carried out *in vitro* by ultracentrifugation techniques using varying concentrations of ^{14}C-neostigmine from 1.79 H 10^{-8} to 1.43 H 10^{-7} mol; ^{14}C-HPTMA, 3.59 H 10^{-8} mol; and ^{14}C-pyridostigmine from 7.66 H 10^{-9} to 3.8 H 10^{-7} mol. Increasing concentrations of NEO, HPTMA, and pyridostigmine bind in an increasing amount to chondroitin sulfate (Somani, unpublished data).

Neostigmine kinetics and metabolism were studied after i.m. administration in 8 patients with myasthenia gravis.[158] The plasma neostigmine level declined monoexponentially from 21 ± 2 to 9 ± 1 ng/ml between 30 and 120 min. Estimates of plasma half-life ($t_{1/2}$) ranged from 51.1 to 90.5 min; apparent volume of distribution varied from 32.0 to 60.6 1 l/kg; and total body clearance from 434 to 549 ml/min. Approximately 80% of the drug was eliminated in urine within 24 h either unchanged or as metabolites. Approximately 50% of the dose was eliminated as the unchanged drug, 15% as 3-hydroxyphenyltrimethylammonium, and 15% as other unidentified metabolites. The neostigmine $t_{1/2}$, based on the urinary excretion of the unchanged drug, ranged from 90.2 to 118.7 min. Therefore, neostigmine is eliminated by renal and extrarenal mechanisms. Calvey et al.[159] have shown that the elimination of neostigmine and its metabolites also occurs via bile, this being the secondary route after urine.

The clinical pharmacology and kinetic interaction of neostigmine and pyridostigmine has been evaluated in patients with myasthenia gravis.[18,39] These two drugs showed similar pharmacokinetic profiles with plasma $t_{1/2}$ of 0.9 and 1.4 h for neostigmine and pyridostigmine, respectively. The oral bioavailability was however higher for pyridostigmine (7.6%) compared to neostigmine (2%). Aquilonius et al.[18] observed no pharmacokinetic interaction between neostigmine and pyridostigmine in five myasthenic patients, when these drugs were given in combination by the oral route. On the contrary, another study suggested that neostigmine might interfere with the bioavailability of pyridostigmine when both drugs are administered orally at the same time.[39] However, a combination of i.m. and oral routes of these quaternary amines may be more advantageous in the treatment of myasthenia gravis.

Neostigmine and PB have similar chemical structure and pharmacological effects. However, neostigmine is more potent and is metabolized extensively to an unusual glucuronide metabolite. Pyridostigmine does not form the glucuronide conjugate and is metabolized in the liver to its major metabolite 3-hydroxy-N-methyl pyridinium. Further, NEO sequesters in the liver, whereas, PB does not. It is thought that the differences in the distribution and metabolism of these two drugs play a role in their duration of action and influence the pharmacodynamic effects after single and multiple dosages.

V. SUMMARY

Pyridostigmine bromide, a peripheral anticholinesterase drug, was used for the first time on a mass scale by military personnel during the Persian Gulf War as a protective measure against possible nerve gas exposure. The soldiers took PB 30 mg tablets three times a day for 2 weeks. Gulf War veterans reported various illnesses, months and years after the war. PB may be implicated in the etiology of Gulf War illnesses. This has led to an increase in research with PB and its interaction with other factors such as the environment, chemicals, and possible low-level exposure to nerve gas. This chapter deals with the pharmacokinetics and pharmacodynamics of carbamates such as PB, physostigmine, and neostigmine. The absorption, distribution, metabolism, and excretion of PB has been reported extensively in various animal species and human beings. The pharmacokinetics of PB plays an important role in determining the pharmacodynamic effects in normal, disease, or stressful conditions, and in the presence of chemicals and low-level nerve gas exposure. The pharmacodynamic effects and toxicity of PB have clearly shown the importance of dose, time of administration, and disposition of this drug, which could determine its protective efficacy against nerve gases. However, there is little information on pharmacokinetics and phamacodynamics of PB under stressful conditions and also with respect to gender and age. This area of research needs further investigation. Although PB was prescribed during the Persian Gulf War, this chapter also discusses PHY, a centrally acting carbamate that also has potential as a pretreatment drug against nerve agents. PHY does not attain an effective concentration in the brain after oral administration due to its first-pass effect. However, this drug was found to be efficacious centrally and peripherally after i.v. and i.m. administration. Physical stress influences the pharmacokinetics and pharmacodynamics of this drug. The pharmacokinetics and metabolism of neostigmine (a close congener of PB) has also been compared with PB. Neostigmine is more potent pharmacodynamically, and its metabolite forms a unique glucuronide of quaternary amine and is sequestered in the liver. The metabolite of PB does not form the glucuronide. Environmental factors play a role in altering pharmacokinetics and pharmacodynamics of PB and PHY. Physical stress seems to enhance the pharmacodynamic effects of PB. Therefore, it may be necessary to titrate the dosage of PB under stressful conditions to enable its safe and effective use.

ACKNOWLEDGMENTS

The authors thank Judith M. Bryan for technical support in preparation of this manuscript. The authors sincerely acknowledge their gratefulness to Drs. James A. Romano, Brian J. Lukey, and Benedict Capacio of the U.S. Army Medical Research Institute of Chemical Defense, 3100 Ricketts Point Road, Aberdeen Proving Ground, MD 21010–5400, and Dr. David A. Gelber, Associate Professor of Neurology, Department of Neurology, Southern Illinois University School of Medicine, Springfield, IL 62702, for their thorough review of this chapter.

REFERENCES

1. Urban, R., Pyridostigmine, U.S. Patent 2, 1951. *Pharmacol. Ther.,* 13(3), 393, 1972.
2. Koelle, G.B., Protection of cholinesterase against irreversible inactivation by DFP *in vitro, J. Pharmacol. Exp. Ther.,* 88, 323, 1946.
3. Koster, R., Synergism and antagonisms between physostigmine and ci-isopropylfluorophosphate in cats, *J. Pharmacol. Exp. Ther.,* 88, 39, 1946.
4. Bergman, F. and Shimona, A., Quaternary ammonium salts as inhibitors of acetylcholine release, *Biochem. Biophys. Acta.,* 8, 520, 1952.
5. Jobst, J. and Hesse, O., Uber die bohne von calabar, *Ann. Chem. Pharm.,* 129, 115, 1864.
6. Vee, M. and LeVen, M., De L'alcaloide de la feve de calabar et experiences physiologiques avec ce meme alcaloide, *J. Pharm Chimie,* 1, 70, 1865.
7. Aeschlimann, J.A. and Reinert, M., Pharmacological action of some analogs of physostigmine, *J. Pharmacol. Exp. Ther.,* 43, 413, 1931.
8. Almog, S., Winkler, E., Amitai, Y., Dani, S., Shefi, M., Tirosh, M., and Shemer, J., Acute pyridostigmine overdose: A report of nine cases, *Isr. J. Med. Sci.,* 27, 659, 1991.
9. Park, K.H., Kim, D.E., Arnold, T.W., Oh, S.J., and Bradley, R., Pyridostigmine toxicity electrophysiological study, *Electromyogr. Clin. Neurophysiol.,* 33, 323, 1993.
10. Kluwe, W.M., Page, J.G., Toft, J.D., Ridder, W.E., and Chung, H., Pharmacological and toxicological evaluation of orally administered pyridostigmine in dogs, *Fundam. Appl. Toxicol.,* 14, 40, 1990.
11. Keeler, J.R., Hurst, C.G., and Dunn, M.A., Pyridostigmine used as a nerve agent pretreatment under wartime conditions, *JAMA,* 266, 693, 1991.
12. Institute of Medicine, Health consequences of service during the Persian Gulf War: Initial findings and recommendations for immediate action, National Academy Press, Washington, D.C., 1992.
13. Cook, J. and Kolka, M., Chronic pyridostigmine bromide administration: Side effects among soldiers working in a desert environment, *Mil. Med.,* 157, 250, 1992.
14. Levine, B.S., Long, R., and Chung, H., Subchronic oral toxicity of pyridostigmine bromide in rats, *Biomed. Environ. Sci.,* 4, 283, 1991.
15. Caldwell, R.W., Lowensohn, H.S., Chryssanthis, M.A., and Nash, C.B., Interactions of pyridostigmine with cardiopulmonary systems and their relationships to plasma cholinesterase activity, *Fundam. Appl. Toxicol.,* 12, 432, 1989.
16. Sorensen, P.S., Flachs, H., Friis, M.L., Hvidberg, E.F., and Paulson, O.B., Steady state kinetics of pyridostigmine in myasthenia gravis, *Neurology,* 34, 1020, 1984.
17. Breyer-Pfaff, U., Maier U., Brinkmann, A.M., and Schumm, F., Pyridostigmine kinetics in healthy subjects and patients with myasthenia gravis, *Clin. Pharmacol. Thera.,* 37, 495, 1985.

18. Aquilonius, S.M., Eckernas, S.A., Hartvig, P., Lindstrom, B., and Osterman, P.O., Clinical pharmacology of neostigmine and pyridostigmine in patients with myasthenia gravis, *J. Neurol. Neurosurg. Psychiat.*, 46, 929, 1983.
19. Cohan, S.L., Drettchen, K.L., and Neal, A., Malabsorption of pyridostigmine in patients with myasthenia gravis, *Neurology,* 27, 299, 1977.
20. Aquilonius, S.M. and Hartvig, P., Clinical pharmacokinetics of cholinesterase inhibitors, *Clin. Pharmacokinetics,* 11, 236, 1986.
21. Aquilonius, S.M., Eckernas, S.A., Hartvig, P., Lindstrom, B., and Osterman, P.O., Pharmacokinetics and oral bioavailability of pyridostigmine in man, *Eur. J. Clin. Pharmacol.,* 18, 423, 1980.
22. Birtley, R.D., Roberts, J.B., Thomas, B.H., and Wilson, A., Excretion and metabolism of ^{14}C-pyridostigmine in the rat, *Br. J. Pharmacol.,* 26(2), 393, 1966.
23. Cohan, S.L., Pohlman, J.L.W., Mikszcwski, J., and O'Doherty, D.S., The pharmacokinetics of pyridostigmine, *Neurology,* 26, 536, 1976.
24. Miller, R.L. and Verma, P., Radio immuno assay of pyridostigmine in plasma and tissues, *Pharmacol. Res.,* 21(4), 359, 1989.
25. Yamamoto, K., Sawada, Y., and Iga, T., Comparative pharmacokinetics of four cholinesterase inhibitors in rats, *Biol. Pharmaceut. Bull.,* 18, 1292, 1995.
26. Eiermann, B., Sommer, N., Winne, D., Schumm, F., Maier, U., and Breyer-Pfaff, U., Renal clearance of pyridostigmine in myasthenic patients and volunteers under the influence of ranitidine and pirenzepine, *Xenobiotica,* 23, 1263, 1993.
27. Telang, F.W., Ding, Y.-S., Volkow, N.D., Molina, P.E., and Gatley, S.J., Letter to the Editor, *Nucl. Med. Biol.,* 26, 249, 1999.
28. Sharma, H.S., Navarro, J.C., and Dey, P.K., Increased blood brain barrier permeability following acute short-term swimming exercise in conscious normotensive rats, *Neurosci. Res.,* 10, 211, 1991.
29. Kornfeld, P., Samuels, A.J., Wolf, R.L., and Osserman, K.E., Metabolism of ^{14}C-labelled pyridostigmine in myasthenia gravis. Evidence for multiple metabolites, *Neurology,* 26, 634, 1970.
30. Somani, S.M., Roberts, J.B., and Wilson, A., Pyridostigmine metabolism in man, *Clin. Pharmacol. Therap.,* 393, 13, 1972.
31. Somani, S.M., Metabolism and pharmacokinetics of pyridostigmine in rat after multiple dosing, *Pharmacologist,* 25, 97, 1983.
32. Howrath, R.D., Lamberton, A.H., and Woodcock, D., Investigations on the influence of chemical constitution upon toxicity. Part II. Compounds related to "prostigmine," *J. Chem. Soc.,* Part 1, 182, 1947.
33. Golomb, B.A., *Pyridostigmine Bromide,* Vol. 2, Rand Corporation, 1999, 1.
34. Somani, S.M., Distribution of neostigmine and its metabolites in rat tissues after acute and chronic administration, *Eur. J. Pharmacol.,* 30, 336, 1975.
35. Somani, S.M. and Anderson, J.H., *In vitro* glucuronization of 3-hydroxy-phenyl-trimethyl ammonium by rat liver microsomes, *Drug Metabol. Dispos.,* 5, 15, 1977.
36. Fromberz, K. and Pellmont, B., Pharmakologische wirkung des mestinon, *"Roche" Schweiz. Med. Wschr.,* 49, 1187, 1953.
37. Husain, M.A., Roberts, J.B., Thomas, B.H., and Wilson, A., The excretion and metabolism of oral ^{14}C-pyridostigmine in the rat, *Br. J. Pharmacol.,* 34(2), 445, 1968.
38. Burdfield, P.A., Calvery, T.N., and Roberts, T.B., *In vitro* metabolism of neostigmine and pyridostigmine, *J. Pharm. Pharmacol.,* 25(4), 428, 1973.
39. Chan, K., Davison, S.C., Dehghan, A., and Hyman, N., The effect of neostigmine on pyridostigmine bioavailability in myasthenic patients after oral administration, *Meth. Find. Exper. Clin. Pharmacol.,* 3, 291, 1981.

40. Calvey, T.N., Chan, K., and Dehghan, A., Kinetics of intravenous pyridostigmine in man, *Br. J. Clin. Pharmacol.,* 11, 406, 1981.
41. Meyer, H.G., Lukey, B.J., Gepp, R.T., Corpuz, R.P., and Lieske, C.N., A radioimmunoassay for pyridostigmine, *J. Pharmacol. Exper. Therapeu.,* 247, 432, 1988.
42. Taylor, T., Hawkins, D.R., Forest, T.J., and Chung, H., Pharmacokinetics of pyridostigmine in dogs. *J. Pharmaceut. Sci.,* 80, 353, 1991.
43. Cronnelly, R., Stanski, D.R., Miller, R.D., and Sheiner, L.B., Pyridostigmine kinetics with and without renal function, *Clin. Pharmacol. Therapeu.,* 28, 78, 1980.
44. Marino, M.T., Schuster, B.G., Brueckner, R.P., Lin, E., Kaminskis, A., and Lasseter, K.C., Population pharmacokinetics and pharmacodynamics of pyridostigmine bromide for prophylaxis against nerve agents in humans, *J. Clin. Pharmacol.,* 38, 227, 1998.
45. Somani, S.M. and Kamemori, G.H., Exercise and absorption, distribution, metabolism, excretion and pharmacokinetics of drugs and chemicals, in *Pharmacology in Exercise and Sports,* Somani, S.M., ed., CRC Press, Inc., Boca Raton, FL, 1996, 1.
46. Hobbiger, F., The action of carbamic esters and tetraethylpyrophosphate on normal and curarized frog rectus muscle, *Br. J. Pharmacol. Chemother.,* 5, 37, 1950.
47. Blaschko, H., Bulbring, E., and Chou, T.C., Tubocurarine antagonism and inhibition of cholinesterase, *Br. J. Pharmacol. Chemother.,* 4, 29, 1949.
48. Desmedt, J.E. and LaGrutta, G., Sur le mode d'action de l'ester dimethyl carbamique de la 3-hydrosy-methyl pyridine (Mestinon), *Rev. Neurol.,* 91, 457, 1954.
49. Tether, J.E., Mestinon in myasthenia gravis (Preliminary report), *Dis. Nervous Syst.,* 15, 227, 1954.
50. Osserman, K.E., Teng, P., and Kaplan, L.I., Studies in myasthenia gravis, Preliminary report on therapy with mestinon, *JAMA,* 155, 961, 1954.
51. Schwab, R.S. and Timberlake, W.H., Pyridostigmine (mestinon) in the treatment of myasthenia gravis, *NEJM,* 251, 271, 1954.
52. Foldes, F.F. and Smith, J.C., The interaction of human cholinesterases with anticholinesterases used in the therapy of myasthenia gravis, *Ann. N.Y. Acad. Sci.,* 135, 287, 1964.
53. Dirnhuber, P. and Green, D.M., Effectiveness of pyridostigmine in reversing neuromuscular blockade produced by soman, *J. Pharm. Pharmac.,* 30, 419, 1978.
54. Dirnhuber, P., French, M.C., Green, D.M., Leadbeater, L., and Stratton, J.A., The protection of primates against soman poisoning by pretreatment with pyridostigmine, *Pharm. Pharmacol.,* 31, 295, 1979.
55. Lennox, W.J., Harris, L.W., Talbot, B.G., and Anderson, D.R., Relationship between reversible acetylcholinesterase inhibition and efficacy against soman lethality, *Life Sci.,* 37, 793, 1985.
56. Anderson, R.J., Chamberlain, W.L., Roesner, M., Dacko, C., and Robertson, D.G., Decreased tetanic contracture of rat skeletal muscle induced by pyridostigmine, *J. Toxicol. Environ. Health,* 18, 221, 1986.
57. Gebbers, J.-O., Lotscher, M., Kobel, W., Portmann, R., and Laissue, J.-A., Acute toxicity of pyridostigmine in rats: Histological findings. *Arch. Toxicol.,* 58, 271, 1986.
58. Bowman, P.D., Schuschereba, S.T., Johnson, T.W., Woo, F.J., McKinney, L., Wheeler, C.R., Frost, D., and Korte, D.W., Myopathic changes in diaphragm of rats fed pyridostigmine bromide subchronically, *Fundam. Appl. Toxicol.,* 13, 110, 1989.
59. Blick, D.W., Murphy, M.R., Brown, G.C., Yochmowitz, M.G., Fanton, J.W., and Hartgraves, S.L. Acute behavioral toxicity of pyridostigmine or soman in primates, *Toxicol. Appl. Pharmacol.,* 126, 311, 1994.

60. Glikson, M., Achiron, A., Ram, Z., Ayalon, A., Karni, A., Sarova-Pinchas, I., Glovinski, J., and Revah, M., The influence of pyridostigmine administration on human neuromuscular functions—Studies in healthy human subjects. *Fund. Appl. Toxicol.,* 16, 288, 1991.
61. Von Bredow, J.D., Adams, N.L., Groff, W.A., and Vick, J.A., Effectiveness of oral pyridostigmine pretreatment and cholinolytic-oxime therapy against soman intoxication in nonhuman primates, *Fund. Appl. Toxicol.,* 17, 761, 1991.
62. Blick, D.W., Kerenyi, S.Z., Miller, S., Murphy, M.R., Brown, G.C., and Hartgraves, S.L., Behavioral toxicity of anticholinesterases in primates: Chronic pyridostigmine and soman interactions, *Pharmacol. Biochem. Behav.,* 38, 526, 1991.
63. Vijayaraghavan, R., Husain, K., Kumar, P., Pandey, K.S., and Das Gupta, S., Time dependent protection by carbamates against inhaled sarin aerosols in rats. *Asia Pac. J. Pharmacol.,* 7, 257, 1992.
64. Koplovitz, I., Harris, L.W., Anderson, D.R., Lennox, W.J., and Stewart, J.R., Reduction by pyridostigmine pretreatment of the efficacy of atropine and 2-PAM treatment of sarin and VX poisoning in rodents, *Fund. Appl. Toxicol.,* 18, 102, 1992.
65. Kluwe, W.M., Efficacy of pyridostigmine against soman intoxication in a private mode, in *Proceedings of the 6th Medical Chemical Defense Bioscience Review,* Aberdeen Proving Ground, MD, U.S. Army Medical Research Institute of Chemical Defense, 1987, 227.
66. Worek, F., Kleine, A., and Szinicz, L., Effect of pyridostigmine pretreatment on cardiorespiratory function in tabun poisoning, *Human Exper. Toxicol.,* 14, 634, 1995.
67. Lintern, M.C., Smith, M.E., and Ferry, C.B., Effects of repeated treatment with pyridostigmine on acetylcholinesterase in mouse muscles, *Human Exp. Toxicol.,* 16, 158, 1997.
68. Husain, K., Vijayaraghavan, R., and Marjit, D.N., Effect of pyridostigmine and physostigmine against acute toxicity of inhaled DFP in rats, *Arch. Ind. Hyg. Toxicol.,* 41, 19, 1990.
69. Hubert, M. and Lison, D., Study of muscular effects of short-term pyridostigmine treatment in resting and exercising rats, *Human Exp. Toxicol.,* 14, 49, 1995.
70. Somani, S.M., Husain, K., Asha, T., and Helfert, R., Interactive and delayed effects of pyridostigmine and physical stress on biochemical and histopathological changes in peripheral tissues of mice, *J. Appl. Toxicol.,* 20, 327, 2000.
71. Augustinsson, K.B. and Nachmansohn, D., Distinction between acetylcholinesterase and other choline ester splitting enzymes, *Science,* 110, 98, 1949.
72. Wade, O.L. and Bishop, J.M., *Cardiac Output and Regional Blood Flow,* Blackwell, Oxford, 1962.
73. Rowell, L.B., Blackmon, J.R., and Bruce, R.A., Indocyanine green clearance and estimated hepatic blood flow during mild to maximal exercise in upright man, *J. Clin. Invest.,* 43, 1677, 1964.
74. Ballard, B.E., Pharmacokinetics and temperature, *J. Pharm. Sci.,* 63, 1345–1357, 1974.
75. Francesconi, R., Hubbard, R., and Mager, M., Effects of pyridostigmine on ability of rats to work in the heat. *J. Appl. Physiol.: Respirat. Environ. Exercise Physiol.,* 56, 891, 1984.
76. Avlonitou, E. and Elizondo, R., Effects of atropine and pyridostigmine in heat-stressed patas monkeys, *Aviat. Space Environ. Med.,* 59, 544, 1988.
77. Friedman, A., Kaufer, D., Shemer, J., Hendler, I., Soreq, H., and Tur-Kaspa, I., Pyridostigmine brain penetration under stress enhances neuronal excitability and induces early immediate transcriptional response, *Nat. Medicine,* 2, 1382, 1996.
78. Verma-Ahuja, S., Husain, K., Verhulst, S., Espinosa, J.A., and Somani, S.M., Delayed effects of pyridostigmine and exercise training on muscle tension in mouse lower extremity, *FASEB J.,* 818, 5, 1999.

79. Kaiser, K.S., Hawksworth, A.W., and Gray, G.G., Pyridostigmine bromide intake during the Persian Gulf War is not associated with postwar handgrip strength, *Mil. Med.,* 165, 165, 2000.
80. Lallement, G., Foquin, A., Baubichon, D., Burckhart, M.-F., Carpentier, P., and Canini, F., Heat stress, even extreme, does not induce penetration of pyridostigmine into the brain of guinea pigs, *Neurotoxicology,* 19, 759, 1998.
81. Epstein, Y., Seidman, D.S., Moran, D., Arnon, R., Arad, M., and Varssano, D., Heat-exercise performance of pyridostigmine-treated subjects wearing chemical protective clothing, *Aviat. Space Environ. Med.,* 61, 310, 1990.
82. Kolka, M.A. and Stephenson, M.S., Human temperature regulation during exercise and after oral pyridostigmine administration, *Aviat. Space Environ. Med.,* 61, 220, 1990.
83. Ram, Z., Molcho, M., Danon, Y.L., Almog, S., Baniel, A.K., and Shemer, J., The effect of pyridostigmine on respiratory function in healthy and asthmatic volunteers, *Isr. J. Med. Sci.,* 27, 664, 1991.
84. Wenger, C.B. and Latzka, W.A., Effects of pyridostigmine bromide on physiological responses to heat, exercise and hypohydration, *Aviat. Space Environ. Med.,* 63, 37, 1992.
85. Wenger, B., Quigley, M.S., and Kolka, M.A., Seven-day pyridostigmine administration and thermoregulation during rest and exercise in dry heat, *Aviat. Space Environ. Med.,* 64, 905, 1993.
86. Forster, E.M., Forster, J.S., Barber, B.A., Parker, Jr., F.R., Whinnery, J.E., Burton, R.R., and Boll, P., Effect of pyridostigmine bromide on acceleration tolerance and performance, *Aviat. Space Environ. Med.,* 65, 110, 1994.
87. Prusaczyk, W.K. and Sawka, M.N., Effects of pyridostigmine bromide on human thermoregulation during cold water immersion, *J. Appl. Physiol.,* 71, 432, 1991.
88. Roberts, D.E., Sawka, M.N., Young, A.J., and Freund, B.J., Pyridostigmine bromide does not alter thermoregulation during exercise in cold air, *Can. J. Physiol. Pharmacol.,* 72, 788, 1994.
89. Sharabi, Y., Danon, Y.L., Berkenstadt, H., Almog, S., Mimouni-Bloch, A., Zisman, A., Dani, S., and Atsmon, Survey of symptoms following intake of pyridostigmine during the Persian Gulf War, *Isr. J. Med. Sci.,* 27, 656, 1991.
90. Brooks, G.A. and Fahey, T.N., *Exercise Physiology,* John Wiley & Sons, New York, 1984, 726.
91. Connolly, R.J., Flow patterns in the capillary bed of rat skeletal muscle at rest and after repetitive tetanic contraction, in *Microcirculation,* Grayson, J. and Zingg, W., Eds., Plenum Press, New York, 1976.
92. Sahlin, K., Intracellular pH and energy metabolism in skeletal muscle of man with special reference to exercise, *Acta Physiol. Scand. Suppl.,* 455, 1, 1978.
93. Hughson, R.L. and Green, H.J., Blood acid-base and lactate relationships studies by ramp work tests, *Med. Sci. Sports Exer.,* 14, 297, 1982.
94. Day, R.E., Effects of exercise performance on drugs used in musculoskeletal disorders, *Med. Sci. Sports Exer.,* 13, 272, 1981.
95. Schwartz, G., Estimating the dimension of a model, *Ann. Stat.,* 6, 461, 1978.
96. Somani, S.M., ed., *Pharmacology in Exercise and Sports,* CRC Press, Inc., Boca Raton, FL, 1996, 1.
97. Powers, S.K., Criswell, D., Lawler, J., Martin, D., Lieu, F., Ji, L.L., and Herb, R.A., Rigorous exercise training increases superoxide dismutase activity in ventricular myocardium, *Am. J. Physiol.,* 34, 2094, 1993.
98. Husain, K. and Somani, S.M., Influence of exercise and ethanol on cholinesterase activity and lipid peroxidation in blood and brain regions of rat, *Prog. Neuro-Psychopharmacol. Biol. Psychiat.,* 21, 659, 1997.

99. Husain K. and Somani, S.M., Effect of exercise training and chronic ethanol ingestion on cholinesterase activity and lipid peroxidation in blood and brain regions of rat, *Prog. Neuro-Psychopharmacol. Biol. Psychiat.,* 22, 411, 1998.
100. Ott, P., Membrane acetylcholinesterases: Purification, molecular properties and interactions with amphyphilic environments, *Biochem. Biophys. Acta,* 822, 375, 1985.
101. Babu, S.R., Somani, S.M., and Dube, S.N., Effect of physostigmine and exercise on choline acetyltransferase and acetylcholinesterase activities in fast and slow muscles of rat, *Pharmacol. Biochem. Behav.,* 45, 713, 1993.
102. Peden-Adams, M.M., Dudley, A.C., EuDaly, J.G., Gilkeson, G.S., and Keil, D.F., Effects of exercise stress on pyridostigmine bromide on immune function parameters in mice, *Toxicologist,* 54, 162, 2000.
103. Abou-Donia, M.B., Wilmarth, K.R., Jensen, K.F., Oehme, F.W., and Kurt, T.L., Neurotoxicity resulting from coexposure to pyridostigmine bromide, DEET, and permethrin: Implications of Gulf War chemical exposures, *J. Toxicol. Environ. Health,* 48, 35, 1996.
104. Husain, K. and Somani, S.M., Influence of physical stress and pyridostigmine on cholinesterase activity in blood and brain regions of male and female mice, *FASEB J.,* 818, 1999.
105. Stone, J.G., Matteo, R.S., Ornstein, E., Schwartz, A.E., Ostapkovich, N., Jamdar, S.C., and Diaz, J., Aging alters the pharmacokinetics of pyridostigmine, *Anes. Analgesia,* 81, 773, 1995.
106. Somani, S.M. and Khalique, A., Pharmacokinetics and pharmacodynamics of physostigmine in the rat after intravenous administration, *Drug Metab. Dispos.,* 15, 627, 1987.
107. Somani, S.M. and Khalique, A., Distribution and pharmacokinetics of physostigmine in rat after intramuscular administration, *Fundam. Appl. Toxicol.,* 6, 327, 1986.
108. Giacobini, E., Somani, S.M., McIlhany, M., Downen, A., and Hallak, M., Pharmacokinetics and pharmacodynamics of physostigmine after i.v. administration in beagle dogs, *Neuropharmacology,* 26, 831, 1987.
109. Hartvig, L., Wiklund, and Lindstrom, B., Pharmacokinetics of physostigmine after intravenous, intramuscular and subcutaneous administration in surgical patients, *Acta Anaesthesiol. Scand.,* 30, 177, 1986.
110. Lukey, B.J., Parrish, J.H., Marlow, D.D., Clark, C.R., and Sidell, F.R., Pharmacokinetics of physostigmine intramuscularly administered to guinea pigs, *J. Pharm. Sci.,* 79, 796, 1990.
111. Unni, L.K. and Somani, S.M., Hepatic and muscle clearance of physostigmine in the rat, *Drug Metab. Dispos.,* 14, 183, 1986.
112. Somani, S.M. and Boyer, A., *Eur. J. Drug Metab. Pharmacokin.,* 10, 343, 1985.
113. Johansson, M. and Nordberg, A., Pharmacokinetic studies of cholinesterase inhibitors, *Acta Neuro. Scand.,* S149, 22, 1993.
114. Asthana, S., Greig, N.H., Hegedus, L., Holloway, H.H., Raffaele, K.C., Schapiro, M.B., and Soncrant, T.T., Clinical pharmacokinetics of physostigmine in patients with Alzheimer's disease, *Clin. Pharmacol. Thera.,* 58, 299, 1995.
115. Somani, S.M., Pharmacokinetics and pharmacodynamics of physostigmine in the rat after oral administration, *Biopharm. Drug Dispos.,* 10, 187, 1989.
116. King, B.F. and Somani, S.M., Distribution of physostigmine and metabolites in brain subcellular fractions of the rat, *Life Sci.,* 41, 2007, 1987.
117. Unni, L.K. and Somani, S.M., Binding of physostigmine to rat and human plasma and crystalline serum albumins, *Life Sci.,* 36, 1389, 1985.
118. Somani, S.M., Unni, L.K., and McFadden, D.L., Drug interaction for plasma protein binding: Physostigmine and other drugs, *Int. J. Clin. Pharmacol. Ther. Toxicol.,* 25, 412, 1987.

119. Whelpton, R. and Hurst, P. R., The binding of physostigmine to human serum albumin, *J. Pharm. Pharmacol.,* 42, 804, 1990.
120. Somani, S.M., Gupta, S.K., Khalique, A., and Unni, L.K., Physiological pharmacokinetic and pharmacodynamic model of physostigmine in the rat, *Drug Metab. Disp.,* 19, 655, 1991.
121. Hemsworth, B.A. and West, G.B., Anticholinesterase activity of some degradation products of physostigmine, *J. Pharm. Sci.,* 59, 118, 1970.
122. Isaksson, K. and Kissinger, P.T., Metabolism of physostigmine in mouse liver microsomal incubations studied by liquid chromatography with dual-electrode amperometric detection, *J. Chromatog.,* 419, 165, 1987.
123. Somani, S.M., Kutty, R.K., and Krishna, G., Eseroline, a metabolite of physostigmine, induces neuronal cell death, *Toxicol. Appl. Pharmacol.,* 106, 28, 1990.
124. Fleisher, J.H. and Harris, L.W., Dealkylation as a mechanism for aging of cholinesterase after poisoning with pinacolyl methylphosphonofluoridate, *Biochem. Pharmacol.,* 14, 641, 1965.
125. Berry, W.K. and Davies, D.R., The use of carbamates and atropine in the protection of animals against poisoning by 1,2,,2-trimethylpropylmethyl phosphonofluoridate, *Biochem. Pharmacol.,* 19, 927, 1970.
126. Heyl, W.C., Harris, L.W., and Stitcher, D.L., Effects of carbamates on whole blood cholinesterase activity: Chemical protection against soman, *Drug Chem. Toxicol.,* 3, 319, 1980.
127. McNamara, B.P., Koelle, G.B., and Gilman, A., The treatment of diisopropyl fluorophosphate (DFP) poisoning in rabbits, *J. Pharmacol. Exp. Ther.,* 88, 27, 1946.
128. Schoene, K., Steinhanses, J., and Oldiges, M., Protective activity of pyridinium salts against soman poisoning *in vivo* and *in vitro, Biochem. Pharmacol.,* 25, 1955, 1976.
129. Gordon, J.J., Leadbeater, L., and Maidment, M.P., The protection of animals against organophosphate poisoning by pretreatment with a carbamate, *Toxicol. Appl. Pharmacol.,* 43, 207, 1976.
130. Ashani, Y., Leader, H., Raveh, L., Bruckstein, R., and Spiegelstein, M., *In vitro* and *in vivo* protection of acetylcholinesterase against organophosphate poisoning by pretreatment with a novel derivative of 1,3,1-diolaphosphoriname 2-oxide, *J. Med. Chem.,* 26, 145, 1983.
131. Harris, L.W., Heyl, W.C., Stitcher, D.L., and Moore, R.D., The effect of atropine and/or physostigmine on cerebral acetylcholine in rats poisoned with soman, *Life Sci.,* 22, 907, 1978.
132. Harris, L.W., Stitcher, D.W., and Heyl, W.C., The effects of pretreatments with carbamates, atropine and mecamylamine on survival and on soman induced alterations in rat and brain acetylcholine, *Life Sci.,* 26, 1885, 1980.
133. Harris, L.W., Lennox, W.J., and Talbot, B.G., Toxicity of anticholinesterase: Interactions of pyridostigmine and physostigmine with soman, *Drug Chem. Toxicol.,* 7, 507, 1984.
134. Inns, R.H. and Leadbeater, L., The efficacy of bispyridinium derivatives in the treatment or organophosphate poisoning in the guinea pig, *J. Pharm. Pharmacol.,* 35, 427, 1983.
135. Karlsson, N., Larsson, R., and Puu, G., Ferrocene-carbamate as prophylaxis against soman poisoning, *Fund. Appl. Toxicol.,* 4, S184, 1984.
136. Leadbeater, L., Inns, R.H., and Pylands, J.M., Treatment of poisoning by soman, *Fund. Appl. Toxicol.,* 5, 225, 1985.
137. Solana, R., Gennings, C., Anderson, D., Lennox, W., and Carter, W., Jr., Absence of effect by pyridostigmine against organophosphate induced lethality and physical incapacitation, *FASEB J.,* 3, 3664A, 1989.
138. Lennox, W.J., Harris, L.W., Anderson, D., and Solana, R., Successful pretreatment/therapy of soman, sarin and VX intoxication, *FASEB J.,* 3, 3683A, 1989.

139. Anderson, D., Harris, L., and Lennox, W., Subacute carbamate plus acute adjunct pretreatment against nerve agent intoxication, *FASEB J.*, 3, 3867A, 1989.
140. Harris, L.W., Anderson, D.A., Lennox, W.J., and Solana, R.P., Effects of subacute administration of physostigmine on blood cholinesterase activity, motor performance and soman intoxication, *Toxicol. Appl. Pharmacol.*, 97, 267, 1989.
141. McDonald, R.B., Hamilton, J.S., Stern, J.S., and Horwitz, B.A., Regional blood flow of exercise-trained younger and older cold-exposed rats, *Am. J. Physiol.*, 256, 41069, 1989.
142. Shand, D.G., Kornhauser, D.M., and Wilkinson, G.R., Effects of route of administration and blood flow on hepatic elimination, *J. Pharmacol. Exp. Ther.*, 195, 424, 1975.
143. Frank, S., Somani, S.M., and Kohnle, M., Effect of exercise on propranolol pharmacokinetics, *Eur. J. Clin. Pharmacol.*, 39, 391, 1990.
144. Somani, S.M., Gupta, S.K., Frank, S., and Corder, N., Effect of exercise on disposition and pharmacokinetics of drugs, *Drug Develop. Res.*, 20, 251, 1990.
145. Dill, D.B. and Costill, D.L., Calculation of percentage changes volumes of blood, plasma and red cells in dehydration, *J. Appl. Physiol.*, 37, 247, 1974.
146. McMaster, S.B. and Foster, R.E., Behavioral and morphological studies of the interaction between exercise and physostigmine, U.S. Army Medical Research and Development Command, *Sixth Ann. Chem. Def. Biosci. Rev.*, August, 629, 1987.
147. Matthew, C.B., Hubbard, R.W., Francesconi, R.P., and Thomas, G.J., Carbamate-induced performance and thermoregulatory decrements restored with diazepam and atropine, *Aviat. Space Environ. Med.*, 58, 1183, 1987.
148. Somani, S.M., Babu, S.R., Arneric, S.P., and Dube, S.N., Effect of cholinesterase inhibitor and exercise on choline acetyltransferase and acetylcholinesterase activities in rat brain regions, *Pharmacol. Biochem. Behav.* 39, 337, 1991.
149. Matthew, C.B., Bowers, W.D., Francesconi, R.P., and Hubbard, R.W., Chronic physostigmine administration in the exercising rat, Report U.S. Army Medical Research and Development Command, Natick, MA, March, 1990.
150. Dube, S.N., Somani, S.M., and Babu, S.R., Concurrent acute exercise alters central and peripheral responses to physostigmine, *Pharmacol. Biochem. Behav.*, 41, 773, 1993.
151. Somani, S.M. and Dube, S.N., Endurance training changes central and peripheral responses to physostigmine, *Pharmacol. Biochem. Behav.*, 41, 773, 1992.
152. Somani, S.M., Giacobini, E., Boyer, A., Hallak, M., Khalique, A., Unni, L., Hannant, M., and Hurley, E., Mechanisms of action and pharmacokinetics of physostigmine in relation to acute intoxication by organofluorophosphates, Reports submitted to U.S. Army Medical Research and Development Command, Fort Detrick, MD, 1988.
153. Somani, S.M., Roberts, J.B., Thomas, B.H., and Wilson, A., Isolation and characterization of metabolites of neostigmine from rat urine, *Eur. J. Pharmacol.*, 12, 114, 1970.
154. Roberts, J.B., Thomas, B.H., and Wilson, A., Distribution and excretion of ^{14}C-neostigmine in the rat and hen, *Br. J. Pharmacol. Chemotherap.*, 25, 234, 1965.
155. Roberts, J.B., Thomas, B.H., and Wilson, A., Metabolism of ^{14}C-neostigmine in the rat, *Br. J. Pharmacol. Chemotherap.*, 25, 763, 1965.
156. Roberts, J.B., Thomas, B.H., and Wilson, A., Excretion and metabolism of oral ^{14}C-neostigmine in the rat, *Biochem. Pharmacol.*, 15, 71, 1966.
157. Somani, S.M. and Anderson, J.H., Sequestration of neostigmine and metabolites by perfused rat liver, *Drug Metabol. Dispos.*, 3, 275, 1975.
158. Somani, S.M., Chan, K., Dehghan, A., and Calvey, T.N., Kinetics and metabolism of intramuscular neostigmine in myasthenia gravis, *Clin. Pharmacol. Ther.*, 28, 64, 1980.
159. Calvey, T.N., Somani, S.M., and Wright, A., Differences between the biliary excretion of tri[^{14}C]methyl-(3-hydroxy-phenyl)ammonium iodide in Wistar and Gunn rats, *Biochem. J.*, 119, 659, 1970.

160. Chan, K. and Calvey, T.N., Renal clearance of pyridostigmine in patients with myasthenia gravis, *E. Neurol.,* 16, 69, 1977.
161. Adler, M., Maxwell, D., Foster, R.E., Deshpande, S.S., and Albuquerque, E.X., *In vivo* and *in vitro* pathophysiology of mammalian skeletal muscle following acute and subacute exposure to pyridostigmine. Studies on muscle contractility and cellular mechanisms. *Proceedings of the Fourth Annual Chemical Defense Bioscience Review,* 1984, 173.
162. Capacio, B.R., Byers, C.E., Anderson, D.R., Matthews, R.L., and Brown, D.E., The effect of ondansetron on pyridostigmine-induced blood acetycholinesterase inhibition in the guinea pig, *Drug Chem. Toxicol.,* 19, 1, 1996.

6 New Approaches to Medical Protection against Chemical Warfare Nerve Agents

Bhupendra P. Doctor, Donald M. Maxwell, Yacov Ashani, Ashima Saxena, and Richard K. Gordon

CONTENTS

I. Introduction 191
II. Stability of Cholinesterases (ChE) *In Vivo* 193
III. Scavenger Protection in Rodents 193
IV. Prophylaxis against Soman Inhalation Toxicity in Guinea Pigs with Human Butyrylcholinesterase (HuBChE) 195
V. Comparison of Antidote Protection against Soman by Pyridostigmine, HI-6, and Acetylcholinesterase (AChE) 196
VI. Experiments with Non-Human Primates 196
VII. Improving the Bioscavenging Capability of ChE 202
A. Amplification of the Effectiveness of ChE for Detoxification of Organophosphates (OP) by Oximes 203
B. Site-Specific Mutagenesis of AChE 203
C. OP Hydrolyzing Enzymes, e.g., OPH, OPAA, Paraoxonase, Parathion Hydrolase, etc. 205
D. Carboxylesterase as a Bioscavenger 206
E. Huperzine A as a Pretreatment Drug 208
F. Immobilized ChE for the Decontamination of OP 208
References 210

I. INTRODUCTION

The acute toxicity of organophosphorus (OP) compounds is usually attributed to their irreversible inhibition of acetylcholinesterase (AChE; EC 3.1.1.7).[1,2] The resultant increase in the level of acetylcholine at cholinergic synapses, particularly in brain and

0-8493-0872-0/01/$0.00+$.50

diaphragm, produces an acute cholinergic crisis characterized by miosis, increased tracheobronchial and salivary secretions, bronchoconstriction, bradycardia, fasciculation, behavioral incapacitation, muscular weakness, and convulsions culminating in death by respiratory failure.[3] Current antidotal regimens for OP poisoning consist of a combination of pretreatment with a spontaneously reactivating AChE inhibitor such as pyridostigmine bromide to protect AChE from irreversible inhibition by OP compounds, postexposure therapy with anticholinergic drugs such as atropine sulfate to counteract the effects of excess acetylcholine, and oximes such as 2-PAM chloride to reactivate OP-inhibited AChE.[4] Although these antidotal regimens are highly effective in preventing lethality of animals from OP poisoning, they do not prevent the post-exposure incapacitation, convulsions, performance deficits, or in many cases, permanent brain damage.[5–7] These symptoms are commonly observed in experimental animals and are likely to occur in humans. An anticonvulsant drug, diazepam, was included as a treatment to minimize convulsions, thereby minimizing the risk of permanent brain damage.[7] The problems intrinsic to these antidotes stimulated attempts to develop a single protective drug devoid of pharmacological effects, which would provide protection against the lethality of OP and prevent post-exposure incapacitation.[7]

One approach to prevent lethality and minimize side effects or performance decrements is through the use of enzymes such as cholinesterases (ChE) as single pretreatment drugs to sequester highly toxic OP before they reach their physiological targets.[8–17] This approach turns the irreversible nature of the OP-ChE interaction from disadvantage to advantage; instead of focusing on the OP as an anti-ChE, one can focus on the ChE as an anti-OP. Using this approach, it was shown that administration of fetal bovine serum (FBS) AChE or human serum butyrylcholinesterase (HuBChE), protected animals from a variety of multiple LD_{50} of highly toxic OP without any toxic effects or performance decrements.[8–17]

The use of enzymes as therapeutic agents is not unique to ChE. In comparison with many drugs, enzymes have many unique advantages; they are specific, catalytically efficient, operate under physiological conditions, and cause essentially no deleterious side effects. Some of the demonstrated uses of enzymes as therapeutic agents include facilitating the digestion of food, wound healing, proteolysis, replacement of defective enzyme in the case of genetic disorders, removal of blood clots, fibrinolysis, and depletion of metabolites in cancer. In almost all instances where enzymes have been employed therapeutically, they have been used for their proteolytic/hydrolytic properties, as replacements for defective or deficient enzymes, or for the improvement or alteration of immune properties. Only recently have enzymes been employed as scavengers cr prophylactic drugs for protection from highly toxic substances or as detoxifying or decontamination agents. Both the enzymes for which the toxic agents are substrates that are catalytically hydrolyzed (e.g., organophosphate hydrolases (OPH) or organophophorous acid anhydride hydrolases (OPAA), and the enzymes which have a very high affinity for these toxic agents and are irreversibly inhibited (e.g., ChE) are potential scavengers for OP compounds. There are requirements for an enzyme to be an effective scavenger for OP toxicity *in vivo.* It should have a relatively high turnover number, a long half-life *in vivo,* be readily available in sufficient quantities, and not be immunoreactive. In addition, for enzymes such as ChE and CaE, the *in vivo* stoichiometry of sequestration of toxic OP agents should approach 1:1.

The contents of this article describe the progress made in the last decade, by several groups of investigators, in exploring the potential use of enzymes to counteract the toxicity of OP. Among the enzymes which hold promise as scavengers of highly toxic OP nerve agents, significant advances have been made using ChE. Since the biochemical mechanism underlying the prophylaxis by exogenous ChE is established and tested in several species, including non-human primates, this concept should enable a reliable extrapolation of results from animal experiments to human application.

II. STABILITY OF CHOLINESTERASES (ChE) *IN VIVO*

ChE purified from animals such as FBS-AChE, equine serum BChE (EqBChE), and HuBChE were selected as appropiate forms of bioscavengers to be tested as pretreatment drugs for OP toxicity. Their selection was based on the fact that all three enzymes are soluble globular forms,[18,19] easily purified in large quantities from serum,[20,21] and have a relatively long half-life *in vivo.*[9,22–25] Figure 6.1 depicts the time courses of three ChE, administered by three different routes, in mice, rats, guinea pigs, and rhesus monkeys. The determination of half-life of all these ChE in mice,[9,23–25] rats,[20,25] guinea pigs,[17] and rhesus monkeys,[16] showed that their mean residence time in circulation was 35–60 h. The route of administration (i.v., i.p., or i.m.) affected the time at which the maximum concentration of enzyme in circulation was reached, but did not affect the mean residence time, and a constant level of enzyme was maintained for a period of approximately 3–10 h. Also, regardless of the route of administration, 60–90 % of administered enzyme was found in the circulation of animals.

All recombinant as well as monomeric forms of native esterases tested so far have a relatively low mean residence time in the circulation of mice.[23,24] Therefore, in their present form they are not suitable as scavengers of OP. This is discussed in detail later in the chapter. In general, only the tetrameric forms of plasma-derived ChE appear to have relatively long residence times in animals. Enzymes isolated from animal species or from plant or bacterial sources may not be suitable for use in humans, for they will cause adverse immune reaction. At the present time, HuBChE appears to be the most suitable bioscavenger enzyme for human use. Notably, the stability of exogenously administered HuBChE was determined in individuals identified as being homozygous "silent" for serum BChE and half-lives of 8–12 days were reported.[26–28]

III. SCAVENGER PROTECTION IN RODENTS

The first successful use of AChE or BChE as pretreatment drugs for OP toxicity was demonstrated in rodents.[8–10] For example, pretreatment of mice with FBS-AChE[8–10] or HuBChE[10,22] successfully protected animals against 2–5 × LD_{50}

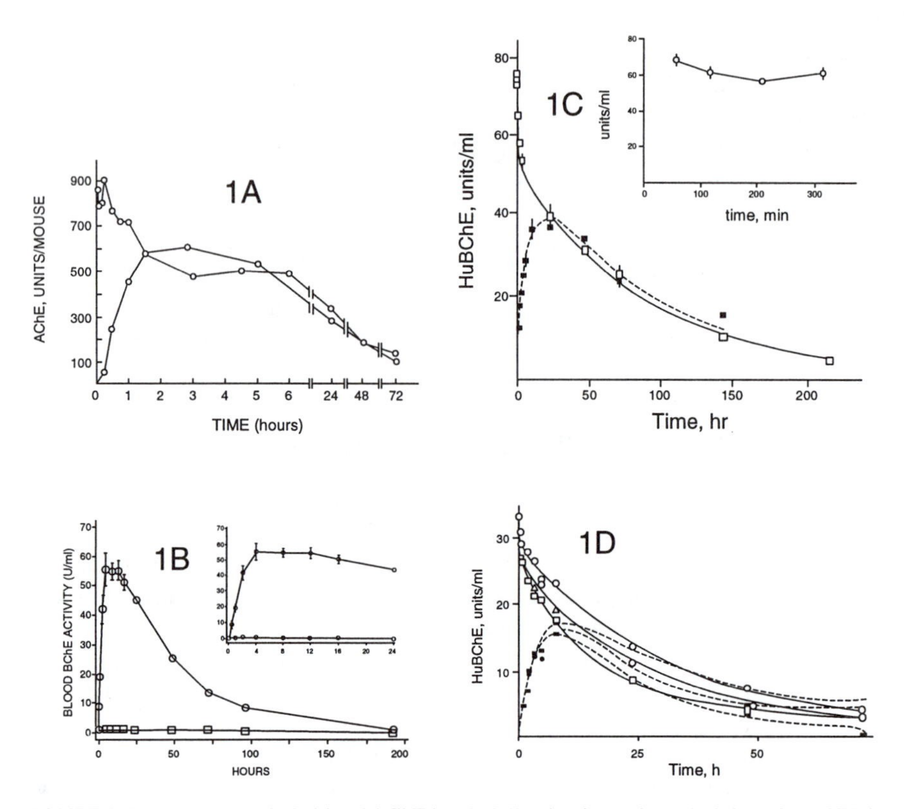

FIGURE 6.1a Average whole blood AChE levels following i.v. or i.p. administration of FBS-AChE: (O) mice administered 920 units/mouse by i.v. route (n = 2), and (□) mice administered 760 units/mouse by i.p. route (n = 3). Variations among individual animals administered the same amount of AChE was assumed 15%. Blood volume was to be 7.5% of body weight. From Raveh, L. et al., *Biochem. Pharmacol.,* 41, 37, 1991. With permission.

FIGURE 6.1b Average BChE levels in rats injected i.p. with 5000 U of EqBChE (n = 6, circles) or vehicle (n = 7, squares) from blood sampled during 192 h (inset 24 h) following administration. Vertical lines about each point represent ± SEM. Points above time 0 represent endogenous BChE levels measured 24 h before injection. From Genovase, R.F. and Doctor, B.P., *Pharmacol. Biochem. Behav.*, 51, 647, 1995. With permission from Elsevier Science.

FIGURE 6.1c Time course of HuBChE in blood of guinea pigs after i.v. (□, 1600 units/animal) and i.m. (2000 units/animal) bolus injections. Each data point is an average from three animals. Endogenous BChE activity (an average 3.4 units/ml of whole blood), was subtracted. (Inset) Expansion of the time between 0 and 5.5 h post-i.v. loading of the enzyme. From Allon et al., *Toxicol. Sci.,* 43, 121, 1998. With permission.

FIGURE 6.1d Individual time course of HuBChE in blood of monkeys following i.v. (open symbols) and i.m. (filled symbols) injections of ~11.5 mg purified enzyme/animal. Each data point is an average of three measurements. Endogenous BChE activity is subtracted. From Raveh, L. et al., *Toxicol. Appl. Pharmacol.,* 145, 43, 1997. With permission.

of VX (ethoxymethyl-S-[2-(diisopropylamino)ethylthiophosphonate), or MEPQ (7-(ethoxymethylphosphinyloxy)-l-methylquinolinium iodide) or soman (pinacoloxymethyl-fluorophosphonate) without requiring any other drug treatment. These studies established a quantitative correlation between the degree of protection against OP compounds and the level of inhibition of administered enzyme, although the protected mice were not evaluated for potential behavioral incapacitation or for any detrimental immunologic response from administering an exogenous enzyme. In addition, these results demonstrated that *in vivo* inhibition of exogenously administered AChE or BChE in blood was proportional to the amount of OP administered as challenge, a result consistent with *in vitro* experiments.

Subsequent studies addressed the question whether pretreatment with a ChE can prevent OP-induced cognitive impairments. Behavioral testing was carried out in rats using the Morris Water Maze Task, evaluating learning, memory, and reversal learning processes. Cognitive functioning in rats was significantly impaired following i.v. administration of 0.9–1.1 $\times$ LD_{50} of soman. HuBChE significantly prevented the development of soman- induced cognitive decrements.[29] These results are consistent with previous conclusions that cognitive functions are sensitive to cholinergic manipulations.[30,31] HuBChE treatment alone was devoid of any impairments in behavioral performance, either motor or cognitive. In that respect, it seems that HuBChE has no undesirable performance decrements. These results further support the concept that pretreatment alone with a scavenger such as HuBChE is sufficient to increase not only survival but also to alleviate deficits in cognitive functioning after exposure to a potent nerve agent such as soman.

IV. PROPHYLAXIS AGAINST SOMAN INHALATION TOXICITY IN GUINEA PIGS WITH HUMAN BUTYRYLCHOLINESTERASE (HuBChE)

The use of a ChE scavenger as a prophylactic treatment against inhalation toxicity, which is a more realistic simulation of exposure to volatile OP, has been described by Allon et al.[17] HuBChE-treated guinea pigs were exposed to a controlled concentration of soman vapors ranging from 417 to 430 μg/l for 45 to 70 s. The correlation between the inhibition of circulating HuBChE and the dose of soman administered by sequential i.v. injections and by respiratory exposure indicated that ~29% of the inhaled dose of soman reached the blood. A HuBChE to soman molar ratio of 0.11 was sufficient to prevent the manifestation of toxic signs following exposure to 2.17 $\times$ LD_{50} of soman (1 LD_{50} inhaled dose = 101 μg/kg). It was noted that protection was far superior to the currently used traditional approach (pyridostigmine and post-exposure therapy). The greater-than-calculated values of protection observed were explained by the fact that unlike an i.v. bolus injection, inhalation exposure allows soman to enter the circulation gradually, which increases the efficacy of soman sequestration to below its toxic levels. The following three important observations are advanced regarding the use of scavengers for OP toxicity:[17] (1) the stoichiometry of protection against inhalation exposure agrees reasonably well with that seen after i.v. challenge, (2) consistent protection is observed across four species of animals, and (3) the pharmacokinetic behavior of HuBChE is similar in mice, rats, guinea pigs, and non-human primates.

V. COMPARISON OF ANTIDOTE PROTECTION AGAINST SOMAN BY PYRIDOSTIGMINE, HI-6, AND ACETYLCHOLINESTERASE (AChE)

Carbamate, oxime, and enzyme scavenger approaches to protection against highly toxic soman were compared by using the prominent example of each type of antidote.[32] Pyridostigmine in combination with atropine, HI-6 in combination with atropine, and FBS-AChE alone were used as examples of carbamate, oxime, and enzyme scavenger antidotes, respectively. Each antidotal regimen produced approximately equal maximal protection against the lethal effects of 952 to 1169 nmol/kg ($8-10 \times LD_{50}$) of soman in mice whose carboxylesterase had been inhibited with CBDP (2-(o-cresyl)-4H-1:3:2-benzodioxaphosphorin-2-oxide). FBS-AChE was much better than either pyridostigmine/atropine or HI-6/atropine in reducing post-exposure incapacitation from soman as measured by lacrimation, motor dysfunction, activity level, and the inverted screen test. A lower dose of pyridostigmine (566 nmol/kg) or FBS-AChE (1150 nmol/kg) was sufficient to protect against 968 nmol/kg ($8 \times LD_{50}$) of soman than was required for HI-6 (200,000 nmol/kg). The circulatory half-life of FBS-AChE (1550 min) was much greater than that of pyridostigmine (48 min) or HI-6 (11 min). These results suggest that FBS-AChE should be considered a superior alternative to either pyridostigmine/atropine or HI-6/atropine antidotal regimens. The major advantages of bioscavengers for protection against OP toxicity are their rapid removal of OP compounds from circulation and the absence of post-exposure incapacitation and toxic effects that are commonly observed in animals protected by traditional antidotal approaches.[32]

VI. EXPERIMENTS WITH NON-HUMAN PRIMATES

The successful demonstration of asymptomatic protection of rodents against a variety of OP by pretreatment with three different ChE prompted the evaluation of sequestration of OP by ChE in non-human primates. The effectiveness of FBS-AChE, EqBChE, and HuBChE as pretreatment drugs was evaluated in rhesus monkeys, which are more sensitive to OP compounds than rodents. Monkeys were exposed to sarin, VX, or soman, the latter OP compound is considered to be the most refractory to current therapy.[7] Behavioral performance was measured by a highly sensitive test of cognitive function, the serial probe recognition (SPR) task.[12,14,15,33–38] This behavioral task was chosen because (a) it is a multiple-item memory task that measures short-term memory capacity and decision-making ability,[36] (b) it has been used extensively to understand human cognitive processing,[37] and (c) it is sensitive to CNS damage in both human and non-human primates.[37,38] For example, rhesus monkeys with damage to the limbic system and humans suffering from amnesia resulting from either Parkinson's or Alzheimer's disease show impaired performance on SPR tasks.[37] This task was also shown to be sensitive to disruption after exposure to doses of soman as low as 1.5–2.0 μg/kg.[38] Following i.v. administration of FBS-AChE, the *in vivo* blood AChE activity was elevated more than 100- to 150-fold after 2 h, yet this treatment had very little effect on the SPR performance. The *in vivo* neutralization of soman by FBS-AChE (Figure 6.2) showed a linear relationship between the progressive inhibition of blood AChE activity and the cumulative dose of soman administered.[15] The percent correct and response latencies of monkeys trained on SPR task to a list length of six items showed complete protection against behavioral incapacitation by soman with no apparent sign of OP toxicity. The monkeys failed to respond within the 10-s interval in only 2 of 4200 trials.

This investigation demonstrated that monkeys displayed minimal adverse reactions from FBS-AChE pretreatment. Following OP exposure, even the best pretreatment/therapy regimen, i.e., pyridostigmine pretreatment and atropine/oxime therapy, does not prevent signs of OP intoxication, such as periods of unconsciousness, respiratory distress tremors, and intermittent convulsions.[6,7] The administration of FBS-AChE prevented the occurrence of all of these signs of OP intoxication. Thus, the ability of FBS-AChE to protect against behavioral incapacitation that results from OP exposure in non-human primates suggests that humans would also be protected.

Concurrently, Broomfield et al. showed in rhesus monkeys that the toxicity of soman ($2 \times LD_{50}$) can be neutralized by administration of an appropriate amount of EqBChE without any performance decrement as measured by SPR.[12] Also, protection of monkeys against 3 to $4 \times LD_{50}$ of soman was obtained with EqBChE pretreatment followed by atropine post-exposure treatment. These animals were able to perform the SPR task about 9 h post-exposure, whereas animals treated with conventional atropine/oxime therapy were not able to perform the same task for 14 days. Animals receiving enzyme alone showed only a subtle transient performance decrement on the SPR task.

A second parameter, the Primate Equilibrium Platform (PEP) task[39–41] was used to demonstrate the protection of rhesus monkeys from the toxicity of as high as $5 \times LD_{50}$ of soman by pretreatment with FBS-AChE or EqBChE without the occurrence of performance deficits.[14] The PEP is a continuous compensatory tracking device that measures the ability of a monkey to compensate for unpredictable perturbations in the pitch induced by a filtered random noise signal. Subjects performed the PEP task for 2.5 h on each soman-challenge testing day, and results were presented for each 5-min block of testing time. During the 6 weeks of long-term follow-up, PEP tests were conducted for 2 h; Φ was computed for each 5-min block of time; and the mean of the 24 resulting data points was calculated to yield one performance score for the entire 2 h.

The i.v. administration of ~0.5 μmol of ChE alone produced a 100-fold increase in blood ChE activity and caused no apparent physiological or neurological effect or deficit, as measured by the PEP task performance. None of the eight monkeys showed any OP toxicity after soman challenges; protection was so complete that there were no fasciculations even at the site of soman injections. Following the first and second soman injections (totaling 25.6 μg/kg, $\sim 4 \times LD_{50}$), the PEP performance of all eight

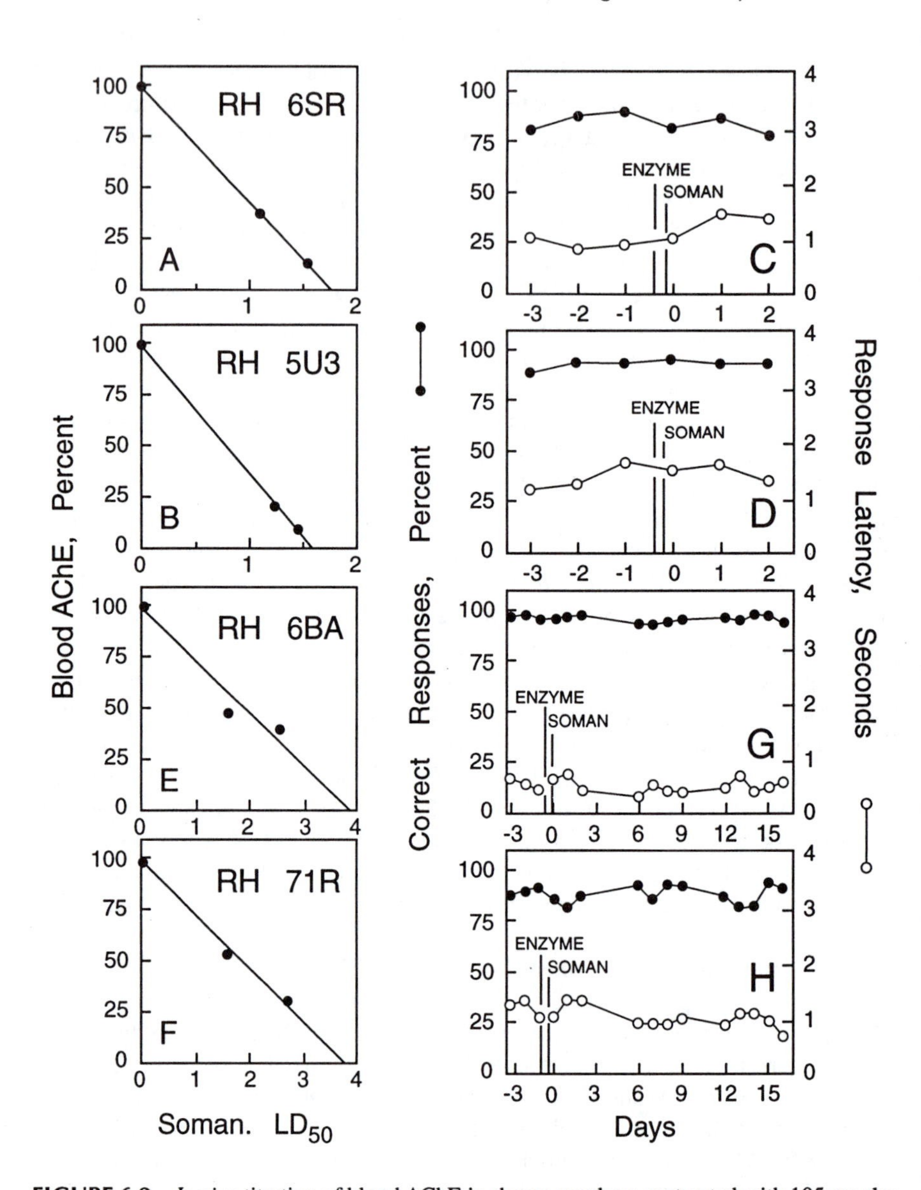

FIGURE 6.2 *In vivo* titration of blood AChE in rhesus monkeys pretreated with 105 nmoles of FBS AChE (ABCD). Soman dose shown is the cumulative LD_{50}. Percent correct responses and response latencies for rhesus monkeys. SPR scores (list length of one item) were obtained at indicated times before administration of 105 nmoles of FBS AChE and after challenge with 1.5 LD_{50} of soman, i.v., in 2 injections. *In vivo* titration of blood AChE in rhesus monkeys pretreated with 210 nmoles of FBS AChE (EFGH). Monkeys were challenged with 2.5–2.7 LD_{50} of soman. Percent correct responses and response latencies for SPR scores (list length of 6 items) before injection of 210 nmoles of FBS AChE and after challenge with 2.5–2.7 LD_{50} of soman, i.v., in two injections. From Maxwell et al., *Toxicol. Appl. Pharmacol.,* 115, 44, 1992. With permission.

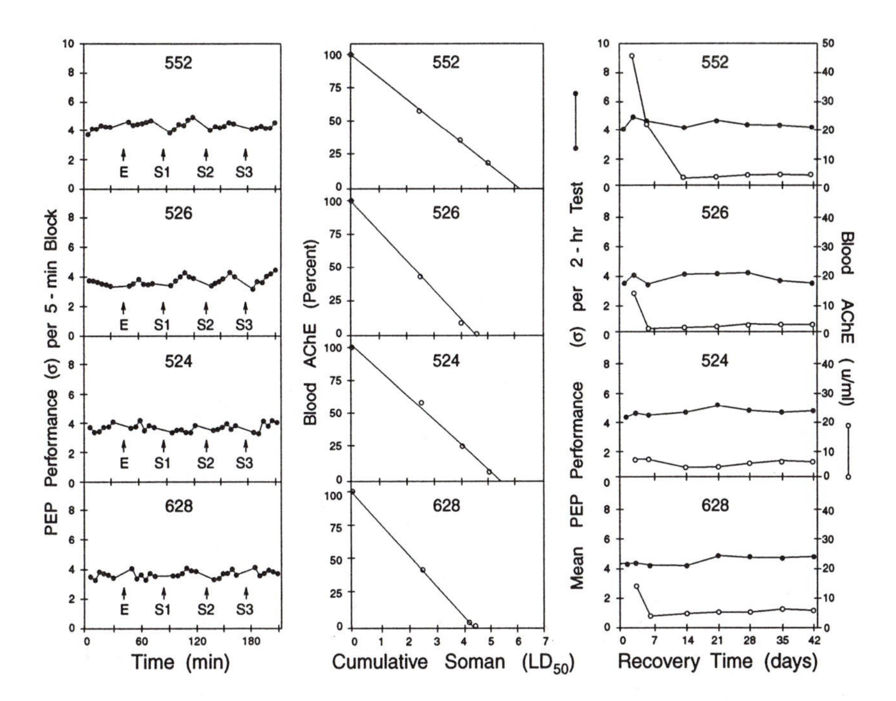

FIGURE 6.3 (Left) Effects of i.v. administered purified FBS AChE on PEP task performance before and after challenge with approximately 2.5, 1.5, and 1.0 LD_{50} of soman. Four male rhesus monkeys (5–8 kg), trained to perform the primate equilibrium platform (PEP) task, each received approximately 0.4–0.5 μmol of AChE i.v. (greater than 1:1 stoichiometry with soman). The sequence of behavioral testing and soman challenges was (a) 30-min PEP task (baseline); (b) AChE injection, E (15 min); (c) 30-min PEP task to determine the effect of administration of AChE alone, followed by a 15-min pause for obtaining blood samples, AChE assay, and soman injection, S1 (16.0 μg/kg, ≈2.5 LD_{50}, i.m.); (d) 30-min PEP testing, followed by a 15-min pause for obtaining blood samples, AChE assay, and soman injection, S2 (9.6 μg/kg, ≈1.5 LD_{50}, i.m.); (e) 30-min PEP testing, followed by a 15-min pause for obtaining blood samples, AChE assay, and the final i.m. soman injection, S3 (6.4 μg/kg, ≈1.0 LD_{50}, was planned but would be reduced if residual AChE activity was judged insufficient); (f) final 30 min of PEP testing. For each 30 min of PEP testing, the data (filled circles) from 6 sequential, 5-min blocks of time are presented. (Middle) *In vivo* titrations of blood AChE in four rhesus monkeys pretreated by i.v. injection with FBS AChE. Details are as described above. The cumulative dose of soman which reduced ChE activity to the indicated final levels exceeded the amount of AChE administered, suggesting involvement of endogenous esterase. (Right) Long-term effects on PEP task performance of i.v. administered FBS AChE and challenge with a total of approximately 5 LD_{50} of soman and residual blood AChE levels. PEP performance and blood AChE levels of 4 monkeys were tested weekly for 6 weeks, filled circles, PEP performance; open circles, enzyme level. PEP performance scores are the mean of data from 24 separate, 5-min blocks that compose the 2-h test. From Wolfe, A.D. et al., *Toxicol. Appl. Pharmacol.*, 115, 44, 1992. With permission.

monkeys was completely normal. The four monkeys pretreated with FBS-AChE (Figure 6.3) or EqBChE (Figure 6.4) continued this level of performance even after the third soman challenge. However, the two remaining monkeys that had been

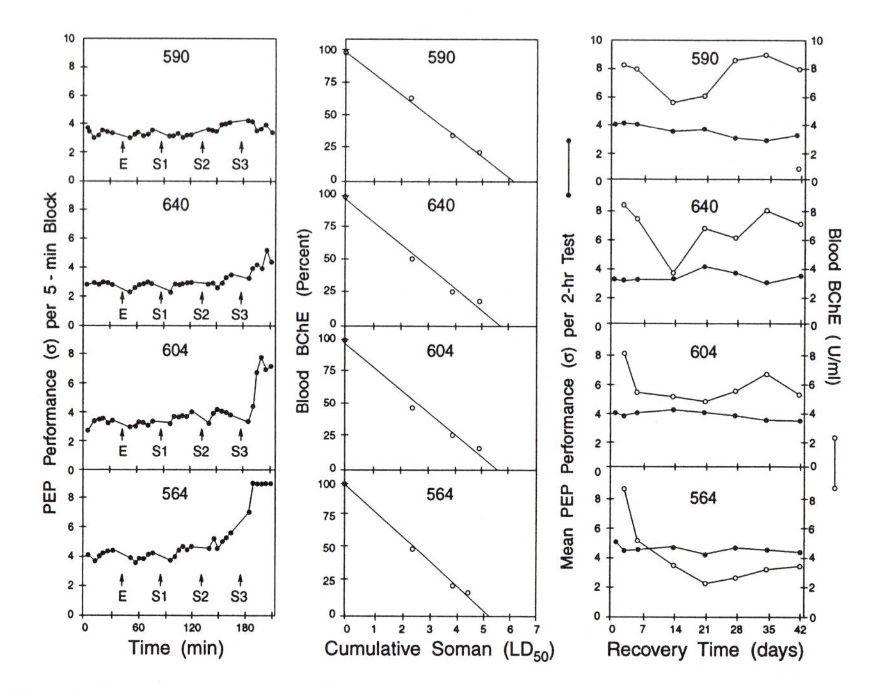

FIGURE 6.4 (Left) Effects of i.v. administered purified horse serum BChE on PEP test performance before and after challenge with approximately 2.5, 1.5, and 1.0 LD_{50} of soman. See legend to Figure 1 for a detailed explanation. (Middle) *In vivo* titrations of blood BChE in four rhesus monkeys pretreated by i.v. injection with horse serum BChE. See legend to Figure 6.1 for details. (Right) Long-term effects on PEP task performance of i.v. administered horse serum BChE and challenge with a total of approximately 5 LD_{50} of soman and residual blood BChE levels. From Wolfe, A.D. et al., *Toxicol. Appl. Pharmacol.*, 115, 44, 1992. With permission.

pretreated with EqBChE exhibited a significant but minor PEP deficit after the third soman injection; this transient PEP performance deficit was similar to that observed after exposure of unprotected monkeys to low doses of soman (<2.8 μg/kg).[40,41] Based upon this comparison, the cumulative protective ratio afforded by ChE pretreatment against soman can be estimated at 10 to 15. During 6 weeks of postsoman testing, none of the monkeys showed any signs of delayed toxicity, convulsions, or other OP symptoms or any abnormality on PEP performance.[14] In non-human primates, the 1:1 stoichiometry between ChE and OP dose, plus the endogenous scavenger (ChE, CaE, and other 3–4 unidentified proteins) present,[42–45] the LD_{50} of soman may be extrapolated to be approximately 4.3 ug/kg.

The ability of HuBChE to prevent toxicity induced by soman and VX was assessed in rhesus monkeys.[16] A molar ratio of HuBChE:OP ~1.2 was sufficient to protect monkeys against an i.v. bolus injection of 2 × LD_{50} of VX, while a ratio of 0.62 was sufficient to protect monkeys against an i.v. dose of 3.3 × LD_{50} of soman, with no additional post-exposure therapy. A remarkable protection was also seen against soman-induced behavioral deficits detected in the performance of a spatial discrimination task (Figure 6.5).

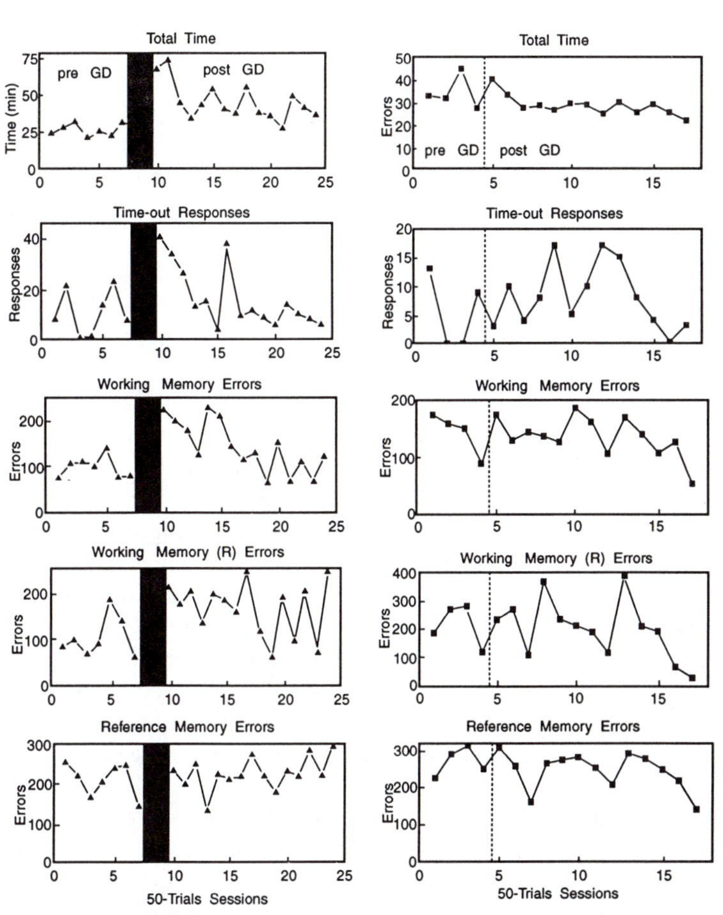

FIGURE 6.5 Effects of soman on performance of the spatial discrimination task in monkeys. (Left) Monkey A, pretreated with 0.1 mg/kg pyridostigmine followed with TAB immediately after exposure to 15 μg/kg soman (2.7 x LD_{50}). The shaded area represents 2 days in which no performance was obtained during the presentation of the behavioral tests. The five panels represent the behavioral parameters. (Right) Effects of soman (18 μg/kg; 3.3 × LD_{50}) on performance of the spatial discrimination task in monkey RQ686, following pretreatment with 26 mg HuBChE. No additional treatment was administered. The animal continued its normal performance with no adverse effects immediately following soman (dotted line). Data points are morning sessions only. Note the different scale on the ordinates of the two panels. From Raveh, L. et al., *Toxicol. Appl. Pharmacol.,* 145, 43, 1997. With permission.

These studies firmly establish that prophylactically administered ChE, with no additional therapy, prevents the toxicity induced by highly toxic OP nerve agents in mice, rats, guinea pigs, and rhesus monkeys. Not only do these bioscavengers prevent lethality, but animals do not show any untoward side effects or performance decrements/deficits determined by the Morris Water Maze task, SPR task, PEP task, or spatial discrimination task.

Of the three ChE investigated so far, only HuBChE appears to be an appropriate candidate for exploration for human use. FBS-AChE and EqBChE are known to induce the production of antibody when administered in heterologous species of animals (unpublished results). The antibody generated by repeated administration of these two enzymes rapidly clears the circulating exogenous ChE from blood, indicating that the use of such enzymes in heterologous species may not be of much value except for a single use. The absence of immunological and physiological side effects following blood and/or plasma transfusions in humans and lack of adverse reaction to partially purified HuBChE administered daily for many weeks,[47] support the contention that HuBChE is the most promising prophylactic antidote. Also, the stability of the exogenously administered HuBChE in humans (half-life of 8 to 12 days)[26–28] suggests a long-lasting therapeutic level even after administration of a single dose of enzyme.

The systematic evaluation of the efficacy of HuBChE in protection of four species against nerve agent toxicity offered an extrapolation model from animal to human[9,10,16,17,29,46] based on the stoichiometry of OP sequestration and pretreatment with HuBChE protection levels in mice, rats, guinea pigs, and monkeys. Further, results show that the stoichiometry of OP sequestration in any given species should depend on the concentration of the circulating enzyme at the time of exposure to challenge. Calculations of protective ratios in humans required quantitative information on the toxicity of OP in humans. These figures were compiled from the literature describing human volunteer studies with non-lethal doses and accidental exposures to nerve agents that enabled an estimate of sign-free doses as well as toxic doses in humans. Predictions were then made by calculating the amount of HuBChE required to reduce toxic levels of OP to below the sign-free doses within one blood circulation time in human (seconds).

It was predicted that 200 mg/70 kg HuBChE would protect against up to $2 \times LD_{50}$ of VX or soman, without the need for immediate post-exposure treatment.[43] Lowering the dose to 50 mg/70 kg is likely to confer protection against long-term exposure to low levels of nerve agents such as soman. It should be noted, however, that the extrapolation from animal-to-human was based on data generated in animals weighing 20 g to 10 kg, and further validation in larger animals may be useful.

VII. IMPROVING THE BIOSCAVENGING CAPABILITY OF ChE

Many approaches have been made to improve the efficacy of stoichiometric bioscavengers. Enzymes that can hydrolyze OP are also being considered as promising bioscavengers. These efforts are summarized below:

1. Amplification of the effectiveness of ChE for detoxification of OP by oximes
2. Site-specific mutagenesis of AChE
3. OP hydrolyzing enzymes, e.g., OPH, OPAA, paraoxonase, parathion hydrolase, etc.

4. Carboxylesterase as a bioscavenger
5. Huperzine A as a pretreatment drug
6. Immobilized ChE for the decontamination of OP

A. Amplification of the Effectiveness of ChE for Detoxification of Organophosphates (OP) by Oximes

A major limitation for use of ChE as pretreatment drugs for OP toxicity is their 1:1 stoichiometry with OP. An approximately 200-fold difference in molecular weight between OP and ChE necessitates the use of large amounts of enzyme to provide protection. To improve the efficacy of ChE as pretreatment drugs, an approach was developed in which the catalytic activity of OP-inhibited AChE was rapidly and continuously restored by having sufficient amounts of appropriate oxime present.[48] In general, OP-inhibited ChE can be reactivated rapidly by mono- or bis-quaternary oximes such as 2-PAM and HI-6 so long as it has not undergone aging. The rate of reactivation of OP-inhibited ChE depends on the type and source of ChE, the structure of OP and oxime, as well as the concentration of oxime used. *In vitro* effectiveness of several oximes in reactivating AChE that has been inhibited by a variety of OP showed that oximes, such as TMB_4, 2-PAM, MMB_4 and HI-6, reactivated AChE inhibited by all OP to some extent, but HI-6 was the most effective in reactivating AChE that was inhibited by soman and sarin. The capacity of AChE in combination with 2mM HI-6 to detoxify large amounts of sarin *in vitro* is shown in Figure 6.6a. One mole of enzyme could detoxify a 3200-fold molar excess of sarin or a 64-fold molar excess of soman in the presence of 2 mM HI-6, as compared to a two-fold excess of sarin or soman in the absence of HI-6. Improved detoxification of OP compounds by AChE in combination with oxime has also been demonstrated *in vivo* (Figure 6.6b). Mice receiving 9 nmol of AChE and 1 mg HI-6 could detoxify a cumulative 57-fold excess of sarin when it was administered by repeated injections at 15-min intervals and as long as the HI-6 level was maintained by repeated injections of 1 mg HI-6.[48] If the level of HI-6 was not maintained, detoxification was less effective as demonstrated by a pronounced decrease in *in vivo* AChE activity.

B. Site-Specific Mutagenesis of AChE

Several recent studies have demonstrated that it is indeed possible to improve the bioscavenging performance of cholinesterases by site-directed mutagenesis.[49] Using this technique, it is possible to obtain mutant enzymes which possess an increased affinity for OP,[50] or are more easily reactivated by oximes,[51] and/or possess a reduced rate of aging.[49,52–54] The kinetics of aging were examined in a soman-inhibited mutant enzyme in which the glutamate E202(*199*), located next to the active-site serine S203(*200*) of AChE, was converted to glutamine.[49] For wild-type enzyme, the soman-AChE conjugate aged very rapidly, giving rise to a form of enzyme resistant to reactivation by oximes. In contrast, the E202(*199*)Q mutant enzyme was largely resistant to aging and could be reactivated by oximes.[49–52] *In vitro* detoxification of soman and sarin by mouse wild-type and E202Q AChE in the presence of 2 mM HI-6

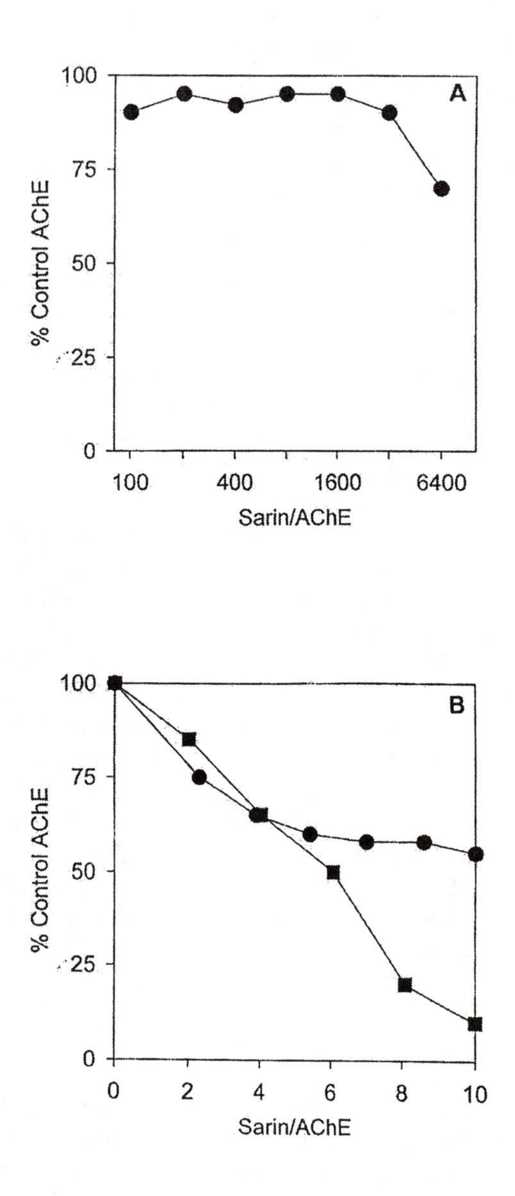

FIGURE 6.6A *In vitro* titration of FBS AChE in the presence of HI-6. Reactivation of FBS AChE (0.125 nmol) in the presence of 2 mM HI-6 at pH 8.0 after repeated additions of sarin at 0.5 h intervals

FIGURE 6.6B *In vivo* detoxification of sarin by FBS AChE in mice. Mice received i.v. FBS AChE (9 nmol) followed by sarin (14 nmol) and 1 mg HI-6. Sarin/HI-6 injections (●) or sarin alone injections (i.v.) were then repeated at 15-min intervals. AChE activity was determined 5 min prior to each sarin injection. All mice survived.

showed that the mutant E202Q AChE was 2–3 times more effective in detoxifying soman and sarin compared to wild-type AChE.[49] These studies show that these recombinant DNA-derived AChE are a great improvement over wild-type AChE as bioscavengers. They can be used to develop effective methods for the safe disposal of stored OP nerve agents and appropriate formulation for medical surgical and skin decontaminants and also for decontamination of materials, equipment, and the environment.

To evaluate the possible use of recombinant ChE as bioscavengers *in vivo,* the mean residence time of five tissue-derived and two rChE (i.v.) injected in mice were compared with their oligosaccharide profiles.[23,24] Monosaccharide composition analysis revealed differences in the total carbohydrate, galactose, and sialic acid contents. The molar ratio of sialic acid to galactose residues on tetrameric HuBChE, rMoAChE, and rHuBChE was found to be ~1.0, suggesting that all the terminal galactose residues were capped with sialic acid. However the mean residence time of HuBChE was 9- and 14-fold greater than that of rMoAChE and rHuBChE, suggesting that the capping of galactose with sialic acid by itself is not sufficient to confer circulatory stability to ChE. For *Torpedo* AChE (mean residence time = 44 min) and monomeric FBS-AChE (mean residence time = 304 min), this ratio was ~0.5, suggesting that only half of the terminal galactose residues were capped with sialic acid, yet these enzymes differed greatly in their circulatory stability. In contrast, a molar ratio of 0.5 for sialic acid-to-galactose was observed for the highly stable tetrameric FBS-AChE and EqBChE. These observations suggest that although the presence of sialic acid appears to be essential for maintaining ChE in circulation, the location rather than the number of the non-sialylated galactose residues may be affecting circulatory stability.

Differences in oligosaccharides of ChE from various sources and the microheterogeneity in glycans on each ChE were elucidated by charge- and size-based separation analyses. However, neither the carbohydrate composition nor the oligosaccharide profile could be completely correlated with the pharmacokinetic parameters of these enzymes. The glycans of recombinant ChE and monomeric FBS-AChE displayed a remarkable heterogeneity in size and consist of hybrid and complex bi-, tri-, and tetra-antennary structures. *Torpedo* AChE also contains high-mannose structures. The three plasma ChE, on the other hand, contain mature glycans which are predominantly of the complex biantennary type, suggesting that these structures are responsible for the extended mean residence times of the enzymes. *Torpedo* AChE, rChE, and monomeric FBS-AChE showed a distinctive shorter mean residence time (44–304 min) compared with tetrameric forms of plasma ChE (1902–3206 min). Differences in the pharmacokinetic parameters of ChE appear to be due to the combined effect of the molecular weight and charge- and size-based heterogeneity in glycans. Site-specific analysis of glycan structures may elucidate the structures responsible for the rapid clearance of non-plasma ChE and suggest suitable manipulations for improving the circulatory stability of rChE.

C. OP Hydrolyzing Enzymes, e.g., OPH, OPAA, Paraoxonase, Parathion Hydrolase, etc.

Lenz et al.'s chapter provides a comprehensive discussion of this topic.

D. Carboxylesterase as a Bioscavenger

Although the development of bioscavenger protection against toxic OP has focused primarily on the use of AChE and BChE, studies of carboxylesterase (CaE) have demonstrated that this esterase has some advantages as an OP bioscavenger. A recent comparison of AChE, BChE, and CaE as bioscavengers has described some of these advantages.[55] AChE, BChE, and CaE are all members of the α/β hydrolase family and have a high degree of overall homology in their amino acid sequences, but they differ in several critical regions that produce distinct differences in their biochemical properties.[56] The most significant biochemical differences in these esterases are related to the extent of aging of the OP-inhibited esterase, the size of the active site, and the ability of the OP-inhibited esterase to undergo spontaneous or oxime-induced reactivation. The ideal OP scavenger would have a fast rate of reactivity for a broad spectrum of OP compounds, a slow rate of aging, and the ability to reactivate to increase its stoichiometry as a scavenger. Evaluation of CaE on these criteria suggests that it is a major candidate as an OP bioscavenger.

One of the primary concerns in the use of esterases as bioscavengers for OP compounds is the 1:1 stoichiometry of their detoxication of OP compounds. The major limitation on the stoichiometry of esterases as OP scavengers is the aging of OP-inhibited esterases that prevents their reactivation. One of the most important advantages of CaE is that OP-inhibited CaE does not undergo the rapid aging that prevents oxime reactivation of OP-inhibited cholinesterases.[57] This means that OP-inhibited CaE can be reactivated to an active enzyme for further sequestration of OP molecules. The effectiveness of this process *in vivo* has been demonstrated by the protection that is produced by diacetylmonoxime, an oxime that reactivates OP-inhibited CaE but does not reactivate OP-inhibited AChE.[58] Oxime reactivation of soman-inhibited CaE by diacetylmonoxime in rats increased soman detoxication enough to produce a two-fold increase in the LD_{50} of soman.[59]

Another advantage of CaE is the size of its active site. Saxena et al. have developed a method to estimate the size of the active site of esterases in which the volume of the active site corresponds to the area defined by the van der Waals surface.[60] The active site volumes of AChE, BChE, and CaE were calculated from the X-ray crystallographic stucture of *Torpedo californica* AChE and models of BChE and CaE that were created from the homology of these enzymes with *Torpedo californica* AChE and *Geotrichum candidum* lipase, respectively. The active site volume of CaE was 10× larger than that of AChE and 6× larger than that of BChE.[60] The larger size of the active site of CaE is important inasmuch as Taylor et al. have demonstrated that substitution of smaller aliphatic amino acid residues for bulky aromatic residues in the active site of AChE increases the volume of the active site and the ease with which oximes can reactivate OP-inhibited AChE.[61] Their site-directed mutagenesis studies showed that changing phenylalanines in the active site of AChE to smaller groups enhanced oxime reactivation 10- to 20-fold. The reasons for this beneficial effect are complex, but a primary factor is that a more spacious active site allows more avenues of nucleophilic attack by oximes on the phosphorylated serine of AChE and increases the probability of a successful reactivation reaction. Jarv discussed the importance of the direction of nucleophilic attack for oxime reactivation of OP-inhibited ChE.[62] His analysis concluded that oxime reactivation is an S_N2 reaction in which the reaction intermediate undergoes inversion of configuration that can be prevented by steric hindrance in a small constrained active site. The large active site volume of CaE, therefore, minimizes steric hindrance in the active site and maximizes the potential for reactivation.

The importance of active site volume is also evident in the decreasing stereospecificity of esterases as the volumes of their active sites increase. In site-specific mutagenesis studies of mammalian AChE, Taylor et al. observed that the stereoselectivity of AChE was reduced 3-fold and 230-fold by substitution of small aliphatic groups for phenylalanine at positions 295 and 297, respectively.[61] Furthermore, in a comparison of the stereoselectivity of AChE, BChE, and CaE, whose relative active site volumes are 3:5:30, the reported ratio of reaction rates of C(+)P(−) and C(+)P(+) stereoisomers of soman for AChE, BChE, and CaE are 17500, 290, and 135, respectively.[63,64] Even though the stereospecificity of CaE is reduced by its larger active site volume in comparison to ChE, it still maintains a 135-fold greater reactivity with the most toxic stereoisomers [i.e., C(+)P(−) soman].

Another advantage of a large active site is that it confers an enzyme specificity for a wider range of OP inhibitors. By measuring the rate constants for esterase inhibition by a spectrum of OP inhibitors, Maxwell et al. compared the structural specificity of AChE, BChE, and CaE.[55] This specificity study found that AChE could accommodate OP inhibitors containing only one bulky group (e.g., isopropyl, pinacolyl, or phenyl); BChE could accommodate OP inhibitors containing two of the smaller bulky groups (i.e., isopropyl); and CaE could accommodate OP inhibitors containing up to two of the largest bulky groups (e.g., phenyl groups). Therefore, CaE had the ability to detoxify the broadest spectrum of OP inhibitors. The only exception to this observation is that the fewer aromatic residues in the active site of CaE in comparison to ChE reduces the affinity of CaE for positively charged OP inhibitors.[52] However, this is not a major deficiency inasmuch as few nerve agents or pesticides are positively charged.

The final advantage of CaE as a bioscavenger is shown by the recent observations that OP-inhibited CaE undergoes spontaneous reactivation. Jokanovic et al. observed that OP-inhibited CaE in rats exhibited spontaneous reactivation after inhibition with dichlorvos, sarin, or soman.[65] In mechanistic studies of this process, Maxwell et al. found that spontaneous reactivation of sarin-inhibited CaE had a pH profile that suggested the involvement of an amino acid residue with a pK_a of 6.1.[55] Subsequent examination of the amino acid sequences of CaE from six mammals and two insects revealed a highly conserved histidine that met this pK_a requirement and was not part of the catalytic triad of CaE. This conserved histidine was not found in any wild-type ChE and was located immediately adjacent to the glycines that comprise the oxyanion hole of AChE, BChE, and CaE. This oxyanion hole region appears to be particularly important for spontaneous reactivation of OP-inhibited esterase and OP hydrolysis since mutagenesis of this region has produced profound changes in these activities. For example, Lockridge et al. produced OP hydrolase activity in rHuBChE by site-directed mutagenesis in which a glycine in this region was changed to a histidine.[66] In addition, Newcomb et al. found a mutant blowfly CaE in which a glycine in the oxyanion region was changed to an aspartate, which converted this CaE to an OP hydrolase.[67] This conversion was so effective that the mutant blowfly was found to be OP resistant, requiring four to five times more diazinon than was necessary to produce lethality in wild-type blowflies.

E. Huperzine A as a Pretreatment Drug

Huperzine A (HUP), an alkaloid isolated from moss Lycopodium *Huperzia Serrata* is a selective, slow reversible inhibitor of mammalian AChE, with K_1 of 20–40 nM.[68–70] HUP has been demonstrated in mice,[71,73] guinea pigs,[72] and monkeys[73] to protect against nerve agent toxicity by treatment of animals prior to challenge with soman. Pretreatment of four monkeys with a sign-free dose of (−)-HUP protected them even 14 h post-loading of HUP. The monkeys displayed only minor toxic signs and survived without the need for post-exposure therapy. Similarly, a protective ratio of 2.0 was obtained 6 h after pretreatment of mice with HUP, with no post-challenge supporting therapy. In both monkeys and mice, the long-lasting antidotal efficacy conferred by HUP correlated well with the time course of blood-AChE inhibition. In guinea pigs, pretreatment with HUP was shown to prevent seizures and neuropathological damage to the hippocampus following exposure to soman. These studies highlighted the superiority of HUP as an antidote against nerve agent toxicity compared to pyridostigmine and physostigmine inasmuch as the duration of protection conferred following administration of a single dose of the prophylactic drug.

HUP is more than 1000-fold less potent inhibitor of BChE than mammalian AChE, a finding that is attributed to the reduced aromaticity of the active site gorge of BChE compared with AChE. It was thought that the combined administration of HUP and BChE would have a synergistic effect in terms of protection and will allow decreasing the amount of protein required for adequate protection. Preliminary observations indeed show that pretreatment with HUP and HuBChE protected mice against soman with the sum of the individual contribution of each drug alone.[73] Thus, a protective ratio of 2.6 was observed while the predicted value from the separate experiments was at 2.5. These studies with HUP as a potential antidote suggest that this slow inhibitor is a promising pretreatment drug that confers protection by a relatively long-lasting reversible inhibition of AChE at physiologically important sites.

F. Immobilized ChE for the Decontamination of OP

It has been demonstrated that a variety of enzymes exhibited enhanced mechanical and chemical stability when immobilized on a solid support, producing a biocatalyst. Munnecke first immobilized a pesticide detoxification extract from bacteria by absorption on glass beads.[74] The absorbed extract retained activity for what was then a remarkable full day. Wood and co-workers, using isocyanate-based polyurethane foams (Hypol®), found that a number of enzymes unrelated to OP hydrolysis could be covalently bound to this polymer.[75] Later, Havens and Rase immobilized a parathion hydrolase.[76] Furthermore, Turner observed that polyurethane foams are excellent adsorption materials for OP such as pesticide vapors.[77]

As described above (see Section VII.A), soluble ChE and oxime together detoxify OP compounds. These features were combined to developed a sponge product composed of ChE (FBS-AChE and EqBChE), organophosphate hydrolase (rabbit or bacterial OPH), oxime (2-PAM or HI-6), and polyurethane foam combinations for the removal and decontamination of OP compounds from medically important biological surfaces such as skin.[78] This is an important extension of the bioscavenger approach to external decontamination and protection against organophosphate toxicity, since currently accepted methods for decontamination of personnel and materials use bleach, which is caustic and harmful and also poses a significant environmental burden. Additionally, the ChE-sponge has unique attributes, making it a biosensor for OP for use in any environmental condition, such as vapor, water, soil, and long-term remote sensing.

ChE, OPH, and other enzymes may be immobilized with a Hypol® toluene diisocyanate polyurethane prepolymer creating the enzyme sponges in less than 20 min at ambient temperature in any desired size or shape. Since the enzymes predominently attach (covalently) at surface lysines to the inert foam at multiple points during the polymerization process, they become an integral part and acquire the structural integrity of the resultant polymerized matrix. This is clearly evident in the enhanced mechanical and chemical stability of immobilized AChE and BChE; they were stable at room temperature for more than two years without any special treatment. The enzymes remained covalently attached to the polymer even after 20 washes over many days; did not wash off; and were very resistent to environmental assaults such as salt water, proteolytic degradation, or saturated organic fumes. Due to the large capacity of the prepolymer for protein, high activity sponges can be synthesized from purified ChE, substantially increasing their efficacy. Multiple OP-hydrolyzing enzymes can be co-immobilized on one sponge, including phosphotriesterases (paraoxonase or OP hydrolases) and/or cholinesterases. The advantage of including OP hydrolases in the multi-enzyme component is that they detoxify all phosphonylated oxime intermediates with little substrate specificity.

Since the enzyme is likely attached to the polymer at multiple points and therefore becomes partially distorted, it is not unexpected that the K_m values for the immobilized ChE and OPH were about 10-fold greater than for the corresponding soluble enzyme, but the combined effects on affinity for substrate and k_{cat} resulted in approximately a 20- to 50-fold decrease in acylation (k_{cat}/K_m). Yet there was no observed shift in the pH profile of the enzymes, and, more important, the bimolecular rate constants for the inhibition of AChE-sponge and BChE-sponge and the soluble enzymes by MEPQ showed no significant difference between soluble and covalently bound enzymes. Therefore, the OP interacts similarly with soluble and immobilized ChE.

OP such as diisopropylfluorophosphate or MEPQ inhibited the activity of ChE-sponges, and the oxime HI-6 restored activity of the AChE-sponge until the molar concentration of MEPQ reached approximately 1000 times that of the cholinesterase active site, demonstrating that the bioscavenger approach works externally as well as *in vivo.* In addition, the AChE-sponge could be recycled many times by rinsing the sponge with HI-6 in the absence of OP. In this case, most of the original cholinesterase activity could then be restored to the sponge for another cycle of detoxification of OP. The ability of the immobilized enzyme-sponge and HI-6 to detoxify the MEPQ was dependent upon the efficiency of the sponge to decontaminate particular surfaces. The sponge alone could decontaminate MEPQ from non-porous plastic and steel surfaces (>97%), and an AChE-sponge with HI-6 detoxified the removed MEPQ. However, the sponge alone (without enzyme) was not more effective than the M291 decontamination kit for removing neat soman applied to a guinea pig (shaved skin). To improve the removal/extraction of OP from skin surfaces, additives were incorporated into the polyurethane matrix both during synthesis and postsynthesis. Liquid additives to the sponges possessing partial organic solubilizing characteristics such as tetraglyme and also oximes such as 2-PAM and HI-6 were particularly effective in protecting guinea pigs from soman exposure, yielding protective ratios about 4-fold (LD_{50} 80 mg/kg) better when compared to the M291 kit (LD_{50} 20mg/kg). Sponges synthesized with activated carbon incorporated into the polymer matrix, a process that did not interfere with the immobilization of ChE, were also useful at removing soman from skin, and might be effective in removing other toxic agents such as vesicants. The sponges should be suitable for a variety of biological surface detoxification and decontamination schemes for both chemical weapons, and for civilians and first-responders exposed to pesticides or highly toxic OP such as sarin or soman.

REFERENCES

1. Karczmar, A.G., Anticholinesterase agents, in *International Encyclopedia of Pharmacology and Therapeutics,* Karczmar, A.G., Ed., Pergamon, Oxford, 1970, 1.
2. Taylor, P., Anticholinesterase agents, in *The Pharmacological Basis of Therapeutics,* Gilman, A.G., Goodman, L.S., Rall, T.W., and Murad, F., Eds., Macmillan, New York, 1985, 110.
3. Brimblecombe, R.W., Drugs acting on central cholinergic mechanisms and affecting respiration, *Pharmacol. Ther.,* B3, 65, 1977.
4. Gray, A.P., Design and structure-activity relationships of antidotes to organophosphorus anticholinesterase agents, *Drug Metab. Rev.,* 15, 557, 1984.
5. Dirnhuber, P., French, M.C., Green, D.M., Leadbeater, I., and Stratton, J.A., The protection of primates against soman poisoning by pretreatment with pyridostigmine, *J. Pharm. Pharmacol.,* 31, 295, 1979.
6. McLeod, C.G., Pathology of nerve agents: Perspectives on medical management, *Fund. Appl. Toxicol.,* 5, S10, 1985.
7. Dunn, M.A. and Sidell, F.R., Progress in medical defense against nerve agents, *JAMA,* 262, 649, 1989.
8. Wolfe, A.D., Rush, R.S., Doctor, B.P., Koplovitz, I., and Jones, D., Acetylcholinesterase prophylaxis against organophosphate toxicity, *Fund. Appl. Toxicol.,* 9, 266, 1987.
9. Raveh, L., Ashani, Y., Levy, D., De La Hoz, D., Wolfe A.D., and Doctor, B.P., Acetylcholinesterase prophylaxis against organophosphate poisoning. Quantitative correlation between protection and blood-enzyme level in mice, *Biochem. Pharmacol.,* 41, 37, 1991.
10. Ashani, Y., Shapira, S., Levy, D., Wolfe, A.D., Doctor, B.P., and Raveh, L., Butyrylcholinesterase and acetylcholinesterase prophylaxis against soman poisoning in mice, *Biochem. Pharmacol.,* 41, 37, 1991.
11. Doctor, B.P., Raveh, L., Wolfe, A.D., Maxwell, D.M., and Ashani, Y., Enzymes as pretreatment drugs for organophosphate toxicity, *Neurosci. Biobehav. Rev.,* 15, 123, 1991.
12. Broomfield, C.A., Maxwell, D.M., Solana, R.P., Castro, C.A., Finger, A.V., and Lenz, D.E., Protection of butyrylcholinesterase against organophosphorus poisoning in nonhuman primates, *J. Pharmacol. Exper. Ther.,* 259, 633, 1991.
13. Maxwell, D.M., Wolfe, A.D., Ashani, Y., and Doctor, B.P., Cholinesterase and carboxyesterase as scavengers for organophosphorus agents, in *Proceedings of the Third International Meeting on Cholinesterase,* Massoulié et al., Eds., ACS Books, Washington, D.C., 1991, 206.
14. Wolfe, A.D., Blick, D.W., Murphy, M.R., Miller, S.A., Gentry, M.K., Hartgraves, S.L., and Doctor, B.P., Use of cholinesterases as pretreatment drugs for the protection of rhesus monkeys against soman toxicity, *Toxicol. Appl. Pharmacol.,* 117, 189, 1992.

15. Maxwell, D.M., Castro, C.A., De La Hoz, D.M., Gentry, M.K., Gold, M.B., Solana, R.P., Wolfe, A.D., and Doctor, B.P., Protection of rhesus monkeys against soman and prevention of performance decrement by pretreatment with acetylcholinesterase, *Toxicol. Appl. Pharmacol.,* 115, 44, 1992.
16. Raveh, L., Grauer, E., Grunwald, J., Cohen, E., and Ashani, Y., The stoichiometry of protection against soman and VX toxicity in monkeys pretreated with human butyrylcholinesterase, *Toxicol. Appl. Pharmacol.,* 145, 43, 1997.
17. Allon, N., Raveh, L., Gilat, E., Cohen, E., Grunwald, J., and Ashani, Y., Prophylaxis against soman inhalation toxicity in guinea pigs by pretreatment alone with human serum butyrylcholinesterase, *Toxicol. Sci.,* 43, 121, 1998.
18. Ralston, J.S., Main, A.R., Kilpatrick, J.L., and Chasson, A.L., Use of procainamide gels in the purification of human and horse serum butyrylcholinesterase, *Biochem. J.,* 211, 243, 1983.
19. Lockridge, O., Eckerson, H.W., and La Du, B.N., Interchain disulfide bonds and subunit organization in human serum cholinesterase, *J. Biol. Chem.,* 254, 8324, 1979.
20. De La Hoz, D., Doctor, B.P., Ralston, J.S., and Wolfe, A.D., A simplified procedure for the purification of large quantities of fetal bovine serum acetylcholinesterase, *Life Sci.,* 39, 195, 1986.
21. Grunwald, J., Marcus, D., Papier, L., Raveh, L., Pittel, Z., and Ashani, Y., Large scale purification and long term stability of human butyrylcholinesterase: A potential bioscavenger drug, *J. Biochem. Biophys. Methods,* 34, 123, 1997.
22. Raveh, L., Grunwald, J., Marcus, D., Papier, Y., Cohen, E., and Ashani, Y., Human butyrylcholinesterase as a general prophylactic antidote for nerve agent toxicity; *in vitro* and *in vivo* quantitative characterization, *Biochem. Pharmacol.,* 45, 2465, 1993.
23. Genovese, R.F. and Doctor, B.P., Behavioral and pharmacological assessment of butyrylcholinesterase in rats, *Pharmacol. Biochem. Behav.,* 51, 647, 1995.
24. Saxena, A., Raveh, L., Ashani, Y., and Doctor, B.P., Structure of glycan moieties responsible for the extended circulatory life of fetal bovine serum acetylcholinesterase and equine serum butyrycholinesterase, *Biochemistry,* 36, 7481, 1997.
25. Saxena, A., Ashani, Y., Raveh, L., Stevenson, D., Patel, T., and Doctor, B.P., Role of oligosaccharides in the pharmacokinetics of tissue-derived and genetically engineered cholinesterases, *Mol. Pharmacol.,* 53, 112, 1998.
26. Jenkins, T., Balinski, D., and Patient, D.W., Cholinesterase in plasma: First reported absence in the Bantu; Half-life determination, *Science,* 156, 1748, 1967.
27. Stovner, J. and Stadsjkeuv, K., Suxamethonium apnea terminated with commercial serum cholinesterase, *Acta Anaesth. Scand.,* 20, 211, 1976.
28. Ostergaard, D., Viby-Mogensen, J., Hanel, H.K., and Skovgaard, L.T., Half-life of plasma cholinesterase, *Acta Anaesth. Scand,* 32, 266, 1988.
29. Brandeis, R., Raveh, L., Grunwald, J., Cohen, E., and Ashani, Y., Prevention of soman-induced cognitive deficits by pretreatment with human butyrylcholinesterase in rats, *Pharmacol. Biochem. Behav.,* 46, 889, 1993.
30. Hunter, A.J. and Roberts, F.F., The effect of pirenzepine on spatial learning in the Morris water maze, *Pharmacol. Biochem. Behav.,* 30, 519, 1988.
31. Smith, G., Animal models of Alzheimer's disease: Experimental cholinergic denervation. *Brain Res. Rev.,* 13, 103, 1988.
32. Maxwell, D.M., Brecht, K.M., Doctor, B.P., and Wolfe, A.D., Comparison of antidote protection against soman by pyridostigmine, HI-6 and acetylcholinesterase, *J. Pharmacol. Exper. Ther.,* 264, 1085, 1993.
33. Sands, S.F. and Wright, A.A., Primate memory: Retention of serial list items by a rhesus monkey, *Science,* 209, 938, 1980.

34. Castro, C. and Finger, A., The use of serial probe recognition in nonhuman primates as a method for detecting cognitive deficits following CNS challenge, *Neurotoxicology,* 125, 125, 1991.
35. Waugh, N.C., Serial position and memory span, *Am. J. Psychol.,* 73, 68, 1960.
36. Wickelgren, W.A. and Norman, D.A., Strength models and serial position in short-term recognition memory, *J. Math. Psychol.,* 3, 316, 1966.
37. Sullivan, E.V. and Sugar, H.J., Nonverbal recognition and recency discrimination deficits in Parkinson's disease and Alzheimer's disease, *Brain,* 112, 1503, 1989.
38. Castro, C.A., Larsen, T., Finger, A.V., Solana, R.P., and McMaster, S.B., Behavioral efficacy of diazepam against nerve agent exposure in rhesus monkeys, *Pharmacol. Biochem. Behav.,* 41, 159, 1991.
39. Farrer, D.N., Yochmowitz, M.G., Mattson, J.L., Lof, N.E., and Bennett, C.T., Effects of benactyzine on an equilibrium and multiple response task in rhesus monkeys, *Pharmacol. Biochem. Behav.,* 16, 605, 1982.
40. Blick, D.W., Murphy, M.R., Fanton, J.W., Kerenyi, S.Z., Miller, S.A., and Hartgraves, S.L., Incapacitation and performance recovery after high-dose soman: Effects of diazepam, *Proceedings of the Medical Chemical Defense Bioscience Review,* Columbia, MD, 1989, 219.
41. Blick, D.W., Kerenyi, S.Z., Miller, S.A., Murphy, M.R., Brown, G.C., and Hartgraves, S.I., Behavioral toxicity of anticholinesterases in primates: Chronic pyridostigmine and soman interactions, *Pharmacol. Biochem. Behav.,* 38, 527, 1991.
42. Boskovic, B., The influence of 2-(*0-cresyl*)-4 H-1:3:2:-benzodioxaphosphorin-2-oxide (CBDP) on organophosphate poisoning and its therapy, *Arch. Toxicol.,* 42, 207, 1979.
43. Sterri, S.H., Lyngaas, S., and Fonnum, F., Toxicity of soman after repetitive injection of sublethal doses in guinea pig and mouse, *Acta Pharmacol. Toxicol.,* 49, 266, 1981.
44. Clement, J.G., Importance of aliesterase as a detoxification mechanism for soman (pinacolylmethylphosphonofluoridate) in mice, *Biochem. Pharmacol.,* 33, 3807, 1984.
45. Maxwell, D.M., Brecht, K.M., and O'Neill, B.L., The effect of carboxyesterase inhibition on interspecies differences in soman toxicity, *Toxicol. Lett.,* 39, 35, 1987.
46. Ashani, Y., Grauer, D., Grunwald, J., Allon, N., and Raveh L., Current capabilities in extrapolating from animal to human the capacity of human BChE to detoxify organophosphates, in *Structure and Function of Cholinesterases and Related Proteins,* Doctor, B.P. et al., Eds., Plenum Press, New York, 1998, 255.
47. Cascio, C., Comite, C., Ghiara, M., Lanza, G., and Popnchione, A., The use of serum cholinesterase in severe phosphorus poisoning, *Minerva Anestesiol.,* 54, 337, 1988.
48. Caranto, G.R., Waibel, K.H., Asher, J.M., Larrison, R.W., Brecht, K.M., Schutz, M.B., Raveh L., Ashani, Y., Wolfe, A.D., Maxwell, D.M., and Doctor B.P., Amplification of the effectiveness of acetylcholinesterase for detoxification of organophosphorus compounds by bis-quaternary oximes, *Biochem. Pharmacol.,* 47(2), 347, 1994.
49. Saxena, A., Maxwell, D.M., Quinn, D.M., Radic, Z., Taylor, P., and Doctor, B.P., Mutant acetylcholinesterases as potential detoxification agents for organophosphate poisoning, *Biochem. Pharmacol.,* 54, 269, 1997.
50. Ordentlich, A., Barak, D., Kronman, C., Ariel, N., Segall, Y., Velan, B., and Shafferman, A., The architecture of human acetylcholinesterase active center probed by interactions with selected organophosphate inhibitors, *J. Biol. Chem.,* 271, 11953, 1996.
51. Ashani, Y., Radic, Z., Tsigelny, I., Vellom, D.C., Pickering, N.A., Quinn, D.M., Doctor, B.P., and Taylor, P., Amino acid residues controlling reactivation of organophosphonyl conjugates of acetylcholinesterase by mono- and bisquaternary oximes, *J. Biol. Chem.,* 270, 6370, 1995.

52. Saxena, A., Doctor, B.P., Maxwell, D.M., Lenz, D.E., Radic, Z., and Taylor, P., The role of glutamate-199 in the acing of cholinesterase, *Biochem. Biophys. Res. Commun.*, 197:1, 343, 1993.
53. Ordentlich, A., Kronman, C., Barak, D., Stein, D., Ariel, N., Marcus, D., Velan, B., and Shafferman, A., Engineering resistance to 'aging' of phosphylated human acetylcholinesterase. Role of hydrogen bond network in the active center, *FEBS Lett.*, 334, 215, 1993.
54. Shafferman, A., Ordentlich, A., Barak, D., Stein, D., Ariel, N., and Velan, B., Aging of phosphylated human acetylcholinesterase: Catalytic processes mediated by aromatic and polar residues of the active centre, *Biochem. J.*, 318, 833, 1996.
55. Maxwell, D.M., Brecht, K., Saxena, A., Feaster, S., and Doctor, B.P., Comparison of cholinesterases and carboxylesterase as bioscavengers for organophosphorus compounds, in *Structure and Function of Cholinesterases and Related Proteins,* Doctor, B.P. et al., Eds., Plenum Press, New York, 1998, 387.
56. Cygler, M., Schrag, J.D., Sussman, J.C., Harel, M., Silman, I., Gentry, M.K., and Doctor, B.P., Relationship between sequence conservation and three-dimensional structure in a large family of esterases, lipases and related proteins, *Protein Sci.*, 2, 366, 1993.
57. Maxwell, D.M., Lieske, C.N., and Brecht, K.M., Oxime-induced reactivation of carboxylesterase inhibited by organophosphorus compounds, *Chem. Res. Toxicol.*, 7, 428, 1994.
58. Myers, D.K., Mechanism of the prophylactic action of diacetylmonoxime against sarin poisoning, *Biochim. Biophys. Acta,* 334, 555, 1959.
59. Maxwell, D.M., Detoxication of organophosphorus compounds by carboxylesterase, in *Organophosphates: Chemistry, Fate and Effects,* Chambers, J.E. and Levi, P.E., Eds., Academic Press, San Diego, 1992.
60. Saxena, A., Redman, M.G., Jiang, X., Lockridge, O., and Doctor, B.P., Differences in active site gorge dimensions of cholinesterases revealed by binding of inhibitors to human butyrylcholinesterase, *Biochemistry,* 36, 14642, 1997.
61. Taylor, P., Wong, L., Radic, Z., Tsigelny, I., Bruggeman, R., Hosea, N.A., and Berman, H.A., Analysis of cholinesterase inactivation and reactivation by systematic structural modification and enantiomeric selectivity, *Chem.-Biol. Interact.*, 3, 119, 1999.
62. Jarv, J., Stereochemical aspects of cholinesterase catalysis, *Bioorg. Chem.*, 12, 259, 1984.
63. De Jong, L.P.A. and Benschop, H.P., Biochemical and toxicological implications of chirality in anticholinesterase organophosphates, in *Stereoselectivity of Pesticides: Biological and Chemical Problems,* Ariens, E.J., van Rensen, J.J.S., and Welling, W., eds., Elsevier, Amsterdam, 1988.
64. Clement, J.G., Benschop, H.P., De Jong, L.P.A., and Wolthuis, O., Stereoisomers of soman: Inhibition of serum carboxylic ester hydrolase and potentiation of their toxicity by CBDP in mice, *Toxicol. Appl. Pharmacol.*, 89, 141, 1987.
65. Jokanovic, M., Kosanovic, M., and Maksimovic, M., Interaction of organophosphorus compounds with carboxylesterase in the rat, *Arch. Toxicol.*, 70, 444, 1996.
66. Lockridge, O., Blong, R.M., Masson, P., Froment, M.-T., Millard, C.B., and Broomfield, C.A., A single amino acid substitution Gly117His confers phosphotriesterase (organophosphorus acid anhydride hydrolase) activity on human butyrylcholinesterase, *Biochemistry,* 36, 786, 1997.
67. Newcomb, R.D., Campbell, P.M., Ollis, D.L., Cheah, E., Russell, R.J., and Oakeshott, J.G., A single amino acid substitution converts a carboxylesterase to organophosphorus hydrolase and confers insecticide resistance on a blowfly, *Proc. Natl. Acad. Sci. U.S.A.*, 94, 7464, 1997.
68. Ashani, Y., Peggins III, J.O., and Doctor, B.P., Mechanism of inhibition of cholinesterases by huperzine A., *Biochem. Biophys. Res. Commun.*, 184, 719, 1992.

69. Ashani, Y., Grunwald, J., Kronman, C., Velan, B., and Shafferman, A., Role of tyrosine 337 in binding of huperzine A to the active site of human acetylcholinesterase, *Mol. Pharmacol.,* 45, 555, 1993.
70. Saxena, A., Qian, N., Kovach, I.M., Kozikowski, A.P., Pang, Y.P., Vellom, D.C., Radic, Z., Quinn, D., Taylor, P., and Doctor, B.P., Identification of amino acid residues involved in the binding of huperzine A to cholinesterases, *Protein Sci.,* 3, 1770, 1994.
71. Grunwald, J., Raveh, L., Doctor, B.P., and Ashani Y., Huperzine A as a pretreatment candidate drug against nerve agent toxicity, *Life Sci.,* 54, 991, 1994.
72. Lallement, G., Veyret, J., Masqueliez, C., Aubriot, S., Burckhart, M.F., and Baubichon, D., Efficacy of huperzine in preventing soman-induced seizures, neuropathological changes and lethality, *Fund. Clin. Pharmacol.,* 11, 387, 1997.
73. Ashani, Y., Grunwald, J., Alakali, D., Cohen, G., and Raveh, L., Studies with huperzine A, a new candidate in the search of prophylaxis against nerve agents, *Proceedings of the Medical Defense Bioscience Review,* Baltimore, MD, 1996, 105.
74. Munnecke, D.M., Chemical, physical and biological methods for the disposal and detoxification of pesticides, *Residue Rev.,* 70, 1, 1979.
75. Wood, L.L., Hardegen, F.J., and Hahn, P.A., Enzyme Bound Polyurethane. U.S. Patent 4,342,834, 1982.
76. Havens, P.I. and Rase, H.F., Reusable immobilized enzyme polyurethane sponge for removal and detoxification of localized organophosphate pesticide spills, *Ind. Eng. Chem. Res.,* 32, 2254, 1993.
77. Turner, B.C. and Glotfelty, D.E., Field air sampling of pesticide vapors with polyurethane foam, *Anal. Chem.,* 49, 7, 1977.
78. Gordon, R.K., Feaster, S.R., Russell, A.J., LeJeune, K.E., Maxwell, D.M., Lenz, D.E., Ross, M., and Doctor, B.P., Organophosphate skin decontamination using immobilized enzymes, *Chemico-Biol. Interact.,* 463, 1999.

7 Nerve Agent Bioscavengers: Protection against High- and Low-Dose Organophosphorus Exposure*

David E. Lenz, Clarence A. Broomfield, Donald M. Maxwell, and Douglas M. Cerasoli

CONTENTS

I. Introduction ... 215
II. Current Therapy for Nerve Agent Exposure ... 217
III. Nerve Agent Bioscavengers: an Alternative to Conventional Approaches ... 218
IV. Stoichiometric Scavengers and the Protection They Offer ... 220
A. Antibodies ... 220
B. Enzymes ... 220
V. Catalytic Bioscavengers ... 228
VI. Behavioral Effects ... 233
VII. Behavioral Effects of Scavengers Alone ... 233
VIII. Summary ... 235
Acknowledgment ... 237
References ... 237

I. INTRODUCTION

Organophosphorous anticholinesterases (OPs), usually acid anhydride derivatives of phosphoric acid, are among the most toxic substances identified.[1] Originally, OP were developed for use as insecticides,[2] but their extreme toxicity toward higher vertebrates has led to their adoption as weapons of warfare.[3] The OPs most com-

* The opinions or assertions contained herein are the private views of the authors, and are not to be construed as reflecting the view of the Department of the Army or the Department of Defense.

0-8493-0872-0/01/$0.00+$.50

monly utilized as chemical weapons (referred to as nerve agents) are anhydrides of hydrocyanic acid, hydrofluoric acid, or of a derivative of thiocholine: tabun (GA), sarin (GB), soman (GD), cyclohexylmethyl phosphonofluoridate (GF), and ethyl-S-diisoproplyaminoethyl methylphosphonothioate (VX). Their molecular weights range from 140 to 267 Daltons (Da) and, under standard conditions, they are all liquids that differ in their degrees of volatility.[4] They have median lethal dose (LD_{50}) values in mammals, including estimates for humans, in the μg/kg dose range for all routes of exposure except dermal, where LD_{50} doses are in the mg/kg range.[3]

OPs produce their acute toxic effects by irreversibly inhibiting the enzyme acetylcholinesterase (AChE, E.C. 3.1.1.7).[5,6] This inhibition leads to an increase in the concentration of acetylcholine in the cholinergic synapses of both the peripheral and central nervous systems. The physiological consequences of elevated acetylcholine include alterations in the function of the respiratory center and over-stimulation at neuromuscular junctions.[7–13] A sufficiently high level of acetylcholine or a sufficiently rapid increase in acetylcholine concentration precipitates a cholinergic crisis, resulting in dimming of vision, headache, shortness of breath, muscle weakness, and seizure. In the extreme, organophosphorus intoxication can be a life-threatening event, with death usually resulting from respiratory failure. This is often accompanied by secondary cardiovascular components including hypotension, cardiac slowing, and arrhythmias.[6] The toxic effects ensuing from low-level exposure have not been well defined and are still the subject of some debate. To provide some context for the ensuing discussion of biological scavengers, the following definitions, suggested by the Gulf War Research Coordination Board, will be adopted to define low-level exposure:

- Level 1: An exposure that results in no clinical signs (and for humans no subjective symptoms) and minimal AChE inhibition (0–20% reduction in red blood cell [RBC] AChE).
- Level 2: An exposure that results in no clinical signs (and for humans no subjective symptoms) and moderate AChE inhibition (>20% reduction in RBC AChE).
- Level 3: An exposure that results in mild clinical signs, such as salivation, miosis, and tachycardia. In humans such an exposure would also be expected to cause symptoms such as shortness of breath.

These low-level exposure definitions refer to the effects observed in a single exposure of less than a 24-h duration.[14] While a level of AChE inhibition is not mentioned for Level 3 exposures, the symptoms described can be considered cholinergic in nature, probably resulting from inhibition of synaptic AChE. This suggests that a prophylactic approach based on the reduction of the concentration of OP toxicant in the blood before it can reach its site of action (synaptic endplates) should be particularly effective; potentially incapacitating or even toxic exposures could be mitigated to Level 3-type outcomes, and lower level exposures could be rendered inconsequential.

II. CURRENT THERAPY FOR NERVE AGENT EXPOSURE

The conventional approach to treatment of OP intoxication involves efforts to counteract the effects of AChE inhibition. Cholinolytic drugs such as atropine are administered at the onset of signs of OP intoxication to antagonize the effects of the elevated acetylcholine levels that result from the inhibition of AChE.[15] Additionally, an oxime nucleophile is given, which reacts with the inhibited (phosphonylated) enzyme to displace the phosphonyl group and restore normal activity.[16] In the United States, the oxime of choice for treatment of nerve agent poisoning is the chloride salt of 2-PAM, usually referred to as 2-PAM Cl, although bis-pyridinium oximes may be more effective depending on the particular OP agent.[17] Anticonvulsant drugs such as diazepam are also administered to control OP-induced tremors and convulsions. In conjunction with therapy, individuals at high risk for exposure to nerve agents are pretreated with a spontaneously reactivating AChE inhibitor such as pyridostigmine, which temporarily masks the active site of a fraction of AChE molecules and thus protects the enzyme from irreversible inhibition by the OP agent.[18]

While these treatment regimens have been the standard for many years, they are not ideal and suffer from a number of disadvantages. The major drawback of current approaches is that, while they can be effective in preventing lethality, they do not prevent performance deficits, behavioral incapacitation, loss of consciousness, or permanent brain damage, all of which can result from acute OP toxicity.[19]

Several nerve agents, including GF, sarin and, in particular, soman, present an additional therapeutic challenge in that after they inhibit AChE, they undergo a second reaction in which the phosphonyl group attached to the inhibited enzyme is dealkylated. This process, known as aging, results in a phosphonylated AChE that is refractory to either spontaneous or oxime-mediated reactivation.[20] The ineffectiveness of therapeutically administered oxime as a treatment for some nerve agents explains the continued research efforts aimed at alternative approaches to protection.[21] In particular, efforts have focused on approaches that prevent the critical enzyme AChE from becoming inhibited in the first place. Although the currently used pretreatment/therapy regimen is able to protect soldiers against the otherwise lethal effects of nerve agents, it does not adequately protect against the incapacitation that results from high levels of nerve agent exposure. Furthermore, it appears that greater than marginal improvement of these pharmacological approaches will not be possible, because stronger drugs or higher doses are likely to produce unacceptable performance decrements by themselves.[21,22]

With respect to low-dose exposures, there are no standardized treatment regimes. Indeed, due to the subtlety of the symptoms and the difficulty in detecting decrements in cholinesterase activity associated with Level 1, 2, or 3 exposures, many such exposures may go unnoticed and unreported. Individuals at risk for low-dose exposures (such as laboratory researchers working with OPs, chemical plant staff, and farmers using pesticides) can be routinely monitored for red blood cell (RBC) AChE activity; should lowered AChE activity be detected, responses include removal of the individual from the work environment, closer monitoring of RBC AChE levels, and reassessment of procedures and practices to reduce the chance of subsequent exposures. The long-term consequences of low-dose exposure to OP, if any exist, remain unknown.

III. NERVE AGENT BIOSCAVENGERS: AN ALTERNATIVE TO CONVENTIONAL APPROACHES

While successful, current treatments for acute nerve agent poisoning always result in the victim suffering a toxic insult that subsequently must be therapeutically managed. In contrast, recent efforts have focused on identifying proteins that can act as biological scavengers of organophosphorus compounds and can remain stable in circulation for long periods of time. The concept of using a protein that can react with a nerve agent, either stoichiometrically or catalytically, to protect against the toxic effects of those compounds, either acute or low level, is not new. As early as 1956 it was shown that injection of exogenous paraoxonase could protect rats against several times the LD_{50} of paraoxon.[23] This approach avoids the side effects associated with current antidotes and the requirement for their rapid administration, by prophylactically inactivating (through sequestration or hydrolysis) anticholinesterase agents before they can react with the target AChE.[21,24–31] The time frame for this inactivation to occur before endogenous AChE is affected is quite narrow (estimated to be approximately 2 min in humans), so especially for situations involving acute exposure, the scavenger function must be very rapid, irreversible, and specific.[32] Ideally, the scavenger would enjoy a long residence time in the bloodstream, would be biologically innocuous in the absence of nerve agent, and would not present an antigenic challenge to the immune system. For these reasons, prime efforts to identify candidate bioscavengers have focused on enzymes of mammalian (usually human) origin.

Candidate bioscavenger proteins, in general, function either by stoichiometrically binding and sequestering the anticholinesterase or by catalytically cleaving the OP substrate into biologically inert products. In the former category are naturally occurring human proteins that bind nerve agents, including enzymes such as cholinesterases (ChE) and carboxylesterases (CaE), as well as antibodies specific for nerve agent haptens. Each of these stoichiometric scavengers has the capacity to bind one or two molecules of nerve agent per molecule of protein scavenger. While this approach has been proven to be effective in laboratory animals, it has the disadvantage that the extent of protection is directly proportional to the concentration of unexposed, active scavenger in the bloodstream at the time of nerve agent exposure. Since the molecular weight of a protein scavenger is in the range of 80,000 Da and the molecular weight of the nerve agents is about 160 Da, the concentration mass ratio of scavenger to nerve agent is 500:1. Thus, a high concentration of scavenger protein in circulation is necessary to protect against exposure to multiples of an LD_{50} dose of nerve agent, although lower concentrations would be sufficient to prevent inactivation of synaptic AChE after a low-dose exposure. It might be possible to mitigate the need for large amounts of scavenger by also administering, either prophylactically or immediately post-exposure, a currently fielded oxime. Oxime treatment would allow for the continual reactivation of the bioscavenger *in vivo,* in effect converting the stoichiometric scavenger into a pseudo-catalytic one.

Candidate enzymes with bona fide catalytic activity against nerve agents include the human organophosphorus acid anhydride hydrolases (OPAH), such as paraoxonase (hu-Pon). Additionally, the ability to generate catalytic antibodies in response to appropriate transition state analogs suggests that nerve agent-specific antibodies that catalyze hydrolysis of their ligands could be effective bioscavengers.[33,34] Finally, the ability to engineer site-specific amino acid mutations into naturally occurring scavenger enzymes can allow investigators to alter the binding and/or catalytic activities of these enzymes. In general, the use of scavengers with catalytic activity would be advantageous because small amounts of enzyme, meaning lower concentrations in circulation, would be sufficient to detoxify both large amounts of nerve agent (as in an acute exposure) or lower amounts of agent associated with low-dose exposure Levels 1, 2, and 3.

By nearly all criteria, the use of biological scavengers, either stoichiometric or catalytic, as a prophylactic approach to providing protection against an exposure to either a low-level or a lethal dose of a nerve agent offers numerous advantages over conventional treatments. In fact, the half-time for reaction of a nerve agent with a biological scavenger can be calculated using some very conservative assumptions. Based on toxicity estimates in humans, the expected concentration of a nerve agent in the blood at an LD_{50} dose would be about 8×10^{-7} M.[35] The bimolecular rate constant for reaction of soman with AChE is ~9×10^{7} M^{-1} min^{-1}.[36,37] If a scavenger were present in blood at a concentration of 1 mg/mL (1×10^{-5} M), then the rate constant for reaction of scavenger with toxicant would be pseudo first order and the $t_{1/2}$ for the reduction of toxicant would be ~3×10^{-4} min. Under those conditions, which assume perfect mixing and that all of the scavenger and all of the toxicant remain in the bloodstream, the concentration of toxicant would be reduced to 1/1000 of its initial concentration within 10 half-times (2×10^{-3} min). In practical terms, the inhibition of AChE by nerve agent essentially would be 0 under Levels 1, 2, and 3 exposure definitions given above. Where actual measurements have been made of the rate of reduction of concentration of soman in animals (guinea pigs), it was found that, in the absence of an exogenous scavenger, the concentration of a $2 \times LD_{50}$ dose of soman in circulation was reduced by 1000-fold in about 1.5 min.[38] These results support our contention that, if a bioscavenger were present in circulation at the time of exposure, the reduction in toxicant concentration to a physiologically insignificant level (with no measurable inhibition of AChE) would be very rapid, and would certainly occur in less than one circulation time at any concentration of OP that could produce a low-level effect in an untreated animal. The need to administer, repetitively, a host of pharmacologically active drugs with a short duration of action at a precise time following exposure is all but eliminated if a scavenger is used. The potential for having to use mission-oriented protective posture (MOPP) gear is greatly reduced. Finally, with the appropriate scavenger(s), such an approach could afford protection against all of the current threat agents, including those that induce rapid aging of AChE and are refractory to treatment by the current atropine and oxime treatment regime.

IV. STOICHIOMETRIC SCAVENGERS AND THE PROTECTION THEY OFFER

A. ANTIBODIES

More than 25 years ago efforts were undertaken to protect animals by actively immunizing them with analogs of paraoxon or soman attached to appropriate protein carrier molecules, to elicit an antibody response against these two highly toxic organophosphorus compounds.[39,40] As summarized in Table 7.1, rabbits that developed antibodies against paraoxon were protected against 2 to 3 times the LD_{50} of paraoxon.[41] The extent of protection was found to be directly related to the concentration of the paraoxon-specific antibodies in circulation. Significantly, the protected animals were essentially asymptomatic and did not require the administration of any additional therapeutic drugs. Rabbits immunized with an analog of soman were not protected against the administration of a lethal dose of that compound. Subsequently it was determined that the polyclonal antibodies induced in these animals were not of sufficiently high affinity to successfully compete with AChE for the binding of soman.[40]

Based on these limited but promising results, efforts were made to generate high-affinity monoclonal antibodies that could be used to afford passive protection from nerve agents. Hunter et al.[42] reported the production of the first anti-soman monoclonal antibodies, which were subsequently shown to be of sufficiently high affinity to compete with AChE for soman binding *in vitro*.[40] When mice were passively immunized with these antibodies they failed to show any protection against the *in vivo* toxicity of soman, although the time to death was almost doubled in the animals pretreated with antibody.[40] Further *in vitro* characterization of the monoclonal antibodies showed that their anti-soman binding constants were only in the micro-molar range, but that they were highly soman-specific, in that they did not bind the structurally related nerve agent sarin.[43] Subsequent calculations suggest that to afford protection on a stoichiometric level against soman or sarin, a monoclonal antibody must have a binding constant in the 50 nano-molar range.[35]

B. ENZYMES

A number of different enzymes that react with OPs but do not catalyze their hydrolysis have been tested for their ability to provide protection against nerve agent poisoning. Wolfe et al.[44] first reported the use of exogenously administered AChE as a bioscavenger (Table 7.2). In that study, fetal bovine serum acetylcholinesterase (FBS-AChE) was administered to mice 20 h before a multiple LD_{50} challenge of VX was administered. Complete protection was afforded against a $2 \times LD_{50}$ dose of VX (100% survival of exposed animals), while moderate protection (80% survival rate) was observed after a challenge of $3 \times LD_{50}$. No protection was observed against higher multiple LD_{50} challenges of VX. When animals pretreated with FBS-AChE were exposed to soman, little protection was afforded. However, FBS-AChE pretreatment in conjunction with post-exposure atropine and 2-PAM treatment protected mice from $2 \times LD_{50}$ of soman. The authors reported that animals displayed no detectable side effects in response to administration of FBS-AChE.

TABLE 7.1
Protection from Organophosphorus Intoxication by Antibody Bioscavengers

Bioscavenger	Test Species	Nerve Agent	Protection (LD_{50})[a]	Serum $T_{1/2}$[b]	Ref.
Polyclonal Antibodies[c]	Rabbit	Paraoxon	2–3	Days to Weeks	41
Polyclonal Antibodies	Rabbit	GD	-	Days to Weeks	41
Monoclonal Antibody[c]	Mouse	GD	−/extended mean survival time	(6–8 days)	40, 77

[a]Values represent multiples of median lethal doses (LD_{50}) of nerve agent survived after antibody administration.

[b]Half-life of antibodies in blood circulation.

[c]Polyclonal antibodies: the endogenous serum titer after priming with nerve agent analogs. Monoclonal antibody: produced *in vitro* by a hybridoma, then passively administered to naïve mice.

TABLE 7.2
Protection from Organophosphorus Intoxication by the Bioscavenger FBS-AChE

Bioscavenger	Test Species	Nerve Agent	Protection (LD_{50})[a]	Serum $T_{1/2}$[b]	Ref.
FBS-AChE	Rhesus Monkey	GD	2–5	30–40 h	45, 48
FBS-AChE	Mouse	GD	2 (w/ Atropine + 2-PAM)	40–50 h	44
FBS-AChE	Mouse	GD	2 (after CBDP treatment)	~24 h	61
FBS-AChE	Mouse	GD	2–8	24–26 h	46, 50, 62
FBS-AChE	Mouse	MEPQ	4	~24 h	50, 61
FBS-AChE	Mouse	VX	2–3.6	~24–50 h	44, 50, 61

[a]Values represent multiples of median lethal doses (LD_{50}) of nerve agent survived after FBS-AChE administration.

[b]Half-life of administered FBS-AChE in blood circulation.

Maxwell and co-workers carried out a similar set of experiments using rhesus monkeys pretreated with the scavenger FBS-AChE.[45] When monkeys pretreated with FBS-AChE were challenged with either 1.5 or 2.5 × LD_{50} of soman, there was protection (Table 7.2) with no decrements in performance on the serial probe recognition (SPR, discussed in Behavioral Effects, below) task as compared with animals treated with FBS-AChE alone. The animals were also monitored for the generation of an antibody response against the administered FBS-AChE, but none was detected. The authors caution, however, that whenever a foreign protein is administered to an animal, the potential for an antibody-mediated immune response must be assessed on a case-by-case basis. Maxwell and co-workers also compared the relative protection against soman afforded to mice by three different treatments: pyridostigmine pretreatment with atropine therapy post-exposure, post-exposure oxime (HI-6) and atropine therapy, or FBS-AChE pretreatment alone.[46] The authors concluded that the FBS-AChE pretreatment offered superior protection against both soman toxicity (survival after 8 to 10 × LD_{50} doses) and behavioral incapacitation. The results of these and other studies using FBS-AChE are summarized in Table 7.2.

Broomfield and co-workers reported that equine butyrylcholinesterase (EqBChE) afforded complete protection against a 2 × LD_{50} challenge dose of soman in rhesus monkeys (Table 7.3) with no supporting therapy and against 3 to 4 × LD_{50} doses when atropine was also administered (post-exposure).[47] Protection against a single LD_{50} dose of sarin was also demonstrated. In all cases (Table 7.3) there were no fatalities. Furthermore, when animals were assessed for behavioral deficits again using an SPR task, they all returned to baseline performance within 9 h after soman exposure (*vide infra*).[22]

In a related study, Wolfe et al. assessed the ability of pretreatment with either FBS-AChE or EqBChE to protect rhesus monkeys against multiple LD_{50} doses of soman (Tables 7.2 and 7.3).[48] Survival and the ability to perform a different behavioral test, the Primate Equilibrium Platform (PEP) task, were the variables assessed. Those animals that received FBS-AChE as a pretreatment were protected against a cumulative exposure of 5 × LD_{50} of soman and showed no decrement in the PEP task. Two of the four monkeys that received purified EqBChE did show some transient decrement in PEP task performance when the cumulative dose of soman exceeded 4 × LD_{50}. All of the experimental animals were observed for an additional 6 weeks, and none displayed any residual or delayed performance decrements suggesting no residual adverse effects. These results were reviewed and expanded upon by Doctor et al., wherein mice pretreated with FBS-AChE were also administered the oxime HI-6 immediately post-exposure to sarin.[49] In theory, the oxime will continuously regenerate the inhibited scavenger enzyme *in vivo;* this approach is predicted to increase the amount of sarin that could be scavenged by a given amount of AChE, making this stoichiometric scavenger pseudo-catalytic. The therapeutic addition of HI-6 after pretreatment with FBS-AChE was found to enhance the efficacy of the scavenger enzyme against sarin *in vivo,* increasing the ratio of neutralized OP compound per FBS-AChE molecule from 1:1 (in the presence of AChE alone) to roughly 65:1.

Maxwell et al. identified carboxylesterase as another enzyme with the potential to be a good anti-organophosphorous scavenger molecule (summarized in Table 7.4).[50]

TABLE 7.3
Protection from Organophosphorus Intoxication by the Bioscavenger EqBuChE

Bioscavenger	Test Species	Nerve Agent	Protection (LD_{50})[a]	Serum $T_{1/2}$[b]	Ref.
EqBChE	Rhesus Monkey	GB	1	620 h	47
EqBChE	Rhesus Monkey	GD	2 (4 w/ atropine)	620 h	47
EqBChE	Rhesus Monkey	GD	5	30–40 h	48

[a]Values represent multiples of median lethal doses (LD_{50}) of nerve agent survived after EqBChE administration.

[b]Half-life of administered EqBChE in blood circulation.

TABLE 7.4
Protection from Organophosphorus Intoxication by Endogenous Plasma CaE

Bioscavenger	Test Species	Nerve Agent	Protection (LD_{50})[a]	Ref.
CaE[b]	Mouse	GD	16	56
CaE	Guinea Pig	GD	3.5	56
CaE	Rabbit	GD	3	56
CaE	Rat	GD	8–9	56, 82
CaE	Rat	GB	8	84
CaE	Rat	GA	4–5	84
CaE	Rat	VX	1	84
CaE	Rat	Paraoxon	2	84

[a]Values represent multiples of median lethal doses (LD_{50}) of nerve agent survived due to the presence of CaE. Because CaE is an endogenous plasma protein in these species, the protection it offers was measured by comparing LD_{50} values in untreated and CBDP-treated animals; 2 mg/kg CBDP completely abolishes endogenous plasma CaE activity.[84]

[b]For each species, the activity of the host's endogenous CaE was tested.

While AChE and BChE were found to be more efficient scavengers for soman in mice than CaE (i.e., they have higher bimolecular rate constants), the latter enzyme was capable of affording equal protection on a molar basis. Carboxylesterases (CaE; EC 3.1.1.1) catalyze the hydrolysis of a wide variety of aliphatic and aromatic esters and amides.[51] As with AChE, catalysis occurs by a two-step process in which the substrate acylates the active site serine of CaE, which subsequently deacylates by the addition of water.[52] CaE can be distinguished from AChE and BChE by the fact that AChE and BChE react with positively charged carboxylesters, such as acetylcholine and butyrylcholine, and are readily inhibited by carbamates, while CaE does not react with positively charged substrates and is inhibited by carbamates only at high concentrations.[52] These differences in substrate specificity also extend to the reaction of CaE with OP compounds. Positively charged OP compounds, such as VX, react poorly with CaE while neutral OP compounds, such as soman, sarin, and paraoxon, react rapidly. Dephosphorylation of the active-site phosphorylated serine of CaE is a slow process compared to deacylation,[53] and therefore CaE has usually been considered to be a stoichiometric detoxification mechanism for OP compounds.

CaE is 60-kDa enzyme that is found in many mammalian tissues—lung, liver, kidney, brain, intestine, muscle, and gonads—usually as a microsomal enzyme. In some species CaE is also found in high concentration in plasma; plasma CaE is probably synthesized in the liver and secreted into the circulation via the Golgi apparatus.[54] Secretion of CaE appears to be controlled by the presence or absence of a retention signal at the carboxy terminal of the enzyme (Figure 7.1). CaE that is retained in the liver has a highly conserved carboxy terminal tetrapeptide sequence (HXEL in single-letter amino acid code, where X represents any amino acid), while the secretory form of CaE has a disrupted version of this retention signal in which the terminal leucine residue is replaced by either histidine-lysine or histidine-threonine.[54] Mammalian species that have high levels of secretory CaE in their plasma require much larger doses of OP compounds to produce toxicity than species with low levels of plasma CaE.[50] For example, the LD_{50} dose for soman in rats is 10-fold larger than the LD_{50} in non-human primates, which correlates with the differences in the plasma concentrations of CaE found in these species (Figure7.2). Although human CaE has been cloned and expressed,[55] there is no commercial source of highly purified CaE for use in *in vivo* testing of protective efficacy. Therefore, the primary evidence demonstrating the effectiveness of CaE as a stoichiometric scavenger against OP, especially sarin and soman, has been by comparison of OP LD_{50} in animals with high endogenous plasma levels of CaE to OP LD_{50} levels in animals of the same species whose plasma CaE has been chemically inhibited.[56] For example, inhibition of plasma CaE prior to the LD_{50} determination of soman in rats reduces its LD_{50} by approximately 8-fold (Table 7.4), strongly suggesting that circulating CaE is an effective bioscavenger against OP compounds.

Recent investigations of the reactivation of OP-inhibited CaE have suggested that it may be possible to increase its potential as an OP scavenger by exploiting its turnover of OP compounds. Maxwell et al. observed that OP-inhibited CaE does not undergo the aging process that prevents oxime reactivation of OP-inhibited cholinesterases,[57] while Jokanovic et al. found that OP-inhibited CaE from plasma

Biochemical Basis for CaE Cellular Trafficking

Enzyme	COO-HTerminal Residues	Reference
Intracellular CaEs		
Rabbit Es-1	..TE**HIEL**	[88]
Rabbit Es-2	..QK**HTEL**	[89]
Hamster AT51p	..GK**HSEL**	[90]
Human CaE-1	..TE**HSEL**	[91]
Human CaE-2	..ER**HTEL**	[92]
Pig CaE	..IK**HAEL**	[93]
Rat Es-10(pI6.1)	..WK**HVEL**	[94]
Rat Es-B	..PH**HNEL**	[95]
Mouse Es-X	..RE**HVEL**	[96]
Mouse Es-22	..TE**HTEL**	[97]
Consensus	..**HXEL**	
Secreted CaEs		
Mouse Es-1	..TE**HTE*HK***	[98]
Rat Es-1	..TE**HTE*HT***	[99-101]

FIGURE 7.1 The carboxy-terminal amino acid residues of carboxylesterase enzymes from disparate species are aligned to show the conserved "HXEL" motif found among intracellular enzymes (shown in bold letters), and the disrupted versions of this retention motif found in the mouse and rat secreted carboxylesterase isoenzymes (alterations to the motif shown in italics). The capacity of the carboxy-terminal "HXEL" motif to act as an endoplasmic reticulum retention signal has been directly demonstrated.[102]

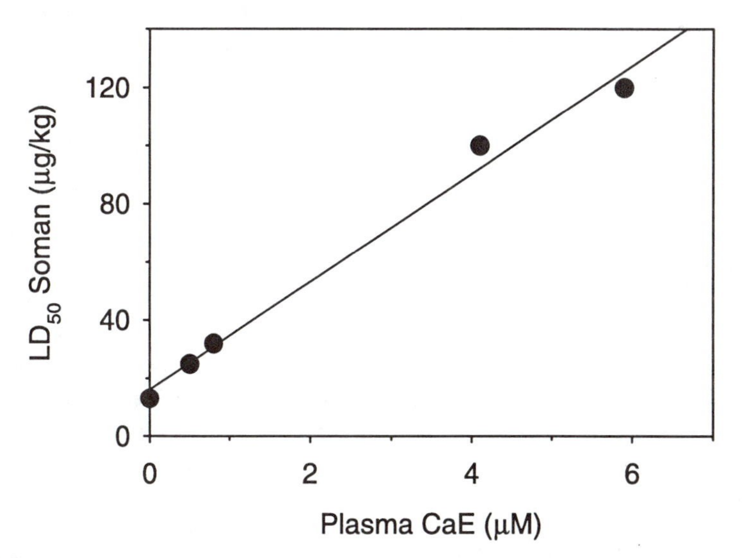

FIGURE 7.2 Effect of plasma CaE concentration on soman LD_{50} (administered s.c.) in different species. Data points (from lower left to upper right of graph) for species were monkey, rabbit, guinea pig, rat, and mouse. Data taken from Maxwell et al.[50]

underwent spontaneous reactivation with a half-time of 1 to 2 h.[58] Comparisons of the amino-acid sequences of CaE, AChE, and BChE are informative with regard to the critical amino acid residues required for occurrence of aging vs. spontaneous reactivation. Of the seven conserved amino acid residues that have been identified by theoretical studies and confirmed by site-directed mutagenesis to be important for aging of OP-inhibited AChE and BChE, only two are conserved in CaE.[59] Conversely, a highly conserved histidine found in CaE from six mammalian species and two insect species, but not in mammalian cholinesterases, correlates with the higher level of spontaneous reactivation of OP-inhibited CaE in comparison to OP-inhibited cholinesterase.[59] Interestingly, introduction of a histidine into BChE at a position nearly identical to the position of the conserved histidine of CaE produces spontaneous reactivation of OP-inhibited BChE.[60]

A more detailed discussion of the relative merits of FBS-AChE, EqBChE, and plasma CaE as scavengers, which describes the extent of protection they offer against a variety of nerve agents, both *in vitro* and *in vivo* in mice, was presented by Doctor et al.[61] The authors note that some of the *in vivo* differences in sensitivity and protection seen may be due to variations in the circulatory pharmacodynamics of the different OP compounds, such that those inhibitors that distribute more slowly from circulation are more readily scavenged. This concept supports the feasibility of using scavengers to protect against low-level exposures of nerve agent. Raveh et al. have provided additional examples that agree with those conclusions.[63,64] The extent of protection afforded by FBS-AChE against soman in marmosets and rhesus monkeys with respect to survival was determined and found to be the same in both species. Significantly, the stoichiometry of the protective dose of FBS-AChE scavenger to OP compound was experimentally determined to be one-to-one on a molar basis in both species of monkey, suggesting that a similar ratio will be maintained in other species, including man. Finally, none of the animals pretreated with scavenger displayed any adverse symptoms following a LD_{100} challenge dose of soman.

Ultimately, the goal of research on scavenger molecules is to generate a means to protect humans from the toxic effects of nerve agents. In an effort to minimize any physiological, immunological, or psychological side effects of scavenger use in humans, research efforts have begun to focus on the use of human BChE (HuBChE), human CaE, and/or FBS-AChE (which does not induce an immune response in rhesus monkeys).[45] In a series of studies, Ashani and his co-workers examined the scavenger properties of FBS-AChE and particularly HuBChE in mice, rats, and rhesus monkeys with respect to several different nerve agents as well as other OP compounds (Table 7.5).[62–64] They found that following administration of exogenous cholinesterase, there was a linear correlation between the concentration of cholinesterase in the blood and the level of protection against OP poisoning. Furthermore, the extent of protection granted to mice was sufficient to counteract multiple LD_{50} doses of soman. When the protective effect of pretreatment with HuBChE was compared in mice and rats, it was found that in both species the same linear correlation existed between blood concentration of HuBChE and protection against soman, sarin, or VX (Table 7.5). They further noted that to be effective, a scavenger had to be present before exposure to the OP compound, because (as discussed above) the nerve agent had to be scavenged within

TABLE 5
Protection from Organophosphorus Intoxication by HuBChE Bioscavengers

Bioscavenger	Test Species	Nerve Agent	Protection (LD_{50})[a]	Serum $T_{1/2}$[b]	Ref.
HuBChE	Rhesus Monkey	GD	2	~30 h	64
HuBChE	Rhesus Monkey	VX	1.5	~30 h	64
HuBChE	Rat	GD	2–3	46 h	63
HuBChE	Rat	VX	2	46 h	63
HuBChE	Mouse	GD	2.1	21 h	64
HuBChE	Mouse	GB	1.6	21 h	64
HuBChE	Mouse	GA	1.8	21 h	64
HuBChE	Mouse	VX	4.9	21 h	64

[a] Values represent multiples of median lethal doses (LD_{50}) of nerve agent survived after HuBChE administration.

[b] Half-life of administered HuBChE in blood circulation.

one blood circulation time period.[63] In the final paper in this series, the authors report similar protection results against a 3.3 LD_{50} dose of soman or a 2.1 LD_{50} dose of VX in rhesus monkeys.[64] They also report considerable protection against soman-induced behavioral deficits in a spatial discrimination task.

V. CATALYTIC BIOSCAVENGERS

While stoichiometric scavengers are able to afford good protection as long as they reside at high levels in the blood stream, they suffer the disadvantage that they are all molecules of high molecular weight (*vide supra*); a comparatively large quantity is required to neutralize a small amount of nerve agent. A catalytic scavenger, even having the same high molecular weight, could be administered in smaller quantities and could produce the same or greater degree of protection. It would also have the advantage of not being consumed in the process of detoxifying the nerve agent, so it would be available to protect against multiple exposures of either high or low dose. Some of these potential bioscavenger proteins along with parameters of their catalytic activities are summarized in Tables 7.6–7.8. As discussed above, in conjunction with an oxime such as HI-6, cholinesterases that have not undergone aging can be continually reactivated to function pseudo-catalytically, eliminating substantially more moles of OP compounds than would be predicted based on binding alone. Furthermore, some enzymes, such as the OPAH from *Pseudomonas diminuta*[65] or the hu-Pon,[66] have intrinsic catalytic anti-organophosphorus activity. The former enzyme has been shown to afford protection against soman lethality in mice and to protect against behavioral side effects (Table 7.6).[67] However, since this bacterially derived enzyme has no known mammalian homologues, it will likely be a potent initiator of immune responses and is therefore unlikely to be appropriate for use as a prophylactic scavenger in humans. Nonetheless, the *Pseudomonas diminuta* OPAH could be used as a one-time pretreatment either in addition to or in place of conventional therapy, since in the short term this enzyme is highly effective against GD, GB, and VX, and alone induces no known behavioral effects. The hu-Pon enzyme has been identified as having a similar potential for affording protection (Table 7.6), but without the complication of inducing an immune response (being an endogenous self-antigen in humans); this enzyme has not yet been tested for efficacy in a mammalian model system.[68]

While the enzymes discussed above possess the desired catalytic activity, none of them is fast enough for use as a nerve agent pretreatment. Since the OP anticholinesterases have been in the environment for only a little over 50 years, it is not likely that any of the enzymes we identify as OPAH have as their primary function the destruction of OP. In fact, an OPAH from an *alteromonas* species has been identified as a prolidase, a dipeptidase that cleaves at a penultimate proline from the carboxyl end of a peptide.[69] Recently, hu-Pon was shown to be a homocysteine thiolactone hydrolase that can protect against protein N-Homocysteinylation.[70] A functional catalytic scavenger must have a lower Km (a measure of the strength of binding of a substrate to the enzyme) and a higher turnover number than has been found to date among these naturally occurring catalytic enzymes, since agent must be

TABLE 6
Kinetic Properties of Naturally Occurring Catalytic Bioscavengers

Bioscavenger	Source Species	Substrate Specificity	km (μM)	Vmax (nmol · min^{-1} · mg^{-1})	Ref.
Phosphotriesterase	*P. Diminuta*	GD	36/500	15/7.3	67/85
Phosphotriesterase	*P. Diminuta*	GB	700	N.D.[a]	85
Phosphotriesterase	*P. Diminuta*	Paraoxon	50	3200	85
Phosphotriesterase	*P. Diminuta*	DFP	100	64	85
			Bimolecular Rate Constant (k_{cat}/km [M^{-1} · (min^{-1}])		
			Q191/R191[b]		
hu-Pon	Human	GD	$2.8 \times 10^6/2.1 \times 10^6$		68
hu-Pon	Human	GB	$9.1 \times 10^5/6.8 \times 10^4$		68
hu-Pon	Human	DFP	3.7×10^4/N.D.		68
hu-Pon	Human	Paraoxon	$6.8 \times 10^5/2.4 \times 10^6$		68

[a] Not determined.

[b] Two naturally occurring allelic variants of hu-Pon (Q191 and R191) have been identified. The activity of each form is shown.

TABLE 7
Kinetic Properties of Catalytic Mutated HuBChE Bioscavengers

Bioscavenger	Substrate Specificity	Spontaneous Reactivation Rate Constant ($\times 10^3$ min^{-1})	Ref.
Wild type HuBChE	GB	<0.05	74
Wild type HuBChE	VX	<0.05	74
Wild type HuBChE	GD	<0.05	74
G117H HuBChE[a]	GB	5	74
G117H HuBChE	VX	7	74
G117H HuBChE	GD	<0.05	74
G117H E197Q HuBChE[b]	GB	62	74
G117H E197Q HuBChE	VX	78	74
G117H E197Q HuBChE	GD (P_SC_R)[c]	6	74
G117H E197Q HuBChE	GD (P_RC_R)[c]	6	74
G117H E197Q HuBChE	GD (P_SC_S)[c]	77	74
G117H E197Q HuBChE	GD (P_RC_S)[c]	128	74

Note: The rate-limiting step in the hydrolysis of organophosphate nerve agents by mutated HuBChEs is the enzyme reactivation step.[74]

[a] A version of HuBChE in which the glycine at amino acid residue 117 has been replaced by histidine.

[b] A double mutant of HuBChE containing both histidine (rather than glycine) at amino acid residue 117 and glutamine in place of glutamic acid at residue 197.

[c] The reactivity with each of the four stereoisomers of GD was determined independently.

TABLE 8
Kinetic Properties of Mouse-Derived Catalytic Antibody Bioscavengers

Bioscavenger	Substrate Specificity	km (μM)	Vmax (nmol · min^{-1} · mg^{-1})	Ref.
Antibody IIA12-ID10	GD, others?	330	25	33
Antibody DB-108Q	GD, others?	110	16	75
Antibody DB-108P	GD, others?	100	53	75

cleared from the bloodstream within the 1 to 2 min before it reaches critical targets.[32] Therefore, it was decided to attempt to create such an enzyme by specific mutation of existing human enzymes. Obvious candidates for such attempts include members of the cholinesterase family (including carboxylesterase) and the paraoxonases, which already possess the desired activity but at insufficient levels. The rationale for the design of mutations in the cholinesterase family was based on the fact that for these enzymes, the OP inhibitors are in reality hemisubstrates; their initial reaction with enzyme is similar to that of normal substrates. However, the subsequent reaction, equivalent to deacylation of the active site serine, is blocked because of the geometry of the active site. The amino acid group responsible for deacylation is not in an appropriate position to effect dephosphorylation.[71]

The perceived solution to this problem was to insert a second catalytic center into the active site specifically to carry out the dephosphorylation step of the reaction.[72] Applying this rationale, the human form of BChE has been mutated (Figure 7.3) to

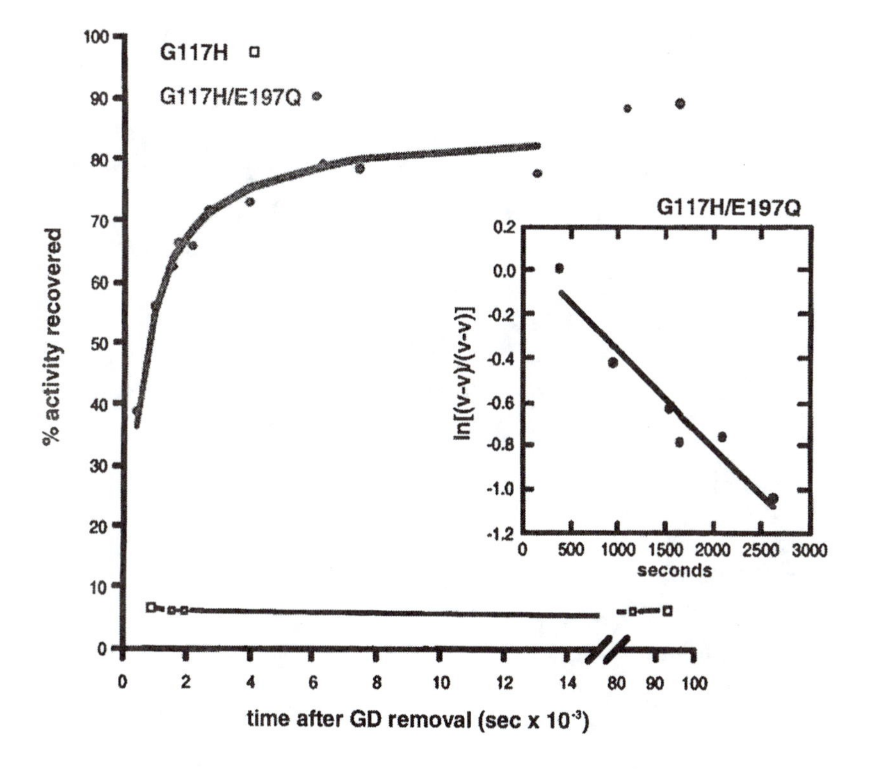

FIGURE 7.3 Comparative reactivation kinetics of soman-inhibited human butyrylcholinesterase single mutant G117H (□) and double mutant G117H/E197Q (●). Note that the recovery rate of the double mutant is very fast (with reaction rates of 77,000 and 128,000 per minute for the P_SC_R and P_SC_R isomers of soman, respectively), while the single mutant does not recover measurably. The insert shows that reactivation of the double mutant with soman can be treated as a first-order reaction for at least 2.5×10^3s.

express an enzyme with the ability to catalyze the hydrolysis of sarin, DFP, paraoxon, VX, and other non-aging nerve agents.[60,73] Aging and reactivation are parallel first-order reactions in phosphorylated enzymes. In the reactivation reaction the phosphoryl group is removed from the active site serine residue, restoring activity, whereas in the aging reaction one of the alkyl groups is removed from the phosphoryl group, rendering the group non-reactivatable. To effect the hydrolysis of rapidly aging nerve agents such as soman, it is necessary to inhibit the aging reaction so that reactivation is faster. This was accomplished by replacing the carboxyl group (glutamic acid) adjacent to the active site serine with an amide (glutamine) (Figure 7.4).[74] Unfortunately, these mutants have catalytic activities that are too slow for practical use (Table 7.7), and thus the search for a faster enzyme continues. For example, human CaE and hu-Pon are currently being subjected to mutation in efforts to generate additional, faster catalytic anti-nerve agent enzymes. It is important to note that

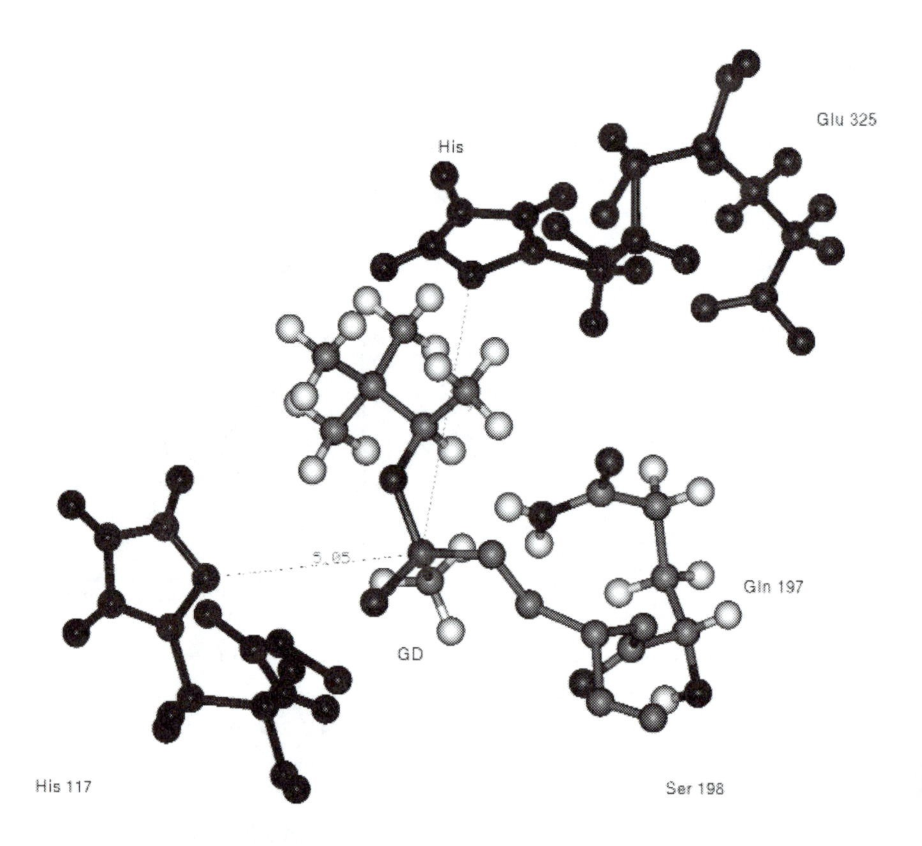

FIGURE 7.4 A ball-and-stick computer model of the active site of the double mutant of butyrylcholinesterase G117H/E197Q. In addition to the His 117 and Gln 197, the active site triad amino acid residues of His 438, Ser 198, and Glu 325 are also depicted with soman at the active site. The distances between the phosphorus atom of soman and His 117 is 5.05 Å and distance between the phosphorus atom of soman and the active site His 438 is 5.94 Å.

in the case of hu-Pon, the desired catalytic activity is present at low levels in the native enzyme; since OP are "accidental" substrates for hu-Pon (see above),[70] it is likely that improvement in activity can be realized through protein engineering.

Finally, through the careful design and synthesis of transition state analogs of the hydrolysis of soman, it has been possible to immunize mice and recover hybridomas whose antibodies display slow catalytic activity (Table 7.8) towards soman.[33,75] Such catalytic antibodies could be "humanized" to reduce their immunologic antigenicity,[76] thereby prolonging their serum half-life into the range of days to weeks, as reported for other mammalian species.[77] While most of these catalytic enzymes and antibodies have not yet been tested in mammalian systems, they are indicative of the types of drugs that may soon be available for use in animals, including humans. Since mutated BChE, CaE, and hu-Pon are based on human proteins, and catalytic antibodies can be rendered predominantly human in structure, the expectation is that these proteins would have no immunological or behavioral side effects.

VI. BEHAVIORAL EFFECTS

Since overt signs, symptoms, or physiological responses may not accompany many low-level exposures, behavioral toxicological measures may be chosen to detect any toxic changes wrought. Under such conditions, it is important to ensure that biological scavengers, either elevated levels of naturally occurring proteins or mutagenized forms thereof, do not elicit behavioral effects of their own after administration. Other considerations are potential behavioral effects that might result after pretreatment with a biological scavenger followed by exposure to a nerve agent, as well as a comparison of the extent of behavioral side effects that ensue from pretreatment with scavenger followed by nerve agent exposure vs. exposure to nerve agent followed by conventional therapy. The discussion here will be limited only to the side effects, if any, resulting from administration of scavengers alone. The other topics, including the ability of scavengers to ameliorate behavioral side effects following nerve agent exposure and the advantages of scavengers vs. conventional therapy, are discussed in detail elsewhere.[78]

VII. BEHAVIORAL EFFECTS OF SCAVENGERS ALONE

Most studies that have examined the behavioral effects of biological scavengers have done so by comparing a behavior before scavenger administration, after scavenger administration, and then after exposure to nerve agents.[78] There are, however, several studies that have examined the behavioral effects of the biological scavengers themselves in the absence of cholinesterase inhibitors. In a study by Genovese and Doctor, rats were trained to perform three behavioral paradigms: a passive avoidance task, a motor activity, and a scheduled-controlled behavior (Table 7.9).[79] The performance of animals before and after administration of purified EqBChE at a dose that would be expected to provide protection against an exposure of several LD_{50} of an OP compound was assessed. They determined the pharmacokinetic profile of EqBChE in rats

TABLE 7.9
Extent of Behavioral Deficits following Bioscavenger Administration or Conventional Therapy

Protection	Species	Behavioral Test(s)	Impairment	Recovery Time	Ref.
Atropine	Rat	Passive Avoidance, VI56 s Schedule	Total	>1 Week	79
EqBChE	Rat	Passive Avoidance, Motor Activity, VI56 s Schedule	None	Immediate	79
HuBChE	Rat	Morris Water Maze	None	Immediate	81
Pyridostigmine	Rhesus Monkey	Primate Equilibrium Platform (PEP)	Substantial	N.D.	86
EqBChE	Rhesus Monkey	Serial Probe Recognition (SPR)	None	Immediate	80
EqBChE	Rhesus Monkey	Observation, SPR	Subtle SPR defect	~6 Days	87
EqBChE	Rhesus Monkey	Observation, SPR	None	Immediate	80
HuBChE	Rhesus Monkey	Spatial Discrimination	Minor (1/4 had errors)	>1 day	64

Note: Impairment, behavioral impairment relative to untreated animals; Recovery Time, time elapsed before performance returns to pretreatment levels; N.D., not determined.

and then examined the behavior of the animals in the passive avoidance task when the levels of administered EqBChE were maximal. Subsequently, the animals were tested after enzyme levels had started to diminish, to enhance the opportunity of detecting any behavioral effects. During the activity tests, individually housed animals were allowed to habituate. Enzyme was given such that maximum levels would be present in circulation about 1 h before the beginning of dark cycle. Motor activity was then monitored for 10 days. As a final test, the effects of excess enzyme were examined in rats trained to perform a VI56 s schedule of food reinforcement. Previously, cholinergic compounds had been shown to disrupt performance of this task. Animals were observed for 10 days to ensure that any prolonged or delayed effects would be noted. In all cases for all test paradigms, the authors report that EqBChE did not disrupt performance of any of the learned tasks, did not upset the circadian cycle of light/dark activity, and had no effect on motor activity. They noted that these outcomes were in contrast to those observed when the standard cholinolytic, atropine, was administered. Finally, they evaluated the protective effects of the levels of enzyme given to the rats in the behavioral studies against MEPQ, a peripherally active OP compound. While the level of protection observed was lower than the theoretical prediction, the authors suggested that the simultaneous administration of scavenger and MEPQ might have reduced the efficacy of the administered EqBChE.

In a separate study also using EqBChE, rhesus monkeys were trained to perform a SPR task.[80] Using a six-object list, the monkeys were tested for same-different discrimination and delayed same-different discrimination. Once the animals became proficient at the task (80% correct for three successive sessions on 3 consecutive days), they received EqBChE in a dose similar to that reported by Broomfield et al. as sufficient to afford protection against 2 or 3 multiples of an LD_{50} soman challenge (*vide supra*).[47] The authors reported that in their study, repeated administration of commercially prepared EqBChE had no effect on the behavior of the monkeys as measured by the SPR studies (Table 7.9). Given the lack of behavioral effects and the relatively long *in vivo* half-life of the EqBChE, they concluded that this biological scavenger was potentially more effective than current chemotherapeutic treatments for OP intoxication. Other studies in rats or monkeys using human BChE also showed virtually no behavioral effects following administration of this enzyme.[64,81]

VIII. SUMMARY

Organophosphorous nerve agents represent a very real threat not only to warfighters in the field but also to the public at large.[82] Nerve agents have already been used by terrorist groups against a civilian population and, due to their low cost and relative ease of synthesis, are likely to be used again in the future.[83] In addition, many commonly used pesticides and chemical manufacturing by-products can act as anticholinesterases, and may be a low-dose exposure threat to workers in a variety of professions. Current therapeutic regimes for acute nerve agent exposure are generally effective at preventing fatalities if administered in an appropriate time frame. While the current therapeutic drugs, atropine and 2-PAM, have not been tested against a low-level exposure, their requirement for timely administration following symptoms makes it unclear whether under low-level exposure conditions these therapeutic interventions could be effectively implemented on a large scale. For acute multi-LD_{50} levels of exposure, pyridostigmine pretreatment coupled with post-exposure administration of an oxime, atropine, and an anti-convulsant does not prevent the substantial behavioral incapacitation or, in some cases, permanent brain damage that can result from OP poisoning. For low-level exposures that result in the Level 1 or Level 2 effects described above, the current therapy will probably not be administered at all since it is to be given at the onset of overt physiological signs. It is therefore important from both military and domestic security perspectives to develop novel defenses against nerve agents, including the use of bioscavenger molecules that avoid many of the difficulties associated with current treatments. While the use of nerve agents on the battlefield may be somewhat predictable, their use in a terrorist situation will be, in all probability, an unanticipatable event. The ability to afford long-term protection for first-responders exposed to toxic, incapacitating or Level 3 doses of OP, thereby reducing the severity of outcomes to Level 2 or Level 1 symptoms (and eliminating the impact of Level 1 and 2 exposures completely), is a notable potential advantage of biological scavengers.

The use of bioscavengers as a defense against OP intoxication has many advantages and few apparent disadvantages. As discussed in detail above, bioscavengers can afford protection against not only mortality, but also most or all of the adverse physiological and behavioral effects of nerve agent exposure. They can be administered prophylactically, precluding the need for immediate post-exposure treatment. In addition, the use of bioscavengers has several psychological benefits that are likely to result in a higher degree of user acceptability than exists for conventional therapy. No post-exposure auto-injectors are necessary, and protection is afforded with little chance of short- or long-term side effects. Of particular significance is the fact that current candidate bioscavenger proteins are, for the most part, enzymes of human origin. From a scientific standpoint, these proteins are good candidates because they are less likely to be recognized by cells of the immune system, and will enjoy prolonged residence times in circulation. From a user point of view, individuals are, in essence, being protected against nerve agents using a substance that their bodies already produce, rather than being injected with drugs and enzyme inhibitors that alone can produce potent side effects; such a distinction may enhance the comfort and compliance of end users.

There are several challenges that must be met in the future before bioscavengers can augment or replace the current therapeutic regimes for nerve agent intoxication. First, scavenger proteins, either alone or in combination, with a range of specificities that encompasses all known nerve agents, must be defined. The immunogenicity and serum half-life of the scavenger(s) must be determined in humans, and efforts may be required to minimize the former and maximize the latter. Finally, appropriate dosages of scavenger(s) must be determined that will, based on animal models, protect against concentrations of nerve agents likely to be encountered under a wide range of scenarios. While the majority of the research to date has focused on stoichiometric scavengers, the use of either naturally occurring or genetically engineered enzymes with catalytic activity holds the greatest theoretical promise for the development of a broad specificity prophylactic scavenger. Future efforts are likely to focus on generating, characterizing, and utilizing such enzymes in rodent and non-human primate models.

ACKNOWLEDGMENT

Dr. Cerasoli was supported by a National Research Council post-doctoral fellowship.

REFERENCES

1. Dacre, J.C., Toxicology of some anticholinesterases used as chemical warfare agents—a review, in *Cholinesterases, Fundamental and Applied Aspects,* Brzin, M., Barnard, E.A., and Sket, D., Eds., de Gruyter, Berlin, Germany, 1984, 415.
2. Ballantyne, B. and Marrs, T.C., Overview of the biological and clinical aspects of organophosphates and carbamates, in *Clinical and Experimental Toxicology of Organophosphates and Carbamates,* Ballantyne, B. and Marrs, T.C., Eds., Butterworth, Oxford, England, 1992, 1.
3. Maynard, R.L. and Beswick, F.W., Organophosphorus compounds as chemical warfare agents, in *Clinical and Experimental Toxicology of Organophosphates and Carbamates,* Ballantyne, B. and Marrs, T.C., Eds., Butterworth, Oxford, England, 1992, 373.
4. Somani, S.M., Solana, R.P., and Dube, S.N., Toxicodynamics of nerve agents, in *Chemical Warfare Agents,* Somani, S.M., Ed., Academic Press, San Diego, 1992, 68.
5. Koelle, G.B., Cholinesterases and anticholinesterases, in *Handbuch der Experimentallen Pharmakologie,* Vol. XV, Ekhler, O. and Farah, A., Eds., Springer-Verlag, Berlin, Germany, 1963.
6. Taylor, P., Anticholinesterase agents, in *The Pharmacological Basis of Therapeutics,* Gilman, A.G., Rall, T.W., Nies, A.S., and Taylor, P., Eds., Macmillan, New York, 1990, 131.
7. de Candole, C.A., Douglas, W.W., Evans, C.L., Holmes, R., Spencer, K.E.V., Torrance, R.W., and Wilson K.M., The failure of respiration in death by anticholinesterase poisoning, *Br. J. Pharmacol. Chemother.,* 6, 466, 1953.
8. Stewart, W.C., The effects of sarin and atropine on the respiratory center and neuromuscular junctions of the rat, *Can. J. Biochem. Physiol.,* 37, 651, 1959.
9. Stewart, W.C. and Anderson, E.A., Effects of a cholinesterase inhibitor when injected into the medulla of the rabbit, *J. Pharmacol. Exp. Ther.,* 162, 309, 1968.
10. Brimblecombe, R.W., Drugs acting on central cholinergic mechanisms and affecting respiration, *Pharmacol. Ther. B.,* 3, 65, 1977.
11. Bajgar, J., Jakl, A., and Hrdina, V., Influence of trimedoxime and atropine on acetylcholinesterase activity in some parts of the brain of mice poisoned by isopropylmethyl phosphonofluoridate, *Biochem. Pharmacol.,* 20, 3230, 1971.
12. Heffron, P.F. and Hobbinger, F., Relationship between inhibition of acetylcholinesterase and response of the rat phrenic nerve-diaphragm preparation to indirect stimulation at higher frequencies, *Br. J. Pharmacol.,* 66, 323, 1979.
13. Chabrier, P.E. and Jacob, J., *In vivo* and *in vitro* inhibition of cholinesterase by methyl-1 (S methyl phosphoryl-3) imidazolium (MSPI), a model of an "instantly" aged phosphorylated enzyme, *Arch. Toxicol.,* 45, 15, 1980.

14. Annual Report to Congress: Federally sponsored research on Gulf War Veterans' Illnesses for 1997, Department of Veteran Affairs, Washington, D.C., 1998, URL http://www.va.gov/resdev/pgrpt97.htm.
15. Heath, A.J.W. and Meredith, T., Atropine in the management of anticholinesterase poisoning, in *Clinical and Experimental Toxicology of Organophosphates and Carbamates,* Ballantyne B. and Marrs, T.C., Eds., Butterworth, Oxford, 1992, 543.
16. Wilson, I.B. and Ginsburg, S., A powerful reactivator of alkyl phosphate-inhibited acetylcholinesterase, *Biochim. Biophys Acta,* 18, 168, 1955.
17. Bismuth, C., Inns, R.H., and Marrs, T.C., Efficacy, toxicity and clinical use of oximes in anticholinesterase poisoning, in *Clinical and Experimental Toxicology of Organophosphates and Carbamates,* Ballantyne, B. and Marrs, T.C., Eds., Butterworth, Oxford, 1992, 555.
18. Gordon, J.J., Leadbeater, L., and Maidment, M.P., The protection of animals against organophosphorous poisoning by pretreatment with a carbamate, *Toxicol. Appl. Pharmacol.,* 43, 207, 1978.
19. Leadbeater, L., Inns, R.H., and Rylands, J.M., Treatment of soman poisoning, *Toxicol. Appl. Pharmacol.,* 5, S225, 1985.
20. Fleisher, J.H. and Harris, L.W., Dealkylation as a mechanism for aging of cholinesterase after poisoning with pinacolyl methylphosphonofluoridate, *Biochem. Pharmacol.,* 14, 641, 1965.
21. Dunn, M.A. and Sidell, F.R., Progress in medical defense against nerve agents, *JAMA,* 262, 649, 1989.
22. Castro, C.A., Larsen, T., Finger, A.V., Solana, R., and McMaster, S.B., Behavioral efficacy of diazepam against nerve agent exposure in rhesus monkeys, *Pharmacol. Biochem. Behav.,* 41, 159, 1991.
23. Main, A.R., The role of A-esterases in the acute toxicity of paraoxon, TEPP, and parathion, *Can. J. Biochem. Physiol.,* 75, 188, 1956.
24. Erdmann, W., Bosse, I., and Franke, P., Zur resorption und ausscheidung von toxigonin nach intramuskularer am menschen, *Dtsch. Med. Wschr.,* 90, 1436, 1965.
25. Wiezorek, W., Kreisel, W., Schnitzlein, W., and Matzkowski, H., Eigenwirkungen von trimedoxin und pralidoxim am menschen Zeitschr, *Militarmedizen,* 4, 223, 1968.
26. Sidell, F. and Groff, W., Toxogonin: Blood levels and side effects after intramuscular administration in man, *J. Pharm. Sci.,* 59, 793, 1970.
27. Vojvodic, V., Blood levels, urinary excretion and potential toxicity of N,N'-trimethylenebis(pyridinium-4-aldoxime) dichloride (TMB-4) in healthy man following intramuscular injection of the oxime, *Pharmacol. Clin.,* 2, 216, 1970.
28. Wenger, G.R., Effects of physostigmine, atropine and scopolamine on behavior maintained by a multiple schedule of food presentation in the mouse, *J. Pharmacol. Exp. Ther.,* 209, 137, 1979.
29. Clement, J.G., HI-6 reactivation of central and peripheral acetylcholinesterase following inhibition by soman, sarin and tabun in vivo in the rat, *Biochem. Pharmacol.,* 31, 1283, 1982.
30. McDonough, J.H. and Penetar, D.M., The effects of cholinergic blocking agents and anticholinesterase compounds on memory, learning and performance, in *Behavioral Models and the Analysis of Drug Action. Proceedings of the 27th OHOLO Conference.* Spiegelstein, M.Y. and Levy, A., Eds., Elsevier, Amsterdam, 1982, 155.
31. Huff, B.B., Ed., *Physician's Desk Reference,* Medical Economics Co. Inc., Oradell, 1986, 1491.

32. Talbot, B.G., Anderson, D.R., Harris, L.W., Yarbrough, L.W., and Lennox, W.J., A comparison of *in vivo* and *in vitro* rates of aging of soman-inhibited erythrocyte acetylcholinesterase in different animal species, *Drug Chem. Toxicol.,* 11, 289, 1988.
33. Brimfield, A.A., Lenz, D.E., Maxwell, D.M., and Broomfield, C.A., Catalytic antibodies hydrolyzing organophosphorus esters, *Chem.-Biol. Interact.,* 87, 95, 1993.
34. Broomfield, C.A., Transition state analogs for catalytic antibodies. *Phosph. Sulf. Silic.,* 109, 110, 233, 1996.
35. Lenz, D.E., Brimfield A.A., and Cook, L.A., The development of immunoassays for detection of chemical warfare agents, in *Development and Applications of Immunoassays for Environmental Analysis,* Aga, D. and Thurman, E.M., Eds., ACS Books, Washington, D.C., 1997, 77.
36. Hanke, D. and Overton, M.A., Phosphylation kinetic constants and oxime-induced reactivation in acetylcholinesterase from fetal bovine serum, bovine caudate nucleus, and electric eel, *J. Toxicol. Environ. Health,* 34, 141, 1991.
37. Ordentlich, A., Kronman, C., Barak, D., Stein, D., Ariel, N., Marcus, D., Velan, B., and Shafferman, E., Engineering resistance to 'aging' of phosphylated human acetylcholinesterase. Role of hydrogen bond network in the active center, *FEBS Lett.,* 334, 215, 1993.
38. Langenberg, J.P., van Dijk, C., Sweeney, R.E., Maxwell, D.M., De Jong, L.P., and Benschop, H.P., Development of a physiologically based model for the toxicokinetics of C(+/−)P(+/−)-soman in the atropinized guinea pig, *Arch. Toxicol.,* 71, 320, 1997.
39. Sternberger, L.A., Sim, V.M., Kavanagh, W.G., Cuculis, J.J., Meyer, H.G., Lenz, D.E., and Hinton, D.M., A vaccine against organophosphorous poisoning, *Army Science Conference Proceedings,* 3, 429, 1972.
40. Lenz, D.E., Brimfield, A.A., Hunter, K.W., Benschop, H.P., De Jong, L.P.A., Van Dijk, C., and Clow, T.R., Studies using a monoclonal antibody against soman, *Fund. Appl. Toxicol.,* 4, S156, 1984.
41. Sternberger, L.A., Cuculis, J.J., Meyer, H.G., Lenz, D.E., and Kavanagh, W.G., Antibodies to organophosphorous haptens: Immunity to paraoxon poisoning, *Fed. Proc.,* 33, 728, 1974.
42. Hunter, K.W., Lenz, D.E., Brimfield, A.A., and Naylor, J.A., Quantification of the organophosphorous nerve agent soman by competitive inhibition enzyme immunoassay using monoclonal antibodies, *FEBS Lett.,* 149, 147, 1982.
43. Brimfield, A.A., Hunter, K.W., Lenz, D.E., Benschop, H.P., Van Dijk, C., and De Jong, L.P.A., Structural and stereochemical specificity of mouse monoclonal antibodies to the organophosphorus cholinesterase inhibitor soman, *Mol. Pharmacol.,* 28, 32, 1985.
44. Wolfe, A.D., Rush, R.S., Doctor, B.P., Koplovitz, I., and Jones, D., Acetylcholinesterase prophylaxis against organophosphate toxicity, *Fund. Appl. Toxicol.,* 9, 266, 1987.
45. Maxwell, D.M., Castro, C.A., De La Hoz, D.M., Gentry, M.K., Gold, M.B., Solana, R.P., Wolfe, A.D., and Doctor, B.P., Protection of rhesus monkeys against soman and prevention of performance decrement by pretreatment with acetylcholinesterase, *Toxicol. Appl. Pharmacol.,* 115, 44, 1992.
46. Maxwell, D.M., Brecht, K.M., Doctor, B.P., and Wolfe, A.D., Comparison of antidote protection against soman by pyridostigmine, HI-6 and acetylcholinesterase, *J. Pharmacol. Exper. Therapeut.,* 264, 1085, 1993.
47. Broomfield, C.A., Maxwell, D.M., Solana, R.P., Castro, C.A., Finger, A.V., and Lenz, D.E., Protection of butyrylcholinesterase against organophosphorus poisoning in nonhuman primates, *JPET,* 259, 633, 1991.

48. Wolfe, A.D., Blick, D.W., Murphy, M.R., Miller, S.A., Gentry, M.K., Hartgraves, S.L., and Doctor, B.P., Use of cholinesterases as pretreatment drugs for the protection of rhesus monkeys against soman toxicity, *Toxicol. Appl. Pharmacol.,* 117, 189, 1992.
49. Doctor, B.P., Blick, D.W., Caranto, G., Castro, C.A., Gentry, M.K., Larison, R., Maxwell, D.M., Murphy, M.R., Schutz, M., Waibel, K., and Wolfe, A.D., Cholinesterases as scavengers for organophosphorous compounds: Protection of primate performance against soman toxicity, *Chem.-Biol. Interact.,* 87, 285, 1993.
50. Maxwell, D.M., Wolfe, A.D., Ashani, Y., and Doctor, B.P., Cholinesterase and carboxylesterase as scavengers for organophosphorous agents, in *Cholinesterases: Structure, Function, Mechanism, Genetics and Cell Biology,* Massoulie, J., Bacou, F., Barnard, E., Chatonnet, A., Doctor, B.P., and Quinn, D.M., Eds., American Chemical Society, Washington, D.C., 1991, 206.
51. Satoh, T., Role of carboxylesterases in xenobiotic metabolism, in *Reviews in Biological Toxicology,* Vol. 8, Hodgsen, E., Bend, J.R., and Philpot, R.M., Eds., Elsevier, New York, 1987, 155.
52. Augustinsson, K.B., Electrophoretic separation and classification of blood esterases, *Nature,* 131, 1786, 1958.
53. Aldridge, W.N. and Reiner, E., *Enzymes Inhibitors as Substrates,* North Holland, Amsterdam, 1972, 53.
54. Satoh, T. and Hosokawa, M., The mammalian carboxylesterases: From molecules to functions, *Ann. Rev. Toxicol.,* 38, 257, 1998.
55. Miller, A.D., Scott, D.F., Chacko, T.L., Maxwell, D.M., Schlager, J.J., and Lanclos, K.D., Expression and partial purification of a recombinant secretory form of human liver carboxylesterase, *Prot. Expr. Purif.,* 17, 16, 1999.
56. Maxwell, D.M., Brecht, K.M., and O'Neill, B.L., The effect of carboxylesterase inhibition on interspecies differences in soman toxicity, *Toxicol. Lett.,* 39, 35, 1987.
57. Maxwell, D.M., Lieske, C.N., and Brecht, K.M., Oxime-induced reactivation of carboxylesterase inhibited by organophosphorous compounds, *Chem. Res. Toxicol.* 7, 428, 1994.
58. Jokanovic, M., Kosanovic, M., and Maksimovic, M., Interaction of organophosphorous compounds with carboxylesterases in the rat, *Arch. Toxicol.* 70, 444, 1996.
59. Maxwell, D.M., Brecht, K.M., Saxena, A., Feaster, S., and Doctor, B.P., Comparison of cholinesterases and carboxylesterases as bioscavengers for organophosphorous compounds, in *Structure and Function of Cholinesterases and Related Proteins,* Doctor, B.P., Quinn D.M., and Taylor, P., Eds., Plenum Press, New York, 1998, 387.
60. Millard, C.B., Lockridge, O., and Broomfield, C.A., Design and expression of organophosphorous acid anhydride hydrolase activity in human butyrylcholinesterase, *Biochem.,* 34, 15925, 1995.
61. Doctor, B.P., Raveh, L., Wolfe, A.D., Maxwell, D.M., and Ashani, Y., Enzymes as pretreatment drugs for organophosphate toxicity, *Neurosci. Biobehav. Rev.,* 15, 123, 1991.
62. Ashani, Y., Shapira, S., Levy, D., Wolfe, A.D., Doctor, B.P., and Raveh, L., Butyrylcholinesterase and acetylcholinesterase prophylaxis against soman poisoning in mice, *Biochem. Pharmacol.,* 41, 37, 1991.
63. Raveh, L., Grunwald, J., Marcus, D., Papier, Y., Cohen, E., and Ashani, Y., Human butyrylcholinesterase as a general prophylactic antidote for nerve agent toxicity, *Biochem. Pharmacol.,* 45, 2465, 1993.
64. Raveh, L., Grauer, E., Grunwald, J., Cohen, E., and Ashani, Y., The stoichiometry of protection against soman and VX toxicity in monkeys pretreated with human butyrylcholinesterase, *Toxicol. Appl. Pharmacol.,* 145, 43, 1997.

65. Serdar, C.M. and Gibson, D.T., Enzymatic hydrolysis of organophosphates: Cloning and expression of a parathion hydrolase gene from *Pseudomonas diminuta, Bio/Technology,* 3, 567, 1985.
66. Gan, K.N., Smolen, A., Eckerson, H.W., and La Du, B.N., Purification of human serum paraoxonase/arylesterase, *Drug. Metab. Dispos.,* 19, 100, 1991.
67. Broomfield, C.A., A purified recombinant organophosphorus acid anhydrase protects mice against soman, *Chem.-Biol. Interact.,* 87, 279, 1993.
68. Masson, P., Josse, D., Lockridge, O., Viguié, N., Taupin, C., and Buhler, C., Enzymes hydrolyzing organophosphates as potential catalytic scavengers against organophosphate poisoning, *J. Physiol.,* 92, 357, 1998
69. Cheng, T.-C., Liu, L., Wang, B., Wu, J., Frank, J.J., Anderson, D.M., Rastogi, V.K., and Hamilton, A.B., Nucleotide sequence of a gene encoding an organophosphorus nerve agent degrading enzyme from *Alteromonas haloplanktis, J. Ind. Microbiol.* 18, 49, 1997.
70. Jakubowski, H., Calcium-dependent human serum homocysteine thiolactone hydrolase, *J. Biol. Chem.,* 275, 3957, 2000.
71. Jarv, S., Stereochemical aspects of cholinesterase catalysis, *Bioorg. Chem.,* 12, 259, 1984.
72. Broomfield, C.A., Millard, C.B., Lockridge, O., and Caviston, T.L., Mutation of human butyrylcholinesterase glycine 117 to histidine preserves activity but confers resistance to organophosphorus inhibitors, in *Enzymes of the Cholinesterase Family,* Quinn D.M., Balasubramanian, A.S., Doctor, B.P., and Taylor, P., Eds. Plenum Press, New York, 1995, 169.
73. Lockridge, O., Blong, R.M., Masson, P., Froment, M.-T., Millard, C.B., and Broomfield, C.A., A single amino acid substitution, Gly117His, confers phosphotriesterase (organophosphorous acid anhydride hydrolase) activity on human butyrylcholinesterase, *Biochemistry,* 36, 786, 1997.
74. Millard, C.A., Lockridge, O., and Broomfield, C.A., Organophosphorus acid anhydride hydrolase activity in human butyrylcholinesterase: Synergy results in a somanase, *Biochemistry,* 37, 237, 1998.
75. Yli-Kauhaluoma, J., Humppi, T., and Yliniemela, A., Antibody-catalyzed hydrolysis of the nerve agent soman, in *NBC Defense 1997,* Proceedings of a Symposium on NBC Defense, Hyvinkaa, Finland, 1997, 164.
76. Hale, G., Dyer, M.J.S., Clark, M.R., Phillips, J.M., Marcus, R., Reichmann, L., Winter, G., and Waldmann, H., Remission induction in non-Hodgkin's lymphoma with reshaped human monoclonal antibody CAMPATH-1H, *Lancet,* 2(8625), 1394, 1988.
77. Vieira, P. and Rajewsky, K., The half-lives of serum immunoglobulins in adult mice, *Eur. J. Immunol.,* 18, 313, 1988.
78. Lenz, D.E. and Cerasoli, D.M., Nerve agent bioscavengers: Protection with reduced behavioral effects, *Mil. Psychol.,* in press.
79. Genovese, R.F. and Doctor, B.P., Behavioral and pharmacological assessment of butyrylcholinesterase in rats, *Pharmacol. Biochem. Behav.,* 51, 647, 1995.
80. Matzke, S.M., Oubre, J.L., Caranto, G.R., Gentry, M.K., and Galbicka, G., Behavioral and immunological effects of exogenous butyrylcholinesterase in rhesus monkeys, *Pharmacol. Biochem. Behav.,* 62, 523, 1999.
81. Brandeis, R., Raveh, L., Grunwald, J., Cohen, E., and Ashani, Y., Prevention of soman-induced cognitive deficits by pretreatment with human butyrylcholinesterase in rats, *Pharmacol. Biochem. Behav.,* 46, 889, 1993.
82. Ember, L., Chemical weapons: Plans prepared to destroy Iraqi arms, *Chem. Eng. News,* 19, 6, 1991.

83. Masuda, A.N., Takatsu, M., Morianari, H., and Ozawa, T., Sarin poisoning in Tokyo subway, *Lancet,* 345, 1446, 1995.
84. Maxwell, D.M., The specificity of carboxylesterase protection against the toxicity of organophosphorus compounds, *Toxicol. Appl. Pharmacol.,* 114, 306, 1992.
85. Dumas, D.P., Durst, H.D., Landis, W.G., Raushel, F.M., and Wild, J.R., Inactivation of organophosphorous nerve agents by the phosphotriesterase from *Pseudomonas diminuta, Arch. Biochem. Biophys.,* 277, 155, 1990.
86. Blick, D.W., Murphy, M.R., Brown, G.C., Yochmowitz, M.G., Fanton, J.W., and Hartgraves, S.L., Acute behavioral toxicity of pyridostigmine or soman in primates, *Toxicol. Appl. Pharmacol.,* 126, 311, 1994.
87. Castro, C.A., Gresham, V.C., Finger, A.V., Maxwell, D.M., Solana, R.P., Lenz, D.E., and Broomfield, C.A., Behavioral decrements persist in rhesus monkeys trained on a serial probe recognition task despite protection against soman lethality by butyrylcholinesterase, *Neurotoxicol. Teratol.,* 16, 145, 1994.
88. Korza, G. and Ozols, J., Complete covalent structure of 60-kDa esterase isolated from 2,3,7,8-tetrachlorodibenzo-p-dioxin-induced rabbit liver microsomes, *J. Biol. Chem.,* 263, 3486, 1988.
89. Ozols, J., Isolation, properties, and the complete amino acid sequence of a second form of 60-kDa glycoprotein esterase. Orientation of the 60-kDa proteins in the microsomal membrane, *J. Biol. Chem.,* 264, 12533, 1989.
90. Sone, T., Isobe, M., Takabatake, E., and Wang, C.Y., Cloning and sequence analysis of a hamster liver cDNA encoding a novel putative carboxylesterase, *Biochim. Biophys. Acta.,* 1207, 138, 1994.
91. Shibata, F., Takagi, Y., Kitajima, M., Kuroda, T., and Omura, T., Molecular cloning and characterization of a human carboxylesterase gene, *Genomics,* 17, 76, 1993.
92. Schwer, H., Langmann, T., Daig, R., Becker, A., Aslanidis, C., and Schmitz, G., Molecular cloning and characterization of a novel putative carboxylesterase, present in human intestine and liver, *Biochem. Biophys. Res. Commun.,* 233, 117, 1997.
93. Matsushima, M., Inoue, H., Ichinose, M., Tsukada, S., Miki, K., Kurokawa, K., Takahashi, T., and Takahashi, K., The nucleotide and deduced amino acid sequences of porcine liver proline-beta-naphthylamidase. Evidence for the identity with carboxylesterase, *FEBS Lett.,* 293, 37, 1991.
94. Robbi, M., Beaufay, H., and Octave, J.N., Nucleotide sequence of cDNA coding for rat liver pI 6.1 esterase (ES-10), a carboxylesterase located in the lumen of the endoplasmic reticulum, *Biochemistry,* 269, 451, 1990.
95. Yan, B., Yang, D., Brady, M., and Parkinson, A., Rat kidney carboxylesterase. Cloning, sequencing, cellular localization, and relationship to rat liver hydrolase, *J. Biol. Chem.,* 269, 29688, 1994.
96. Ellinghaus, P., Seedorf, U., and Assmann, G., Cloning and sequencing of a novel murine liver carboxylesterase cDNA, *Biochim. Biophys. Acta.,* 1397, 175, 1998.
97. Ovnic, M., Swank, R.T., Fletcher, C., Zhen, L., Novak, E.K., Baumann, H., Heintz, N., and Ganschow, R.E., Characterization and functional expression of a cDNA encoding egasyn (esterase-22): The endoplasmic reticulum-targeting protein of beta-glucuronidase, *Genomics,* 11, 956, 1991.
98. Ovnic, M., Tepperman, K., Medda, S., Elliott, R.W., Stephenson, D.A., Grant, S.G., and Ganschow, R.E., Characterization of a murine cDNA encoding a member of the carboxylesterase multigene family, *Genomics,* 9, 344, 1991.
99. Long, R.M., Satoh, H., Martin, B.M., Kimura, S., Gonzalez, F.J., and Pohl, L.R., Rat liver carboxylesterase: cDNA cloning, sequencing, and evidence for a multigene family, *Biochem. Biophys. Res. Commun.,* 156, 866, 1988.

100. Robbi, M. and Beaufay, H., Topogenesis of carboxylesterases: A rat liver isoenzyme ending in -HTEHT-COOH is a secreted protein, *Biochem. Biophys. Res. Commun.*, 183, 836, 1992.
101. Murakami, K., Takagi, Y., Mihara, K., and Omura, T., An isozyme of microsomal carboxyesterases, carboxyesterase Sec, is secreted from rat liver into the blood, *J. Biochem.* (Tokyo), 113, 61, 1993.
102. Medda, S. and Proia, R.L., The carboxylesterase family exhibits C-terminal sequence diversity reflecting the presence or absence of endoplasmic-reticulum-retention sequences, *Eur. J. Biochem.*, 206, 801, 1992.

8 Chronic Effects of Acute, Low-Level Exposure to the Chemical Warfare Agent Sulfur Mustard*

Charles G. Hurst and William J. Smith

CONTENTS

I. Introduction 245
II. Clinical Effects of Sulfur Mustard 246
A. Carcinogenesis 247
B. Chronic Pulmonary Disease 248
C. Chronic Eye Disease 248
D. Scarring, Pigmentation Changes, and Cancer of Epithelial Surfaces 248
E. Central Nervous System 249
F. Summary for Symptomatic Exposures 249
III. Acute Subclinical Exposure 251
A. Carcinogenesis 251
B. Radiation 252
IV. *In Vitro* Studies of Sulfur Mustard Toxicity 253
V. Dose Dependency of the Mustard Lesion 254
VI. Summary 255
Acknowledgments 255
References 255

I. INTRODUCTION

Chemical warfare agents have been around for at least 4000 years and probably were originally used as poisons on individuals. The use of chemical weapons dates from at least 423 B.C. when allies of Sparta in the Peloponnesian War took an Athenian-held fort by directing smoke from lighted coals, sulfur, and pitch through a hollowed-out beam into the fort. Other conflicts during the succeeding centuries saw the use of smoke and flame. During the seventh century A.D., the Greeks invented "Greek fire,"

*The opinions or assertions contained herein are the private views of the authors, and are not to be construed as reflecting the view of the Department of the Army or the Department of Defense.

0-8493-0872-0/01/$0.00+$.50

a combination probably of rosin, sulfur, pitch, naphtha, lime, and saltpeter that floated on water and was particularly effective in naval operations. During the fifteenth and sixteenth centuries, Venice employed unspecified poisons in hollow explosive mortar shells and sent poison chests to its enemy to poison wells, crops, and animals.[1–3] Finally, World War I and the Iran-Iraq War saw the advent of modern chemical warfare.

Mustard has been stockpiled in the arsenals of various countries since it was first used on July 12, 1917, when the Germans fired shells containing mustard at British troops entrenched near Ypres, Belgium. When a single agent was identified as the source of injury, it was estimated that mustard caused about 80% of the chemical casualties in World War I; other agents such as chlorine and phosgene caused the remaining 20%. The British had 180,983 chemical casualties; the injuries of 160,970 (88%) were caused solely by mustard. Of these casualties, 4,167 (2.6%) died. Of the 36,765 single-agent United States (U.S.) chemical casualties, the injuries of 27,711 (75%) were caused solely by mustard. Of the casualties who reached a medical treatment facility, 599 (2.2%) died. Just as disconcerting was the fact that mustard survivors required lengthy hospitalizations: the average length of stay was 42 days.[4]

Since the first use of mustard as a military weapon, there have been a number of isolated incidents in which it was reportedly used. In 1935, Italy probably used mustard against Abyssinia (now Ethiopia); Japan allegedly used mustard against the Chinese from 1937 to 1944; and Egypt was accused of using the agent against Yemen in the mid-1960s.

Chemical agents were not used during World War II. It is thought that Germany did not use mustard because Hitler had been a mustard victim during World War I and was loathe to use it.[4]

Sidell and Hurst have described the long-term effects produced from acute symptomatic clinical dose exposure to mustard, but less is known about the clinical effects from chronic, sometimes symptomatic, low-dose exposure.[5,6] Acute is defined here as an exposure lasting less than 24 h. The term chronic refers to an exposure lasting for days, weeks, months, and even years. Clinical means producing a recognizable illness directly related to mustard exposure. Symptomatic means having either the acute or chronic clinical illness produced by sulfur mustard. Asymptomatic, of course, is without symptoms at all.

The argument can be made that acute, subclinical asymptomatic injury causing long-term effects does not exist. The pros and cons will be considered in this chapter, and parallels will be drawn. Certainly chronic, subclinical asymptomatic exposures do exist and there are parallels to other harmful situations. Obviously, workers in the manufacture of sulfur mustard, who were asymptomatic for part or all of their employment, fall into this category.

II. CLINICAL EFFECTS OF SULFUR MUSTARD

The organs most commonly affected by mustard are the skin, eyes, and airways: the organs which mustard contacts directly. After a substantial amount of mustard has been absorbed through the skin or inhaled, the hemopoietic system, gastrointestinal tract, and CNS are also damaged. Mustard may also affect other organs but rarely do these produce clinical effects.[4] After an asymptomatic latent period of hours, mustard causes erythema and blisters on the skin. This response ranges in severity from mild redness resembling sunburn to severe third-degree burns. Eye damage ranges from mild irritation-conjunctivitis, to corneal opacity, to perforation of the eye, and blindness. In the lung, the injury extends from mild upper respiratory signs to marked airway damage, bronchitis, and pneumonia. On rare occasions, acute laryngospasm can result in rapid death. Gastrointestinal effects vary from nausea and vomiting to severe hemorrhagic diarrhea. In the bone marrow, severe stem cell suppression can result in profound pancytopenia. In the CNS, at least in laboratory animals, seizures and death have been produced at high concentration exposures. The worst possible outcome from mustard exposure is death. However, mortality from mustard is uncommon. Less than 5% mortality from mustard gas was observed in allied troops in World War I.

Laboratory animal studies have shown that mustard is mutagenic and carcinogenic, and thus, it is not surprising that it is carcinogenic in man.[7–9] Both Morgenstern et al. and Buscher and Green emphasize that chronic low-dose exposure over months to years in occupationally exposed workers leads to chronic bronchitis, bronchial asthma, hoarseness, aphonia, and hypersensitivity to smoke, dust, and fumes.[10,11] Such individuals typically show persistent disability, with increased susceptibility to respiratory tract infections and evidence of bronchitis and bronchiectasis.[10–12]

All human studies dealing with chronic mustard disease processes are retrospective and fraught with the problems inherent in retrospective studies. These problems include bias in the sampling populations; lack of epidemiological controls from the effects of smoking, lifestyle, race, gender, age, or exposure to other chemicals; differential quality of available health care; and incorrect diagnosis.[12] These limitations make absolute interpretation of the studies difficult.

A. Carcinogenesis

Mustard is an alkylating agent similar to drugs that have been used in cancer chemotherapy, such as nitrogen mustards, Cytoxan, and cis-platin. Since DNA is one of mustard's most sensitive targets, it is not surprising that carcinogenesis and radiomimetic effects are seen.[5]

Human data on the carcinogenicity of mustard are from (a) battlefield exposures, (b) accidents, and (c) workers in chemical factories. Both British and American studies have investigated the increased incidence of pulmonary carcinoma arising from World War I battlefield exposure. All are difficult to interpret, owing to the lack of controls for age, chronic pulmonary disease, cigarette smoking, and other factors that might affect the outcome.[13–15]

In contrast to battlefield exposures, studies of factory workers involved in the production of mustard have shown a definite link between prolonged exposure to low doses of mustard and cancer.[12] Several studies have provided evidence of an increased risk of respiratory tract cancers in factory workers.[6,16–20] Easton et al. found a 45% increase in death due to lung cancer, a 170% increase in death from cancer of the larynx, and a 450% increase in death from cancer of the pharynx, compared with expected deaths in the general population.[17] The risks from cancer of the pharynx and lung were significantly related to the duration of employment at the factory.

B. Chronic Pulmonary Disease

Inhalation of mustard vapor primarily affects the laryngeal and tracheobronchial mucosa.[12] Evidence exists to suggest that mustard inhalation causes sustained respiratory difficulties even after the acute lesions have healed. Clinical follow-ups on 200 Iranian soldiers who were severely injured by mustard during the Iran-Iraq War indicate that about one third had experienced persistent respiratory effects 2 years or more after initial exposure. Reported problems included chronic bronchitis, asthma, rhinopharyngitis, tracheobronchitis, laryngitis, recurrent pneumonia, bronchiectasis, and, in some cases, severe, unrelenting tracheobronchial stenosis.[21–25]

Of the British soldiers exposed to mustard in World War I, 12% were awarded disability compensation for respiratory disorders that were believed to be due to mustard exposures during combat.[26]

Little contemporary information regarding the pathogenesis of the respiratory lesions is available, and few data from people or animals exposed to nonlethal concentrations of mustard vapor exist. Even fewer studies investigate the histopathology of the recovery process in animals exposed to mustard.[9] However, two studies conducted during World War I suggest that low-level exposure or survivable exposures in dogs and rabbits may produce scar tissue following small ulcerations in the trachea and larynx, causing contractions of these areas.[27,28] The more severe respiratory tract lesions described in animals exposed to mustard vapor appear to be quite similar in type and location to those described in humans.[12]

C. Chronic Eye Disease

Individuals who sustain acute ocular injury due to high-dose mustard exposure may experience difficulties even after the initial effects of the injury have subsided.[29–32] Recurrent or persistent corneal ulceration can occur after latent periods of 10 to 25 years. Chronic conjunctivitis and corneal clouding may accompany this delayed keratopathy.[31–32] Anecdotal accounts suggest that low-dose exposure also causes increased sensitivity to later exposure to mustard, although the existence of increased sensitivity is difficult to substantiate with available scientific evidence.[12, 33]

D. Scarring, Pigmentation Changes, and Cancer of Epithelial Surfaces

Skin cancer occurring at the site of old scar formation is an acknowledged biological phenomenon.[34,35] Cutaneous cancers resulting from acute mustard exposure usually localize in scars, whereas those caused by chronic exposure can occur on any exposed site.[36]

In a prospective study of delayed toxic effects from mustard exposure, Balali-Mood followed a group of Iranian solders exposed to mustard gas during the Iran-Iraq War.[24] After 2 years, 41% of the exposed victims were experiencing pigmentary disorders.

In the absence of melanocyte destruction, hyperpigmentation predominates. If melanocytes are locally destroyed, and inward migration from destroyed adnexal structures does not occur, depigmentation predominates.[5]

In its study of mustard and Lewisite effects, the Institute of Medicine concluded that, following mustard exposure:

- The evidence indicates a causal relation between acute, severe exposure to mustard agents and increased pigmentation and depigmentation in human skin.
- Acute and severe exposure can lead to chronic skin ulceration, scar formation, and the development of cutaneous cancer.
- Chronic exposure to minimally toxic and even subtoxic doses can lead to skin pigmentation abnormalities and cutaneous cancer.[9]

E. Central Nervous System

Excitation of the CNS after mustard exposure, resulting in convulsions and followed by CNS depression, has been reported by the U.S. Army.[37] Convulsions and cardiac irregularities appear to occur only after extremely acute, high doses, which are probably attainable only in laboratory settings.[12, 38] Mustard casualties of the Iran-Iraq War did not display severe CNS or cardiac abnormalities.[21]

F. Summary for Symptomatic Exposures

The organs most commonly affected by mustard are the skin, eyes, and airways; the organs mustard comes in direct contact with. After a substantial amount of mustard has been absorbed through the skin or inhaled, the hemopoietic system, gastrointestinal tract, and CNS are also damaged. Mustard may also affect other organs, but rarely do these produce clinical effects.[4] After an asymptomatic latent period of hours, mustard causes erythema and blisters on the skin. This ranges in severity from mild redness resembling sunburn, to severe third-degree burns. Eye damage ranges from mild irritation-conjunctivitis to corneal opacity, or even perforation of the eye and blindness. In the lung, the injury ranges from mild upper respiratory signs to marked airway damages, bronchitis, and pneumonia. On rare occasion, acute laryngospasm can result in rapid death. Gastrointestinal effects range from nausea and vomiting to severe hemorrhagic diarrhea. And in the bone marrow, severe stem cell suppression can result in profound pancytopenia. In the CNS, at least in laboratory animals, seizures and death have been produced at high concentration exposures. The worst outcome from all these organs systems, except for possibly the eye, is death. Death, however, is not the usual outcome from mustard exposure.

Studies of English and Japanese mustard factory workers establish repeated symptomatic exposures to mustard over a period of years as a causal factor in an increased incidence of airway cancer. The association between a single exposure to mustard and airway cancer is not as well established as the association between one-time mustard exposure and other chronic airway problems, such as chronic bronchitis (based on World War I data). In some cases, the long-term damage was probably a continuation of the original insult resulting from insufficient therapy in the pre-antibiotic era. Morgenstern et al. give the following graphic description of symptoms and injuries incurred by some of the mustard factory workers.

> Less widely known is the fact that many persons employed in the handling of mustard gas and exposed to small quantities of the vapor over a prolonged period of time may sustain damage to the respiratory mucosa, which may leave them partially or totally disabled. This statement is based on two and one-half years of observation in the medical department of an industrial plant where over 200 patients have been treated for both the acute symptoms and the residual effects of mustard gas exposure. The evolution of chronic mustard bronchitis may be traced as follows:
>
> A young, white male previously engaged in farming or some other nonindustrial occupation with no history of any previous chronic lung disease goes to work on the mustard filing line. There is a varying concentration of mustard vapor in the air during a good part of the working day. After a period of time ranging anywhere from 3 weeks to 6 or 12 months he begins to show signs of definite irritation of the conjunctival and respiratory mucous membranes. He develops symptoms. He is given sick time off with his condition improving and returns to work. After a number of such episodes it becomes apparent that this man is not suitable for work in mustard and he is transferred out to another department free of toxic fumes.
>
> After removal from mustard, his eyes and throat gradually heal. The conjunctivitis recedes and the vision returns to normal. The sore throat and hoarseness subside. The sense of taste returns, but the sense of smell may remain impaired. The appetite improves and he regains some of his lost weight with overall improvement. But he remains troubled by a persistent hacking cough, which come in paroxysms. It is most common in the morning but also occurs on lying down at night. It is often precipitated by physical exertion or when the man walks from the cold into a warm room or comes into contact with fumes of smoke. The cough is productive of anywhere from a teaspoon to a cupful of white or yellow mucoid or mucopurulent sputum, which may have a foul odor on occasion. There may be a troublesome wheezing and chest tightness most marked during damp weather. The patient seems to be more susceptible to respiratory infections than he was prior to exposure to mustard and the infections tend to last longer. Definite clinical bronchiectasis may develop as a result of repeated attacks of acute infectious bronchitis. He is hypersensitive to fumes and dust of any kind. He may develop dyspnea on slight or moderate exertion and therefore cannot perform any arduous labor.[10]

This description of chronic bronchitis developing in factory workers in the setting of World War II is both accurate and quite convincing.

Several eye diseases, such as chronic conjunctivitis, appear after an acute, usually severe, insult to the eye. In particular, delayed keratitis has appeared more than 25 years after the acute, severe lesion. Similarly, skin scarring, pigment changes, and even cancer have either followed the initial wound as a continuation of the process (scarring) or later appeared at the site of the lesion.

The production of nonairway cancer by mustard has been demonstrated in animals, but scant evidence exists to implicate mustard as a causative factor in nonairway cancer in humans.[5]

III. ACUTE SUBCLINICAL EXPOSURE

We are not convinced that acute, asymptomatic injuries that result in clinical disease truly exist. It is conceivable that certain synergistic situations can develop, such as would happen if co-factors or preexisting conditions (immunosuppression, genetic deficiency, or an additional chronic subclinical exposure) were triggered by some otherwise uneventful insult. This is certainly an unknown for mustard exposure at this time. The best that can be done is to draw analogies to other circumstances.

A. Carcinogenesis

Genotoxic substances usually have a direct effect on DNA and are occasionally effective after a single exposure. This helps to explain why they are frequently carcinogenic at subtoxic doses. These toxic compounds often act in a cumulative manner and synergistically with other DNA-reactive carcinogens. They usually produce neoplasms in more than one target organ and have a variable latency.[39]

Strong promoters also possess weak intrinsic carcinogenicity. This experimental evidence comes from two test systems: (1) continued high-level administration of promoters such as croton oil to mouse skin or (2) oral administration of DDT or phenobarbital to rats. Both systems yield a small but definite crop of benign and malignant neoplasms in the absence of any obvious genotoxic carcinogen. An explanation is needed because promoters, by definition, do not have intrinsic properties of altering the genetic apparatus. One example has supplied an explanation that could apply to the others: when croton oil was applied to the skin of mice, it appeared to induce by itself a high incidence of papillomas and carcinomas.[40] Careful analysis revealed that the mice used were purchased from a supplier who housed the mice in creosoted cages. Thus, the mice had been exposed to genotoxic carcinogens in the creosote prior to the application of croton oil. Similarly, the weak carcinogenicity of DDT or phenobarbital may stem from prior exposure of the animals to small amounts of carcinogens, possibly mycotoxins or certain nitrosamines in the diet.[41–43]

In mice, chemical enzyme-inducing substances have produced liver tumors, but were not found to be genotoxic using *in vitro* testing. Such chemicals have shown promoting activity in these systems.[44–46] The liver of these animals appears to respond as if a DNA gene structure change has occurred. The role mycotoxins or nitrosamines play in the diet of these animals is open for speculation. Complex polychlorinated aliphatic and cyclic hydrocarbons also fall into the class of enhancing substances.[47,48] They exhibit a nonlinearity in their dose-time response curves that differs from the genotoxic carcinogens.[49]

The mechanism of promotion is subject to scientific research and considerable conjecture. The probability of multiple mechanisms is strong. A sequence of steps may be involved, leading to proliferation and differentiation. A parallel example might be the prostaglandin and cyclic nucleotide membrane effector systems. An early biochemical indicator of promotion is the induction of ornithine decarboxylase (ODC), and its presence has been used to discover new promoters.[50–55] Increased levels of ODC appear to be associated with increased liver cell proliferation and might be related to the mechanisms of the promotion.[56,57]

Promoters and carcinogens involved in human cancer induction were first discovered through carefully conducted epidemiological studies, and then tested in animal models or in *in vitro* systems. Newer techniques use exfoliated cells or cells in culture to ascertain exposure to these genotoxic materials.[57–60]

Also, monoclonal antibody techniques may be able to trace carcinogen-macromolecular adducts in human tissues, verifying exposure to specific carcinogens.[61,62] The gold standard will be to identify human hazards and carcinogens before exposure, "an ounce of prevention is worth a ton of cure."[63–70]

B. Radiation

Ultraviolet light is a form of nonionizing radiation exposure to man, while X-rays are an ionizing radiation. Both forms have the effect of damaging DNA similar to sulfur mustard, especially in rapidly dividing cells. It may never be possible to determine whether an individual can go through life without manifesting symptoms of acute sun damage: erythema, blisters, pigmentation, etc. or, at the very least, mild erythema. But, the chronicity of exposure producing the skin damage and cancers is undeniable. It is the acute symptomatic vs. chronic asymptomatic mix of exposures that will always remain a mystery.

Sulfur mustard has been called "radiomimetic" primarily because it appears to target DNA in rapidly dividing cells. There is a latent onset to its acute clinical symptoms very much paralleling those of ultraviolet sunlight damage, and a delayed onset to pulmonary cancers from chronic exposure. It is well established that the consequence of lifelong exposure to sunlight significantly enhances the development of skin cancers in fairer skinned individuals. The chronic damage to fair-skinned individuals is seen as premature aging, pigmentary changes, solar elastosis, solar keratosis, basal and squamous cell cancers, and malignant melanoma.[71,72] The threshold for developing these lesions in man is multifactoral: heredity (skin type, individual's ability to repair DNA), environment, and lifestyle. Thus, the cumulative dose responsible for these affects has great variability. Dark-skinned individuals may never experience any of the clinical entities mentioned above while individuals with the genetic disorder, Xeroderma pigmentosa, will experience marked acceleration of sun-related skin damage and cancers. This is caused by the genetic defect, which impairs the DNA repair produced by ultraviolet damage.[73]

Likewise, it is well recognized that certain dyes, pigments, and drugs (i.e., tetracycline, psoralens, etc.) enhance sunlight and artificial light effects on skin. Existing genetic defects and specific drugs have not been identified that enhance the damage caused by sulfur mustard, but this possible synergism may be acute, subacute, or chronic.

Radiation-induced cancer is due to a nonlethal mutation of somatic cells. The latent period between irradiation and the development of cancer varies from 4 to 40 years, the average being 7 to 12 years. Even relatively low doses of X-rays increased the risk of cancer. Of school children epilated with 300 to 400 r of unfiltered 100-kV radiation, 1.6 percent had skin, thyroid, or parotid tumors 20 years later, while untreated control group showed only 0.2 percent such tumors. Apparently the mutated cells can survive 10 to 20 years before proliferating.[74]

Evidence that various derivatives of tar and oil cause squamous cell carcinoma of the skin is both environmental and experimental. Experimental production of skin cancer in rodents with the various carcinogenic hydrocarbons has been well demonstrated. In fact, the effect of a chemical on animal skin is currently regarded as the best method of testing the carcinogenicity of the chemical.[75]

IV. *IN VITRO* STUDIES OF SULFUR MUSTARD TOXICITY

Sulfur mustard is an alkylating agent that acts through cyclization of an ethylene group to form a highly reactive sulfonium electrophilic center. This reactive electrophile is capable of combining with any of the numerous nucleophilic sites present in macromolecules of cells. The products of these reactions are stable adducts which can modify the normal function of the target macromolecule. Since nucleophilic areas exist in peptides, proteins, RNA, DNA, and membrane components, extensive efforts have been underway to identify the most critical biomolecular reactions leading to mustard injury.

While the chemistry of mustard interactions with cellular components is well defined, the correlation of these interactions with injury has not been made. Over the past few decades, scientists have made major advances in understanding the cellular and biochemical consequences of exposure to mustard. While not the only target for alkylation by mustard, DNA is presumed to be an early reactant in the pathogenic cascades leading to the mustard lesion. Alkylation of nucleotides can result in apurinic site formation, disruption of normal DNA replication, activation of DNA repair pathways and eventually, to cytotoxic or mutagenic events. At high-exposure doses, such as those that lead to vesication *in vivo* or above 50 μM *in vitro,* the exposed cells sustain so much damage that they will die. The cells show activation of the nuclear enzyme poly(ADP-ribose) polymerase (PARP), disruption of cellular metabolism, loss of cell energetics, and total cellular breakdown. Many of these cells initially respond to the agent insult by activation of apoptotic death pathways, but in the absence of sufficient energy stores, quickly shift to a necrotic pattern of death. One could visualize this response as a cell suicide to eliminate the threat of long-term genotoxic sequelae such as mutation or cancer.

At low-dose exposure, the pattern could switch to one in which there is little to no immediate demonstration of injury but does set the stage for long-term consequences. As discussed previously, as a genotoxic agent, mustard can function in carcinogenesis in concert with, or playing the role of, promoter or inducer. The presence of nucleotide adducts or incorrect base replacement following DNA repair attempts coincidental with exposure to a second chemical insult could result in genesis of a cancer. But the question we wish to address is, "Can a low-dose, acute, asymptomatic exposure to mustard lead, by itself, to disease in later years?" Mustard is a mutagen and the mutation rate following *in vitro* exposure to cell culture systems has been studied. In the early 1970s, our laboratories reported that *in vitro* exposure of mouse lymphoma cells to submicromolar concentrations of mustard could increase the reversion rate for asparagines independence. *In vivo,* using a host-mediated assay in mice, comparable reversion rates were seen with a single subcutaneous dose of 100 mg/kg.[76] In a concurrent study of chronic exposure over 37 weeks, mice were exposed in an inhalation chamber to 0.1 mg/m^3 for 6 h per day, 5 days/week. No statistically significant increase in reversion rate was detected.[77] It is also interesting to note that the bifunctional alkylating agent sulfur mustard is less mutagenic than many monofunctional agents presumably because the cross-links formed by sulfur mustard lead to death of the affected cell.

Besides genotoxicity, sulfur mustard is known to affect other cellular parameters. Concentrations of mustard above 50 μM result in a marked reduction of cellular NAD+, a critical glycolytic cofactor, within 4 h of *in vitro* exposure to human keratinocytes.[78] However, no alterations in NAD+, ATP, or mitochondrial dehydrogenase activity are seen at 10-fold less concentrations in this model. One of the central cellular enzymes involved in NAD+ turnover is the nuclear enzyme PARP. As mentioned earlier, this enzyme undergoes a large and rapid activation in human epithelial cells exposed *in vitro* to vesicating equivalent doses (i.e., >50 μM). At concentrations of mustard below 10 μM, however, a completely different pattern of PARP response is seen.[79] It appears that even though significant DNA damage is detected at these low doses, the repair response is not as aggressive and the net metabolic disruption is transient.[80]

As one studies the response to *in vitro* exposure to mustard in cell systems, there appears to be a threshold level above which death processes are initiated that are rapid and totally destructive to the cells. This appears to be in the range of 50–100 μM for most mammalian cells. If concentrations below 10 μM are studied, one can observe toxic processes occurring, but depending on the cell system employed, these are often reversible.

V. DOSE DEPENDENCY OF THE MUSTARD LESION

In 1946, Renshaw reviewed the understanding of the mechanisms of mustard injury to that point.[81] He defined three dose ranges based on μg of mustard fixed per cm^2 of human skin. From 0.1–1.0 μg fixed/cm^2 the result was "mild erythema, occasional vesication with a histology that showed hyperemia and edema *without sufficient epidermal injury to cause death of more than occasional isolated basal cells*" (italics added). For 1.0–2.5 μg fixed/cm^2, the result was moderate injury with routine blister formation, and at doses >2.5 μg fixed/cm^2, the resulting injury was described as severe with central necrosis and circumferential vesication. Furthermore, he went on to state that minimal reversible injury was seen at 0.1 μg fixed/cm^2. One can say, therefore, at exposures resulting in less than 0.1 μg mustard fixed per cm^2 of skin, the outcome will be minimal clinical symptoms and fully reversible changes with no long-term effects. This exposure level is less than one-tenth that required for full demonstration of vesication and very close to what we refer to as a mild erythematic exposure.

Finally, for many years, sulfur mustard was used topically in the treatment of psoriasis in the form known as Russian Ointment (0.005% mustard-vaseline). This was also known as Psoriasin. In the 1970s, Illig reviewed the clinical information from these studies and evaluated the potential skin carcinogenicity and off-gassing problems associated with cutaneous exposure to low-dose mustard.[82,83] The following is a quote from his 1976 paper: "It is extremely improbable that the carcinogenic risk of the external S-mustard treatment is higher than that of a parenteral Methotrexate therapy, carried out at many clinics and in many cases over a period of many years, especially in the U.S.A.; it is rather to be expected that the carcinogenic risk in the case of external application of Psoriasin is also substantially lower over a longer period of time than in the Methotrexate treatment."

VI. SUMMARY

Genotoxic agents have the potential of long-term consequences, especially when synergistically coupled with promoters, immunosuppression or genetic deficiencies. Our contention, that acute subclinical asymptomatic injury causing long-term effects does not exist, is based on the following:

1. Lack of reliable clinical cases
2. *In vitro* observations
3. Renshaw's suggestion that, once the dose of applied mustard drops below that which yields observable, sustainable injury, no untoward consequences will become evident in the patient
4. More than 30 years' experience with Russian Ointment

Based on our scientific and medical experience, we should never-say-never, but the probability of chronic illness developing from an acute asymptomatic exposure to sulfur mustard appears to be extremely low.

ACKNOWLEDGMENTS

The authors wish to acknowledge the assistance provided by the following: Patricia Little for editorial assistance and manuscript preparation, Cynthia Martinez and Bethany Toliver for collecting and organizing the references, and Clark Gross for reminding us of the papers on Russian Ointment.

REFERENCES

1. Joy, R.J.T., Historical aspects of medical defense against chemical warfare, in *Textbook of Military Medicine—Medical Aspects of Chemical and Biological Warfare,* Zajtchuk, R. and Bellamy, R.F., Eds., Office of The Surgeon General, Department of the Army, Washington, DC, 1997, chap. 3.
2. Smart, J.K., History of chemical and biological warfare: An American perspective, in *Textbook of Military Medicine—Medical Aspects of Chemical and Biological Warfare,* Zajtchuk, R. and Bellamy, R.F., Eds., Office of the Surgeon General, Department of the Army, Washington, DC, 1997, chap. 2.

3. *Medical Management of Chemical Casualties Handbook,* 3rd Ed., U.S. Army Medical Research Institute of Chemical Defense, Aberdeen Proving Ground, MD, 1999.
4. Sidell, F.R., Urbanetti, J.S., Smith, W.J., and Hurst, C.G., Vesicants, in *Textbook of Military Medicine—Medical Aspects of Chemical and Biological Warfare,* Zajtchuk, R. and Bellamy, R.F., Eds., Office of the Surgeon General, Department of the Army, Washington, DC, 1997, chap. 7.
5. Sidell, F.R. and Hurst, C.G., Long-term health effects of nerve agents and mustard, in *Textbook of Military Medicine—Medical Aspects of Chemical and Biological Warfare,* Zajtchuk, R. and Bellamy, R.F., Eds., Office of the Surgeon General, Department of the Army, Washington, DC, 1997, chap. 8.
6. Manning, K.P., Skegg, D.C.G., Stell, P.M., and Doll, R., Cancer of the larynx and other occupational hazards of mustard gas workers, *Clin. Otolaryngol.,* 6, 165, 1981.
7. Prokes, J., Svovoda, V., Hynie, I., Hroksova, M., and Keel, K., The influence of x-radiation and mustard gas on methionine-35-S incorporation in erythrocytes, *Neoplasma,* 5, 393, 1968.
8. Heston, W.E., Induction of pulmonary tumors in strain A mice with methyl-bis(beta-chloroethyl)amine hydrochloride, *J. Natl. Cancer Inst.,* 10, 125, 1949.
9. *Veterans at Risk: The Health Effects of Mustard Gas and Lewisite,* Pechura, C.M. and Rall, D.P., eds., The Institute of Medicine, Washington, DC, 1993.
10. Morgenstern, P., Koss, F.R., and Alexander, W.W., Residual mustard gas bronchitis: Effects of prolonged exposure to low concentrations of mustard gas, *Ann. Intern. Med.,* 26, 27, 1947.
11. Buscher, H. and Conway, N., *Green and Yellow Cross,* Cincinnati, OH, Kettering Laboratory of Applied Physiology, University of Cincinnati, OH, 1944.
12. Papirmeister, B., Feister, A.J., Robinson, S.I., and Ford, R.D., *Medical Defense against Mustard Gas: Toxic Mechanisms and Pharmacological Implications,* CRC Press, Boca Raton, FL, 1991.
13. Case, R.A.M. and Lea, A.J., Mustard gas poisoning, chronic bronchitis, and lung cancer: An investigation into the possibility that poisoning by mustard gas in the 1914–1918 war might be a factor in the production of neoplasia, *Br. J. Prev. Soc. Med.,* 9, 62, 1955.
14. Norman, J.R., Lung cancer mortality in World War I veterans with mustard gas injury: 1919–1965, *J. Natl. Cancer Inst.,* 54, 311, 1975.
15. Fletcher, C., Peto, R., Tinker, C., and Speizer, F.E., *The Natural History of Chronic Bronchitis and Emphysema,* Oxford University Press, Oxford, England, 1976.
16. Wada, S., Miyanishi, M., Nashimoto, Y., Kambe, S., and Miller, R.W., Mustard gas as a cause of respiratory neoplasia in man, *Lancet,* 1, 1161, 1968.
17. Easton, D.F., Peto, J., and Doll, R., Cancers of the respiratory tract in mustard gas workers, *Br. J. Ind. Med.,* 45, 652, 1988.
18. Minoue, R. and Shizushiri, S., Occupationally-related lung cancer—Cancer of the respiratory tract as sequentia from poison gas plants, *Jpn. J. Thorac. Dis.,* 18, 845, 1980.
19. Albro, P.W. and Fishbein, L., Gas chromatography of sulfur mustard and its analogs. *J. Chromatogr.,* 46, 202, 1970.
20. Yanagida, J., Hozawa, S., and Ishioka, S., Somatic mutation in peripheral lymphocytes of former workers at the Okunojima poison gas factory, *Jpn. J. Cancer Res.,* 79, 1276, 1988.
21. Willems, J.L., Clinical management of mustard gas casualties, *Ann. Med. Mil. Belg. 3(Suppl),* 1, 1989.
22. Urbanetti, J.S., Battlefield chemical inhalation injury, in *Pathophysiology and Treatment of Inhalation Injuries.,* Loke, J., Ed., Marcel Dekker, New York, 1988.

23. Balali-Mood, M., Clinical and laboratory findings in Iranian fighters with chemical gas poisoning, in *Proceedings of the 1st World Congress on New Compounds in Biological and Chemical Warfare: Toxicological Evaluation,* 21–23 May 1984, Heyndrickx B., Ed., State University of Ghent, Ghent, Belgium, 254, 1984.
24. Balali-Mood, M., First report of delayed toxic effects of yperite poisoning in Iranian fighters, in *Proceedings of the 2nd World Congress on New Compounds in Biological and Chemical Warfare: Toxicological Evaluation, Industrial Chemical Disasters, Civil Protection and Treatment,* 24–27 August 1986, Heyndrickx, B., Ed., University of Ghent, Ghent, Belgium, 489, 1986.
25. Freitag, L., Fizusian, N., Stamatis, G., and Greschuchna, D., The role of bronchoscopy in pulmonary complications due to mustard gas inhalation, *Chest,* 100, 1436, 1991.
26. Gilchrist, H.L., *A Comparative Study of World War Casualties from Gas and Other Weapons,* Government Printing Office, Washington, DC, 1928.
27. Warthin, A.S. and Weller, C.V., The lesions of the respiratory and gastrointestinal tract produced by mustard gas (dichloroethyl sulphide), *J. Clin. Lab. Med.,* 4, 229, 1919.
28. Winternitz, M.C., Anatomical changes in the respiratory tract initiated by irritating gases, *Mil. Surg.,* 44, 47, 1919.
29. Rimm, W.R. and Bahn, C.F., Vesicant injury to the eye, in *Proceedings of the Vesicant Workshop,* U.S. Army Medical Research Institute of Chemical Defense, Aberdeen Proving Ground, MD, 1987.
30. Hughes, W.F., Jr., Mustard gas injuries to the eyes, *Arch. Ophthalmol.,* 27, 582, 1942.
31. Blodi, F.C., Mustard gas keratopathy, *Int. Ophthalmol. Clin.,* 2, 1, 1971.
32. Duke-Elder, W.S. and MacFaul, P.A., Chemical injuries, in *System of Ophthalmology,* Duke-Elder, W.S. and MacFaul, P.A., eds., CV Mosby, St. Louis, MO, 1994.
33. Otto, C.E., *A Preliminary Report on the Ocular Action of Dichlorethyl Sulfide (Mustard Gas) in Man as Seen at Edgewood Arsenal, Edgewood, MD,* Edgewood Arsenal, Chemical Warfare Service, EAL 539, 1946.
34. Novick, M., Gard, D.H., Hardy, S.B., and Spira, M., Burn scar carcinoma: A review and analysis of 46 cases, *J. Trauma,* 17, 809, 1977.
35. Treves, N. and Pack, G.T., Development of cancer in burn scars: Analysis and report of 34 cases, *Surg. Gynecol. Obstet.,* 51, 749, 1930.
36. Inada, S., Hiragun, K., Seo, K., and Yamura, T., Multiple Bowen's disease observed in former workers of a poison gas factory in Japan with special reference to mustard gas exposure, *J. Dermatol.,* 5, 49, 1978.
37. U.S. Army, U.S. Navy, and U.S. Air Force, Vesicants (blister agents), Section I—Mustard and nitrogen mustard, in *NATO Handbook on the Medical Aspects of NBC Defensive Operations,* U.S. Army, U.S. Navy, U.S. Air Force, Washington, DC, AMedP-6, 1973.
38. Anslow, W.P. and Houch, C.R., Systemic pharmacology and pathology of sulfur and nitrogen mustards, in *Chemical Warfare Agents and Related Chemical Problems,* Office of Scientific Research and Development, Washington, DC, 1946.
39. Weisburger, J.H. and Williams, G.M., Bioassay of carcinogens: *In vitro* and *in vivo* tests, in *Chemical Carcinogens,* Searle, C. E., ed., ACS Monograph 182, Vol. 2, American Chemical Society, Washington DC, 1984, chap. 22.
40. Boutwell, R.K. and Bosch, D.K., The carcinogenicity of creosote oil: Its role in the induction of skin tumors in mice, *Cancer Res.,* 18, 1171, 1958.
41. Grice, H.C., Cleff, D.J., Coffin, D.E., Lo, M.T., Middleton, E.J., Sandi, E., Scott, P.M., Sen, N.P., Smith, B.L., and Withey, J.R., in *Carcinogens in Industry and the Environment,* Marcel-Dekker, New York, 1981, 439.

42. Walker, E.A., Castegnaro, M., Griciute, L., Börzönyi, M., and Davis, W., Eds., in *N-Nitroso Compounds: Analysis, Formation, and Occurrence,* IARC Scientific Publication No. 31, Lyon, France, 1980.
43. Silverman, J. and Adams, J.D., N-nitrosamines in laboratory animal feed and bedding, *Lab. Anim. Sci.,* 33, 161, 1983.
44. Berenblum, I., *Carcinogenesis as a Biological Problem,* Frontiers of Biology, North Holland, Amsterdam, 34, 1974.
45. Ford, J.O. and Pereira, M.A., Short-term *in vivo* initiation/promotion bioassay for hapatocarcinogens, *J. Environ. Pathol. Toxicol.,* 4, 39, 1980.
46. Williams, G.M., Phenotypic properties of preneoplastic rat liver lesions and applications to detection of carcinogens and tumor promoters, *Toxicol. Pathol.,* 10, 3, 1982
47. Stott, W.T. and Watanabe, P.G., Differentiation of genetic vs. epigenetic mechanisms of toxicity and its application to risk assessment, *Drug Metab. Rev.,* 13, 353, 1982.
48. Stott, W.T., Reitz, R.H., Schumann, A.M., and Watanabe, P.G., Genetic and nongenetic events in neoplasia, *Food Cosmet. Toxicol.,* 19, 567, 1981.
49. Tennekes, H.A., Edler, L., and Kunz, H.W., Dose-response analysis of the enhancement of liver tumor formation in CF-1 mice by dieldrin, *Carcinogenesis,* 3, 941, 1982.
50. Hecker, E., Fusenig, N.E., Kunz, W., Marks, F., and Thielmann, H.W., Eds., *Cocarcinogenesis and Biological Effects of Tumor Promoters; Carcinogenesis—A Comprehensive Survey,* Raven Press, New York, 1982, 7.
51. Astrup, E.G. and Boutwell, R.K., Ornithine decarboxylase activity in chemically induced mouse skin papillomas, *Carcinogenesis,* 3, 303, 1982.
52. O'Brien, T.G., in *Polyamines in Biomedical Research,* Gaugas, J. M., Ed., Wiley Interscience, New York, 1980, 237.
53. Russell, K.H. and Haddox, M.K., Cyclic AMP-mediated induction of ornithine decarboxylase in normal and neoplastic growth, *Adv. Enzyme Regul.,* 17, 61, 1979.
54. Scalabrino, G. and Ferioli, M.E., Polyamines in mammalian tumors. Part I, *Adv. Cancer Res.,* 35, 151, 1981.
55. Fujiki, H., Suganuma, M., Nakayasu, M., Hoshino, H., Moore, R.E., and Sugimura, T., The third class of new tumor promoters, polyacetates (debromoaplysiatoxin and aplysiatoxin), can differentiate biological actions relevant to tumor promoters, *Gann,* 73, 495, 1982.
56. Izumi, K., Reddy, J.K., and Oyasu, R., Induction of hepatic ornithine decarboxylase by hypolipidemic drugs with hepatic peroxisome proliferative activity, *Carcinogenesis,* 2, 623, 1981.
57. Ide, F., Ishikawa, T., Takagi, M., Umemura, S., and Takayama, S., Unscheduled DNA synthesis in human oral mucosa treated with chemical carcinogens in short-term organ culture, *J. Natl. Cancer Inst.,* 69, 557, 1982.
58. Bruce, W.R. and Heddle, J.A., The mutagenic activity of 61 agents as determined by the micronucleus, salmonella, and sperm abnormality assays, *Can. J. Genet. Cytol.,* 21, 319, 1979.
59. Jenssen, D. and Ramel, C., The micronucleus test as part of a short-term mutagenicity test program for the prediction of carcinogenicity evaluated by 143 agents tested, *Mutat. Res.,* 75, 191, 1980.
60. Stich, H.F. and Rosin, M.P., Quantitating the synergistic effect of smoking and alcohol consumption with the micronucleus test on human buccal mucosa cells, *Int. J. Cancer,* 31, 305, 1983.
61. Perera, F.P., Poirier, M.C., Yuspa, S.H., Nakayama, J., Jaretski, A., Curnen, M.M., Knowles, D.M., and Weinstein, I.B., A pilot project in molecular cancer epidemiology: Determination of benzo [a] pyrene-DNA adducts in animal and human tissues by immunoassays, *Carcinogenesis,* 3, 1405, 1982.

62. Groopman, J.D., Haugen, A., Goodrich, G.R., Wogan, G.N., and Harris, C.C., Quantitation of aflotoxin B1-modified DNA using monoclonal antibodies, *Cancer Res.,* 42, 3120, 1982.
63. Gibson, J.L., Symposium: Peer review and scientific decision making, *Fundam. Appl. Toxicol.,* 2, 271, 1982.
64. Campbell, T.C., A decision tree approach to the regulation of food chemicals associated with irreversible toxicities, *Regul. Toxicol. Pharmacol.,* 1, 193, 1981.
65. Munro, I.C. and Krewski, D.R., Risk assessment and regulatory decision making, *Food Cosmet. Toxicol.,* 109, 549, 1981.
66. Starr, C. and Whipple, C., Risks of risk decisions, *Science,* 208, 1114, 1980.
67. Brown, S.M., The use of epidemiologic data in the assessment of cancer, *J. Environ. Pathol. Toxicol.,* 4, 573, 1980.
68. Vogt, T.M., Risk assessment and health hazard appraisal, *Ann. Rev. Public Health,* 2, 31, 1981.
69. Lave, L.B., Balancing economics and health in setting new standards, *Annu. Rev. Public Health,* 2, 183, 1981.
70. Scientific Committee Food Safety Council, Proposed system for food safety assessment, *Food Cosmet. Toxicol.,* 16, 1, 1978.
71. Urbach, F., *The Biologic Effects of Ultraviolet Radiation (with Emphasis on the Skin),* Pergamon Press, Oxford, England, 1969.
72. Epstein, J.H. and Forbes, F.D., Ultraviolet carcinogenesis: Experimental, global and genetic aspects, in *Sunlight and Man,* Pathak, M.A., Harber, L.C., Leifik, M., and Kukita, A., Eds., University of Tokyo Press, Tokyo, Japan, 1974, 259.
73. Fornace, A.J., DNA single-strand breaks during repair of UV damage in human fibroblasts and abnormalities of repair in Xeroderma pigmentosum, *Proc. Nat. Acad. Sci., U.S.A.,* 73, 39, 1976.
74. Menon, I.A. and Haberman, H.F., Mechanisms of actions of melanins. *Br. J. Dermatol.,* 97, 109, 1977.
75. Poel, W.E., Skin as test site for the bioassay of carcinogens and carcinogen precursors, *Natl. Cancer Inst. Monogr.,* 10, 611, 1963.
76. Capizzi, R.L., Smith, W.J., Field, R.J., and Papirmeister, B., A host-mediated assay for chemical mutagens using the L5178Y/Asn(−) murine Leukemia, *Mutat. Res.,* 21, 6, 1973.
77. Rozmiarek, J., Capizzi, R.L., Papirmeister, B., Furman, W.H., and Smith, W.J., Mutagenic activity in somatic and germ cells following chronic inhalation of sulfur mustard, *Mutat. Res.,* 21, 13, 1973.
78. Smith, W.J., Gross, C.L., Chan, P., and Meier, H.L., The use of human epidermal keratinocytes in culture as a model for studying the biochemical mechanisms of sulfur mustard induced vesication, *Cell Biol. Toxicol.,* 6, 285, 1990.
79. Clark, O.E. and Smith, W.J., Activation of poly(ADP-ribose) polymerase by sulfur mustard in HeLa cell cultures, in *Proceedings of the 1993 Medical Defense Bioscience Review,* U.S. Army Medical Research Institute of Chemical Defense, Aberdeen Proving Ground, MD, DTIC Accession # A275667, 1, 199, 1993.
80. Smith, W.J., Toliver, B.S., Nealley, E.W., Guzman, J.J., and Gross, C.L., Effects of low dose sulfur mustard on growth and DNA damage in human cells in culture, *Toxicol. Sci.,* 54(1-S), 152, 2000.
81. Renshaw, B., Mechanisms in production of cutaneous injuries by sulfur and nitrogen mustard, in *Chemical Warfare Agents and Related Chemical Problems,* Bush, V., Ed., Office of Scientific Research and Development, National Defense Research Committee, Division 9, Parts 1–6, Washington DC, 1946, chap. 23.

82. Illig, L., The treatment of psoriasis vulgaris with S-mustard vasoline externally with special consideration to the possible carcinogenic risk (First continuation and conclusion): On the carcinogenicity of S-mustard in animal tests and in humans, *Z. Hautkrankh.,* 52, 1035, 1976.
83. Illig, L., Paul, E.L., Eyer, P., Weger, H., and Born, W., The treatment of psoriasis vulgaris with S-mustard-vaseline externally, taking especially into consideration the possible carcinogenic risk: III-Communication. Clinical and experimental studies on the extent of percutaneous and inhalative intake of S-mustard-vaseline, *Z. Hautkrankh.,* 54, 941, 1979.

9 Gulf War Syndrome: Questions, Some Answers, and the Future of Deployment Surveillance

Coleen Baird Weese

CONTENTS

I. Introduction 262
II. The Population at Risk 265
III. Outcomes in the Population at Risk 266
  A. Mortality 267
  B. Morbidity 269
  C. Symptom Prevalence 270
  D. Reproductive Outcomes 271
IV. Subsets of the Population at Risk: Symptom-Based Clusters 272
  A. 123rd ARCOM 273
  B. Air National Guard 274
  C. Seabees 274
  D. Reproductive Effects 275
V. Subsets of Population at Risk: Common Exposures 276
  A. Infectious Diseases 276
  B. Immunizations, Pesticides, and Occupational Exposures 277
  C. Depleted Uranium 278
  D. Oil Well Fires 279
  E. Chemical and Biological Warfare Agents 279
  F. Khamisiyah, Iraq 280
  G. Mixed Exposures and Synergistic, Additive, or Other Combined Effects 283
VI. Subsets Enrolled in Registries 283
VII. Outcomes in Subpopulations in Registries 285
VIII. Is There a Single PGW Syndrome? The Problem with Case Definition 289

0-8493-0872-0/01/$0.00+$.50

IX. Future Research Directions 290
X. Association vs. Causation in Environmental Epidemiology 291
Acknowledgments 295
References 295

I. INTRODUCTION

Ten years following the Persian Gulf conflict, uncertainty remains regarding potential exposures, health risks, and adverse outcomes in the 697,000 U.S. troops deployed to Operations Desert Shield/Desert Storm. While this was not the first wartime cohort to report medically unexplained symptoms, it is certainly the most studied. Somatic complaints such as fatigue, shortness of breath, headache, sleep disturbance, forgetfulness, and impaired concentration have been reported following armed conflicts since the Civil War (Table 9.1).[1] The authors described two general categories of war-related illness—a poorly understood group thought to be associated with physiological disease, and another group of psychological illnesses attributed to wartime stress. "War syndromes have not been consistently defined or identified by a pathognomonic physical sign or laboratory abnormality. As a result, the diagnosis of a physiological or psychological illness in individual patients has been imprecise and has depended on self-reported symptoms and the impression of the examining physician."

Past wartime deployments have resulted in concerns over specific potential exposures as well. Following the Vietnam War, uncertainty relating to exposure to herbicides ultimately led to the Congressional passage of Public Law 102–4 (the "Agent Orange Act of 1991"). This legislation directed the National Academy of Sciences (NAS) to conduct a comprehensive review and evaluation of scientific and medical information regarding the health effects of exposure to Agent Orange, other herbicides used in Vietnam, and the various chemical components of these herbicides, including dioxin. The review was intended to determine, to the extent that available data permitted, whether there was: (1) a statistical association between herbicide exposure and disease outcomes, (2) an increased risk of the disease among those exposed to herbicides during Vietnam service, and (3) whether there was a plausible biological mechanism or other evidence of a causal relationship between herbicide exposure and disease.[2] The NAS committee faced considerable issues of cohort reconstruction and dose estimation in the absence of quantified exposure information, as well as difficulties in assessing causality. Ultimately, epidemiological studies were reviewed, and specific health outcomes were assigned to one of four categories of evidence based on "statistical associations," not on causality.

Similarly, following the Persian Gulf War (PGW), the Department of Defense (DoD) and the Department of Veterans Administration (DVA) faced basic questions of exposure, outcome, and association. These questions address exposures that were known or possible for the deployed cohort, the potential outcomes of importance that might be associated with such exposures, and the studies and actions undertaken to evaluate these associations. Multiple expert boards and committees have studied PGW veterans and health consequences of service in the Gulf (Table 9.2).[3–8] The

TABLE 9.1
Somatic Symptoms Commonly Associated with War-Related Medical and Psychological Illnesses

Symptom	War and Illness					
	U.S. Civil War DaCosta Syndrome	World War I	World War II Combat Stress Reaction	Vietnam Agent Orange Exposure	Vietnam Post-Traumatic Stress	Persian Gulf Unexplained Illness
Fatigue and exhaustion	+	+	+	+	+	+
Shortness of breath	+	+	+		+	+
Palpitations and tachycardia	+	+	+		+	
Precordial pain	+	+			+	+
Headache	+	+	+	+	+	+
Muscle or joint pain				+	+	+
Diarrhea	+		+	+	+	+
Excessive sweating	+	+	+			
Dizziness	+	+	+	+	+	
Fainting	+	+				
Disturbed sleep		+	+	+	+	+
Forgetfulness		+	+	+	+	+
Difficulty concentrating		+	+	+	+	+

Note: A plus sign indicates a commonly reported symptom.

Source: Hyams, K.C., Wignall, S.W., and Roswell, R., *Ann. Intern. Med.*, 125, 398, 1996. With permission.

Defense Science Board (DSB) panel was originally charged to evaluate the scientific and medical evidence relating to long-term health effects of low levels of neurotoxic agents, but expanded its scope to the full range of exposures to low levels of chemicals, as well as environmental pollutants, biological agents, and other health hazards.[3] The task force was unable to define the medical nature and cause or causes of a Gulf War Syndrome, and did not identify any cause-and-effect relationships between putative exposures and an undefined illness. The panel did not find evidence to suggest that illnesses suffered by PGW veterans were related to chemical or biological weapons.[3] In April 1994, the National Institutes of Health Technology Assessment Workshop was held to consider the evidence for increased incidence of unexpected illness attributable to service in the PGW and the components of a practical case definition. They further considered the plausible etiologies and biological explanations for any unexpected illness and future research deemed necessary.[4] The panel

TABLE 9.2
Expert Panels Evaluating Health Effects of Gulf War Service

Panel	Funding	Report
Task Force on PGW Health Effects, Defense Science Board	DoD	Report to the Under Secretary of Acquisition, DSB, 1994
National Institutes of Health Technology Assessment Workshop Panel	Interagency	NIH Technology Assessment Panel, 1994
Institute of Medicine Committee to Evaluate the Comprehensive Clinical Evaluation Program	DoD	IOM 1996, Evaluation of the U.S. Department of Defense Persian Gulf Comprehensive Clinical Evaluation Program
Presidential Advisory Committee on Gulf War Veteran's Illnesses	DoD	Presidential Advisory Committee on Gulf War Veteran's Illnesses Interim Report 1996, Final Report 1996
Institute of Medicine Committee to Review the Health Consequences of Service during the Persian Gulf War	DoD	Health Consequences of Service During the Persian Gulf War, IOM, 1996

was unable to formulate a case definition to determine whether plausible exposures were associated with outcomes (unexplained illnesses). The panel did note the lack of available data on exposures and made a series of recommendations regarding future research. Both the Institute of Medicine (IOM) and the Presidential Advisory Committee (PAC) noted that the formalized registries established by the DoD and the DVA, which provide free medical evaluation to concerned PGW veterans, served an important purpose but were not designed to answer epidemiological questions.[5–8] The PAC noted that the current scientific evidence did not support a causal link between the symptoms and illnesses reported by PGW veterans and exposures while in the Gulf to pesticides, chemical warfare agents, biological warfare agents, vaccines, pyridostigmine bromide, infectious diseases, depleted uranium, oil well fires and smoke, and petroleum products.[8] The PAC determined, however, that the investigation of possible exposures of troops to chemical and biological agents was "superficial and inadequate." They made a series of recommendations regarding improved communication, better data on baseline health conditions of troops, locations and exposures on deployments, and better services to veterans. The IOM reviewed the studies that were available to date and reported that the scope and focus was "of uneven depth and quality" and noted a series of potential biases.[5,6] They considered the initial research efforts "poorly organized both strategically and tactically." The committee identified a lack of reference population for many data collection and analysis activities, and noted that predeployment demographic information on health and medical interventions such as vaccinations was incomplete and possibly inaccurate. In the evaluation of health outcomes, there was little standardization and operationalization of data on disease symptoms and signs. Further, follow-up was difficult and incomplete, and

DoD and DVA databases did not communicate effectively. It was noted that very little personalized exposure information was available, and defining relevant control groups and obtaining data for them were very difficult. The committee also noted that the "full range of potential biases (selection bias, follow-up bias, dropout bias, observation bias, ascertainment bias, and recall bias) was operating. These problems further limit the ability of even the most expert and well-funded investigation to identify health outcomes linked to specific exposures or risk factors."[6] The Government Accounting Office recommended a re-examination of research emphasis in 1997.[9] They noted that the majority of research focused on the prevalence and cause of Gulf War illnesses, rather than diagnosis, treatment, and prevention. "While this epidemiological research will provide descriptive data on veterans' illnesses, methodological problems are likely to prevent researchers from providing precise, accurate, and conclusive answers regarding the causes of veterans' illnesses. Without accurate exposure information, the investment of millions of dollars in further epidemiological research on the risk factors or potential causes for veterans' illnesses may result in little return."[9]

In summary, the panels and committees evaluating the available data noted a significant lack of exposure data on the population or populations at risk. Given these constraints, limited conclusions could be drawn concerning exposure and outcome relationships. The following sections discuss the population at risk and various subsets and provide an overview of outcomes reported in these populations.

II. THE POPULATION AT RISK

Although early epidemiological studies typically focused on infectious diseases and death, current epidemiology has a broader application as "the study of the distribution and determinants of health-related states and events in specified populations and the application of this study to the control of health problems."[10] The population under study is logically dependent on the study question. Ideally, the two groups should differ only with respect to the exposure under study and have equal opportunity for the outcome under consideration. Differences between the two groups with respect to other relevant factors, for example, age, sex, general state of health, or smoking habits, known as confounding variables, should be addressed. Studies of a population exposed to a factor under study and the comparable unexposed control group must recognize known confounders for the outcome or outcomes of interest and measure the rate of exposure to confounders in both groups so that adjustment can be performed in the analysis. Apart from any relevant specific factors, consideration should be given to whether or not the exposed population has more frequent occupational or other relevant exposures, higher rates of disease or is otherwise more at risk for the outcome of interest due to reasons totally unrelated to the exposure under question. This information is typically not available when populations are studied at the community level—by county cancer rates, for example—and thus, differences in individual factors and their impact on the findings of the study cannot be known. Making inferences about individuals from studies of groups can be subject to error known as the "ecological fallacy."[11]

The population at risk for adverse health outcomes associated with service during the war in the Persian Gulf is, in the broadest sense, the cohort of troops deployed to the Gulf. This population has been identified and considered the "exposed" population in a number of studies attempting to assess whether or not Gulf War veterans were at increased risk for adverse health outcomes, as compared with veterans from the same era who did not deploy to the Persian Gulf.[12–17]

Reconstructing this cohort required integration of a number of data sets. As one investigator reports, the data on service in the Gulf War and demographic variables were obtained from the Defense Manpower Data Center. Data on Gulf War Service were compiled from Army, Navy (including Marine Corps), and Air Force records of unit-deployment locations and pay for exposure to hostile fire.[13] Demographic data were obtained from routine data files on U.S. military personnel. Military personnel deployed to serve in the Gulf War for one or more days between August 8, 1990 and July 31, 1991 were considered Gulf War veterans. Approximately 83% of the approximate 697,000 U.S Gulf War veterans served on regular active duty in the Army, Navy (including Marine Corps), or Air Force. The Defense Manpower Data Center also provided personal identifiers for the Gulf War veterans as well as data on their sex, age, race or ethnic group, marital status, branch of service, occupation, rank, pay grade, and total number of months of active duty service. The demographics of this population are provided in Table 9.3.

Most of the studies of outcomes in this deployed or exposed cohort utilized the entire cohort of deployed troops for which data was complete.[12–17] The cohort thus identified as Persian Gulf veterans often excludes reservists and civilians. The completeness of the outcome data varied with the outcomes assessed (for example, reproductive outcomes vs. mortality outcomes). For most of the studies of outcomes in this cohort, rates were compared to a random sample of roughly an equal number of all personnel on active duty but not deployed to the Gulf, with the sampling percentage from each service proportional to the numbers from each service sent to the Gulf. The underlying assumption is that this non-deployed cohort represents a group with the same opportunity for outcomes of interest, apart from deployment, and as a group has no more or less of attributes that may be associated with these outcomes. Issues relating to whether or not this is a valid assumption are discussed specifically in the sections on health outcomes. A more basic question is whether or not the "population at risk" identified as the deployed cohort has any real meaning. In all reality, it is not a homogeneous group whose collective exposure was "the Gulf." It is in actuality a composite of groups with differential experiences, exposures, and duties from locations throughout the Gulf theater. This is further discussed in the section addressing subsets of exposures.

III. OUTCOMES IN THE POPULATION AT RISK

A variety of studies compared PGW veterans with non-deployed military personnel of the same era as a control group.[12–17] The selection of non-deployed military attempts to address the assumption that active duty military personnel are likely to be healthier than typical U.S. workers due to the physical demands of the military.[8] The

TABLE 9.3
Demographic Characteristics of the Persian Gulf War Population at Risk

Characteristic	PGW Participants
Gender (%)	
Male	93
Female	7
Race (%)	
White	70
Black	23
Hispanic	5
Other/no data	2
Age	
Mean	26
Median	24
Rank (%)	
Enlisted	89
Officer	10
Other/no data	1
Branch	
Air Force	12
Army	50
Marines	15
Navy	23
Status (%)	
Active	83
Reserve component	17

Note: N = 697,000.

Source: Desert Shield/Storm Participation Reports Vol. 1 & 2, Defense Manpower Data Center, DoD, 1994

ideal comparison population would be identical to the Gulf War veteran population in every aspect except deployment to the Gulf region. Therefore, concerns relate to whether some aspect of health resulted in non-deployed status and thus would be over-represented in the non-deployed group. Selected studies conducted to ascertain the frequency and scope of outcomes in troops and subpopulations of troops who served in the Gulf are summarized below.

A. Mortality

Vital status was determined for all of the approximately 700,000 military personnel who served in the Gulf and compared with a roughly equal group of personnel on active duty who did not deploy to the Gulf for the period August–September 1990 to April 1991.[12] Potential confounders such as age, sex, race, and military variables were controlled. A 9% higher death rate in PGW veterans (exposed) was demonstrated as compared with other veterans of the same era, or "unexposed" (relative risk = 1.09, CI = 1.01–1.16) (Table 9.4). The excess mortality was entirely attributable to external causes with an excess of deaths from motor vehicle injuries (relative risk = 1.31, CI = 1.14–1.49) and unintentional injuries (RR = 1.25, CI =

TABLE 9.4
Deaths, Mortality Rates, and Mortality-Rate Ratios among the Study Subjects According to Cause of Death and Sex

	Gulf War Veterans		Other Veterans		Mortality-Rate Ratios	
Cause of Death	No. of Deaths	Mortality Rate[a]	No. of Deaths	Mortality Rate[a]	Crude	Adjusted (95% CI)[b]
All Causes						
Men	1437	10.7	1084	9.8	1.10	1.09 (1.01–1.18)
Women	70	5.8	84	4.1	1.41	1.32 (0.95–1.83)
Disease-related causes						
Men	238	1.8	286	2.6	0.69	0.87 (0.73–1.04)
Women	14	1.2	26	1.3	0.92	0.89 (0.45–1.78)
All external causes						
Men	1110	8.3	732	6.6	1.26	1.17 (1.07–1.29)
Women	47	3.9	41	2.0	1.95	1.78 (1.16–2.73)
All accidents						
Men	689	5.1	422	3.8	1.34	1.26 (1.11–1.42)
Women	25	2.1	22	1.1	1.91	1.83 (1.02–3.28)
Motor Vehicle Accidents						
Men	457	3.4	269	2.4	1.42	1.27 (1.09–1.48)
Women	21	1.7	19	0.9	1.89	1.81 (0.96–3.41)
Suicide						
Men	211	1.6	191	1.7	0.94	0.88 (0.72–1.08)
Women	11	0.9	12	0.6	1.50	1.47 (0.63–3.43)
Homicide						
Men	116	0.9	101	0.9	1.00	0.80 (0.61–1.05)
Women	11	0.9	6	0.3	3.00	2.66 (0.96–7.36)

Note: Data for men are based on 544,270 Gulf War veterans and 456,726 controls assigned to active units. Data for women are based on 49,919 Gulf War veterans and 84,517 controls assigned to active duty.

[a]Crude rates shown are per 10,000 person-years

[b]Adjusted rate ratios (and 95 percent confidence intervals [CI] were derived from the Cox proportional-hazards model after adjustment for age, race, branch of service, and type of unit.

Source: Kang, H.K. and Bullman, T.A., *N. Engl. J. Med.*, 335, 1498, 1996. With permission.

1.13–1.39). No excess of deaths from suicide, homicide, or specific disease was observed. Risk of death from infectious diseases was reduced in the deployed population (RR = 0.21, CI = 0.11–0.43). Mortality for both groups was less than half that of the U.S. general population. Precise reasons for the excess of deaths due to external causes among war veterans are not well understood. The findings of the extension of this study through 1997 indicated that, while the risk of disease-related deaths (RR = 0.65, CI = 0.60–0.71) in deployed veterans as compared to controls did not increase over time, the excess deaths from motor vehicle accidents persisted (RR = 1.32, CI = 1.23–1.41). Post-war mortality from external causes, including suicides and homicides, was greater among female veterans than males, and the risk of suicide and homicide was even greater among married female Gulf War veterans.[18]

This was a large study that compared broad outcomes in large populations without respect to specific risk factors. While it was critically important to describe mortality in the cohort deployed and have a comparison population, in the sense that "risk" was related to deployment, this makes the study ecological in nature. No attempt to differentiate exposure to any specific location or hazard was made due to lack of data. In this instance, risk was equated to deployment. Additionally, mortality represents an infrequent and rather serious outcome, and patterns of mortality might not reflect patterns of morbidity. Risks under consideration might not be substantial enough to lead to significant detectable changes in mortality. Nonetheless, all PGW veterans and almost half of all military personnel not deployed to the Gulf were included to minimize sampling biases. Interpretation of the study is somewhat limited by the possibility that the two populations are not comparable. Military personnel who were ill or recovering from surgery, and perhaps more at risk for morbidity outcomes, might be differentially represented in the non-deployed population, and this would confound the results. This further extension of the healthy worker effect, coined the "healthy warrior effect," noted that lower mortality should be expected in those healthy enough to deploy; although the magnitude of the expected difference is not known, it is relevant to the interpretation of the results of morbidity and mortality studies.[19,20] This was a very complete study with several sources of mortality data. Death certificates were utilized that may vary in quality and completeness. In general, mortality data are suitable for some assessments of outcomes in broad categories, but less useful for diseases that are difficult to diagnose. Writer et al. reported similar findings for a comparison of deployed and non-deployed service members during the period of the Gulf War and shortly thereafter.[17]

B. Morbidity

A study of hospitalization in military hospitals of 547,076 PGW veterans who had remained on active duty in the 2 years following the war compared 618,335 veterans who served elsewhere. All diagnostic categories were evaluated.[13] The study adjusted for the possibility that disease rates should be lower in the cohort that was deployed. This considers that individuals with pre-existing illness might differentially not deploy, and thus the rate of hospitalization in this cohort might be expected to differ. The study noted that PGW veterans were at a slightly lower risk for hospitalization for any cause than the non-deployed cohort before the war, but not after the war.

Higher rates of hospitalization for alcohol and drug use and adjustment disorders were noted in PGW veterans in 1992 and 1993. Gulf War veterans were at increased risk of hospitalization for benign neoplasms in 1991, diseases of the genitourinary system in 1991, and diseases of the blood and blood-forming organs in 1992. Further analysis indicated that most of these were anemias associated with pregnancy. This study did not demonstrate an emerging illness requiring increased hospitalization in troops deployed to the Gulf. However, this study also equated deployment to the Gulf to exposure and did not attempt to differentiate hospitalization rates among various subpopulations of deployed troops with different exposures or experiences. Outcomes were obtained from computerized military hospital discharge data. While hospitalization of virtually all active-duty troops takes place in military hospitals, hospitalizations after discharge may occur in private, public, or DVA facilities.[6] For certain outcomes of interest such as obstetrical outcomes, civilian sector care may be frequently utilized. Also to be considered is that hospitalization outcomes on those who remain on active duty may be biased if health-status-specific discharges differ for the two cohorts.[19,21]

C. Symptom Prevalence

The Iowa Persian Gulf Study Group evaluated symptom prevalence in a cross-sectional telephone interview survey of 3,695 PGW and non-PGW military personnel from the state of Iowa.[22] The study tool was a validated questionnaire which attempted to gather information on the prevalence of self-reported medical and psychiatric conditions in PGW veterans compared to military personnel on active duty at the same time but non-deployed. The PGW veterans reported a significantly higher prevalence of depression, post-traumatic stress disorder, fatigue, cognitive dysfunction, bronchitis, asthma, alcohol abuse, anxiety, and sexual discomfort. The relationship between self-reported exposures and conditions suggested that no single exposure was related to the medical and psychiatric conditions among PGW personnel (Table 9.5). The most commonly reported exposures in symptomatic PGW personnel were solvents, smoke, pesticides, pyridostigmine bromide, and chemical warfare agents. However, this study suffered from recall bias related to self-reporting, and the number and variety of units involved precluded the focus on any subpopulations with specific exposures. Telephone interviewing may result in a select group who is willing to participate, and generalizability is also somewhat limited by restricting eligibility to those from Iowa.

The Veterans Administration's "National Health Survey of Gulf War Era Veterans and Their Families" is a three-phased project currently underway.[23] Phases I and II collected self-reported health data using mail and telephone interviews with about 21,000 veterans. Phase III involves a comprehensive, in-person examination on a stratified random sample of Phase I and II participants and their families. Participants are currently being recruited. The study's aim is to test the hypothesis that the prevalence of chronic fatigue syndrome, fibromyalgia, post-traumatic stress disorder, selected neurological abnormalities, and general health status of deployed and non-deployed veterans and their families are not significantly different.

TABLE 9.5
Reported Exposures among Persian Gulf Military Personnel

	Regular Military (N = 985)	National Guard/Reserve (N = 911)
Estimated days in theater Mean (SE)	167.8 (2.5)	138.1 (1.2)
Number of assigned units	820	137
Number of vaccinations (injections and oral) % of subjects		
0	1.5	1.1
1–5	28.3	26.8
6–10	31.1	35.8
>10	33.6	27.1
missing data	5.5	9.2
Number of pyridostigmine tablets used % of subjects		
0	45.7	40.8
1–10	17.7	27.0
11–30	14.6	15.0
>30	33.6	27.1
missing data	2.3	4.3
Smoking history % of subjects		
Never	44.9	45.1
Former	21.0	22.4
Current	34.1	32.5
Agent, % of subjects		
Solvents/petrochemicals	88.7	91.2
Smoke/combustion products	85.2	96.0
Sources of infectious agents	84.0	92.6
Psychological stressors	82.6	96.3
Sources of lead from fuels	78.2	88.5
Pesticides	43.8	63.4
Ionizing/nonionizing radiation	27.2	16.0
Chemical warfare agents	4.6	6.4
Physical trauma	3.7	5.6

Source: JAMA, 277, 238, 1997. With permission.

D. Reproductive Outcomes

Birth outcomes for 579,931 active duty military personnel deployed for at least one day to Operations Desert Shield/Desert Storm from August 8, 1990, to July 31, 1991, were compared to that for 700,000 service members occurring in a similar time frame.[15] Information was obtained from hospital-recorded, International

Classification of Disease codes coded to five digits and included up to eight diagnoses. The primary outcome assessed in the study was the occurrence of birth defects with the number of live births per 1,000 population and the ratio of male to female babies as secondary outcome. Birth defects were defined in two ways—a sensitive definition of "any birth defect" and a second "severe birth defect" as defined by the Centers for Disease Control (CDC). Birth rates were comparable in the two populations. The hypothesis that children born to PGW veterans were at increased risk of overall birth defects was not supported. For male service members, no positive association was noted between PGW service and any birth defects. For female PGW veterans, the risk of any birth defect was slightly higher, but appeared to be the result of confounding by race, ethnicity, marital status, or length of service, and did not persist after adjustment. The risk of birth defects in both the deployed and the non-deployed military populations approximated the risk in a civilian population. No linear trend of increasing risk with increasing length of time spent in the Gulf was demonstrated. The study was limited by gross classification of exposure simply as service in the Gulf. Further, only 68% of all births to military personnel occurred in military hospitals and were studied. Births to reserve component members or individuals who left active duty after the study period were excluded. Also, only defects evident at birth and coded before discharge were included. Nonetheless, this study provides substantial evidence that PGW veterans do not have decreased fertility or increased risk of birth defects. Another study identified 17,182 live births to military personnel in the state of Hawaii between 1989 and 1993. The Hawaii Birth Defects Program records were utilized to identify birth defects. A total of 3,717 infants were born to PGW veterans and 13,465 to non-deployed veterans. Of these, 367 infants (2.14%) were identified with one or more of 47 major birth defects diagnoses. The prevalence of birth defects was similar between both groups and was similar among infants conceived prior to and after the Gulf War. Although the number of infants in the birth defects categories was small, this study eliminates some of the limitations of previous studies that utilized information only from military hospitals and included diagnoses made during the first year of life.

IV. SUBSETS OF THE POPULATION AT RISK: SYMPTOM-BASED CLUSTERS

Another subset of the deployed population at risk was identified on the basis of symptoms. Within a few months of the return of troops in 1991, complaints of fatigue, headaches, joint pains, rashes, sleep disturbances, and other cognitive difficulties began to arise.[25] A concern was raised that perhaps there was something unique to the Gulf or the war fought there that was linked to a specific illness. These concerns were raised initially by individuals and then by other outbreak or cluster investigations that reported a high prevalence of a cluster of symptoms later proposed to be a characteristic of a "Gulf War Syndrome."[26–29]

Cluster investigations are typically initiated to evaluate reported rates of symptoms in a group of individuals. The "cluster" is first identified on the basis of symptoms or conditions. Typically, the individuals are linked by a common workplace, community, or experience and have some concern about an "excess" of symptoms or findings that they believe are linked to a poorly defined exposure. A key difficulty with a systematic approach to cluster evaluations is identifying the correct denominator, that is, the population at risk. "Clusters sometimes arise, are publicized, generate interest, and often lead to the collection of cases (numerator data) that are poorly defined with little knowledge of the population at risk (denominator data)."[6] Selection of the participants for study may be problematic, as is the selection of a control group for comparison, if one is used, because it depends on correct classification of exposure, although the exact exposure of concern might not be identified. "Even if the disease is well defined and its diagnosis properly operationalized, clusters cannot be used to evaluate causation because it is virtually impossible to identify a reference population. Clusters will arise in the absence of causation; indeed, they are inevitable in any large and complex collection of study participants and data. It is the task of the investigating analyst to sort out clusters that occur by chance from those that occur as a result of some exposure of interest." These studies are also limited by sample size and have a significant potential for bias resulting from respondent awareness of the underlying concern.

The initial reports of symptom clusters (fatigue, headache, muscle aches) helped to formulate hypotheses for subsequent studies and served as a starting point for survey questions. Clusters of disease have been reported among various units deployed to the Gulf.

A. 123RD ARCOM

Early in 1992, the staff of the 123rd ARCOM Surgeon's Office became aware of symptomatic complaints among reservists belonging to the 123rd ARCOM, Lafayette, Indiana. Similar complaints were later reported from the members of the 417th Quartermaster Company in Scottsburg, Indiana. A team from the Walter Reed Army Institute of Research evaluated 79 reservists with medical questionnaires; 78 completed a brief symptom inventory and a detailed interview. Other components of the evaluation included a brief psychiatric intake interview, a dental exam, vital signs, a laboratory evaluation that included a complete blood count with differential, an erythrocyte sedimentation rate, and liver function studies. All sera were tested for antibodies to Leishmania tropica, and sera from selected individuals were tested for antibodies to brucellosis. The most common complaint was fatigue (70%); other less-common symptoms included fever, abdominal pain, and diarrhea. The onset of fatigue and other associated symptoms were related to redeployment from the Gulf, although diarrhea was more frequent during the deployment. Blood testing revealed no cases of leishmaniaisis, brucellosis, Lyme disease, nor any characteristic pattern of other laboratory measures. There was no documented exposure to microwaves, chemicals, radiation, or other suspected environmental hazards. High levels of stress were reported, although Post-Traumatic Stress Disorder (PTSD) was present in few, if any, of the reservists. No common pattern of illness was noted among study group members. This study provided no basis for identifying a "case" of disease, and no evidence of a common exposure was found.[25]

B. Air National Guard

In 1994, an evaluation of unexplained illness among PGW veterans of the Pennsylvania Air National Guard unit was conducted.[26] The initial cluster investigation expanded into a three-stage study. To identify and characterize the signs and symptoms of disease in these veterans, 59 identified symptomatic PGW veterans received standardized interviews and physical examinations. Gulf War veterans reported a higher prevalence of symptoms identified as "moderate" or "severe" (Table 9.6). Overall, the patients reported that symptoms began in the Gulf or 2 to 3 months following return and persisted for greater than 6 months. No consistent abnormalities were noted on physical examination or medical records' review. To establish the frequency of reported symptoms in PGW veterans as compared to guardsmen who had not deployed to the Gulf, a second stage surveyed 3,927 members of the index unit and three comparison units. In all units, the prevalence of 13 symptoms lasting greater than 6 months was higher among deployed personnel. The index unit had a higher prevalence of chronic diarrhea, other gastrointestinal complaints, difficulty remembering, "trouble finding words," and fatigue, but also had twice the deployment rate of comparison units. Deployment rate appeared to account for the higher rates of symptoms. All outcomes or symptoms were self-reported in this stage of the evaluation. Similar increases in symptom prevalence were noted in a study of Navy Seabees.[27]

C. Seabees

Haley et al. studied 249 (61%) of the members of a Reserve Naval Mobile Construction Battalion that served in the Gulf. Illness had been common in this group, and some members had previously undergone evaluation of cognitive function.[28–30] The 249 veterans were evaluated through the administration of a detailed

TABLE 9.6
Ten Most Frequently Reported Symptoms in 59 PGW Veterans, Air National Guard

Symptom	% Reported
Fatigue	61%
Joint Pain	51%
Nasal/Sinus Congestions	51%
Diarrhea	44%
Joint Stiffness	44%
Unrefreshing Sleep	42%
Excessive Gas	41%
Difficulty Remembering	41%
Muscle Pains	41%
Headaches	39%

questionnaire that included anatomic distribution of symptoms, wartime exposures, and a standard personality assessment inventory. Seventy percent of the veterans reported "serious health problems" that most attributed to the war; 30% reported no such problems. The results were subjected to factor analysis, and the authors identified six clusters of self-reported symptoms that were grouped into syndromes. Sixty-three of the 249 veterans were classified as having one of the six syndromes. The three syndromes with most strongly clustered syndromes were characterized as impaired cognition, confusion ataxia, and arthromyoneuropathy.[28] Twenty-three veterans with clinical symptoms were further evaluated with detailed neuropsychological studies, as were 10 PGW veterans without symptoms, and 10 non-deployed controls. Thirteen veterans identified as having the "confusion ataxia syndrome" had significantly higher mean brain dysfunction scores than the 20 controls. These scores were based on the Halstead impairment index, General Neuropsychological Deficit Scale, and Trail Making Test Part B. Individuals classified as having the other two syndromes demonstrated impairment more frequently on other scales. The author concluded that individuals found by factor analysis to have one of these three syndromes "consistently scored in the abnormal direction on objective tests of neurological function than control veterans of the same battalion who were matched for age, sex, and education and were either deployed to the war zone and remained well, or who were not deployed." The study was not population-based, but limited to a single battalion that was the focus of a cluster evaluation and whose experiences and exposures may not be widely generalizable. The rate of participation (41%) may indicate some selection bias, and exposure and outcome information was self-reported. Finally, the study involved very few participants for the detailed neuropsychological evaluations.[28]

D. Reproductive Effects

A cluster investigation evaluated a perceived excess of birth defects and health problems in children born to two National Guard units from Southwest Mississippi. The two units, both deployed to the Gulf, consisted of 282 veterans. Initial contact was by telephone, and it was learned that 67 pregnancies had occurred since return from the Gulf. Medical records on 54 of the children were available and reviewed. The children ranged in age from 3–26 months at the time of the review. Records were reviewed for evidence of serious birth defect, minor birth defects, low birth weight, or premature birth. Baseline rates for comparison were obtained from three major U.S. birth defect surveillance systems. The rate of birth defects of all types in children born to this group of veterans was similar to that expected in the general population. The small size of the study population and the occurrence of only one case of each of five different types of birth defects (three major and two minor) made calculation of individual rates for the purpose of comparison difficult. Clustering of any one type or affected system was not noted. The amount of morbidity observed during the first year of life was not excessive.[16]

Taken as a whole, these studies support claims that deployed troops reported high rates of a variety of non-specific symptoms. However, they were initiated in essentially "self-selected" groups who had reported concerns and, as such, provide little information about the larger cohort of troops who deployed to the Gulf. Self-reported symptoms may in some part result from recall bias. This bias has been commonly reported in epidemiological investigations of health effects associated with exposure to hazardous waste sites. In the context of an ill-defined exposure possibly linked to health effects, concerned individuals tend to report more symptoms, or differentially recall exposures.[31,32] Troops may have been aware of the general public debate regarding Gulf War Syndrome and medical concerns of others in their units. Additionally, concerns about service-connected disorders may have sensitized troops to report conditions for fear that the symptoms might progress in severity.[6] Thus, defining a general population to compare the prevalence of symptoms may be inappropriate if the general population does not share the same general concerns.

V. SUBSETS OF POPULATION AT RISK: COMMON EXPOSURES

As previously discussed, the identification of an exposed population to compare with an unexposed population is fundamental to cohort studies attempting to evaluate differential rates of outcomes. While there have been many assessments of potential and known exposures that are related to PGW service, quantitative data with which to distinguish the exposed from the unexposed is lacking.[3–8] The environment of the Gulf was described as hostile, with uncomfortable temperatures and extreme rainfalls, and desert conditions with blowing sand, insects, animals, fumes, and smoke.[6] Troops were exposed to vaccines to protect against biological warfare and other infectious diseases, pyridostigmine bromide to protect against chemical warfare, and pesticides to protect against insects carrying diseases such as sandfly fever and leishmaniasis. Depleted uranium was used in munitions and tank armor, and subsets of troops faced occupational exposures to fuel, solvents, chemical-agent-resistant coating (CARC) paint, and vehicle exhaust fumes. Additionally, oil wells set on fire south of Kuwait City created a superplume of smoke. Hazardous exposures have been considered by many of the expert panels evaluating consequences of PGW service[6] (Table 9.7).

As noted by the IOM, "Although a wide range of possible exposures might be associated with adverse health outcomes in PGW veterans, data on these exposures are often not available; when they are available, they are poorly documented. This lack of exposure information is at the core of the frustration in obtaining answers from epidemiological studies. Self-reports of exposure and estimates of individual exposure from unit level measurements will be subject to so much error that they are likely to yield inconclusive results and additional questions."[6]

A. Infectious Diseases

With respect to individual exposures of interest, infectious diseases such as shigellosis, malaria, sandfly fever, and cutaneous leishmaniasis were a known threat in the region.[33–35] However, infectious disease did not exert a major toll on deployed troops. This success has been attributed to preventive medicine efforts and the timing of

TABLE 9.7
Exposures of Interest in the Gulf War and Availability of Exposure Information

Exposure	Available Data
Infectious Diseases	Case reports and summaries
Pyridostigmine Bromide	No centralized record of recipients
	Toxicological studies of effects
	Toxicological studies of interactions
Immunizations	No centralized database of recipients
Pesticides	Amount shipped to theater
Chemical Agent Resistant Coating Paint	No industrial hygiene monitoring data
Depleted Uranium	Clinical follow-up of soldiers with imbedded shrapnel
	Health risk assessment in progress based on modeled exposures
Petroleum Products	No industrial hygiene monitoring data
Oil-Well Fires	Human Health Risk Assessment based on monitoring data
Biological Warfare Agents	No data
Chemical Warfare Agents	Modeled data based on Khamisayah

major troop strengths in the region when insect populations were decreased due to cooler weather. Short-term diarrhea was common initially, but gastroenteritis rates decreased from 4% per week early in the deployment to less than 0.5% per week once controls over food sources, particularly locally grown produce, were instituted.[7] Seven cases of malaria and one case of West Nile fever were diagnosed, but there were no cases of sandfly fever, rickettsial illnesses, or arthropod-borne viral illness diagnosed. Visceral leishmaniasis and cutaneous leishmaniasis appeared to be the only endemic infectious disease associated with chronic morbidity in deployed troops.[35] Twelve cases of visceral leishmaniasis and 20 cases of cutaneous leishmaniasis were reported.[36]

B. Immunizations, Pesticides, and Occupational Exposures

Distinguishing the exposed from the unexposed with respect to immunizations is not possible due to incomplete documentation and the lack of a centralized database. Although it is estimated that at least 250,000 troops took at least some pyridostigmine bromide, no records of self-administered medications were kept.[8] Pesticide volumes shipped to theater are known, but estimating an individual's exposure is not possible, nor is it possible to identify those who used topical repellents (DEET) and impregnated their uniforms with permethrin or to what degree. Certain subsets of soldiers performed occupational duties that exposed them to unique hazards such as fuels, solvents, metals, and chemical-agent-resistant coatings applied to vehicles. Industrial operations performed in field settings are not subjected to strict industrial hygiene oversight. Modifications to procedure, lack of fixed ventilation, and the lack of recommended protective equipment may lead to exposures of individuals in excess of permissible workplace standards. One episode of overexposure to CARC paint was reported, but reliable monitoring information is not available to define a subset of exposed individuals.[37]

C. Depleted Uranium

Depleted uranium (DU) is a heavy metal that contains decreased amounts of the most radioactive isotopes of uranium. It is 40% less radioactive than naturally occurring uranium, but chemically and toxicologically similar to natural uranium. The health effects are considered to be generally comparable to other heavy metals such as lead and tungsten.[38–41] Depleted uranium is created as a by-product of the nuclear energy industry. The U.S employed steel-encased DU for increased armor protection, and the M2/3 Bradley Fighting Vehicle, M1 Abrams Tank and the M60 series tank can fire penetrating munitions containing DU.[38] There are no additional safety procedures required for intact DU and armor beyond those required for all munitions. When a DU munition pierces a target, it pyrolyzes, resulting in high concentrations of airborne oxides of uranium and metallic shards. Concerns arose over exposure to DU in the Gulf relating to proximity to a vehicle at the time of impact by DU munitions or a DU-armored vehicle at the time of impact by munitions. Other scenarios of concern involved proximity to actively burning fires involving DU, or routinely entering vehicles with penetrated DU armor or vehicles that had been struck by DU munitions. Exposure may occur through retained fragments, inhalation, or wound contamination. A health risk assessment addressing human health risk to modeled exposures is currently near completion. Toxicity related to DU exposures is expected to be related to heavy-metal-like effects on the kidneys. Renal toxicity occurs only at very substantial doses but was at least a theoretical concern relating to soldiers exposed in the Gulf. Surveillance was conducted on several small populations of troops considered at risk for exposure.[39–41] Twelve military personnel who helped salvage disabled tanks were studied by whole body counts and eight received urine analysis for uranium approximately 1 year after exposure. None were found to have increased body burdens of uranium.[39] A surveillance program was initiated in 1993 for referral of soldiers identified at increased risk from DU exposure. Thirty-three personnel were evaluated in 1993–1994. Testing included complete history to include medical, reproductive, occupational components, laboratory examinations to include complete blood count (CBC), chemistries, renal function tests, urinary uranium levels, and neuroendocrine measures. These individuals also received detailed physical examinations, neuropsychological tests, and radiologic tests. These initial evaluations demonstrated some persistent health problems related to wounds. Those with evidence of retained fragments (shrapnel) had increased urinary uranium, but no association between uranium excretion and clinically detectable adverse health effects was documented. Twenty-nine of this original cohort were re-evaluated in 1997. Controls without DU exposure were included at this time to provide a point of comparison for some of the clinical parameters and to assess the range of urinary and other uranium from natural sources. The evaluation was expanded to include genotoxicity assessments, neurocognitive evaluations, psychiatric and psychosocial evaluations, and risk communication. An additional focus at this point was to identify the most sensitive and relevant biologic measures of uranium such as spot and 24-h urines, seminal fluid, and whole-body radiation. It has been determined that 24-h urine collection and analysis is the most sensitive biologic exposure indice for uranium.[40] Thus far, the highest urinary uranium values were found in those with retained fragments; individuals who had fragments removed still had uranium levels somewhat above controls. On clinical examination, the exposed and unexposed groups were similar, although the unexposed had more genitourinary (GU) and nervous system complaints. Psychiatric complaints were similar. Exposed individuals were more likely than controls to have normal laboratory parameters such as CBC, urinalysis (UA), semen parameters, and blood chemistries. No renal abnormalities were noted. The most common abnormality in both groups was triglyceride levels. Only prolactin levels were found to be more elevated in exposed individuals as a group. This surveillance will continue, and the statistically significant differences between the two groups with respect to reproductive hormone and neurocognitive function will be further investigated.

D. Oil Well Fires

Of all of the exposures in the Gulf, the oil well fires are the most studied. A U.S. Interagency Air Assessment team of scientists studied the potential health effects of the oil well fires.[42,43] The U.S. Army Environmental Hygiene Agency collected nearly 4,000 ambient air and soil samples from May to December 1991.[44] These data were collected after a number of the fires had been extinguished, but were utilized in a health-risk assessment performed to assess the potential for health effects from exposure. Analyses were performed for criteria pollutants such as particulates, nitrogen oxides, sulfur dioxide, carbon monoxide, and lead. Other pollutants measured included volatile organic compounds and semi-volatile organic compounds, such as polycyclic aromatic compounds and metals. Results indicated that most contaminants did not exceed findings in a typical U.S. industrialized city, and the risk of long-term adverse health effects was minimal. The total predicted excess carcinogenic risks did not exceed 3 excess cancers per 1,000,000 attributable to this exposure.[44] The Environmental Protection Agency considers this level of risk to be *de minimus,* or indistinguishable from background. Non-carcinogenic risks were also predicted to be minimal following health risk assessment methodology. While particulate levels were high, analysis indicated that they resulted from sand-based material typical for the Gulf, and levels of associated metals and organic compounds were low.

E. Chemical and Biological Warfare Agents

While there was no available evidence from the Gulf to indicate a subpopulation of troops exposed to biological warfare agents, data reconstruction was performed to identify troops potentially exposed to chemical weapons.[6–8] This effort and the associated study are discussed in some detail here.

F. Khamisiyah, Iraq

While Iraq was known to possess chemical weapons, review of the available exposure and medical data from the Gulf concluded that there was no evidence that these weapons were used during the conflict.[3] However, in June 1996, the U.S. Department of Defense announced the United Nations' findings that U.S. forces near Khamisiyah, Iraq, had destroyed chemical agents in March 1991. Attempts were made to identify a possibly exposed population and compare their hospitalization experience with that of PGW veterans who were not likely exposed.[45]

Khamisiyah was a large ammunition storage facility located in southern Iraq and contained numerous ammunition bunkers, storage buildings, and pits and sand mounds to protect stored weapons. During March 1991, engineers operating from remote sites destroyed much of this. On March 10, 1991, a cache of 1,250 rockets stored in an open pit was destroyed. At the time, it was not known that any of the munitions at Khamisiyah contained chemical agents. In May 1996, the United Nations Special Commission inspectors determined from debris that some of the destroyed rockets contained the nerve agents sarin and cyclosarin. Although quantitative exposure data was not available, concerns about the possible health implications to troops were raised. Utilizing available data regarding numbers of rockets and nerve agent concentrations, the DoD and Central Intelligence Agency jointly conducted destruction testing of simulated rockets containing simulated nerve agent. The simulations and intelligence data led to an estimate that 342 gallons (1,294.57 liters) of nerve agent were released on March 10, 1991. Further analysis estimated the percentages released instantaneously and over time by evaporation. Meteorological, transport, and diffusion modeling were performed by an expert panel of federal and nonfederal experts using meteorological data from a number of sources and three transport and diffusion models. These were combined to generate five estimates of simulations of daily plume coverage.[46]

Although no U.S. personnel casualties were associated with the event of March 10, the DoD defined two nerve agent concentrations to be used in modeling to estimate the population potentially at risk. "The first noticeable effects concentration, 1 mg-minute/m^3, was defined as the dosage expected to cause mild symptoms such as rhinorrhea, muscle twitching, chest tightness, and headache."[45] The general population limit concentration is defined as "The dosage below which the general population, including children and the elderly, could endure for at least 72 hours without symptoms."[45] Following an independent review panel, a notification plume was determined combining the five meteorological/dispersion model simulations. These model simulation contours "represent a 99% probability that persons exposed to the general population limit dosage would fall within that perimeter."[45] Another independent panel recommended the construction of an epidemiological plume from the "best" meteorological and dispersion models for unit-specific dose estimates. This plume "enabled epidemiologists to estimate nerve agent concentration at specific troop locations over time." The troop location data was obtained from a geographical information system that contained all available daily unit locations in latitude and longitude. This data was not available for all units and was reconstructed after the war, and therefore subject to some limitations. Plumes were estimated for each day from March 10 to 13, 1991 and overlaid on the geographic information system troop unit location map (Figure 9.1).

Although no units were identified as having been exposed to the first noticeable effects of vapor concentration or higher in vapor plume modeling, 124,487 Army PGW veterans were identified as having the possibility of at least low-level exposure under either the notification or epidemiological plumes. This group was stratified into four dose groups: uncertain low dose (n = 75,717); exposure 1 defined as 0.0–0.01256 mg-minute/m3 (n = 18,952); exposure 2 defined as 0.01257–0.09656 mg-minute/m3 (n = 23,0610); and exposure 3 defined as 0.09657–0.51436 mg-minute/m3 (n = 6,757). These U.S. Army personnel were compared with 224,804 other Army PGW veterans who were deployed to the Gulf at the same time but were not under the vapor plumes. Hospitalization data was obtained from all DoD hospitals for the period of March 10, 1991 to September 30, 1995 for all study participants. Data included date of admission, up to eight individual International Classification of Diseases, Ninth Revision discharge codes, and disposition. Diagnoses with the same major diagnostic category codes were considered the same. Further, specific diagnoses determined by an expert panel to be possible manifestations of subtle, nerve-agent-induced neurophysiologic effects were examined. These included mononeuritis, peripheral neuropathy, toxic neuropathy, and myoneural disorders and myopathies. Cox proportional hazard modeling was performed for each of the 15 diagnostic categories over 54 months. Possible nerve-agent exposure was not associated with post-war hospitalizations. Analysis of the specific diagnoses between dose groups did not reveal increased risk for personnel possibly exposed to vapor plume. Further analysis of a dichotomous yes/no exposure and yes/no hospitalizations for any cause and for the 15 major codes found only a slight risk for adjustment reaction and nondependent drug abuse. The authors concluded that "These data do not support the hypothesis that PGW veterans who were possibly exposed to nerve agent plumes . . . experienced unusual post-war morbidity." This study was a unique effort to combine operational data (temporal and geographic) with dispersion and meteorological data to estimate an exposure. The accuracy of these models and estimates cannot be known, but represent considerable effort to determine the magnitude and scope of possible nerve agent exposure to deployed troops. The study also compared PGW veterans possibly exposed to all other PGW veterans, eliminating some of the concerns raised regarding the inappropriateness of comparing PGW veterans with non-deployed cohorts of the same era. Finally, dose gradients were estimated to enable dose-response trends to be evaluated. Limited only to hospitalization data, the study could not address symptoms or complaints not resulting in hospitalizations, and available information was limited to personnel remaining on U.S. Army active duty. Nonetheless, this innovative use of available information serves to rule out significant morbidity in a subset of the population at risk when quantitative exposure information was not available.

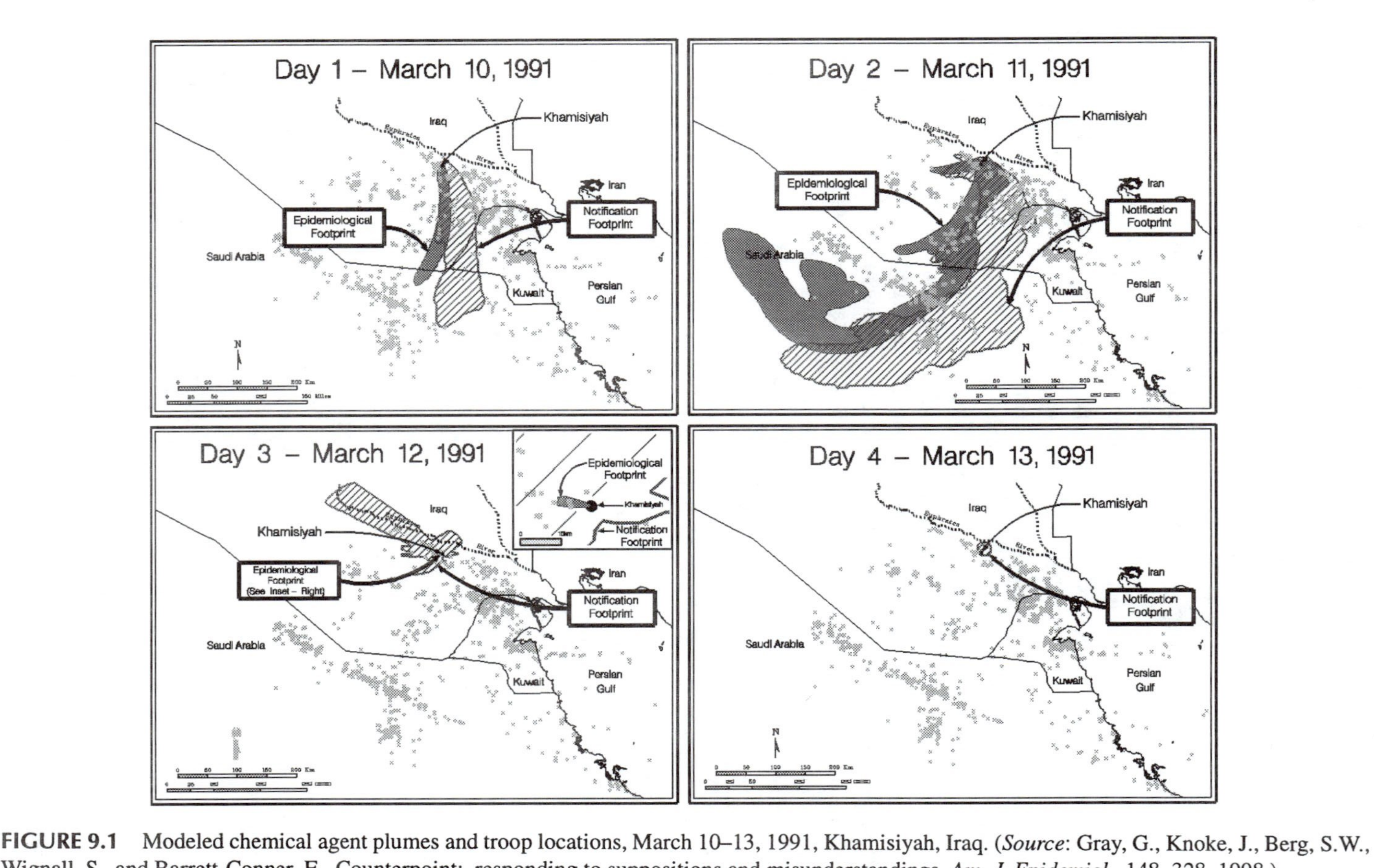

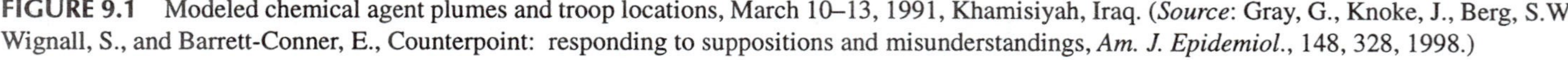

FIGURE 9.1 Modeled chemical agent plumes and troop locations, March 10–13, 1991, Khamisiyah, Iraq. (*Source*: Gray, G., Knoke, J., Berg, S.W., Wignall, S., and Barrett-Conner, E., Counterpoint: responding to suppositions and misunderstandings, *Am. J. Epidemiol.*, 148, 328, 1998.)

G. Mixed Exposures and Synergistic, Additive, or Other Combined Effects

It has been noted by many of the review panels that a further issue of consideration is the potential for additive or synergistic or other complex effects from exposures. This has been a particular focus of study with respect to the combination of neurotoxic compounds such as pyridostigmine bromide, DEET, and permethrin, as well as possible nerve-agent exposure.[47–48] While these interactions are complex and not discussed here, the issue of additive effects has importance for a number of reasons. Most of the evaluations of individual exposures and their likelihood of producing health effects of significance addressed individual exposures. Therefore, the conclusions of such evaluations are often questioned because they do not address additive or other combined effects. As noted by the IOM, "Service personnel stationed in the Gulf were exposed to an extraordinary array of environmental conditions. Their complex experiences combined to yield what is truly a varied and sometimes confusing picture of exposure that has proven difficult to understand, much less reconstruct."[6] It was noted that the exposures combined to produce an environment are not easily described or evaluated. Environmental exposure reports for a variety of hazards encountered during the Persian Gulf conflict can be obtained from http://www.gulflink.osd.mil/cia_092297. The Office of the Secretary of Defense, Special Assistant to the Deputy Secretary of Defense for Gulf War Illnesses, also commissioned eight literature reviews on potential health hazards in the Gulf. The series addresses infectious diseases, pyridostigmine bromide, immunizations, stress, chemical and biological warfare agents, oil well fires, depleted uranium, and pesticides. These reviews describe available data on exposure to these hazards in the context of the Gulf War and published literature on possible health effects.[38,49–55]

VI. SUBSETS ENROLLED IN REGISTRIES

Other populations of Gulf War veterans have been described and have provided data with respect to health outcomes in Gulf War veterans. However, they are not random samples of deployed veterans; rather, they are subsets of the population at risk. Two such subsets are the DoD Comprehensive Clinical Evaluation Program (CCEP) registrants and the Department of Veterans Affairs Persian Gulf Health Registry (PGHR) participants.[56,57] To address the health concerns of PGW veterans, to enable them to receive a clinical evaluation, and to assemble information regarding patterns of illness on a large scale, health registries and referral services were developed. The Department of Veterans Affairs established the National Referral Center and PGHR. Similarly, the Department of Defense created the CCEP. Participation in CCEP is open to Gulf War veterans who are active duty, military retirees, full-time National Guard personnel, members of the reserve units who are placed on orders, family members of above categories who are eligible beneficiaries for military health care, and DoD civilians (current and former) who were in the Persian Gulf between August 1990 and July 1991. Gulf War veterans who have separated or are in a Reserve Component are not eligible but may request a medical evaluation through the

DVA PGHR. The PGHR offers a free complete physical examination with basic laboratory studies to every PGW veteran. In 1996, Public Law 103–446 established a special program to fund health examinations for some spouses and children of PGW Veterans Registry participants. The results of these examinations are included in the PGHR. The demographics of these registry participants are noted in Table 9.8.

The most obvious indication that the registries do not represent a random sample of those veterans who deployed to the Gulf is that the inclusion of family members does not restrict participation to those who actually deployed to the Gulf. Further, the group is self-selected, in that participation is not mandatory. Reasons for requesting a medical evaluation might include diagnosis and treatment of a symptom or illness or perhaps a desire for a complete medical evaluation. Others might wish to obtain information about the health of other Gulf War veterans or to register in case of future health problems that might be compensated. Other individuals might not

TABLE 9.8
Demographic Characteristics of Gulf War Participants Enrolled in the DoD CCEP and DVA Registry

	Total Gulf War Participants (N = 697,000)	CCEP Participants (N = 18,075)	Registry Participants (N = 52,216)
Gender			
Male	93	88	90
Female	7	2	10
Race			
White	70	57	64
Black	23	32	23
Hispanic	5	6	10
Other/unknown	2	5	3
Branch			
Air Force	12	10	7
Army	50	81	72
Marines	15	4	12
Navy	23	1	1
Other/unknown	NA	1	1
Status (in 1991)			
Active Duty	83.3	83	54
Reserve	10.4	13*	20*
National Guard	6.3	NA	19*
Age (years) Mean (in 1991)	26	30	29

Note: NA = Not available; *denotes Reserve/National Guard combined.

Source: PAC, Final Report, 1996.

request a medical evaluation due to a lack of health problems or a lack of individual association between any health problem and service in the Gulf. Some have raised the concern over whether or not participation will adversely impact one's career. An analysis of demographic risk factors for participation in the two registries identified service branch and type were strongly associated with registry participation with Army and National Guard personnel most likely to participate.[58]

Service in the Gulf during the fighting, age, enlisted rank, and construction work were also associated with participation. Other variables associated with participation included female sex and hospitalization during the 12-month period before the war. The significance of the overrepresentation of these demographic groups in the registries remains unknown. Nonetheless, participation rates in the registries have sometimes been presented as a surrogate for the rate of illness in a unit, as in a comparison of CCEP participation as a function of proximity to oil well fires or Khamisayah, the location of the chemical agents' release. Participation in CCEP does not automatically imply illness. Participation in CCEP may be a measure of the tendency to seek health evaluation. While these registries serve a purpose of responding to troops desiring evaluation and creating a centralized repository for information, conclusions drawn from the data are limited in generalizability to the entire cohort of deployed troops. Information or findings relative to a sample of a population is generalizable back to the population from which it came to the degree that any member of the population has a random chance of being included in the sample. To the degree that inclusion in the sample is not random, selection bias limits the ability to generalize the observations and conclusions about the sample to the entire population from which it came. Other limitations to the ability to generalize from this data are that symptoms and exposures were self-reported, and control groups were not utilized. The registries serve as an important source of entry into the medical system for veterans who need clinical services, and provide a source of hypothesis regarding the nature and extent of health problems experienced by PPGW veterans who enrolled.[6] They do not necessarily reflect new conditions or conditions related to Gulf service.

VII. OUTCOMES IN SUBPOPULATIONS IN REGISTRIES

Participants in both the DVA Persian Gulf Health Registry and the CCEP registry represent a broad cross-section of service members who deployed to the Gulf, although the demographics of participants as a group differ from the deployed population in some respects as discussed above. At the time the comprehensive reports were published, 18,075 individuals had participated in CCEP, and 52,216 individuals had been evaluated through the PGHR.[56–60] The Presidential Advisory Committee (PAC) combined the data from both sources in their evaluation of the findings of the registries.[59] As stated, not all registry participants are ill; 10% of CCEP participants are asymptomatic, while 12% of PGHR participants report no symptoms. Symptomatic participants in both registries reported a broad range of symptoms spanning a variety of organ systems. The most common symptoms reported in CCEP participants were joint pain, fatigue, headache, and skin rash. Most commonly reported symptoms for the PGHR were almost identical.

The CCEP report included prevalence data from three studies of outpatient practice in the U.S. for common symptoms.[56] The prevalence of fatigue reported in the general population ranged from 25–58%, whereas in the combined registry data, fatigue was listed as 1 of the top 7 symptoms in 47% of participants. Joint pain prevalence in the general population ranged from 32–59%; 49% of registry participants reported fatigue to be one of their top seven complaints. Headaches reported in the general population ranged from 24–38% as compared to 39% of registry participants. Finally, sleep disturbances were reported in the general population with a prevalence of 15–35% as compared with 32% in the registry participants. While this data indicates that the types of symptoms reported by registry participants are not uncommon, it is noted that community outpatient surveys include populations estimated to be 20–25 years older, and the percentage of women is higher than in the PPGW registries.[56] Table 9.9 lists the ten most frequent symptoms in the combined registries as well as the percentage of participants reporting the symptom in the top seven and top three of their complaints.

For diagnosed conditions, the distribution of major diagnostic categories was similar in the two registries (Table 9.10).[56,57] Approximately 10% of registry participants are healthy. The most common primary diagnostic categories are psychological conditions, musculoskeletal system diseases, and symptoms, signs and ill-defined conditions (SSIDC). These three categories represent greater than 50% of the diagnoses made. Apart from these categories, diagnoses do not center in any single organ system.

In the PGHR population, relative rank-order of major diagnostic categories was the same for men and women, with the exception of digestive system disease ranking

TABLE 9.9
Frequency of the Ten Most Common Symptoms Reported by DoD CCEP Participants (N = 18,075) and the DVA Registry Participants (N = 52,216)

Reported Symptoms	Chief Complaint	Any of Top Seven Symptoms	Any of the Top Three Symptoms
No symptoms	10%	10%	10%
Joint pain	11%	49%	17%[a]
Fatigue	10%	47%	20%
Headache	7%	39%	18%
Memory loss	4%	34%	14%
Sleep disturbance	2%	32%	6%
Rash/dermatitis	7%	31%	18%
Difficulty concentrating	<1%	27%	NA
Depression	1%	23%	NA
Muscle pain	1%	21%	[a]

Note: NA = Not available.

[a]In the VA registry, muscle and joint pain combined are 17%.

Source: PAC, Final Report, 1996.

TABLE 9.10
Frequency Distribution of Major Diagnostic Categories (ICD-9-CM) in Participants in DoD CCEP Participants (N = 18,075) and the DVA Registry Participants (N = 47,624)

ICD-9-CM Diagnostic Code	Primary Diagnosis	Any of the Top Seven Diagnosis	Any of the Top Three Diagnosis
Psychological conditions	18%	36%	15%
Muscular system disease	18%	47%	25%
Symptoms, signs, ill-defined conditions	17%	43%	20%
Healthy	10%	10%	NA
Respiratory system diseases	7%	18%	14%
Digestive system diseases	6%	20%	11%
Skin diseases	6%	20%	14%
Nervous system diseases	6%	18%	8%
Infectious diseases	3%	9%	7%
Circulatory system diseases	2%	8%	7%
Endocrine disorders	2%	8%	NA
Genitourinary system diseases	1%	5%	3%
Injury and poisoning	1%	3%	5%
Neoplasms	<1%	3%	<1%
Blood and blood organ diseases	<1%	3%	NA

Note: NA, not available.

Source: PAC, 1996.

sixth in women and fifth in men. Infectious disease was diagnosed more commonly in men, and genitourinary disease was more common in women. One hypothesis is that the relative lack of gynecological care in the Gulf resulted in increased diagnoses on return.[60] Overall, for all participants, 69% of women reported their health to be all right, good, or very good, compared with 73% of men. For the CCEP population, the rank-listing of major diagnoses for men is the same as in Table 10. For women in the CCEP, the top three categories retained their rank order. Nervous system were the fourth most frequent diagnosis, followed by healthy, respiratory conditions, skin disorders, digestive system disorders, genitourinary diseases, endocrine disorders, infectious diseases, blood diseases, and circulatory diseases, with the remaining categories unchanged in order. To provide a point of comparison, as was done with data on the prevalence of symptoms in the general U.S. population, the CCEP report provided data on the frequency of primary diagnoses for the CCEP as compared to the National Ambulatory Medical Care Survey (NAMCS) for individuals aged 20–40 years.[56] The NAMCS population, although random, is said to differ from the CCEP population in that individuals captured in the NAMCS represent those seeking care for unknown conditions, as well as routine examinations in the absence of any adverse condition (Table 9.11).

TABLE 9.11
Frequency of Primary Diagnosis for CCEP and NAMCS, by Sex, for Subjects 20–40 Years of Age

	Men Aged 20–40 Percent Primary Diagnosis		Women Aged 20–40 Percent Primary Diagnosis	
Primary Diagnosis	**CCEP**	**NAMCS**	**CCEP**	**NAMCS**
Psychological conditions	18	7	18	5
Muscular system disease	19	9	16	5
Symptoms, signs, ill-defined conditions	18	3	17	3
Healthy	10	11	9	27
Respiratory system diseases	7	11	6	10
Digestive system diseases	6	5	5	3
Skin diseases	6	9	6	7
Nervous system diseases	5	9	9	7
Infectious diseases	3	5	3	4
Circulatory system diseases	2	3	2	2
Endocrine disorders	2	2	3	3
Genitourinary system diseases	1	5	4	10
Injury and poisoning	1	17	1	7
Neoplasms	1	2	1	3
Blood and blood organ diseases	<1	<1	2	<1

Source: CCEP Report, DoD, 1996.

Comparing the frequencies of diagnoses between the two age-matched populations, men in the CCEP were two to five times more likely to receive a diagnosis in the categories of psychological conditions, signs, symptoms and ill-defined conditions, and musculoskeletal conditions. The proportion with a diagnosis of healthy did not differ substantially, and CCEP participants were less likely to receive a diagnosis in the respiratory, nervous, infectious disease, and skin categories. For women, CCEP participants were three or more times as likely to receive a diagnosis of psychological conditions, signs, symptoms and ill-defined conditions, and musculoskeletal conditions. They were much less likely to receive a diagnosis of healthy, or in the genitourinary group. While these differences are interesting, they are not readily interpretable with respect to risk factors or significant differences between the two populations. Similar symptom prevalence has been documented in Canadian and British forces, and preliminary results from a Danish study of troops deployed to the Gulf following the war for peacekeeping and humanitarian tasks indicate "a pattern of diseases and symptoms in some respects comparable to the findings in U.S. Gulf War veterans."[61,62]

Both the CCEP and PGHR serve an important purpose as an access to care for concerned individuals and a centralized database of information on those seeking to register. Interpretation of the actual significance of the findings has been limited, as the biases associated with a voluntary, self-referred registry without a comparison population have been noted. The Institute of Medicine committee evaluated the data from the initial 10,000 DoD participants and noted that the CCEP evaluations "were not, however, designed to answer epidemiological questions. Instead, it was designed as a medical evaluation and treatment program. Although useful to bound and explain the problem in a subgroup of veterans, the information is of limited value for determining the prevalence and incidence of illnesses in the full cohort of PGW veterans because they are not necessarily representative of the troops who did not participate, and they do not include comparison populations."[60]

VIII. IS THERE A SINGLE PGW SYNDROME? THE PROBLEM WITH CASE DEFINITION

Given that both large registries found a frequency of unexplained, as yet undiagnosed conditions in about 20–25% of participants, a basic question asked whether or not the symptoms represented a new and unique syndrome. Examinations of large numbers of individuals in a systematic fashion would seemingly provide a reasonable opportunity to diagnose a new definitive condition. A series of six expert panels evaluated the available scientific data but did not identify a single, coherent syndrome, although many illnesses reported by veterans might be attributable to Gulf War service.[3–8] The 1994 NIH Workshop Panel found that no single disease or syndrome is apparent, but rather found evidence for multiple illnesses with overlapping symptoms and causes.[4] Symptomatic veterans were found to be ill due to a wide diversity of health problems, but no specific previously unknown disease was identified, and no case definition related to unexplained symptoms emerged. The NIH panel concluded that "An evolving case definition might be more appropriately used in developing a research strategy." The PAC noted that many veterans were interested in possible links between unexplained illness and symptom-based conditions such as chronic fatigue syndrome (CFS), fibromyalgia (FM), and multiple chemical sensitivity (MCS).[7,8] These conditions lack specific diagnostic tests, but are based on symptoms reported by patients, rather than physical abnormalities or laboratory tests. Chronic fatigue syndrome was defined by a 1994 Centers for Disease Control (CDC) and Prevention as new and unexplained fatigue of 6 months duration accompanied during the six months by persistent and recurrent symptoms. At least four of the following should also be present: memory impairment significant enough to impair function, sore throat, tender cervical or axillary lymph nodes, muscle pain, multi-joint pain without redness or swelling, headaches, unrefreshing sleep or post-exertional malaise lasting more than 24 h. Chronic fatigue syndrome is considered a disease of exclusion in that many conditions must be ruled out before a diagnosis is made.[63]

The PAC noted that the DoD reported 42 of the first 10,020 registry participants met the CDC case definition, but that the VA has not reported the proportion of veterans with this diagnosis.[59]

Fibromyalgia is defined by the American College of Rheumatology as chronic, widespread pain in all 4 quadrants of the body and pain in at least 11 of 18 tender point sites on digital palpation.[64]

Patients with FM also report sleep disturbance, fatigue, morning stiffness, anxiety, headache, and depression. Patients can be diagnosed with other conditions simultaneously, and no specific laboratory test exists. The DVA has not reported the prevalence of FM in its registry participants, but DoD noted that 1.5% of CCEP participants received a primary or secondary diagnosis of FM.[56] Multiple chemical sensitivity does not have a consensus case definition and thus, the frequency in the registry participants cannot be estimated. Hyams discussed symptom-based diagnoses in the context of the Gulf War.[65] Federal funds have been awarded to researchers in the area of CFS, FM, and MCS.

IX. FUTURE RESEARCH DIRECTIONS

On August 31, 1993, in response to Section 707 of Public Law 102–585, President William J. Clinton named the Secretary of Veterans Affairs to coordinate research funded by the Executive Branch of the Federal Government into health consequences of service in the Gulf War. Section 104 of Public Law 105–368 (1998) expands the responsibilities. The DVA carries out the coordinating role through the auspices of the Research Working Group (RWG) of the Persian Gulf Veterans' Coordinating Board (PGVCB). The Secretaries of the Department of Defense, Health and Human Services and DVA chair the PGVCB and have representatives on the RWG, as does the Environmental Protection Agency.[66]

The RWG has developed a strategic plan for research, to include a plan for research on the health effects of exposure to low levels of organophosphorous nerve agents. It also established a programmatic review of peer-reviewed, completed research proposals leading to funding recommendations for more than $100 million in research projects. The strategic plan for the conduct of research on Gulf War veterans' illnesses aims to: (1) determine the nature and prevalence of symptoms, diseases, and other conditions among Gulf War veterans; (2) identify risk factors for symptoms, diseases, and other conditions; and (3) identify diagnostic tools, treatment methods, and prevention/intervention strategies. The plan contains about 20 research questions in broad areas of exposure and outcome posed by Gulf War veterans' illnesses. In 1996, new factual and conceptual knowledge about exposures and outcomes during and after the Gulf War led to a revised set of short-term and long-term research recommendations. Short-term recommendations include epidemiological follow-up on Gulf War veterans' mortality experience at appropriate time intervals and longitudinal follow-up studies of Gulf War veterans' health status. Also to be addressed is peer-review of the atmospheric exposure models for pollutants such as the oil well fires and chemical warfare agent releases at Khamisiyah, Iraq. Long-term research recommendations include research on risk factors for stress-related disorders, excess mortality due to accidents, biomarkers of chemical warfare agents, a strategic plan for investigation of the health effects of low-level chemical warfare agent exposures, and a test for *L. tropica* infection.[67] Since 1994, the Federal Government has sponsored 145 research projects and committed $133.5 million in resources. Non-governmental researchers conduct more than half of these studies. Through 1998, 40 projects have been completed, 103 are ongoing, and 2 are pending

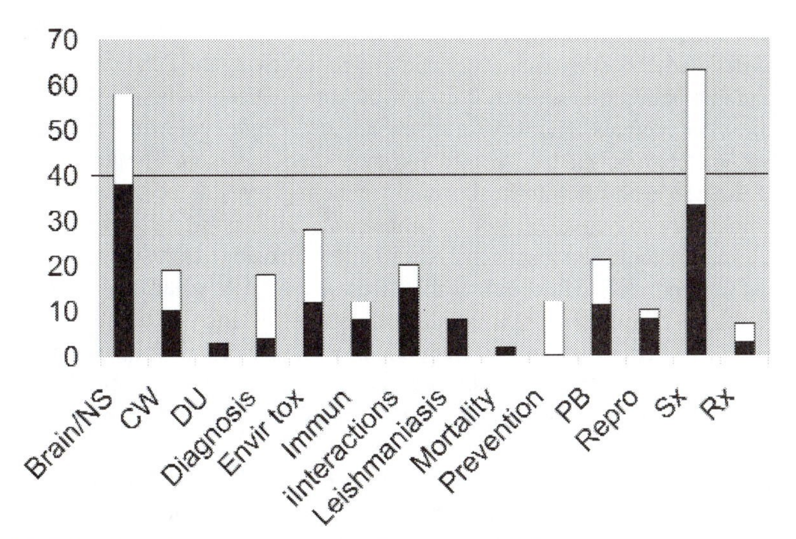

FIGURE 9.2 Cumulative number of research projects funded, by focus area.

start-up. Since 1994, the proportion of research projects funded in epidemiology has remained constant, while funding for products related to the toxicology of chemical weapons has markedly increased. Beginning in 1998, new research on treatment has received increased funding, and $10 million has been invested in two clinical trials. The first is a large multi-center trial to address the effectiveness of behavioral and cognitive therapy and exercise on symptomatic veterans. The other is multi-center trial of the effectiveness of doxycycline in reducing symptomatic complaints. This research was prompted by a growing number of veterans receiving this treatment without evidence of infection or known efficacy for this purpose. The cumulative numbers of research projects across various areas of research focus are shown in Figure 9.2.

X. ASSOCIATION VS. CAUSATION IN ENVIRONMENTAL EPIDEMIOLOGY

While research into the outcomes in PGW veterans continues, a corollary effort is to utilize the lessons learned from the Gulf to improve data collection on future deployments. Currently, the DoD has asked the National Academy of Sciences and the Institute of Medicine to recommend ways in which the DoD can enhance or improve its protection of the health of deployed U.S. military forces in the future.[68] Specific issues to be addressed include: assessing health risks during deployments in hostile environments through the use of an analytical framework; assessing technology and methods for detection and tracking of exposures to a subset of harmful agents; and assessing past, current, and potential future approaches for developing, evaluating and fielding protective equipment, clothing, and technologies related to decontamination.

Given that a major limitation of all epidemiological studies to date has been the lack of detailed exposure data, every committee reviewing the Persian Gulf has recommended that broad-based exposure and outcome data collection be conducted on all future deployments. The IOM recommended "A single, uniform, continuous and retrievable electronic medical record for each service person. The uniform record should include each relevant health item (including baseline personal risk factors, every inpatient and outpatient medical contact and all health-related interventions."[6] It was also recommended that such a record allow linkage to exposure and other data sets and incorporate medical information from the DVA, civilian, and other healthcare facilities. Presumably, this would enable the tracking of outcomes or events resulting in medical interface on all individuals with service time for their entire lives. If realized and perfected, this would address the concerns regarding complete capture of all medical outcomes (as opposed to self-referral for entry into a registry or symptom-based cluster evaluations), and would also address incomplete follow-up for individuals who leave the service, and capture of events that occur outside DoD/DVA health care facilities. If functional and amenable to epidemiological analysis, this would represent a close to perfect data set with respect to outcome capture. With respect to exposure information, the other critical component of the exposure-outcome association question, recommendations have been made as well.[6,7] "The DoD should ensure that military medical preparedness for deployments includes detailed attempts to monitor natural and man-made environmental exposures and to prepare for rapid response, early investigation and accurate data collection, when possible, on physical and natural environmental exposures that are known or possible in the specific theater of operations."[6]

One difficulty with this recommendation is that no specific level of threat has been identified, and the range of possible exposures is broad. For some hazards, guidance for acceptable levels for occupational exposure exist but may not be applicable for extended work shifts or continuous exposure possible in a deployed setting. Screening levels derived for application in risk assessment to represent "No adverse effect levels" for the general population are not suitable because they are meant to protect sensitive members of the population for lifetime exposures and utilize very conservative assumptions at each step of the derivation. Exceedances of such screening levels may be suitable as a basis for determining whether or not a remedial action should be considered, but do not serve as a threshold useful to predict the frequency or magnitude of a health effect. Health effects, if they occur at all, might be subtle and not discernable without specific, tailored, outcome-based medical surveillance, apart from waiting for and tallying specific outcomes. With respect to cancer outcomes, values are derived based on a non-threshold model that may not be appropriate for all hazards. Exceeding a screening level derived to address a cancer endpoint based on a theoretical model may result in anxiety, and consideration of latency would leave the issue unresolved for many years. In actuality, monitoring on recent deployments has been troubled by a time lag between measurement and available results such that information cannot be utilized in any preventive sense to reduce exposure, but may raise questions with respect to significance and prognostic interpretation for those exposed. This raises questions regarding the value of such information for any purpose other than after-the-fact epidemiological analysis, which is not useful for the commander in the field who has the responsibility to complete the mission managing competing risks. Commanders are currently trained to manage risk in accordance with FM 100–14, Risk Management, which applies a probability/severity of health outcome matrix to hazards[69] (Figure 9.3). Obvious catastrophic events such as a release of highly toxic materials would have severe health risks, although the probability of such a release can only be estimated (Figure 9.4). However, since the most profound preventive action is avoidance, troop locations can be selected with regard to proximity and plume direction from industrial facilities. With respect to exposure to low ambient levels of chemicals, health effects may be delayed or produce little obvious and measurable impact on the immediate mission, but the probability of occurrence is high. Even if monitoring information were available immediately, uncertainties relating to actual health impact would make decision making difficult. One approach adopted by the U.S. Army Center for Health Promotion and Preventive Medicine provides concentrations of chemicals of interest representing high, medium, and low risk for short-term exposure.[70] A companion document is under development to address the more problematic long-term exposures.[71] These documents can be viewed at http://chppmwww.apgea.army.mil/hracp/pages/caw/index.html. A major consideration relates to the degree of conservatism to apply to the available toxicological reference values to fit the scenario of long-term exposure of a healthy population on a continuous basis. Appreciated, but not well quantifiable, are issues relating to mixtures of compounds and potentially additive or synergistic effects, interaction with other biologicals such as vaccines and medications, and the effects of stress, reduced sleep, and other considerations in a deployed setting.

RISK ASSESSMENT MATRIX

SEVERITY	*PROBABILITY*				
	Frequent	Likely	Occasional	Seldom	Unlikely
Catastrophic	E	E	H	H	M
Critical	E	H	H	M	L
Marginal	H	M	M	L	L
Negligible	M	L	L	L	L

E- Extremely High Risk

H - High Risk

M- Moderate Risk

L- Low Risk

Figure 2-4, FM 100-14, Risk Management

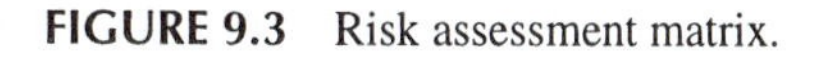

FIGURE 9.3 Risk assessment matrix.

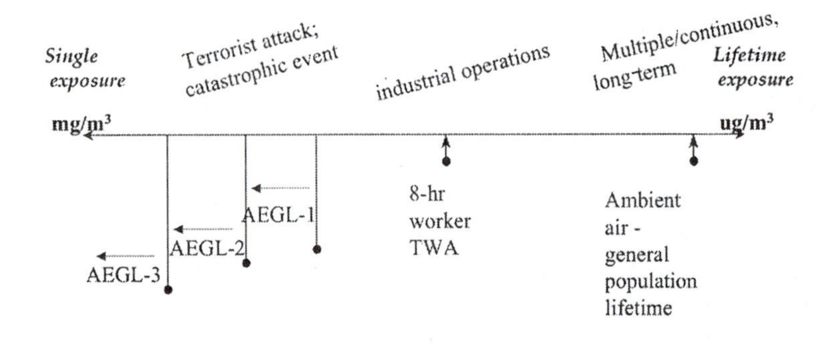

FIGURE 9.4 Air exposure concentration "continuum."

Given that such exposure and outcome data systems come to fruition, will they eliminate or alleviate questions regarding exposure and outcome associations following future deployments? Questions such as addressing whether or not a specific deployed cohort is experiencing statistically significant excesses of certain adverse outcomes would conceivably be answerable. Questions relating to the association of such outcomes with specific exposure on a deployment may not. Given measurable and measured exposures to a known hazard in the range known to produce health effects in humans, the question should be easy to answer. Given measured concentrations of a broad variety of hazards with unclear, but possible health effects ("gray-zone" concentrations), much more sophisticated methods will be required to evaluate the association. Additionally, much more exposure data is needed. The Government Accounting Office, in its review of Gulf War Illness efforts, stated that "The need for accurate, dose-specific information is particularly critical when low-level or intermittent exposure to drugs, chemicals or air pollutants is possible. It is important not only to assess the presence or absence of exposure, but to characterize the intensity and duration of the exposure."[9] This essentially calls for continuous monitoring on a broad range of low-level hazards on deployments, but, in actuality, comparable data would be needed on a control population, unless sufficient data is collected on a large enough population with frequent enough outcomes to assess for a trend in dose-response. Further, adequate information on confounding variables would be required. Identifying the confounding variables up-front may be somewhat difficult without knowledge of which exposures or outcomes will be a concern and subject to analysis. Will sufficient data ever be available following a deployment to evaluate an exposure/outcome relationship in terms of causation? To avoid an ecological fallacy, quite specific information is required at the individual level. Adequate baseline on conditions and/or symptoms pre-deployment is necessary to establish the critical chronological relationship (exposure must precede the disease to be considered causal). Current predeployment questionnaires are too simplistic, although the "seamless medical record," which has been proposed, may alleviate this problem. Causality is supported by the strength of the association, in that the greater the magnitude of the demonstrated association, the more likely the significance.[70] Low-level exposures, such as those evaluated with respect to hazardous waste and health effects have largely been determined to pose low-level risks with broad confidence intervals.[11] Causality is also supported if a dose-response trend can be demonstrated, that is, that those with the most intense and longest duration exposure have a greater chance of developing the outcome. Given enough data points of exposure magnitude and/or duration, and a sufficiently large population, this would be a possibility. Another criteria that supports causality relates to the specificity of the association. If the effect or outcome is specific and/or unusual, associated with the particular potential cause, the relationship between exposure and outcome is more likely to be causal. Whether or not this factor will be relevant depends to some degree on the potential exposures and mechanisms of toxicity. If an unusual outcome is identified, relating it to a particular exposure would be dependent upon the toxicological research associated with that outcome, or more specifically, potential exposures associated with that outcome. Another criterion that supports causation is consistency of the association. If many observers in many studies or settings have replicated the finding, then the role of chance as an explanation for the finding is minimized. Abundant data on particulates and respiratory effects exist, and so, for example, if a finding related to particulate levels and respiratory disease outcomes is noted, causality would be much less of a question. The remaining criterion for causality is biological plausibility. The connection between the potential cause and the possible effect must make biological sense. Documented exposures and documented outcomes may be associated statistically, but causality requires that some plausible mechanism link the two. Much of the basic research currently conducted aims to elucidate the mechanisms of neurotoxic damage to provide support for hypotheses related to exposures in the Gulf and neurological outcomes.[66]

"Recent military deployments, especially in Vietnam and the Persian Gulf, have demonstrated that concerns about the health consequences of participation in military action arise long after deployment has ended and that the evaluation of those concerns and the provision of health care to affected personnel may represent formidable challenges to both epidemiologists and to medical caregivers. Although some of these challenges can be attributed to the intrinsic difficulty of evaluating poorly understood clusters of events that were not among the expected consequences of combat or of environmental conditions, they also may be attributed in part to limitations of the systems used to collect and manage data regarding the health and service-related exposures of military personnel. No system of record keeping can be expected to provide the information needed to address every unanticipated research issue, including the health consequences of military service."[6]

ACKNOWLEDGMENTS

The author wishes to thank Dr. Robert DeFraites, Dr. John Brundage, Dr. K. Craig Hyams, and Dr. Donald MacCorquodale for their thoughtful review and comments on this chapter.

REFERENCES

1. Hyams, K.C., Wignall S.W., and Roswell, R., War syndromes and their evaluation: From the U.S. Civil War to the Persian Gulf War, *Ann. Intern. Med.*, 125, 398, 1996.
2. Institute of Medicine, *Veterans and Agent Orange,* National Academy of Sciences, National Academy Press, Washington DC, 1996.
3. Defense Science Board, Report of the Defense Science Board Task Force on Persian Gulf War Health Effects, Office of the Under Secretary of Defense for Acquisition and Technology, Washington, DC, 1994.
4. NIH Technology Assessment Workshop Panel, The Persian Gulf Experience and Health, *JAMA,* 272, 391, 1994.
5. Institute of Medicine, Committee to Review the Health Consequences of Service during the Persian Gulf War, Medical Follow-up Agency, *Health Consequences of Service during the Persian Gulf War: Initial Findings and Recommendations for Immediate Action,* National Academy Press, Washington, DC, 1995.
6. Institute of Medicine, Committee to Review the Health Consequences of Service during the Persian Gulf War, Medical Follow-up Agency, *Health Consequences of Service during the Persian Gulf War: Recommendations for Research and Information Systems,* National Academy Press, Washington, DC, 1996.
7. Presidential Advisory Committee on Gulf War Veterans' Illnesses, Special Report, U.S. Government Printing Office, Washington, DC, October 1997.
8. Presidential Advisory Committee on Gulf War Veteran's Illnesses: Final Report, U.S. Government Printing Office, Washington, DC, December 1996.
9. Government Accounting Office, *Gulf War Illnesses: Improved Monitoring of Clinical Progress and Reexamination of Research Emphasis Are Needed,* GAO/NSAID-97–163, June 23, 1997.
10. Tyler, C.W., Jr. and Last, J.M., Epidemiology, in *Maxcy-Rosenau-Last Public Health and Preventive Medicine,* Last, J.M. and Wallace, R.B., Eds., 13th Ed., Appleton and Lange, 1991, 12.
11. National Research Council, *Environmental Epidemiology: Public Health and Hazardous Wastes,* Vol. 1, National Academy Press, Washington, DC, 1991.
12. Kang, H.K. and Bullman, T.A., Mortality among U.S. veterans of the Persian Gulf War, *N. Eng. J. Med.,* 335, 1498, 1996.
13. Gray, G.C., Coate, B.D., Andeson, C.M., et al., The post-war hospitalization experience of U.S. veterans of the Persian Gulf War, *N. Engl. J. Med.,* 335, 1505, 1996.
14. The Iowa Persian Gulf Study Group, Self-reported illness and health status among Gulf War veterans: A population based study, *JAMA,* 277, 238, 1997.
15. Cowan, D.N., DeFraites, R.F., Gray, G., Goldenbaum, M., and Wishik, S.M., The risk of birth defects among children of Gulf War veterans, *N. Eng. J. Med.,* 336, 1650, 1997.
16. Penham, A.D. and Tarver, R.S., No evidence of increase in birth defects and health problems among children born to Persian Gulf War veterans in Mississippi, *Mil. Med.,* 161(1), 1, 1996.
17. Writer, J.V., DeFraites, R.F., and Brundage, J.F., Comparative mortality among U.S. military personnel in the Persian Gulf and worldwide during Operations Desert Shield and Desert Storm, *JAMA,* 275, 18, 1996.
18. Kang, H.K. and Bullman, T.A., Mortality among U.S. veterans of the Gulf War: Update through December 1997, presented at the Research Working Group Persian Gulf Veterans Coordinating Board Conference on Federally Sponsored Gulf War Veteran's Research, June 23–25, 1999.

19. Haley, R., Commentaries, Point: Bias from the "healthy warrior effect" and unequal follow-up in three government studies of health effects of the Gulf War, *Am. J. Epidemiol.,* 148, 315, 1998.
20. Kang, H. and Bullman, T., Counterpoint: Negligible "healthy warrior effect" on Gulf War veterans mortality, *Am. J. Epidemiol.,* 148, 324, 1998.
21. Gray, G.C., Knoke, J., Berg, S.W., Wignall, S., and Barrett-Conner, E., Counterpoint: Responding to suppositions and misunderstandings, *Am. J. Epidemiol.,* 148, 328, 1998.
22. The Iowa Persian Gulf Study Group, Self-reported illness and health status among Gulf War veterans: A population-based study, *JAMA,* 277, 238, 1997.
23. Eisen, S.A., Kang, H.K., and Murphy, F., Update on VA Cooperative Study #458, National Health Survey of Gulf War Era veterans and their families, presented at the Research Working Group Persian Gulf Veterans Coordinating Board Conference on Federally Sponsored Gulf War Veterans' Research, June 23–25, 1999.
24. Araneta, M.R., Destiche, D.A., Schlangen, K.M., Merz, R.D., Forrester, M.B., and Gray, G.C., Birth defects prevalence among infants of Gulf War veterans born in Hawaii, 1989–1993, presented at the Research Working Group Persian Gulf Veterans Coordinating Board Conference on Federally Sponsored Gulf War Veterans' Research, June 23–25, 1999.
25. DeFraites, R.F., Wanat, E.R., and Norwood, A.E., Report, investigation of a suspected outbreak of an unknown disease among veterans of Operation Desert Shield/Storm, 123d Army Reserve Command, Fort Benjamin Harrison, IN, April 1992, Epidemiology Consultant Service (EPICON), Division of Preventive Medicine, Walter Reed Army Institute of Research, Washington, DC, June 15, 1992.
26. Centers for Disease Control and Prevention, Unexplained illness among Persian Gulf War veterans in an Air National Guard Unit: Preliminary report—August 1990–March 1995, *MMWR Morb Mortal Wkly Rep.,* 44, 443, 1995.
27. Berg, W., Post-Persian Gulf Medical findings: Naval mobile construction Battalion 24, presentation to the NIH Technology Assessment Workshop Panel on the Persian Gulf Experience and Health, Bethesda, MD, April 1994.
28. Haley, R.S., Kurt, T.L., and Hom, J., Is there a Gulf War Syndrome? Searching for syndromes by factor analysis of symptoms, *JAMA,* 277, 1997.
29. Haley, R.W., Hom, J., Roland, P.S., et al., Evaluation of neurological function in Gulf War veterans: A blinded case-control study, *JAMA,* 277, 223, 1997.
30. Kotler-Cope, S., Milby, J.B., et al., Neuropsychological deficits in Persian Gulf War veterans: A preliminary report, presented at the annual meeting of the International Neuropsychological Society, Chicago, IL, 1996.
31. Roht, L.R., Vernon, S.W., Wier, F.W., Pier, S.M., Sullivan, P., and Reed, L.J., Community exposure to hazardous waste disposal sites: Assessing reporting bias, *Am. J. Epidemiol.,* 122 (3), 418, 1985.
32. Hopwood, D.G. and Guidotti, T.L., Recall bias in exposed subjects following a toxic exposure incident, *Arch. Environ. Health,* 43, 234, 1988.
33. Baker, M.S. and Strunk, H.K., Medical aspects of Persian Gulf operations: Serious infections and communicable diseases of the Persian Gulf and Saudi Arabian Peninsula, *Mil. Med.,* 156, 385, 1991.
34. Gasser, R.A., Magill, A.J., Oster, C.N., et al., The threat of infectious disease in Americans returning from Operation Desert Storm, *New Eng. J. Med.,* 324(12), 859, 1991.
35. Hyams, K.C., Hanson, K., Wignall, F.S., et al., The impact of infectious diseases on the health of U.S. troops deployed to the Persian Gulf during Desert Shield and Desert Storm, *Clinic Infect. Dis.,* 20, 1497, 1995.

36. Magill, A.J., Grogl, M., and Gasser, R.A., Visceral infection caused by Leishmania Tropica in veterans of Operation Desert Storm, *New Eng. J. Med.,* 328(19), 1383, 1993.
37. *Operation Desert Shield/Desert Storm, History of Participation of the U.S. Army Environmental Hygiene Agency,* Aberdeen Proving Ground, MD, August 7, 1990–December 31, 1991.
38. National Defense Research Institute, *RAND, A Review of the Scientific Literature as it Pertains to Gulf War Illnesses,* Vol 7, *Depleted Uranium,* 1999.
39. Mather, S., Depleted uranium, presented at the Research Working Group Persian Gulf Veterans Coordinating Board Conference on Federally Sponsored Gulf War Veteran's Research, June 23–25, 1999.
40. McDiarmid, M.A., Hooper, F.J., Squibb, K., and McPhaul, K., The utility of spot collection for urinary uranium determinations in depleted uranium exposed Gulf War veterans, *Health Physics,* 77, 261, 1999.
41. McDiarmid, M.A., Keogh, J., Hooper, F.J., McPhaul, K., Squibb, K., et al., unpublished data on health effects of depleted uranium on exposed Gulf War veterans.
42. U.S. Interagency Air Assessment Team, *Kuwait Oil Fires,* Interagency Interim Report, April 1991.
43. World Meteorological Organization, *Report of the Second WMO Meeting to Assess the Response to and Atmospheric Effects of the Kuwait Oil Fires,* WMO/TD-No 512, Geneva, Switzerland, May 1992.
44. U.S. Army Environmental Hygiene Agency, *Final Report: Kuwait Oil Fire Risk Assessment Project, 5 May–3 December 1991,* Report No. 39–26-L192–91, February 1994.
45. Gray, G.C., Smith, T.C., Knoke, J.D., and Heller, J.M., The postwar hospitalization experience of Gulf War veterans possibly exposed to chemical munitions destruction at Khamisiyah, Iraq, *Am. J. Epidemiol.,* 150, 532, 1999.
46. Walpole, R.D. and Rostker, B., *Modeling the Chemical Warfare Agent Release at the Khamisiyah Pit,* Department of Defense, September 4, 1991, URL http://www.gulflink.osd.mil/cia{&lm}092297/.
47. Abou-Donia, M.B., Wilmarth, K.R., Jenson, K.F., Oehme, F.W., and Kurt, T.L., Neurotoxicity resulting from coexposure to pyridostigmine bromide, DEET, and permethrin: Implications of Gulf War chemical exposures, *J. Toxicol. Environ. Health.,* 48, 35, 1996.
48. McCain, W.C., Lee, R., and Johnson, M.S., Acute oral toxicity study of pyridostigmine bromide, DEET and permethrin in the laboratory rat, *J. Toxicol. Environ. Health,* 50, 101, 1996.
49. National Defense Research Institute, *RAND, A Review of the Scientific Literature as it Pertains to Gulf War Illnesses,* Vol. 1, *Infectious Diseases,* 1999.
50. National Defense Research Institute, *RAND, A Review of the Scientific Literature as it Pertains to Gulf War Illnesses,* Vol. 2, *Pyridostigmine Bromide,* 1999.
51. National Defense Research Institute, *RAND, A Review of the Scientific Literature as it Pertains to Gulf War Illnesses,* Vol. 3, *Immunizations,* 1999.
52. National Defense Research Institute, *RAND, A Review of the Scientific Literature as it Pertains to Gulf War Illnesses,* Vol. 4, Stress, 1999.
53. National Defense Research Institute, *RAND, A Review of the Scientific Literature as it Pertains to Gulf War Illnesses,* Vol. 5, *Chemical and Biological Warfare Agents,* 1999.
54. National Defense Research Institute, *RAND, A Review of the Scientific Literature as it Pertains to Gulf War Illnesses,* Vol. 6, *Oil Well Fires,* 1999.

55. National Defense Research Institute, *RAND, A Review of the Scientific Literature as it Pertains to Gulf War Illnesses,* Vol. 8, *Pesticides,* 1999.
56. U.S. Department of Defense, Comprehensive Clinical Evaluation Program for Ants, Persian Gulf War Veterans: CCEP Report on 18,598 Participants, U.S. Department of Defense, Washington, DC, 1996.
57. Kang, H.H., Dalager, N.A., and Lee, K.Y., *Health Surveillance of Persian Gulf War Veterans—A Review of the DVA Persian Gulf Registry Data,* Department of Veterans Affairs, Washington, DC.
58. Gray, G.C., Hawksworth, A.W., Smith, T.C., Kang, H.K., Knoke, J.D., and Gackstetter, G.D., Gulf War Veterans' health registries. Who is most likely to seek evaluation, *Am. J. Epidemiol.,* 148, 343, 1998.
59. Presidential Advisory Committee of Gulf War Veteran's Illnesses. *Final Report,* U.S. Government Printing Office, Washington, DC, December 1996.
60. Institute of Medicine, Committee on the DoD Persian Gulf Syndrome Comprehensive Clinical Evaluation Program. Evaluation of the U.S. Department of Defense Persian Gulf Syndrome Comprehensive Clinical Evaluation Program, Committee on the DoD Persian Gulf Comprehensive Clinical Evaluation, National Academy Press, Washington, DC, 1996.
61. Guldager, B., The Danish Gulf War study, presented at the Research Working Group Persian Gulf Veterans Coordinating Board Conference on Federally Sponsored Gulf War Veterans' Illnesses Research, June 23–25, 1999.
62. Graham, J.T., Gulf War veterans' health concerns in the United Kingdom, presented at the Research Working Group Persian Gulf Veterans Coordinating Board Conference on Federally Sponsored Gulf War Veterans' Research, June 23–25, 1999.
63. Fukuda, K., Straus, S., Hickie, I., et al., The chronic fatigue syndrome: A comprehensive approach to its definition and study, *Ann. Int. Med.,* 121, 953, 1994.
64. Wolfe, S., Smythe, H., Yunus, M., et al., The American College of Rheumatology 1990 criteria for fibromyalgia, arthritis and rheumatism, *Arthritis Rheum.,* 33, 160, 1990.
65. Hyams, K.C., Developing case definitions for symptom-based conditions: The problem of specificity, *Am. J. Epidem,* 20, 148, 1998.
66. Gerrity, T.R. and Fuessner, J.R., Emerging research on the treatment of Gulf War veterans' illnesses, *J. Occup. Environ. Med.,* 41(6), 440, 1999.
67. Gerrity, T. R., Update on federal research program, presented at the Research Working Group Persian Gulf Veterans Coordinating Board Conference on Federally Sponsored Gulf War Veteran's Research, June 23–25, 1999.
68. Gulf War Illnesses and Related Research at the Institute of Medicine, National Academy of Sciences, An overview of research and recommendations, presented at the Research Working Group Persian Gulf Veterans Coordinating Board Conference on Federally Sponsored Gulf War Veteran's Research, June 23–25, 1999.
69. Department of the Army, Field Manual 100–14, *Risk Management.*
70. U.S. Army Center for Health Promotion and Preventive Medicine, Technical Guide 230 A: *Short Term Chemical Exposure Guidelines for Deployed Military Personnel.*
71. U.S. Army Center for Health Promotion and Preventive Medicine, Technical Guide 230 A: *Short Term Chemical Exposure Guidelines for Deployed Military Personnel.*
72. Hill, A.B., The environment and disease: Association or causation? *Proc. R. Soc. Med.,* 58, 295, 1965.

10 Acute and Chronic Cyanide Toxicity*

Joseph L. Borowitz, Gary E. Isom, and Steven I. Baskin

CONTENTS

I. Introduction ... 301
II. Cyanide Exposure ... 302
III. Symptoms Produced by Cyanide ... 303
IV. Chemical Reactivity of Cyanide ... 305
V. Metabolism of Cyanide ... 306
VI. Effects of Cyanide on Neural Tissue ... 308
A. Elevated Cell Calcium ... 308
B. Effects of Cyanide on Metabolism of Neurons ... 308
C. Oxidative Stress in Neuronal Cells and Cyanide ... 308
D. Hyperpolarization by Cyanide ... 309
E. Neuronal Activation by Cyanide ... 310
VII. Effects of Cyanide on the Heart ... 310
VIII. ADP-Ribosylation by Cyanide ... 312
IX. Production of Cyanide in Neural Tissue ... 312
X. Cyanide Antidotes ... 313
XI. Summary ... 313
Acknowledgments ... 314
References ... 314

I. INTRODUCTION

Since the previous report in 1992, important new findings have provided fresh insight into CN mechanisms of action in both neural and cardiac tissue, the primary targets of CN intoxication.[1] Most studies use CN to produce chemical hypoxia or to mimic conditions caused by stroke and myocardial infarction. Generally, actions of CN resemble those of ischemia and hypoxia, so information gained from CN studies is as important for the analysis of the chemical itself as for study of common pathological conditions.[2–5]

*The opinions or assertions contained in this paper are the private views of the authors and are not to be construed as official or as reflection of the views of the Army or Department of Defense.

0-8493-0872-0/01/$0.00+$.50

Not only is CN acutely toxic in high (mg) doses, serious neurological problems are associated with chronic exposure at lower levels.[6] This review also includes recent observations on this timely issue.

Finally, there is strong evidence that mammalian tissues actually produce CN,[7,8] and it has been proposed that CN may serve as a neuromodulator.[6] Much work needs to be accomplished to determine mechanisms by which neural and other tissues produce CN, and also the physiological and pathological significance of endogenous CN.

II. CYANIDE EXPOSURE

Recordings from antiquity show that Egyptians and Romans utilized CN-containing poisonous plant extracts as a chemical instrument for suicide or murder. Preparations of cherry laurel water containing cyanogenic glycosides distilled from the bark of the tree were utilized by Nero to dispose of individuals who displeased him.[9]

Napoleon III proposed the use of CN tipped bayonets during the Franco-Prussian war. Lord Playfair also sought to implement its use during the Crimean War. The brilliant German chemists such as Michaelis and Haber studied the kinetics of CN in the laboratory and their lessons were applied to the field. World War I experiences taught that CN could produce rapid death in the field, but the slow WWI munition delivery and manufacture of impure product did not allow for dependable dispersal of HCN as a munition.[10] However, introduction of vincennite mixtures (shell No. 4) at the Somme and the method of rapid firing made it almost impossible to put masks on in time to protect against CN. Subsequent reports showed CN mixtures were far more effective than realized.[10] More efficient delivery systems and improved methods of CN synthesis and storage may overcome the technical problems experienced in WWI. Magnum and Skipper reported from observations made during convict execution that man is incapacitated (onset of convulsions) by approximately 10 mg/L within 10 to 18 s.[11]

At the beginning of WWII, CN was used by the Japanese forces on the Bataan Peninsula in the form of a hand grenade and in Manchuria and China for poisoning wells.[11] The Nazis used the poison at the beginning of WWII to kill entrenched Yugoslav partisans in caves (Adjimushkaiskye) and during WWII to exterminate over 2 million concentration camp inmates. In a chaotic 3-day period with the Russian forces approaching, Höss, the commandant of Auschwitz, increased the Zyklon B (hydrocyanic acid adsorbed onto a dispersible pharmaceutical base) concentration to accelerate the normal killing rate for inmates and to exterminate over 10,000 Russian soldiers. In the1980s, several Middle Eastern sites were reported to be CN targets. The inhabitants of Hama, Syria were gassed as a part of a political solution, as were inhabitants of Halabja, Iraq, and possibly in Shahabad, Iran, during the Iran-Iraq war.[12–14]

Certain parameters of CN-induced lethality in man and other mammals have been examined for many years; however, because of its highly toxic and rapid-acting nature, much less is known about sublethal CN toxicity.[15] It has been suggested that the central nervous system (CNS), in particular, is highly sensitive to the toxic effects of CN, and may be the primary target system.[16,17] CNS changes due to non-CN-induced (e.g., hypoxic) hypoxia resemble those induced by CN, although the latter also produce enzyme and neurotransmitter changes.[15]

It is interesting to note that CN is formed, exchanged with every breath we take, and exhaled at concentrations much less than what is considered toxic. The observations that man is constantly exchanging endogenous CN,[7] and studies suggesting its function as a central modulator in rats in its gaseous state, similar to what has been seen for carbon monoxide,[18] suggest that CN may be providing a biological role as a neuromodulator in addition to that of an exogenous synthetic poison.

Today, poisonings have taken place as a result of contact with or breathing of cleaning products for silver, which contain CN. Cyanide is also used in many industrial applications, such as electroplating, case hardening steel at ~900°C, mining, and agricultural fumigation. It is also used in products related to hydrogen cyanide [HCN] (hydrogen nitrile) by incorporation of other nitriles such as industrial solvents (acetonitrile [methylcyanide], for example) or nitrile polymers such as nylon. Thus, CN polymers have become useful items in everyday life. However, these same products can, like nylon, depolymerize in fire and release short-chain monomers or CN resulting in serious CNS toxicity or death. Cyanide is found in many forms and precursors that can be taken into the body. It was noted since antiquity that certain plants could produce a CNS respiratory gasp followed by anoxia, convulsions, occasionally culminating in death. (This breathing reflex appears to be one of the most sensitive responses to CN exposure.)

A wide variety of plant life incorporates nitrile-containing substances that are metabolically or chemically converted to CN, toxic to both animal and man.[19] For example, a large number of plants in the Rosecea genus (e.g., cherry, peach, and bitter almond) are known to contain cyanogenic glycosides. Many of the behavioral and CNS effects of CN were originally observed after the ingestion of CN-containing plant products. Plants can contain cyanogenic lipids (for example, in *Sapandous drummondii*) or cyanogenic glycosides (for example, in cassava, sorgum, flax, white clover). Cassava (*Manihot esculenta*) is a common crop utilized as a foodstuff (manioc) in parts of Asia, South America, and Africa. If not properly processed, it can pose a serious cyanogenic hazard. The plant stores a cyanogenic glycoside, linamarin that is degraded by the enzyme linamarase to cyanohydrins and subsequently to hydrocyanic acid.

III. SYMPTOMS PRODUCED BY CYANIDE

High doses of CN are rapidly fatal, probably due to respiratory arrest. Severe CN poisoning disrupts neural mechanisms controlling consciousness and breathing, though the heart continues to beat (at a much slower rate which is probably incompatible with normal or life-sustaining function).[9] Animals given high but sublethal doses of CN (e.g., mice with 5-mg/kg KCN s.c.; see Figure 10.1) become quiescent a few minutes after injection but may remain conscious and can respond to physical stimulation. After another few minutes, the animals appear to resume normal locomotor activity. These important symptoms reflect transient actions of CN on different neural

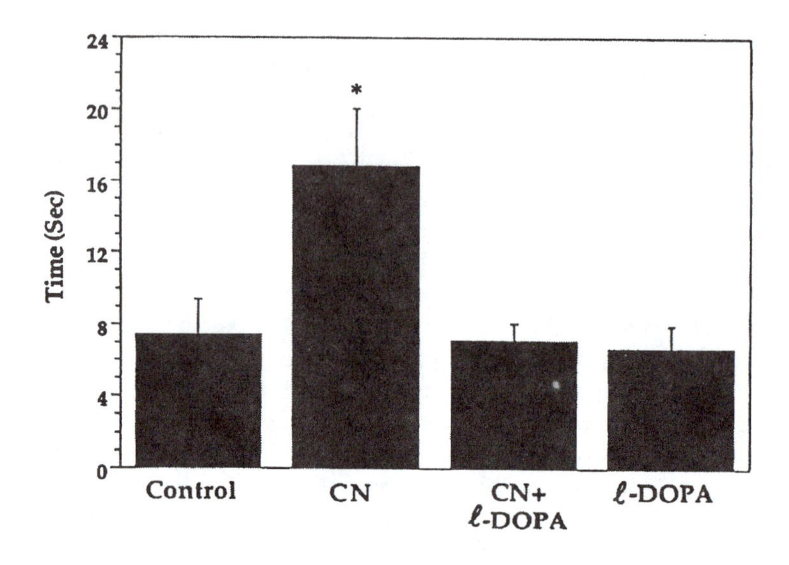

FIGURE 10.1 Production of catalepsy in mice by repeated CN treatment: Mice were treated with 6 mg/kg (s.c.) KCN twice a day for 7 days. Sixteen hours after the last dose, catalepsy was quantitated. In *l*-DOPA experiments, 100 mg/kg (i.p.) *l*-DOPA was administered 1 h prior to quantitation of the degree of catalepsy. Values are the means ± SEM from four determinations and each determination consisted of three animals in each group. Asterisk indicates a significant difference from control at $p < 0.05$. (Reproduced with permission)

systems. Many recent studies discussed below reveal the complex nature of CN's effects on different neural pathways.

Cyanide stimulates chemoreceptor reflexes.[20] Denervation of the carotid sinus removes the respiratory stimulation and bradycardia. Using isolated cat carotid bodies or sinuses, dopamine produces a transient depression of the frequency of chemoreceptor discharges. The effect of dopamine can partially or totally antagonize the excitation of chemoreceptor discharges evoked by acetylcholine or CN.[21] The effect of dopamine is long acting. Also, changes in the intracellular sodium and calcium concentration influence the excessive depolarization and sensory discharge of the cat carotid body and nerve produced by CN.[22] Concentrations as low as 10–50 nM CN reduce cytochromes in the carotid body reflecting the extreme sensitivity of this tissue to CN. It is this site that appears to be responsible for the respiratory gasp. Fluorometry revealed reduction of NADH as well.[23,24] Thus, the interaction with CN in the peripheral nervous system appears to be the most sensitive at the chemoreceptor site.

Studies are currently ongoing to identify the primary oxygen-sensing (and perhaps the CN-sensing) protein controlling transmitter release and electrical activity of the carotid sinus nerve. It is also suggested that this primary oxygen- and CN-sensing receptor is a hemeprotein that does not participate in mitochondrial energy production. A cytochrome b (558) was described for the NAD(P)H oxidase.[25] These results suggest that there may be a specific molecular site for the sensing of CN.

Non-lethal chronic exposure to CN can lead to neurological problems; some neurodegenerative diseases are associated with chronic CN treatment.[6] Mice that are given potassium cyanide (6-mg/kg s.c. twice daily for 7 days) exhibit Parkinsonian symptoms of decreased motor activity and akinesia[26] (Figure 10.1). The reviews by Isom et al. and Baskin and Rockwood cover the relationship between CN ingestion and both the conditions of tropical ataxic neuropathy and the upper motor neuron disease "Konzo."[6,27] Furthermore, evidence of abnormal CN metabolism has been reported for tobacco amblyopia, Leber's optic atrophy, and amyotrophic lateral sclerosis.[27] Symptoms seen in the epidemic of optic neuropathy in Cuba between 1991 and 1993, discussed by Isom et al. resemble those of tobacco amblyopia, and many cases have peripheral neuropathies as well, such as painful dysthesias and decreased ankle reflexes.[27] Vitamin B_{12} is used to treat tobacco amblyopia according to the concept that B_{12} deficiencies increase susceptibility to CN in tobacco smoke. The optic neuropathy in Cuba affected 50,000 people and was also effectively treated with B_{12}.[27] Sudan suggests that vitamin deficiencies, exposure to methanol, and CN contributed to the Cuban epidemic, and that defective mitochondrial function impairs ATP production even to the extent of interfering with axonal transport of mitochondria to nerve endings.[28]

Epidemiological studies reveal a high incidence of Parkinsonism occurring in rural areas.[29,30] More recently, Hobson et al. have noted a direct relationship between use of calcium CN dust and Parkinsonism in beekeepers.[31] Cyanide as an environmental factor appears to be important in some neurological disorders.[27]

IV. CHEMICAL REACTIVITY OF CYANIDE

Cyanide (hydrocyanic acid, HCN) is a small molecule with good lipid and water solubility. Physically, it can exist as a gas or liquid; it is miscible with water and slightly soluble in ether. Like nitric oxide and carbon monoxide, it easily penetrates biological membranes and acts intracellularly.[9] At physiological pH, over 98% of the molecule is in the form of HCN and only a small fraction occurs as CN. The major biological effects are most likely due to the undissociated molecule. Cyanide strongly interacts with iron in protein molecules, inhibiting enzymes including carbonic anhydrase and succinic dehydrogenase.[32] Formation of cyanhemoglobin by interaction of CN with ferric iron abolishes the ability of hemoglobin to carry oxygen. Interaction of CN with the ferric iron in mitochondrial cytochrome oxidase blocks cellular respiration; this has long been considered an important toxic action of CN.[33] Sun et al. also suggest that interaction of CN with disulfide groups on the NMDA receptor regulatory sites enhances receptor function.[34] Arden et al. reported that CN acts on the NMDA receptor as a reducing agent to potentiate NMDA-induced electrical activity in rat cortical neurons, though an oxidizing agent reverses this action.[35] Cyanide is thought to potentiate glutamate neurotoxicity by this mechanism; however, how glutamate-CN interactions relate to CN's *in vivo* toxicity is not completely established. Thus, the primary chemical interactions of CN are thought to involve ferric iron and disulfide bonds.

V. METABOLISM OF CYANIDE

In contrast to other chemical warfare agents, CN appears biologically in blood, urine, and expired breath.[7] It is actually generated in small amounts in neuronal tissue, and researchers have proposed that CN functions as a neuromodulator similar to nitric oxide.[27] Cyanide contrasts with nitric oxide in that it is chemically more stable and is not immediately broken down. Enzymes exist that regulate CN concentrations and two sulfurtransferases, rhodanese and 3-mercaptopyruvate sulfurtransferase, as well as thiosulfate reductase, convert CN to thiocyanate, which is about seven times less toxic.[36] These enzymes account for 60–70% of the metabolism of non-toxic concentrations of CN and may act in concert since they have different tissue distributions. Rhodanese occurs in highest concentration in the liver with high levels also in kidneys, adrenals, and thyroid, whereas mercaptopyruvate sulfurtransferase has a broad tissue distribution with high levels in the liver, kidneys, and heart. Being lipid soluble and relatively stable, CN probably accumulates in lipoid depots throughout the body, and is also bound to an albumin-binding site.[37] Mobilization from lipid and

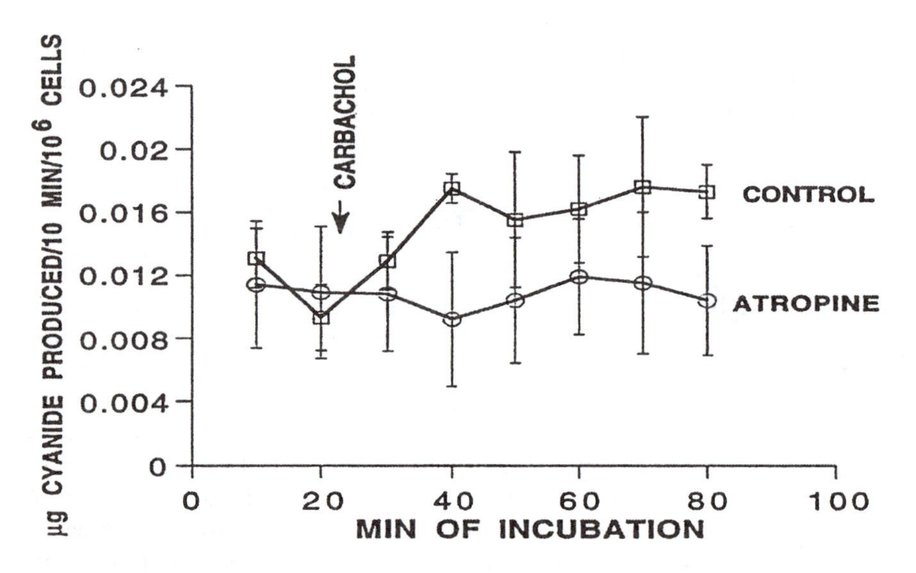

FIGURE 10.2 Blockade of carbachol-induced CN production in undifferentiated rat pheochromocytoma cells by atropine: Atropine 500 µM was added at the beginning of the experiment and carbachol (100 µM) was added after 20 min to both atropine and control samples. Air 95%, CO_2 5% was passed over the cells and bubbled through 0.1 M NaOH to trap the CN. Aliquots of the NaOH were taken to measure CN colorimetrically (Lambert, J., Ramasamy, J., and Pakstelis, J., *Anal. Chem.* 47, 916, 1975. With permission). Note atropine completely blocked the response to carbachol but basal CN production was not affected by atropine. Apparently the cells generate CN from an atropine insensitive source which includes release from lipoid depots and from proteins.

release from protein binding is suggested to account for some of the generation of CN detected in neural tissue (Figure 10.2).

A minor (approximately 20% under non-toxic conditions) but toxicological metabolic pathway (that may increase during CN poisoning) for CN involves the disulfide cystine. 2-ICA, or its tautomer 2-aminothiazolidine-4-carboxylic acid (2-ACA), is a detoxification product of CN that is formed by what is thought to be a non-enzymatic reaction of CN with cystine.[38] Cyanide reacts with cystine producing β-thiocyanoalanine, which spontaneously undergoes ring closure to form 2-ICA and its tautomer 2-ACA (Figure 10.3). These tautomers are in rapid chemical equilibrium and exist in equal concentration in solution. The formation of 2-ICA may increase with increased exposure to CN. One mechanism for this increase may be the decreased pH in the cells that favor the formation of 2-ICA compared with the maximal activity of the sulfurtransferases at a much higher pH.

Only limited research has been conducted to study *in vivo* formation of 2-ICA following systemic administration of CN.[39,40] Depending on species, sensitivity of the assay and CN exposure conditions, the reported percentage of CN converted to 2-ICA ranges from 5–15% of delivered CN dose.[40,41] 2-ICA does not appear to be metabolized, but is excreted slowly in the urine and saliva.[38,41] We have studied its biological activity (i.e., memory loss, convulsions, loss of consciousness) and concluded 2-ICA contributes to the CNS actions of CN.[42–44] The toxicokinetics of 2-ICA formation and its elimination (half-life) have not been determined. However, in a preliminary study of 2-ICA as a CN biomarker, Lundquist et al. showed 2-ICA was detectable in the urine by HPLC assay up to 4 weeks after administration of acetonitrile, a cyanogenic compound that is metabolized to CN.[39] In smokers or human subjects ingesting cyanogenic compounds, 2-ICA was detected in urine. In isolated rat hepatocytes, Huang et al. prevented cell death by 400 μM CN using 1 mM cystine.[45] They found thiocyanate levels were also increased under these conditions, so the cystine may provide sulfur for thiocyanate formation as well as for 2-ICA production.

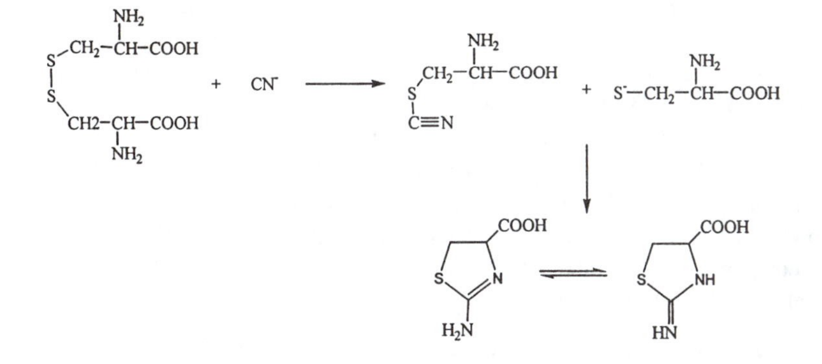

FIGURE 10.3 Conversion of cyanide to 2-aminothiazolidine-4-carboxylic acid or 2-iminothiazolidine-4-carboxylic acid.

VI. EFFECTS OF CYANIDE ON NEURAL TISSUE

A. Elevated Cell Calcium

Since the previous review, several significant papers further implicating elevated cellular calcium in CN-induced neurotoxicity have appeared.[1] Ferger and Krieglstein exposed chick telencephalic neurons to 1 mM NaCN for up to 2 h.[46] Increases in $[Ca]_i$ were measured with Fura-2, and viability was estimated by trypan blue exclusion. Elevation of $[Ca]_i$ paralleled neuronal damage.[46] On the other hand, insertion into PC12 cells of a herpes simplex vector expressing cDNA for calbindin did not prevent the rise in calcium or cell survival after exposure to 1–5 mM sodium CN (18 h) even though these calbindin-containing cells were protected against the effects of glutamate.[47] Compared to the neurotoxicity of cyanide, glutamate-induced neurotoxicity may be more intimately related to increases in cell calcium.

Two other reports suggest that calcium must be taken up into mitochondria to mediate toxicity. Thus, glutamate is not toxic to cultured rat forebrain neurons when uptake of calcium into mitochondria is inhibited.[48] An associated increase in cytosolic calcium occurs however despite the lowered toxicity. The authors suggest calcium is toxic only when it enters mitochondria and that high levels of cytosolic calcium do not appear to be toxic.[48] In support, Sengpiel et al. report that 1 mM sodium CN (admittedly a high concentration) prevented mitochondrial calcium uptake and reduced both neurotoxicity of NMDA in cultured rat hippocampal neurons and the associated NMDA-induced superoxide production.[49]

B. Effects of Cyanide on Metabolism of Neurons

CA1 hippocampal neurons are preferentially susceptible to hypoxia and ischemia. In CA1, CA3, and dentate gyrus neurons dissected from fresh rat hippocampal slices, CN specifically enhanced release of acid metabolic products from CA1 cells but had little effect on the other cells.[50] By contrast, kainate, which has CA3-specific effects, increased acid metabolite release only in CA3 neurons.[50] Actions of CN appear to be metabolic in nature and not all neuronal cell types are equally affected.

Zu and Krnjevic studied CN in hippocampal slices.[51] They found 300 μM CN did not block electrical responses to field stimulation as long as glucose levels were elevated to 10 mM, but in 4 mM glucose (physiological level), CN caused a characteristic hypoxic injury potential followed by a blockade of the response to electric fields. Intracellular recordings reveal a continued hyperpolarization in response to CN in 10 mM glucose, but in 4 mM glucose only a brief hyperpolarization occurred, followed by a major and usually irreversible depolarization. The authors suggested a reduced supply of ATP impairs restoration of membrane potential and causes the irreversible depolarization.

C. Oxidative Stress in Neuronal Cells and Cyanide

Isom et al. reviewed mechanisms of apoptotic or necrotic neural damage caused by CN.[27] Cyanide-induced calcium entry by way of voltage-sensitive calcium channels or NMDA receptors has three main actions. First, activation of lipases in the cell membrane increases arachidonic acid release, which leads to increases in reactive oxygen species. Calcium then activates nitric oxide synthase to increase nitric oxide levels. Finally, calcium activates proteases, lipases, and endonucleases that can damage structural and functional elements in neuronal cells. Reactive oxygen species and nitric oxide also can form peroxynitrite by reacting with superoxide. Peroxynitrite is a powerful oxidant which has many and varied effects in neurons and other cell types, including depletion of cell thiol groups, lipid peroxidation, mobilization of cell calcium, impaired mitochondrial function correlated with muscular contractile failure in rat diaphragms, and modification of synaptic proteins.[52–57] Uric acid is a peroxynitrite scavenger and protects cells against this powerful oxidant. In granule cells of the cerebellum, uric acid protects against CN-induced apoptotic death indicating that peroxynitrite is an important mediator of cell damage by CN.[58,59]

D. Hyperpolarization by Cyanide

Cyanide causes either a hyperpolarization or a depolarization when tested on neuronal tissue depending on conditions and type of neurons involved. Hippocampal CA1 neurons usually hyperpolarize on exposure to hypoxia, but hypoglossal neurons depolarize under the same conditions.[60–62] The hyperpolarization may be a protective mechanism to prevent activation of the cell in a time of stress.[63] Usually the potassium channels involved are ATP regulated (K_{ATP}), but this also varies with the cell type. In undifferentiated rat pheochromocytoma cells, hyperpolarization occurs due to opening of K_{Ca} channels subsequent to an increase in $[Ca^{2+}]_i$.[64] In dissociated rat locus coerulus neurons, the hyperpolarization caused by sodium CN involves both IK_{ATP} and IK_{Ca}.[65] Studying neurons in rat locus coerulus slices, Yang et al. found 61% of the neurons hyperpolarized when treated with 2 mM CN (albeit, a large amount of CN) but 39% of the neurons depolarized.[66] Thus, in neurons responsible for sending noradrenergic impulses throughout the CNS from the same tissue, the response to histotoxic anoxia is variable. Yang et al. suggest that distribution of K_{ATP} channels among neurons of the locus coerulus is variable since the K_{ATP} channel-opener diazoxide could mimic the hyperpolarizing effect of CN in 61% of the neurons, but not in the 39% depolarized by CN.[66]

In a test of the concept that hyperpolarization protects neurons from toxic damage, a potassium channel opener, bimakalim, was employed and was found to protect embryonic chick telencephalic neurons from 1 mM CN-induced injury. The protective effect of bimakalim was canceled by the K_{ATP} blocker tolbutamide.[63] Apparently the extent of the hyperpolarization caused by CN is not sufficient to give optimal protection and a further increase in neuronal polarity provides even more damage control.

Also in hippocampal slices using high glucose (11 mM), Zhu and Krenjevic report that the inhibitory effect of 100 μM KCN was blocked by adenosine antagonists, potentiated by the adenosine uptake blocker dipyridamole but was not affected by glyburide, a K_{ATP} channel blocker.[67] They suggest that adenosine release may be a major cause of the early depression of CNS function caused by CN. Adenosine is known to be released from nerve cells by CN, and to cause hyperpolarization by a

G protein effect on potassium channels.[68–71] Adenosine release by CN must be considered a factor in CN-induced neural injury.

E. Neuronal Activation by Cyanide

Exposure of freshly excised rat CA1 hippocampal neurons to 5 mM CN increased the $I_{Na,P}$ sodium current but had no significant effect on the amplitude of the more transient current $I_{Na,T}$.[72] Bubbling 100% N_2 into the medium similarly increased $I_{Na,P}$ indicating that CN and hypoxia have similar mechanisms. Persistent increase in sodium current probably explains the increased $[Na^+]_i$ seen in cortical neurons during hypoxia.[73] The $I_{Na,P}$ caused by CN was blocked by tetrodotoxin or lidocaine. Persistent flow of sodium through sodium channels may activate voltage-sensitive calcium channels or activate the Na^+/Ca^{++} exchanger, to increase $[Ca^{2+}]_i$. Thus an increase in $I_{Na,P}$ may be the initial event caused by hypoxia leading to cell death. In fact sodium channel blockers can block the $[Ca^{+2}]_i$ increase and prevent cell damage during hypoxia.[74] In support, procaine protects mice against the lethal effects of CN.[75] Combination of procaine with sodium nitrite and sodium thiosulfate enhanced the effectiveness of the nitrite/thiosulfate treatment. Furthermore, the CN-induced increase in whole mouse-brain calcium from 28 to 48 mg/g dry weight was also blocked by procaine pretreatment.[75] Abnormal sodium channel function may be a primary event in CN-induced neuronal damage.

VII. EFFECTS OF CYANIDE ON THE HEART

The previous review mentioned CN-induced changes in myocardial, calcium, and H^+ as factors in myocardial depression caused by this agent. Marked CN-induced increases in circulating catecholamine stimulate the heart,[75] but, at the same time, energy metabolism is impaired and heart failure results.[76]

How CN decreases cardiac contractility is important and has been studied by several groups. Hydrogen ion accumulation contributes to the lack of effectiveness of $[Ca^{++}]_i$ in activating the contractile process.[77] Blockade of oxidative metabolism by CN increases glycolysis and therefore increases lactic acid production. Because ATP is continuously broken down, inorganic phosphate (P_i) accumulates since less is being used to make ATP. Increases from 4 to 10.5 mM P_i have been measured in CN-treated perfused ferret hearts.[78] Essentially, this provides heart cells with added phosphate buffer to minimize pH changes. Changes of only 0.2 unit were noted in ferret hearts perfused with 1 mM CN, or 0.08 units in rat hearts perfused with 1 mM KCN.[79,80] Even though hydrogen ion accumulation is not large, it explains some of the decreased myocardial contractility caused by CN. Hydrogen ion is a strong competitor with calcium for binding sites in tissues.[81] Effects of pH may be more noticeable in intact hearts compared to isolated myocytes because of differences in the rate at which lactic acid can leave the tissue.[80]

Cytosolic calcium overload is generally associated with cell injury and energy deprivation increases intracellular calcium.[82,83] Kondo et al. measured 2 mM

CN-induced increases in systolic (104% above control) and diastolic (37%) calcium in paced rat myocytes.[77] Despite the increase in calcium, contractile function decreased to 58% of control. Doubling extracellular calcium, restored contractility to 123% of control and increased systolic (225% above control) and diastolic (73%) $[Ca^{++}]_i$. However no increase in cell damage was noted over a period of 40 min (25% of the cells went into contracture when exposed to normal $[Ca^{++}]_o$ and 2 mM CN and incidence of damage was the same in high calcium). These observations have important implications. First, the decrease in contractility was related to the relative ineffectiveness of $[Ca]_1$ to activate the contractile machinery, partly due to elevated hydrogen ion. When calcium was further increased, contraction was fully restored. Second, reduced energy availability does not appear to be a problem at least when an abundance of glucose (19.5 mM) was provided. The hearts functioned well when $[Ca^{++}]_o$ was increased despite the continued presence of CN, and remarkably no greater increase in contracture or increase in cell destruction occurred.

Kupriyanov et al. perfused rat hearts with 1 mM KCN and showed decreased heart rate and perfusion pressure associated with an increase in osmolarity.[79] Increases in P_i occur when ATP is broken down, 3 P_i are formed, and the nucleoside leaves the cell; phosphocreatine is also broken down to further increase P_i levels. Breakdown of glycogen to lactate also contributes to the overall increase in osmolarity estimated to be about 26 mM. Some increase in intracellular water (~10%) would be expected in CN-treated heart and this cellular edema may affect function.

Kupriyanov et al. also noted an increase in $[Na^+]$ and a decrease in $[K^+]_i$ in rat hearts perfused with 1 mM KCN.[79] Decreased Na^+K^+ATPase activity due to decreased ATP levels could explain this change. However, it is reported that even a 20-fold decrease in cytoplasmic ATP/ADP does not decrease Na^+K^+ATPase activity in perfused rat heart,[84] so the 5-fold decrease observed by Kupriyamov et al. cannot explain the increased $[Na^+]_i$.[79] These authors suggest that Na^+K^+ATPase is inhibited by the increased P_i, which can form a ternary abortive complex with the enzyme and ADP.

Cyanide activates K_{ATP} channels in the brain and also in the heart.[63,79] The K_{ATP} channel inhibitor glibenclamide blocked the effect of KCN in the Langendorf perfused rat heart.[79] However part of the effect of glidenclamide and that of KCN on cell potassium is due to inhibition of Na^+K^+ATPase. An increase in K^+loss through the K^+/lactate co-transporter by KCN was also demonstrated by use of a blocker of this transport system, α-cyano-4-hydroxycinnamic acid. Thus the effect of KCN on K^+ efflux in the heart involves three factors: activation of the K_{ATP} channel, blockage of Na^+K^+ATPase, and activation of the K^+ lactate cotransporter.

The diaphragm is similar to the heart in that it also responds rhythmically to stimulation. After a brief potentiation of muscle twitch, CN (0.1–1 mM) causes a slow progressive depression of contractility of the rat diaphragm.[85] Potentiation is due to an increase in pH from replenishment of ATP by phosphocreatine (creatine kinase mediated transphosphorylation of ADP to ATP). Inhibition of muscle twitch is due to lactate accumulation as well as increased P_i and increased $[Mg^{2+}]_i$ from breakdown of magnesium phosphocreatine.[85] No decreases in ATP or action potential generation were caused by CN treatment in rat diaphragms.[85] Because skeletal muscle, including the diaphragm, is less active than heart muscle, it is also less sensitive to metabolic inhibition by CN.

VIII. ADP-RIBOSYLATION BY CYANIDE

Proteins may be modified posttranslationally by transfer of the ADP-ribose moiety of nicotinamide adenine dinucleotide to an amino acid. Five mammalian ADP-ribosyl transferases (ART-I-ART-5) have been cloned and expression is limited to certain tissues including heart and brain.[86] These transferases are regulated by ADP-ribosylation factors (ARF) which are small monomeric G proteins activated by combination with GTP.[87] The system is stimulated by reactive oxygen species and may protect cells from oxidative damage or may influence the type of death a cell undergoes.[88–90]

It was reported in 1988 that CN increases ADP ribosylation of mitochondrial proteins.[91] Surprisingly, this interesting effect has not been studied further. Some of the observed actions of CN such as enhanced neurotransmitter release and alignment of chromaffin granules along the plasma membrane may be explained by ADP-ribosylation of certain proteins, since protein ribosylation can affect exocytosis from chromaffin cells and membrane recycling in the Golgi apparatus.[92,93]

A similar process involves poly (ADP-ribose) polymerase (PARP), which catalyses attachment of multiple ribose units from NAD to nuclear proteins. Genetic disruption of PARP protects against ischemic insults *in vitro* and limits infarct volume after reversible middle cerebral artery occlusion in mice.[94] Apparently excessive PARP activation in ischemia depletes NAD and ATP (which regenerates NAD) and causes cell death by energy depletion.[94] It would seem that PARP is certainly involved in the action of CN on neural tissue but no such work has been reported.

IX. PRODUCTION OF CYANIDE IN NEURAL TISSUE

Isom et al. mentioned endogenous generation of CN and the possibility that CN may function as a neuromodulator in a manner similar to nitric oxide.[27] Brain CN levels are increased by hydromorphone and the effect is blocked by naloxone.[27] Undifferentiated rat pheochromocytoma cells also show increased CN production in response to hydromorphone or morphine.[27] Since PC12 cells have mainly kappa opiate receptors and no mu receptors, hydromorphone probably acts through kappa receptors to increase CN release.[95]

If CN is indeed a neuromodulator, it contrasts with nitric oxide. Except for conversion to thiocyanate by sulfurtransferase enzymes, CN is relatively stable in biological systems and exists to the extent of about 3 μM in human blood.[96,97] Those who smoke have elevated blood CN levels. Nitric oxide, on the other hand, spontaneously breaks down in biological fluids, having a half-life of a few seconds.[98] Thus CN can accumulate in biological materials, collecting in lipoid depots since it is lipid soluble. Cyanide also forms complexes with albumin through addition to disulfide bonds, and one study proposed this interaction to be a mechanism to remove CN from blood.[99]

Cyanide may interact with proteins in other ways by forming hydrogen bonds or salt bridges with appropriate sites on protein molecules. Cyanide in lipoid membranes or bound to protein may be in equilibrium with free CN in biological fluids.

Whether disturbances in CN generation or metabolism can cause disease is controversial, although CN imbalance is implicated in Leber's optic atrophy and amyotrophic lateral sclerosis.[27] Important work remains to be done to determine the role of endogenous CN in physiological systems and in disease states.

X. CYANIDE ANTIDOTES

Cyanide is a powerful intracellular poison that acts rapidly due to its good lipid and water solubility, and can quickly cause profound hypoxia in vital organs resulting in death. Prompt diagnosis and timely, effective use of antidotes is critical for the severely poisoned patient.

In the United States, the only Food and Drug Administration-approved antidote is the Cyanide Kit currently manufactured by Taylor Pharmaceutical Co. It actually contains three antidotes: amyl nitrite, sodium nitrite, and sodium thiosulfate. The nitrites form methemoglobin, which is an avid scavenger of CN. They also may give rise to nitric oxide, which is an effective CN antidote independent of methemoglobin formation.[100] Amyl nitrite is a volatile liquid; the glass vial containing the drug is crushed in gauze to allow inhalation by the comatose patient. Sodium nitrite is then given slowly (i.v.) for more extensive methemoglobin generation. Thiosulfate is a sulfur donor aiding the sulfur transferase enzymes, rhodanese and 3-mercaptopyruvate sulfurtransferase, which convert CN to thiosulfate, a much less toxic substance.

Cobalt diedetate (Kelocyanor) is well known in Europe and popular as a CN antidote. It is not available in the United States. Adverse effects of the antidote are seizures, angioedema, cardiovascular instability, and gastrointestinal problems. However, cobalt is a rapid-acting antidote and effective even in the severely poisoned patient.

Thiosulfate enhances the antidotal effect of many substances other than the nitrites. As mentioned in the 1992 review, α-ketoglutarate is a potential antidote with few side reactions and good effectiveness against the toxic effects of CN.[1] Its activity is markedly enhanced when given in combination with thiosulfate.[101]

XI. SUMMARY

In conclusion, low-level acute exposure to CN has been characterized by a respiratory gasp, which is believed to be caused by stimulation of chemoreceptors in the aortic arch. The chronic consequences of this type of acute exposure to CN are largely unknown. Since there are normal cellular mechanisms that maintain the balance between CN and sulfur, the equilibrium of the systems is thought to be well controlled.

Low-level chronic exposure to CN has not been fully characterized. It is believed that enzymes modulate and regulate CN and sulfur turnover at the cellular level to try to maintain homeostasis. Studies of overload of the regulatory balance systems need to be systematically undertaken to determine which enzymes compensate as feedback compensation.

CN does not uniformly affect all brain cells. CA1 neurons in the hippocampus are more susceptible than CA3 cells to metabolic inhibition by CN. Certain neuronal type cells, e.g., those in the carotid body, are highly sensitive to the actions of CN. Thus, CN's actions on the neural systems are complex and depend on the type of neuron involved. Most likely, some nerve pathways are activated while others are inhibited or unaffected when an individual is exposed to CN.

ACKNOWLEDGMENTS

The authors thank Mr. Pinal C. Patel and the library staffs at the U.S. Army Medical Research Institute of Chemical Defense and Purdue University.

REFERENCES

1. Borowitz, J.L., Kanthasamy, A.G., and Isom, G.E., *Chemical Warfare Agents,* Somani, S., Ed., Academic Press, New York, 1992, 209.
2. Ballanyi, K. and Kulik, A., Intracellular Ca^{2+} during metabolic activation of K_{ATP} channels in spontaneously active dorsal vagal neurons in medullary slices, *Eur. J. Neurosci.,* 10, 2574, 1998.
3. Hammerstrom, A.K. and Gage, P.W., Inhibition of oxidative metabolism increases persistent sodium current in rat CA1 hippocampal neurons, *J. Physiol.,* 510, 935, 1998.
4. Yang, J.J., Chou, Y.C., Lin, M.T., and Chiu, T.H., Hypoxia-induced differential electrophysiological changes in rat locus coeruleus neurons, *Life Sci.,* 61, 1763, 1997.
5. Inoue, M., Fujishiro, N., and Imanaga, I., Hypoxia and CN induce depolarization and catecholamine release in dispersed guinea-pig chromaffin cells, *J. Physiol.,* 507, 807, 1998.
6. Isom, G.E., Gunasekar, P.G., and Borowitz, J.L., *Chemicals and Neurodegenerative Disease,* S. Bondy, Ed., Prominent Press, Scottsdale, AZ, 1999, 101.
7. Lundquist, P., Rosling, H., and Sorbo, B., The origin of hydrogen CN in breath, *Arch. Toxicol.,* 61, 270, 1988.
8. Borowitz, J.L., Gunasekar, P.G., and Isom, G.E., Hydrogen cyanide generation by μ opiate receptor activation: Possible neuromodulatory role of endogenous CN, *Brain Res,* 768, 294, 1997.
9. Sollmann, T., *A Manual of Pharmacology and Its Applications to Therapeutic and Toxicology,* 7th ed., W. B. Saunders Co., Philadelphia, PA, 1948.
10. Macy, R., *Hydrocyanic Acid: Its Military History and a Summary of Its Properties,* Edgewood Arsenal, MD, Chemical Warfare Service, War Department. (DTIC No. AD-B957 032), 1937.
11. Magnum, G.H. and Skipper, H.E., Hydrocyanic acid: The toxicity and speed of action on man, *Edgewood Arsenal Memorandum Report* (Project A 3.5–1), Aberdeen Proving Ground, MD, (T.D.M.R. 471), 1942.
12. Lang, J.S., Mullin, D., Fenyvesi, C., Rosenberg, R., and Barnes, J., Is the "protector of lions" losing his touch? *U.S. News World Rep.,* 10, 29, November 1986.
13. Anonymous, Medical experts use of chemical weapons in Iran-Iraq war, *UN Chronicle,* 22, 24, 1985.

14. Heylin, M., Ed., U.S. decries apparent chemical arms attack, *Chem. Eng. News,* 66, 23, 1988.
15. D'Mello, G.D., Neuropathological and behavioral sequelae of acute CN toxicosis in animal species, in *Clinical and Experimental Toxicology of Cyanides,* Ballantyne, B. and Marrs, T.C., eds., Bristol, UK, Wright, 1985, 156.
16. Way, J.L., Mechanism of CN intoxication and its antagonism, *Fund. Appl. Toxicol.,* 3, 339, 1983.
17. Way, J.L., Sylvester, D., Morgan, R.L., Isom, G.E., Burrows, G.E., Tamulinas, C.B., and Way, J.L., Recent perspectives on the toxicodynamic basis of CN antagonism, *Fund. Appl. Toxicol.,* 4, S231, 1984.
18. Borowitz, J., Gunasekar, P., and Isom, G., Hydrogen cyanide generation by mu opiate receptor activation: Possible neuromodulatory role of endogenous cyanide, *Brain Res.,* 768, 294, 1997.
19. Evered, M.D., Robinson, M.M., and Rose, P.A., Effect of arterial pressure on drinking and urinary responses to angiotensin II, *Am. J. Physiol.,* 254, R69, 1988.
20. Heymans, C., Bouckaert, J.J., and Dautrebande, L., Sinus carotidien et reflexes respirtores. III: Sensibilité des sinus carotidiens aux substances chimiques. Action stimulante respiratoire reflexe du sulfure de sodium, du cyanure de potassium, de la nicotine et de la lobeline, *Arch. Int. Pharmacodyn. Théra.,* 40, 54, 1931.
21. Zapata, P., Effects of dopamine on carotid chemo- and baroreceptors in vitro, *J. Physiol.,* 244, 235, 1975.
22. Eyzaguirre, C. and Nishi, K., Effects of different ions on resting polarization and on the mass receptor potential of carotid body chemosensors, *J. Neurobiol.,* 7, 417, 1976.
23. Acker, H., Eyzaguirre, C., and Goldman, W.F., Redox changes in the mouse carotid body during hypoxia, *Brain Res.,* 330, 158, 1985.
24. Lahiri, S., Ehleben, W., and Acker, H., Chemoreceptor discharges and cytochrome redox changes of the rat carotid body: Role of heme ligands, *Proceedings of the National Academy of Science USA,* 96, 9427, 1999.
25. Acker, H., Mechanisms and meaning of cellular oxygen sensing in the organism, *Resp. Physiol.,* 95, 1, 1994.
26. Kanthasamy, A.G., Borowitz, J.L., Pavlakovic, G., and Isom, G.E., Dopaminergic neurotoxicity of CN: Neurochemical, histological and behavioral characterization, *Toxicol. Appl. Pharmacol.,* 126, 156, 1994.
27. Baskin, S.I. and Rockwood, G.A., *Neurotoxicological and Behavioral Effects of Cyanide and Its Potential Therapies,* in press.
28. Sudan, A., Acquired mitochondrial impairment as a cause of optic nerve disease. Transaction Am., *Ophth. Soc.,* 96, 881, 1998.
29. Lanston, J.W., Epidemiology versus genetics in Parkinson's disease, *Ann. Neurol.,* 44 (suppl), S45, 1998.
30. Rajput, A.H., Uitti, R.J., and Rajput, A., Neurological disorders based on provincial health case records, *Neuroepidemiology* 7, 145, 1998.
31. Hobson, D.E., Del Bigio, M.R., and Stoss, B.J., Beekeeper Parkinsonism: A consequence of chronic intermittent CN poisoning, submitted for publication.
32. Ballantyne, B., *Clinical and Experimental Toxicology of Cyanides,* Ballantyne, B. and Marrs, T.C., eds., IOP Publishers, Bristol, England, 1987, 41.
33. Way, J.L., Cyanide intoxication and its mechanism of antagonism, *Ann. Rev. Pharmacol. Toxicol.,* 24, 51, 1984.
34. Sun, P.W., Rane, S.G., Gunasekar, P.G., Borowitz, J.L., and Isom, G.E., Cyanide interaction with redox modulatory sites enhances NMDA receptor responses, *J. Biochem. Molec. Toxicol.,* 13, 253, 1999.

35. Arden, S.R., Sinov, J.D., Potthoff, W.K., and Aizenman, E., Subunit specific interactions of CN with the N-methyl-D-aspartate receptor, *J. Biol. Chem.*, 293, 21505, 1998.
36. Isom, G.E. and Baskin, S.I., *Comprehensive Toxicology,* Vol. 3, Sipes, G., McQueen, C.A., and Gandolfi, A.J., Eds., Pergamon Press, Cambridge, UK, 1997, 477.
37. Lieske, C.N., Clark, C.R., Zoeffel, L.D., von Tersch, R.L., Lowe, J.R., Smith, C.D., Broomfield, C.A., Baskin, S.I., and Maxwell, D.M., Temperature effects in cyanolysis using elemental sulfur, *J. Appl. Toxicol.,* 16(2), 171, 1996.
38. Wood, J.L. and Cooley, S.L., Detoxification of cyanide by cystine, *J. Biol. Chem.,* 218, 449, 1956.
39. Lundquist, P., Kagedal, B., Nilsson, L., and Rosling, H., Analysis of the cyanide metabolite 2-aminothiazoline-4-carboxylic acid in urine by high-performance liquid chromatography, *Anal. Biochem.,* 228, 27, 1995.
40. Swenne, I., Eriksson, U.J., Christoffersson, R., Kagedal, B., Lundquist, P., Nilsson, L., Tylleskar, T., and Rosling, H., Cyanide detoxification in rats exposed to acetonitrile and fed a low protein diet, *Fund. Appl. Toxicol.,* 32, 66, 1996.
41. Ruzo, L.O., Unai, T., and Casida, I.E., Decamethrin metabolism in rats, *J. Agric. Food Chem.,* 26, 918, 1978.
42. Weuffen, W., Jess, G., Julich, W.D., and Bernhardt, D., Untersuchungen zur Beziehuny zwischen der 2-Iminothiazolidin-4carbon säure und dem thiocyanatstoffwechsel des Mecrschweinschens, *Die Pharmazie,* 35, 221, 1980.
43. Bitner, R.S., Kanthasamy, A., Isom, G.E., and Yim, G.K.W., Seizures and selective CA-1 hippocampal lesions induced by an excitotoxic cyanide metabolite, 2-iminothiazolidine-4-carboxylic acid, *Neurotoxicology,* 16, 115, 1995.
44. Bitner, R.S., Yim, G.K.W., and Isom, G.E., 2-Iminothiazolidine-4-carboxylic acid produces hippocampal CA-1 lesions independent of seizure excitation and glutamate receptor activation, *Neurotoxicology,* 18, 3215, 1997.
45. Huang, J., Niknahad, H., Kahn, S., and O'Brien, P.J., Heptocyte-catalysed detoxification of CN by L- and D-cysteine, *Biochem. Pharmacol.,* 55, 1983, 1998.
46. Ferger, D. and Krieglstein, J., Determination of intracellular Ca^{2+} concentration can be a useful tool to predict neuronal damage and neuroprotection properties of drugs, *Brain Res.,* 932, 87, 1996.
47. Meier, T.J., Ho, D.Y., Parks, T.S., and Sapolsky, R.M., Gene transfer of calbindin A28K < DNA via herpes simplex virus amplicon vector decreases cytoplasmic calcium ion response and enhances neuronal survival following glutamatergic challenge but not following CN, *J. Neurochem.,* 71, 1013, 1998.
48. Stout, A.K., Raphael, H.M., Kanterewicz, B.I., Klann, E., and Reynolds, I.J., Glutamate-induced neuron death requires mitochondrial calcium uptake, *Nat. Neurosci.,* 1, 366, 1998.
49. Sengpiel, B., Dreis, E., Krieglstein, J., and Prehn, J.H., NMDA-induced superoxide production and neurotoxicity in cultured rat hippocampal neurons: Role of mitochondria, *Eur. J. Neurosci.,* 10, 1903, 1998.
50. Adjilore, O.A. and Sapolsky, R.M., Application of silicon microphysiometry to tissue slices: Detection of metabolic correlates of selective vulnerability, *Brain Res.,* 752, 99, 1997.
51. Zhu, P.J. and Krnjevic, K., Persistent block of CA1 synaptic function by prolonged hypoxia, *Neuroscience,* 90, 759, 1999.
52. Ozetecan, T., Kocak-Toker, N., and Aykag-toker, G., *In vitro* effects of peroxynitrite on human spermatozoa, *Andrologia,* 31, 195, 1999.

53. Violi, F., Marino, R., Milite, M.T., and Loffredo, L., Nitric oxide and its role in lipid peroxidation, *Diabetes/Metab. Res. Rev.,* 15, 283, 1999.
54. Virag, L., Scott, G.S., Antal-Szalmas, P., O'Connor, M., Ohshima, H., and Szabo, C., Requirement of intracellular calcium mobilization for peroxynitrite-induced poly (ADP-ribose) synthetase activation and cytotoxicity, *Molec. Pharmacol.,* 56, 824, 1999.
55. Bockowski, J., Lisdero, C.L., Lanone, S., Samb, A., Carreras, M.C., Boveris, A., Aubier, M., and Poderoso, J.J., Endogenous peroxynitrite mediates mitochondrial dysfunction in rat diaphragm during endotoxemia, *FASEB J.,* 13, 1637, 1999.
56. Supinski, G., Stotan, D., Callahan, L.A., Nethery, D., Nosek, T.M., and DiMarco, A., Peroxynitrite induces contractile dysfunction and lipid peroxidation in the diaphragm, *J. Appl. Physiol.,* 87, 743, 1999.
57. Distasi, A.M., Mallozzi, C., Macchia, G., Petrucci, T.C., and Minetti, M., Peroxynitrite induces tyrosine nitration and modulates tyrosine phosphorylation of synaptic proteins, *J. Neurochemistry,* 93, 927, 1999.
58. Yu, Z.F., Bruce-Keller, A.J., Goodman, Y., and Mattson, M.P., Uric acid protects neurons against excitotoxic and metabolic insults in cell culture and against focal ischemic brain injury *in vivo, J. Neurosci. Res.,* 53, 613, 1998.
59. Gunasekar, P.G., Borowitz, J.L., and Isom, G.E., Cyanide-induced apoptosis involves NMDA receptor-mediated oxidative stress and NF-$_k$B linked activation of caspase-3 protease, submitted for publication.
60. Fujiwara, N., Higashi, H., Shimoji, K., and Yoshimura, M., Effects of hypoxia on rat hippocampal neurons *in vitro, J. Physiol.,* 384, 131, 1987.
61. LeBlond, J. and Krnjevic, K., Hypoxic changes in hippocampal neurons, *J. Neurophysiol.,* 62, 1, 1989.
62. Haddad, G.G. and Donnelly, D.F., O_2 deprivation induces a major depolarization in brain stem neurons in the adult but not in the neonatal rat, *J. Physiol.,* 429, 411, 1990.
63. Wind, T., Prehn, J.H., Peruche, B., and Krieglstein, J., Activation of ATP-sensitive potassium channels decreases neuronal injury caused by chemical hypoxia, *Brain Res.,* 751, 295, 1997.
64. Latha, M.V., Borowitz, J.L., Yim, G., Kanthasamy, A., and Isom, G.E., Plasma membrane hyperpolarization by cyanide: Role of potassium channels, *Archiv. Toxicol.,* 68, 37, 1994.
65. Koyama, S., Jin, Y., and Akaike, N., ATP-sensitive and Ca^{2+}-activated K^+ channel activities in the rat locus coeruleus neurons during metabolic inhibition, *Brain. Res.,* 828, 189, 1999.
66. Yang, J.J., Chou, Y.C., Lin, M.T., and Chiu, T.H., Hypoxia-induced differential electrophysiological changes in rat locus coeruleus neurons, *Life. Sci.,* 61, 1763, 1997.
67. Zhu, P.J. and Krnjevic, K., Adenosine release mediates cyanide-induced suppression of CA1 neuronal activity, *J. Neurosci.,* 17, 2355, 1997.
68. Maire, J., Medilanski, J., and Straub, R., Release of adenosine, inosine, and hypoxanthine from rabbit non-myelinated nerve fibers at rest and during activity, *J. Physiol.,* 357, 67, 1984.
69. Kurbat, J., Buchanan, R., Wolff, S. and Yoon, K. W., Cyanide mediated adenosine release from rat hippocampal neurons, *Soc. Neurosci.,* Abstr. 19, 1961.
70. Green, R. and Haas, H., Adenosine actions on CA1 pyramidal neurons in rat hippocampal slices, *J. Physiol.,* 366, 119, 1985.
71. Trussel, L. and Jackson, M., Dependence of an adenosine-activated potassium current on a GTP-binding protein in mammalian central neurons, *J. Neurosci.,* 7, 3306, 1987.
72. Hammerstrom, A.K. and Gage, P.W., Inhibition of oxidative metabolism increases persistent sodium current in rat CA1 hippocampal neurons, *J. Physiol.,* 510, 935, 1998.

73. Friedman, J.E. and Haddad, G.G., Anoxia induces an increase in intracellular sodium in rat central neurons *in vitro, Brain Res.*, 663, 329, 1994.
74. Haigney, M.C., Lakatta, E.G., Stern, M.D., and Silverman, H.S., Sodium channel blockade reduces hypoxic sodium loading and sodium-dependent calcium loading, *Circulation,* 90, 391, 1994.
75. Jiang, S., Liu, Z., and Zhuang, X., Effect of procaine hydrochloride on CN intoxication and its effect on neuronal calcium in mice, *Toxicol. Appl. Pharmacol.,* 150, 32, 1998.
76. Baskin, S.I., Wilkerson, G., Alexander, K., and Blitstein, A.G., *Clinical and Experimental Toxicology of Cyanides,* Ballantyne, B. and Marrs, T. C., Eds., IOP Publishing Ltd., Bristol, England, 1987, 138.
77. Kondo, R.P., Apstein, C.S., Eberli, F.R., Tillotson, D.L. and Suter, T.M., Increased calcium loading and inotropy without greater cell death in hypoxic rat cardiomyocytes, *Am. J. Physiol.,* 275, H2292, 1998.
78. Elliott, A., Smith, G., Eisner, D., and Allen, D., Metabolic changes during ischemia and their role in contractile failure in isolated ferret heart, *J. Physiol.,* 454, 467, 1992.
79. Kupriynov, V., Yang, L., and Deslauriers, R., Cytoplasmic phosphates in Na^+-K^+ balance in KCN-poisoned rat heart: a ^{87}Rb- ^{23}Na- and ^{31}P-NMR study, *Am. J. Physiol.,* 270, H1303, 1996.
80. Smith, G., Donoso, P., Bauer, C., and Eisner, D., Relationship between intracellular pH and metabolite concentrations during metabolic inhibition in isolated ferret heart, *J. Physiol.,* 492, 11, 1993.
81. Shanbaky, N. and Borowitz, J., Effect of pH on the response of adrenal medulla to various agents, *J. Pharmacol. Exp. Ther.,* 207, 998, 1978.
82. Maduh, E., Borowitz, J., Turek, J., Rebar, A., and Isom, G., Cyanide-induced neurotoxicity: calcium mediation of morphological changes in neuronal cells, *Toxicol. Appl. Pharmacol.,* 103, 214, 1990.
83. Lee, J. and Allen, D., Mechanisms of acute ischemic contractile failure of the heart: Role of intracellular calcium, *J. Clin. Invest.,* 88, 361, 1991.
84. Stewart, L., Deslauriers, R., and Kupriyanov, V., Relationships between cytosolic [ATP], [ATP]/[ADP] and ionic fluxes in the perfused rat heart, a ^{31}P, ^{23}Na, ^{87}Rb NMR study, *J. Mol. Cell Cardiol.,* 26, 1377, 1994.
85. Adler, M., Lebeda, F., Kaufmann, F., and Deshpande, S., Mechanism of action of sodium cyanide on rat diaphragm muscle, *J. Appl. Toxicol.,* 19, 411, 1999.
86. Okazaki, I.J. and Moss, J., Characterization of glycosysphosphatidyl inositol-anchored, secreted and intracellular vertebrate mono-ADP-ribosyltransferases, *Ann. Rev. Nutrition,* 19, 485, 1999.
87. Moss, J. and Vaughan, M., Activation of toxin ADP-ribosyltransferases by eukaryotic ADP-ribosylation factors, *Molec. Cell Biochem.,* 193, 153, 1999.
88. Mayer-Kuckuk, P., Ullrich, O., Ziegler, M., Grune, T., and Schweiger, M., Functional interaction of poly (ADP-ribose) with the 20S proteasome *in vitro, Biochem. Biophys., Res. Comm.,* 259, 576, 1999.
89. Stout, A.K. and Woodward, J.J., Mechanism for nitric oxide's enhancement of NMDA-stimulated [^{3}H] norepinephrine release from rat hippocampal slices, *Neuropharmacology,* 34, 923, 1995.
90. Lee, Y.J. and Shacter, E., Oxidative stress inhibits apoptosis in human lymphoma cells, *J. Biol. Chem.,* 274, 19792, 1999.
91. Masmoudi, A., Mandel, P., and Maluiya, A., Unexpected stimulation of mitochondrial ADP ribosylation by CN, *FEBS Lett.,* 237, 150, 1988.

92. Tsuyama, S., Fujita, H., Hijikata, R., Okamoto, H., and Takanaks, S., Effects of mono-ADP-ribosylation on cytoskeletal actin in chromaffin cells and their release of catecholamine, *Int. J. Biochem. Cell Biol.,* 31, 601, 1999.
93. Jones, D.H., Bax, B., Fensome, A., and Crockcroft, S., ADP ribosylation factor 1 mutants identify a phospholipase D effector region and reveal that phospholipase D participates in lysosomal secretion but is not sufficient for recruitment of coatomer 1, *Biochem. J.,* 341, 185, 1999.
94. Eliasson, M., Samper, K., Mandir, A., Hurn, P., Traystman, R., Bao, J., Peiper, A., Wang, Z., Dawson, T., Snyder, S., and Dawson, V., Poly (ADP-ribose) polymerase gene disruption renders mice resistant to cerebral ischemia, *Nat. Med.,* 3, 1089, 1997.
95. Venihaki, M., Gravanis, A., and Margioris, A., Opioids inhibit dopamine secretion from PC12 rat pheochromocytoma cells in a naloxone-reversible manner, *Life Sci.,* 58, 75, 1996.
96. Anderson, R. and Harland, W., *Forensic Toxicology,* Oliver, J. S., Ed., Droon Helan, London, 1989, 289.
97. Maehly, A. and Swensson, A., Cyanide and thiocyanate levels in blood and urine of workers with low grade exposure to CN, *Int. Arch. Arbeitsmed.,* 27, 195, 1970.
98. Moncala, S., Palmer, R.M.J., and Higgs, E.H., Nitric oxide: Physiology, pathophysiology and pharmacology, *Pharmacol. Rev.,* 43, 109, 1991.
99. Westley, J., *Cyanide in Biology,* Vennesland, B., Conn, E., Knowles, C., Westley, J., and Wissing F., Eds., Academic Press, New York, 1981, 6l.
100. Sun, P., Borowitz, J., Kanthasamy, A., Kane, M., Gunasekar, P., and Isom, G.E., Antagonism of cyanide toxicity by isosorbide dinitrate: Possible role of nitric oxide, *Toxicology,* 104, 105, 1995.
101. Moore, S., Norris, J., Ho, I., and Hume, L., The efficacy of α-ketoglutaric acid in the antagonism of cyanide intoxication., *Toxicol. Appl. Pharmacol.,* 82, 44, 1986.

11 Riot-Control Agents

Harry Salem, Eugene J. Olajos, and Sidney A. Katz

CONTENTS

I. Introduction 322
II. Historical Perspectives 323
III. Chemistry of Selected Riot-Control Agents 326
A. Chlorobenzylidene Malononitrile (CS) 326
B. Dibenz(b,f)1:4-Oxazepine (CR) 327
C. Chloroacetophenone (CN) 327
D. Oleoresin Capsicum (OC) 328
E. Adamsite (DM) 328
IV. Clinical Aspects of Riot-Control Agents 328
V. Toxicology of Riot-Control Agents 329
VI. Ocular and Cutaneous Effects of Riot-Control Agents 329
VII. Specific Riot-Control compounds 332
A. o-Chlorobenzylidene Malononitrile (CS) 332
1. Mammalian Toxicology 333
2. Ocular and Cutaneous Effects 335
3. Reproductive and Developmental Effects 335
4. Genotoxicity and Carcinogenicity 336
5. Metabolism, Metabolic Fate, and Mechanisms 336
6. Human Toxicology 338
B. Dibenz(b,f)1:4-Oxazepine (CR) 341
1. Mammalian Toxicology 342
2. Ocular and Cutaneous Effects 343
3. Reproductive Toxicity and Developmental Effects 344
4. Genotoxicity and Carcinogenicity 344
5. Clinical Chemistry 345
6. Metabolism, Metabolic Fate, and Mechanisms 345
7. Human Toxicology 346
8. Ocular and Cutaneous Effects (Human) 347
C. Chloroacetophenone (CN) 349
1. Mammalian Toxicology 349
2. Ocular and Cutaneous Effects 351
3. Genotoxicity and Carcinogenicity 351
4. Metabolism, Metabolic Fate, and Mechanisms 351
5. Human Toxicology 352

0-8493-0872-0/01/$0.00+$.50

D. Oleoresin Capsicum (OC) 352
1. Mammalian Toxicology 353
2. Ocular and Cutaneous Effects.... 355
3. Mutagenicity and Carcinogenicity 356
4. Metabolism, Metabolic Fate, and Mechanisms 357
5. Human Toxicology 357
E. Diphenylaminochloroarsine (Adamsite) 359
1. Toxicology and Physiological Effects.... 359
2. Human Toxicology 360
VIII. Summary 361
References.... 362

I. INTRODUCTION

Riot-control agents, chemicals that produce disabling physiological effects when they come in contact with the eyes or skin or when inhaled, are a subset of a larger group of chemicals known as "harassing agents." These compounds have the capability of causing intense sensory irritation and marked irritation of the skin and mucous membranes of the eye and respiratory tract. Riot-control agents are peripheral sensory irritants and are collectively referred to as lacrimators. In common parlance they are known as "tear gases." Peripheral sensory irritants are substances that pharmacologically interact with sensory nerve receptors in skin and mucosal surfaces at the site of contamination, resulting in local sensation (discomfort or pain) with associated reflexes. This is a normal biological response giving warning and protective functions. For example, in the eye, sensory irritation results in pain in the eye (warning) and excess reflex lacrimation and blepharospasm (protection). The response is usually concentration-related and disappears on removal of the sensory irritant stimulus. The intense lacrimation best typifies the biological response to such compounds; however, it must be kept in mind that riot-control compounds have multiple physiological effects. A lacrimatory compound may also elicit pulmonary irritation and/or nausea and vomiting. Generally, classification of military chemicals and chemical agents is based on a salient physiologic action although classification may also be based on use, physical state, or persistency.[1–4] Sartori was of the opinion that the physiological classification of chemical agents and military chemicals, although widely used, was less exact than other classification schemes.[4] He long ago suggested that classification should be based according to the mechanism of action on the organism.

Physiologically, riot-control agents may be classified as to type: lacrimators, which primarily cause eye irritation and lacrimation; vomiting agents, which additionally cause vomiting; and sternutators, which mainly cause uncontrollable sneezing and coughing. Riot-control agents have also been referred to as irritants or irritating agents,[5,6] harassing agents,[7–10] and incapacitating agents or short-term incapacitants.[9–11] The aforementioned categories are general classifications or have special meaning in terms of military usage and may not represent useful equivalents. As a case in point, Cookson and Nottingham are of the opinion that vomiting agents are incorrectly described as riot-control agents and should be considered as a separate category of military chemicals.[12] Furthermore, it must be recognized that physiologically based classification of chemical agents and compounds of military interest is by no means a rigid one—i.e., the classifying of a military compound, as a lung irritant for instance, does not mean it cannot act as a lacrimatory compound. The issue of classification may never be fully resolved; however, a system of classification nevertheless serves to provide some sort of basis for comparison of chemical agents and compounds. The reader is referred to an excellent overview by Verwey concerning criteria to distinguish riot-control agents from chemical warfare agents, as well as a discussion focusing on the concepts of "harassing," "irritating," and "incapacitating."[13] Characteristics common to riot-control agents are: (1) a rapid onset of effects; (2) a relatively short duration of effects after cessation of exposure; and (3) a relatively high safety ratio. Ideally, riot-control agents should produce "harassing effects" that are relatively benign with a low incidence of casualties in riot-control situations. They should have very low acute toxicity and possess physical and toxicological properties that ensure minimal risks.

A distinction has been made between chemical warfare agents and military chemicals and is recognized in military field and technical manuals and in chemical warfare literature. The term military chemical compound excludes chemical warfare agents. Chemical warfare agents include the following categories: nerve agents [e.g., sarin (GB), soman (GD), and VX]; blister agents [e.g., mustard (HD) and lewisite (L)]; choking agents/lung irritants [i.e., phosgene (CG)]; blood agents [e.g., hydrogen cyanide (AC) and cyanogen chloride (CK)]; and incapacitating agents [e.g., adamsite (DM) and 3-quinuclidinyl benzilate (BZ)]. Military chemical compounds include the following groupings: riot-control agents [e.g., chloroacetophenone (CN), dibenz (b,f)-1:4 oxazepine (CR), and o-chlorobenzylidene malononitrile (CS)]; training agents [e.g., CN]; smoke materials [e.g., fog oil (SGF) and white phosphorus (WP)]; and herbicides [e.g., 2,4,5-trichlorophenoxy acetic acid (2,4,5-T) and arsenic trioxide]. Further to this discussion, it should be stated that the United States does not consider riot-control agents to be chemical weapons; however, some other countries do not draw such a distinction. Official American sources such as military field and technical manuals (i.e., Army FM 8–285) provide definitions for chemical agent, military chemical, and riot-control agent.[14] Sidell, in writing about riot-control agents, refers to the United States' position on these compounds and states the following: "The United States does not recognize riot-control agents as chemical warfare agents as defined in the Geneva Convention of 1925."[15] Despite considerable focus and debate on the definition and classification of riot-control agents, recently published literature on the subject matter has not provided clear distinctions on the classification of chemical warfare agents and riot-control compounds.[11,16] Nonetheless, the currently held official policy on riot-control agents by the United States is that riot-control agents are not chemical warfare agents.[17]

II. HISTORICAL PERSPECTIVES

Lacrimatory and irritant compounds, with a history dating from World War I, have been used in riot-control and civil disturbances, military exercises and training, and as chemical warfare agents. A listing of these chemicals and their use application is presented in Table 11.l. Chloropicrin (trichloronitromethane, Green Cross, PS) was a well-known chemical substance prior to World War I, having been first synthesized circa 1850. It was used both as a harassing agent and lethal chemical in the First World War. In fact, chloropicrin was one of several lethal agents—the others being chlorine, phosgene, and trichlorethylchloroformate. Adamsite (DM, diphenylaminochlorarsine), an arsenic-based compound, was developed as a chemical variation of diphenylchloroarsine for use during World War I. It is classified militarily as a vomiting agent and as a sternutator and was used as a riot-control agent after the war. According to Swearengen, ethyl bromoacetate was the first riot-control agent, based on its use in Paris in 1912.[18] This tear gas was again utilized in the 1970s.[19] Tear gases used in World War I included such chemicals as acrolein (papite), bromoacetone (BA,B-stoff), bromobenzyl cyanide (BBC,CA), chloroacetone (A-stoff), diphenylaminochloroarsine (DM), and xylyl bromide (T-stoff). Xylyl bromide was an early war gas, and bromoacetone, a highly potent lacrimator, was the most widely used lacrimatory agent in World War I. Chloroacetophenone ("mace"),* discovered in 1871, was not used during World War I; however, American investigators were certain of its potential utility as a tear gas and worked out a satisfactory process of manufacture.

Military experience with harassing agents encouraged the utilization of these compounds in law enforcement operations. However, many of the military harassing agents are not suited to law enforcement use either, because the risk of fatalities or the likelihood of total incapacitation is too great. The development of modern riot-control agents has been driven by the need to develop safe and effective compounds that can be easily disseminated. Riot-control agents are intended to simply temporarily disable—the intense irritant effects lead to a more or less pronounced incapacitation. Further discussion on "incapacitating" effects of riot-control agents can be found in the literature.[12,20–22] A systematic search of candidate compounds suitable for riot control and temporary incapacitation was in place at the conclusion of World War I. Despite the evaluation of a considerable number of candidate compounds, interest still centered on CN, DM, and a handful of promising compounds such as CR and CS. The war gas bromobenzyl cyanide (BBC, CA) saw early use as a riot-control agent. However, CN and DM were the harassing agents of choice and, at the time of World War II, considerable stockpiles of CN and DM existed. Although adamsite (DM) has been used as a riot-control agent,[1] chloroacetophenone (CN) became the lacrimator of choice for police use. Chlorobenzylidene malononitrile (CS), synthesized in the late 1920s by Corson and Stoughton, was not developed as a riot-control agent until the 1950s.[23] CS has largely replaced CN and is the tear gas (lacrimator) most widely used by law enforcement personnel. Dibenz(b,f)1:4-oxazepine (CR), a riot-control agent of relatively recent origin, is used only to a very limited extent. However, it may see greater use because CR has greater potency and lower toxicity than some of the other riot-control agents. The compound 1-methoxy-,3,5-cycloheptatriene (tropilidene, CHT), a highly volatile and unstable

*Mace® is a liquid mixture containing CN (active ingredient), hydrocarbons, and freon propellant in 1,1,1-trichloroethane.

TABLE 11.1
Lacrimatory Agents and Irritants: Application/Use Information

Chemical	Synonyms	Code[a]	Application/Use Current	Application/Use Former
σ-Chlorobenzylidene malononitrile	2-Chlorobenzalmalononitrile	CS	Riot control	Riot control
Dibenz (b,f)-1:4 oxazepine	CR	CR	Riot control	Riot control
ω-Chloroacetophenone	2-Chloroacetophenone	CN	Riot control	War gas
Diphenylaminochloroarsine	10-Chloro-5,0-dihydro-phenarsazine	DM	Obsolete	War gas
Acrolein	2-Propenal	Papite	Intermediate[b]	War gas
Benzyl bromide	1-Bromotoluene	(−)	Intermediate[c]	Intermediate
Benzyl iodide	1-Iodotoluene	(−)	Reagent	Experimental tear agent
Bromoacetone	1-Bromo-2-propanone	BA	Reagent	War gas
Bromobenzyl cyanide	α-Bromo- α-tolunitrile	BBC	Agricultural chemical	Riot control
Chloroacetone	1-Chloro-2-propanone	A-stoff	Intermediate[b]	War gas
Chloropicrin	Trichloronitromethane	PS	Fumigant war gas	
Ethyl bromoacetate	Ethyl 2-bromoacetate	EBA	Intermediate[d]	Riot control
Ethyl iodoacetate	Iodoacetic acid, ethyl ester	KSK	Reagent	Experimental tear gas
Iodoacetone	1-iodo-2-propanone	(−)	Reagent	Experimental tear gas
Oleoresin of capsicum	OC pepper spray	(−)	Food additive incapacitant	Food additive
Phenyl carbylamine chloride	Phenylimidocarbonyl chloride	(#)[e]	Reagent	War gas
Tropilidene	1-Methoxy-1,3,5-cycloheptatriene _cycloheptatriene	CHT	Experimental tear gas	Experimental tear gas
Xylyl bromide	α-Bromoxylene	T-stoff	Reagent	War gas

[a]Military code or identifier.

[b]Chemical intermediate for various industrial chemicals and pharmaceuticals.

[c]Chemical intermediate for certain industrial chemicals.

[d]Chemical intermediate for pharmaceuticals.

[e](#) Military designation = Green Cross I.

liquid, has also been studied and evaluated as a riot-control agent. Tropilidene has been demonstrated to be a potent irritant with physiological effects characteristic of riot-control agents. Its toxicity is generally similar to that of CR. The naturally occurring compound capsaicin may have potential use as a riot-control agent—"pepper spray" is currently available over the counter for personal protection and is used by postal carriers for repelling animals, and by campers as a bear repellant.

III. CHEMISTRY OF SELECTED RIOT-CONTROL AGENTS

A considerable number of chemicals have been developed for riot control and law enforcement use. The most commonly available riot-control agent is chlorobenzylidene malononitrile (CS), which replaced chloroacetophenone (CN), the latter agent having replaced adamsite (DM). Oleoresin capsicum (OC), in various formulations, has gained popularity in law enforcement and riot-control use. The structures of riot-control agents CS, CR, CN, and DM are depicted in Figure 11.1, and Table 11.2 summarizes selected physicochemical properties of several lacrimatory agents. The common riot-control agents are all solids in pure form, although lacrimatory agents such as acrolein, chloroacetone, and tropilidene, which have been considered and/or used for riot control, are liquids. Of the modern riot-control agents, CS hydrolyzes rather rapidly; however, other compounds such as dibenz (b,f) 1:4-oxazepine (CR) are particularly stable and persist for prolonged periods. The common riot-control agents are alkylating agents that react with nucleophilic sites of macromolecular moieties. A brief description of the chemicophysical properties of the common riot-control agents is presented in Table 11.2.

A. Chlorobenzylidene Malononitrile (CS)

Chlorobenzylidene malononitrile has the military designation CS. It is also known as β,β-dicyano-ortho-chlorostyrene, 2-chlorophenylmethylenepropanedinitrile, and o-chlorobenzalmalononitrile. CS is a white solid with a molar mass of 188.5 corresponding to a molecular formula of $C_{10}H_5N_2Cl$. The molar solubility in water at 20°C

dibenz (b,f) - 1:4 - oxazepine (CR)

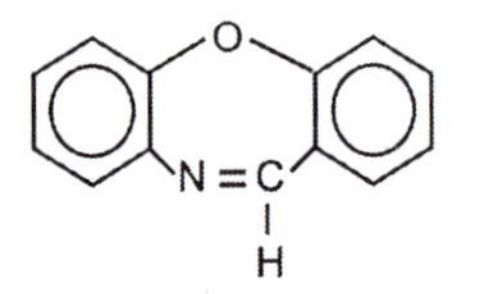

2 - chlorobenzylidene malononitrile (CS)

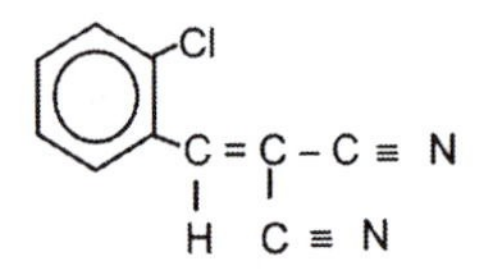

1 - chloroacetophenone (CN)

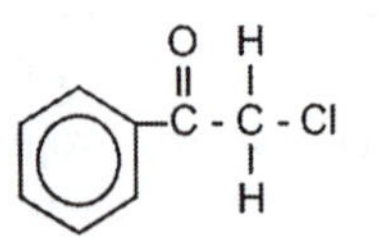

10 - chloro - 5,10 - diphenylarsazine (DM)

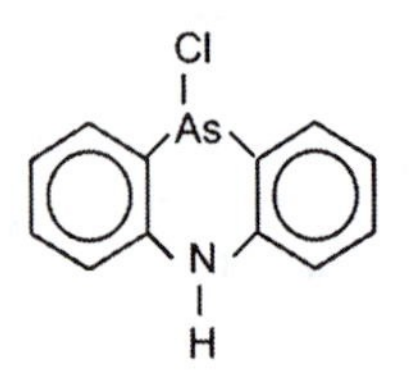

FIGURE 11.1 Structures of CS, CR, CN, and DM.

TABLE 11.2
Selected Chemical and Physical Properties of Lacrimatory Agents

Compound	CS	CR	CN	OC[a]	DM	CA	PS
Molecular Wt	188.5	195.3	154.5	(−)	277.5	196.0	164.5
Melting Point	93°C	72°C	54°C	(−)	195°C	25.5°C	−69°C
Vapor Pressure	0.00034	0.00059	0.0054	(−)	2×10^{-13}	0.011	18.3
Volatility	0.71/25	0.63/25	1.06/52	(−)	(−)	271/30	(−)
Solubility	IOC	IOC	IOC	(−)	IO	IO	IO
Hydrolysis	Slow	V. slow	Slow	(−)	Inhibited	Slow	None

Note: CS = o-chlorobenzylidene malononitrile; CR = dibenz-(b,f)-1:4 oxazepine; CN = chloroacetophenone; OC = oleoresin capsicum; DM = adamsite; CA = bromobenzylcyanide; PS = chloropicrin. Vapor pressure at 20°C (68°F) (mmHg). Volatility, mg/m^3/°C for other than 20°C. Solubility, I = limited in water, O = soluble in organics, C = soluble in chlorinated organics. Hydrolysis (rate of hydrolysis). (−) Denotes no value.

[a]Oleoresin capsicum is a mixture—no values.

is 2.0×10^{-4} mol/l (= ~4 mg/100 ml). Dissolved CS is rapidly hydrolyzed; however, CS may persist in the environment because its solubility in water is limited. The melting and boiling points are 93–96°C and 310–315°C, respectively. The vapor is several times heavier than air, and the vapor pressure of the solid is 0.00034 mm Hg at 20°C.

B. Dibenz (b,f)1:4-Oxazepine (CR)

The military designation for dibenz (b,f) 1:4-oxazepine is CR. This compound is a pale yellow solid with a molar mass of 195.3 corresponding to a molecular formula of $C_{13}H_9ON$. The molar solubility in water at 20°C is 3.5×10^{-4} mol/l (= ~7 mg/100 ml). The melting and boiling points are 72°C and 335°C, respectively. The vapor is 6.7 times heavier than air, and the vapor pressure of the solid is 0.00059 mm Hg at 20°C. CR is a stable chemical and may persist for prolonged periods in the environment.

C. Chloroacetophenone (CN)

Chloroacetophenone is also referred to as ω-chloroacetophenone, α-chloroacetophenone, phenacyl chloride, 2-chloro-l-phenylethanone, and phenyl chloromethyl ketone. It has the military designation CN. Chloroacetophenone is a white solid with a molar mass of 154.5 corresponding to a molecular formula of C_8H_7OCl. The molar solubility at 20°C is 4.4×10^{-3} mol/l (= 68 mg/l00 ml). Melting and boiling points are 54°C and 247°C, respectively. Density of the solid is 1.318 g/cm^3 at 0°C, and density of the liquid is 1.187 g/m^3 at 58°C. The vapor is 5.3 times heavier than air. The vapor pressure of the solid is 2.6×10^{-3} torr at 0°C, 4.1×10^{-3} torr at 20°C, and 15.2×10^{-3} torr at 50°C.

D. Oleoresin Capsicum (OC)

Oleoresin capsicum is a reddish-brown, oily liquid obtained by extracting dried, ripe fruit of chili peppers, usually *Capsicum annuum* or *Capsicum frutescenes*. Oleoresin capsicum is a mixture of many compounds. Its composition is variable and depends on factors such as maturity of the fruit and the environment in which the plants are grown, as well as the conditions of the extraction. More than 100 compounds have been identified in oleoresin capsicum. Among the branched- and straight-chain alkyl vanillyamides isolated from oleoresin capsicum, capsaicin (8-methyl-N-vanillyl-6-noneanamide) is the major constituent. Capsaicin is the major pungent component in many peppers, and it is particularly noted for its irritant properties. Depending on the variety of chili pepper, oleoresin capsicum contains from 0.01 to 1.0% capsaicinoids on a dry mass basis. Some of the capsaicinoids found in oleoresin capsicum are capsaicin (~70%), dihydrocapsaicin (~20%), norhydrocapsaicin (~7%), homocapsaicin (~1%), and monodihydrocapsaicin (~1%). Other components of oleoresin capsicum may also possess irritant properties (e.g., phenolic compounds, acids, and esters).

E. Adamsite (DM)

Diphenylaminochloroarsine (phenarsazine chloride, adamsite) has the military designation, DM. Adamsite is a yellowish and odorless solid that is very stable in pure form. The melting point is 195°C, and the vapor pressure is negligible (2×10^{-13} mm Hg at 20°C). As a solid, the rate of hydrolysis is not significant, owing to the formation of an oxide coating; however, the rate of hydrolysis is rapid when as an aerosol. DM has a molecular weight of 277.5 with the formula $C_6H_4(AsCl)(NH)C_6H_4$.

IV. CLINICAL ASPECTS OF RIOT-CONTROL AGENTS

Riot-control agents exert their effects on eyes and skin and can enter the body via the respiratory tract, skin, and gastrointestinal tract. The clinical symptoms following exposure to riot-control agents are the consequence of these agents' ability to cause intense sensory irritation. Most of the symptoms are felt within 10 to 30 s. The eyes are affected almost immediately with copious lacrimation, blepharospasm, conjunctivitis, and pain. Nasal effects include rhinorrhea, itching, and pain. A stinging or burning sensation of the mucosal surfaces is also experienced. Sneezing, coughing, and increased respiratory tract secretions are accompanied by a burning sensation and chest tightness. There is a burning sensation of the skin followed by erythema. The more severe effects such as marked coughing, retching, and vomiting may occur if an individual remains in a riot-control agent atmosphere following the onset of irritation. Anxiety and panic are reactions that are commonly noted on exposure to these compounds. The intense physical discomfort and anxiety can produce cardiovascular changes such as increased blood pressure. After cessation of exposure, most symptoms persist for a brief period, and by 30 min, most symptoms have completely abated. Conjunctivitis can remain for up to 30 min. On exposure to massive doses, which can be achieved with aggressive use of certain riot-control agents such as CN, severe effects involving the eyes (i.e., corneal damage) and lungs (e.g., hemorrhaging, edema, and congestion) can result. These agents may also complicate and exacerbate existing conditions such as bronchitis and asthma.

V. TOXICOLOGY OF RIOT-CONTROL AGENTS

Riot-control agents are potent sensory irritants of low toxicity that produce dose- and time-dependent acute, site-specific toxicity (refer to Figure 11.2 and Tables 11.3 and 11.4). These agents have been described as non-lethal. Exposures to these compounds involve the ocular, inhalation, and cutaneous routes and indirectly via the oral route. These compounds primarily act on the eye, which is the most sensitive target organ; however, most of these compounds will also cause effects involving the respiratory tract and skin. These agents can cause several or all of the effects on these target organs to a greater or lesser extent. The immediate effects on exposure to riot-control agents are: intense irritation of the eyes; marked irritation of the nose, throat, and lungs; and irritation of the skin. The margin of safety between the amount eliciting an intolerable effect and that which may cause serious adverse effects is large. For example, the lethal amount for the riot-control agent CS is estimated to be 2600 times as great as the dosage required to cause temporary disabling, and that of bromobenzyl cyanide is 3000 times as great. Riot-control agents are not usually accompanied by permanent toxic effects, although the risks for deleterious effects, longer-term sequelae, or even death increase with higher exposure concentrations and greater exposure duration. Overall, the acute and short-term repeated toxicity of riot-control agents is well characterized; however, the extent of our knowledge regarding long-term and chronic effects on exposure to some of these compounds is somewhat limited. The animal and human toxicology of the main riot-control agents (CN and CS), along with CR, DM, and capsaicin is presented; each agent will be considered separately. Topics covered are comparative toxicology, dose-effect relationships, target organ effects, low-dose toxicity, biochemistry, and mechanism(s), as well as consideration of the effects in susceptible subpopulations.

VI. OCULAR AND CUTANEOUS EFFECTS OF RIOT-CONTROL AGENTS

Many compounds possess more or less lacrimatory properties that vary in intensity from mild to severe irritation, with copious flow of tears. The most characteristic feature of riot-control agents is their ability to cause immediate stinging sensation in the eyes with tearing (stimulatory effect) at low concentrations that results in a temporary disabling effect. These compounds produce stinging and lacrimation and reversible and non-injurious effects at low concentrations; however, at high concentrations, ocular damage can result with some irritants. Moderate injury to the eyes following exposure to riot-control agents consist of corneal edema, which is reversible. More serious injurious action of riot-control agents may include corneal opacification, vascularization and scarring of the cornea, and corneal ulceration. Lacrimatory agents that have been associated with ocular injury, for example, include chloroacetophenone (CN), chloracetone, and bromobenzyl cyanide. Ocular injuries are more prevalent

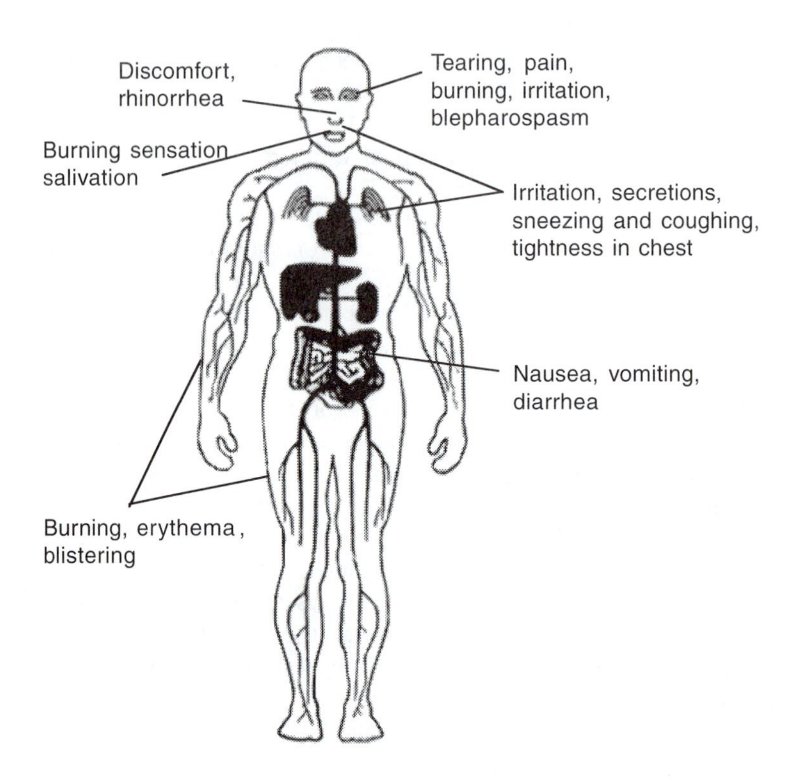

FIGURE 11.2 Site-specific toxicity of riot control agents (human body).

TABLE 11.3
Comparative Toxicity of Lacrimatory Compounds: Human Estimates[1,2,4,6,12,15,105]

Compound	LCt_{50} (mg-min/m^3)	ICt_{50} (mg-min/m^3)	Minimal Irritant ConC(mg-min/m^3)
CN	8500—25000	20—50	0.3—1
CR	>100000	~1	0.002
CS	25000—150000	5	0.004
DM	11000—35000	20—150	1—5
Acrolein	3500—7000	(−)	2—7
Bromobenzyl cyanide	8000—11000	30	0.3
Chloroacetone	>3000	(−)	18
Chloropicrin	2000	(−)	2—9
Xylyl bromide	5600	(−)	~5
Capsaicin	(−)	(−)	(−)

Note: When more than one estimate has been reported, a range is given; (−) denotes value not determined.

TABLE 11.4
Comparative Toxicity (LD_{50}) and LCt_{50}) of CR[a], CS[b], and CN[b]

Route	Species	LD_{50} (mg/kg)[c]		
		CR	CS	CN
i.v.	Mouse	112	48	81
	Rat	68	28	40
	Rabbit	47	27	29
i.p.	Rat	766	48	38
	Guinea pig	463	73	17
Oral	Mouse	4000	(−)	(−)
	Rat	5900	1284	52
	Rabbit	1760	142	118
	Guinea pig	629	212	157
		LCt_{50}(mg–min/m^3)[c]		
		CR	CS	CN
Inhalation				
Pyrotechnically generated	Mouse	203,600	76,000	(−)
	Rat	139,000	68,000	23,000
	Rabbit	160,000	63,000	15,800
	Guinea pig			
Aerosol	Mouse	169,500	67,200	18,200—73,500[d]
	Rat	428,400	88,460	3,700—18,800[d]
	Rabbit	169,000	54,100	5,840—11,480[d]
	Guinea pig	169,500	50,010	3,500—13,140[d]

Note: (−) Denotes no data; i.v. = intravenous; i.p. = intraperitoneal.

[a] Data from several sources as reported by Ballantyne.[24]

[b] Data from several sources as documented in a report by the NAS.[6]

[c] Lowest value reported.

[d] Range of values from several sources.

following use of explosive (thermal type) tear gas devices, as contrasted to solvent spray-type tear gas devices. A description of the differences between thermal and solvent spray-type devices has been provided by MacLeod.[25] Reviews regarding riot control agent-induced ocular injury have been published.[25–30] The comparative ocular irritancy of various lacrimogenic compounds is presented in Table 5. Ocular effects are described in greater detail for each of the main riot-control agents.

Although the eyes and respiratory tract are the primary organs affected by riot-control agents, the skin is also often involved. Riot-control agents are primary irritants that in low concentrations produce tingling or burning sensation and transient

TABLE 11.5
Human Ocular Irritancy and Toxicity of Lacrimatory Compounds [1,2,4,6,10,12,15,24,105]

Compound	Ocular Irritancy	Onset of Action Threshold (mg/m^3)	Irritancy[a]	Intolerable Conc. (mg/m^3)	Lethal Conc. (mg/m^3)
CN	Profound	Immediate	0.3	5–30	850
CR	Profound	Immediate	0.002	~1	10000
CS	Profound	Immediate	0.004	~3	2500
DM	High	Rapid	~1	5	650
Acrolein	High	Rapid	2–7	50	350
Benzyl bromide	High	Rapid	4	50	4500
Bromobenzyl cyanide	Profound	Rapid	0.15	0.8	350
Chloroacetone	High	Rapid	18	100	2300
Chloropicrin	High	Rapid	2–9	50	2000
Xylyl bromide	High	Rapid	~5	15	5600
Capsaicin	High	Rapid	(−)	(−)	(−)

[a]A range is given when more than one value has been reported.

[b]Minimum lethal concentration for 10-min exposure.

erythema. At higher concentrations, agents such as CN, CS, and DM can cause edema and blistering. In addition, riot-control agents can produce allergic contact dermatitis after an initial exposure. The effects of riot-control agents on the skin are successfully treated with topical steroid preparations and oral antihistamines for itching. Appropriate antibiotics are administered to treat secondary infection.

VII. SPECIFIC RIOT-CONTROL COMPOUNDS

A. O-CHLOROBENZYLIDENE MALONONITRILE) (CS)

The riot-control agent o-chlorobenzylidene malononitrile, commonly known as CS, is named after the initials of the two British chemists who prepared it in 1928, Corson and Stoughton.[23] In the 1950s, CS was developed as a potent and safe riot-control agent. The United States Army adopted CS as its standard riot-control agent in 1959. CS has been extensively studied in animals and humans, and has been widely used around the world with no verified deaths in humans following its use. CS, like CN and DM, is a crystalline, solid substance that is soluble in organic solvents, but poorly soluble in water. These compounds can be disseminated as dry powders, by thermal or explosive methods, via spraying of the molten materials or in solution with organic solvents. CS2, a micronized formulation of CS, consists of 95% CS, 5% Cab-o-Sil® (Cabot Corp) and 1% hexamethyldisilzane. The additives prevent agglomeration and produce a free-flowing powder, which can be dispersed in the dry form.[31]

1. Mammalian Toxicology

CS is a sensory irritant, highly irritating to mucous membranes that cover or line tissues of the eyes, nose, throat, and stomach. Irritation of the eyes may cause pain, excessive tearing, conjunctivitis, and uncontrolled blinking (blepharospasm). The nose and mouth may perceive a stinging or burning sensation with excessive rhinorrhea or discharge of nasal mucous. Irritation of the respiratory tract may cause tightness of the chest, sneezing, and cough, as well as increased respiratory secretions. Severe lung injury and, consequently, respiratory and circulatory failure. characterize death in experimental animals after inhalation of CS. Irritation of the gastrointestinal tract may cause vomiting and/or diarrhea. When the skin is exposed, a burning sensation may be experienced, which may be followed by inflammation and redness. In hot and humid environments, the skin effects may be more severe and result in blistering. Some or all of these effects may occur, usually within 30 s of exposure, and disappear within minutes after the exposure. The irritation during exposure is so intensive that it causes the exposed individual to seek escape from the exposure. The lethal effect of CS in animals by inhalation is caused by lung damage leading to asphyxia and circulatory failure, or from bronchopneumonia secondary to respiratory tract injury. Furthermore, pathologic changes involving the liver and kidneys following exposure to high concentrations of CS are secondary to respiratory and circulatory failure. The reader is referred to numerous publications regarding the animal and human toxicity of CS.[24,32–38]

Prior to testing in humans, chemicals and drugs must undergo extensive animal testing in multiple species and by many routes of administration. including the expected route of exposure. For CS, toxicity studies included eye and skin irritation, as well as incapacitating and lethality studies by aerosol or vapor exposure. The airborne dosage is expressed as Ct, which is the product of the concentration (C) in mg/m^3 multiplied by the exposure time (t) in minutes. The product is described as the inhalation exposure dosage in $mg \cdot min/m^3$. The terms LCt_{50} and ICt_{50} describe the airborne dosages lethal (L) or incapacitating (I) to 50% of the exposed population. Some of the animal studies on CS have been summarized by McNamara et al.[31] CS aerosols were generated by various methods, and various species were exposed to a single exposure from 5 to 90 min. The toxic signs observed in mice, rats, guinea pigs, rabbits, dogs, and monkeys on acute exposure to CS were immediate and included hyperactivity followed by copious lacrimation and salivation within 30 s in all species except the rabbit. The initial level of heightened activity subsided, and by 5 to 15 min from start of the exposure, the animals exhibited lethargy and pulmonary stress, which continued for about 1 h on cessation of exposure. All other signs abated within 5 min on removal from the exposure atmosphere. Goats, pigs, and sheep did not exhibit hyperactivity on exposure. When toxic signs were noted, these occurred following exposure via all dispersion methods. Lethality estimates, expressed as LCt_{50}, from acute exposures to CS dispersed from a 10% CS in methylene dichloride are as follows: rats, 1,004,000 $mg \cdot min/m^3$; mice, 627,000 $mg \cdot min/m^3$; and guinea pigs, 46,000 $mg \cdot min/m^3$. No deaths occurred in rabbits exposed to CS dosages of up to

47,000 mg · min/m^3. CS at dosages of up to 30,000 mg · min/m^3 did not kill any of the monkeys that had associated pulmonary dysfunction (i.e., pulmonary tularemia). The combined LCt_{50} for CS dispersed from methylene dichloride for rats, mice, guinea pigs, and rabbits was calculated to be 1,230,000 mg · min/m^3. The results from acute exposures to CS, sprayed as molten agent, were as follows:

Species	LCt_{50} (mg · min/m^3)
Rat	32,000
Mouse	42,000
Guinea pig	8,000
Rabbit	17,000
Dog	34,000
Monkey	50,000

LCt_{50} values have been rounded off.

Because of their resistance to the lethal effects of CS, LCt_{50} values could not be calculated for swine, sheep, and goats. However, the combined LCt_{50} for mice, rats, guinea pigs, rabbits, dogs, monkeys, swine, sheep, and goats was estimated to be 300,000 mg · min/m^3.

The results (LCt_{50}) from acute exposures to CS dispersed from M18 thermal grenades were as follows: rats, 164,000 mg · min/m^3 and guinea pigs, 36,000 mg · min/m^3. Results from acute exposure to CS dispersed from M7A3 thermal grenades are provided below.

Species	LCt_{50} (mg · min/m^3)
Rat	94,000
Guinea pig	66,000
Rabbit	38,000
Goat	48,000
Swine	17,000
Dog	30,000
Monkey	120,000

LCt_{50} values have been rounded off.

When the results from all of the acute exposures are combined, the LCt_{50} values are as follows: all non-rodents combined, 36,000 mg · min/m^3; all rodents combined, 79,000 mg · min/m^3; and all species combined, 61,000 mg · min/m^3. The inhalation toxicity of CS2, which is comprised of 95% CS, 5% Cal-o-Sil® , and 1% hexamethyldisilazane, has also been evaluated. The LCt_{50} results from acute exposure to CS2 are as follows: rats, 68,000 mg · min/m1^3; guinea pigs, 49,000 mg · min/m^3; dogs, 70,000 mg · min/m^3; and monkeys, 74,000 mg · min/m^3.

Repeated-exposure studies on CS via the inhalation and oral routes have been conducted and the findings reported.[39] The inhalation studies conducted on rats and dogs are highlighted. In these studies, rats and dogs were exposed to thermally dispersed CS for 4 to 5 min per day, 5 days per week for 5 weeks. The total accumulated dosage (Ct) for dogs was 17,000 mg · min/m^3 (daily Ct of 680 mg · min/m^3), and for rats the dosage was 91,000 mg · min/m^3 (daily Ct of 3640 mg · min/m^3). During the exposure, rats manifested considerable hyperactive and aggressive behavior. In CS-exposed rats, accumulated dosages of 25,000 and 68,000 mg · min/m^3 resulted in mortalities. No gross pathology was evident in any of the rats that died or the surviving animals that were sacrificed following completion of the exposures. Body weight losses in the CS-exposed animals were minimal, and no significant difference was noted in organ-to-body weight ratios following the 5-week exposure. Based on the findings, it was concluded that repeated exposure did not increase the susceptibility to the lethal effects of CS. Marrs and co-workers studied the effects of repeated inhalation doses (1 h/d, 5d/wk, for 120 days) of neat CS aerosol in rats, mice, and guinea pigs.[40] High concentrations of CS were fatal to the animals after only a few exposures. Mortality in the low- and mid-dose animals was not significantly different from controls. It was concluded that CS concentrations below 30 mg/m^3 were without deleterious effects. These concentrations are about 10 times the IC_{50} of an exposed human population in 1 min.

2. Ocular and Cutaneous Effects

The effects of CS on the rabbit eye have been examined after topical application of CS in methylene chloride.[32] All animals manifested conjunctivitis, which had completely subsided within a few hours. Moderate injury involving the cornea was not observed. Application of more concentrated solutions of CS also had no effect on the cornea.

CS is a primary irritant that elicits injurious action on the skin when topically applied either as a powder or a solution or on exposure to CS aerosol. Excessive perspiration at areas of clothing contact may contribute to the development of dermal lesions. Gutentag et al. and Bowers et al. reported the occurrence of erythema and vesiculation in human subjects topically exposed to CS powder or CS solution.[41,42] Skin exposure to CS aerosols at a concentration of 300 mg/m^3 for 45 min produced erythema and vesiculation, whereas skin lesions were not evident at an exposure duration of 30 min.[43] Workers in a CS manufacturing and processing plant developed rashes, pruritis, vesicles, and wheals, which may have been representative of sensitization and reaction to re-exposure. Rothberg[44] confirmed that both CS and CN could produce skin sensitization in guinea pigs when administered topically and intradermally.

3. Reproductive and Developmental Effects

The developmental toxicity of CS was studied by Upshall in rats and rabbits exposed via inhalation to test article at concentrations most likely to exist in riot-control situations (~10 mg/m^3).[45] Fetuses were examined for abnormalities, and no significant increase in the numbers of abnormal fetuses or resorptions were noted. However, it should be noted that the exposure conditions (low dosages and short exposure duration [5 min]) may not have been adequate to assess the fetotoxic and teratogenic potential of CS. No data were given on maternal systemic toxicity or mortality. Teratology studies are routinely conducted at dosages that produce maternal toxicity. Based on the findings of the Upshall study, it is impossible to conclude definitively that CS would not be fetotoxic and/or teratogenic under other exposure conditions.

4. Genotoxicity and Carcinogenicity

The mutagenic potential of CS and CS2, a formulation containing CS in a mixture of 5% Cab-o-Sil® and 1% hexamethyl disilizane, have been studied in microbial and mammalian bioassays. CS was positive for mutagenicity in the Ames assay, as reported by von Daniken et al.;[46] however, subsequent findings by Zeiger et al.[47] indicated questionable genotoxicity for *S. typhimurium* and those of Reitveld[48] and Wild[49] non-mutagenic for *S. typhimurium.* CS2 was negative when tested in *S. typhimurium* strains TA98, TA 1535, and TA 1537 with or without metabolic activation.[50] The mutagenic potential of CS and CS2 was also assessed in mammalian genotoxicity assays, namely, the Chinese hamster ovary (CHO) assay for induction of sister chromatid exchange (SCE) and chromosomal aberration (CA), and the mouse lymphoma L5178Y assay for induction of trifluorothymidine (Tft) resistance.[50–52] The results of the cytogenetic tests indicated that CS2 induced sister chromatid exchanges, chromosomal aberrations, and induction of Tft resistance.

CS2 was evaluated for carinogenicity in 2-year rodent bioassays.[50] Compound-related non-neoplastic lesions of the respiratory tract were noted. Pathological changes observed in CS2-exposed rats included squamous metaplasia of the olfactory epithelium and hyperplasia and metaplasia of the respiratory epithelium; hyperplasia and squamous metaplasia of the respiratory epithelium were observed in mice exposed to CS2. Neoplastic effects were not observed in either rats or mice exposed to test article. Conclusions drawn from these findings suggest that CS2 is non-carcinogenic for rats and mice.

5. Metabolism, Metabolic Fate, and Mechanisms

CS is absorbed very rapidly from the respiratory tract, and the half-lives of CS and its principal bioconversion products are reported to be extremely short.[53] The disappearance of CS follows first-order kinetics over the dose range examined. CS spontaneously hydrolyzes to malononitrile,[54] and the latter is transformed to cyanide in animal tissues.[55,56] Metabolically, CS undergoes conversion to 2-chlorobenzyl malononitrile (CSH_2), 2-chlorobenzaldehyde (oCB), 2-chlorohippuric acid, and thiocyanate.[24,53,57–60] CS and its metabolites can be detected in the blood after inhalation exposure, but only after large inhalation doses. Following inhalation exposure of rodent and non-rodent species to CS aerosol, CS and two of its metabolites 2-chlorobenzaldehyde and 2-chlorobenzyl malononitrile were detected in the blood.[53,59] Brewster and co-workers studied the fate of CS in rats following intravenous and intragastric doses.[61] Findings demonstrated that in most cases the majority of the administered dose was eliminated in the urine. *In vivo,* CS is converted to 2-chlorobenzaldehyde, which can undergo various metabolic pathways, namely oxidation to 2-chlorobenzoic acid with subsequent glycine conjugation or reduction to 2-chlorobenzyl alcohol with ultimate excretion as 2-chlorobenzyl acetyl cysteine or 1-O-(2-chlorobenzyl) glucuronic acid (refer to Figure 11.3). The principal urinary metabolites are 2-chlorohippuric acid, l-O-(2-chlorobenzyl) glucuronic acid, 2-chlorobenzyl cysteine, and 2-chlorobenzoic acid.[58] Lesser amounts of 2-chlorophenyl acetyl glycine, 2-chlorobenzyl alcohol, and 2-chlorophenyl 2-cyano propionate were also identified. In the study by Leadbeater on the uptake of CS by the human respiratory tract, 2-chlorobenzyl malononitrile was detected in trace amounts in the blood;[53] however, CS and 2-chlorobenzaldehyde were not detected after exposure to a very high dose of CS (Ct = 90 mg · min/m^3). This finding is consistent with the CS uptake studies in animals and with the maximum tolerable concentration in humans, which is below 10 mg/m^3. It is unlikely that significant amounts of CS would be absorbed via inhalation at or near the tolerable concentration.

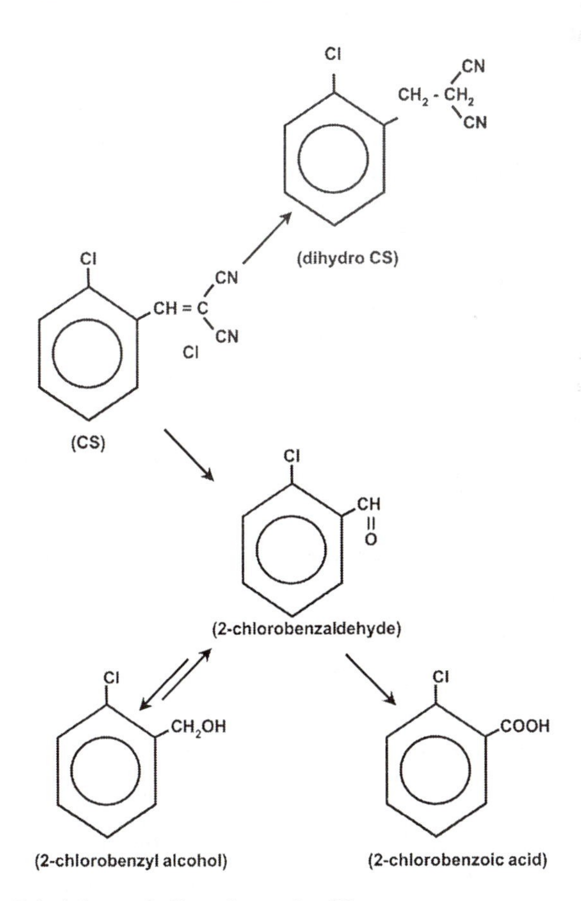

FIGURE 11.3 Principle metabolic pathways for CS.

The formation of cyanide from CS has been the subject of studies in laboratory animals and humans. Free cyanide has been detected following i.v. administration of CS in dogs exposed to lethal doses of CS, but little experimental data have been presented.[57] Studies by Frankenberg and Sorbo were conducted to determine thiocyanate excretion, blood cyanide levels, and the relationship between cyanide levels and symptomatology.[62] They determined blood cyanide levels and thiocyante excretion in mice following intraperitoneal dosing and inhalation exposure to CS. Mice were exposed to CS aerosol (20,000 mg · min/m^3), corresponding to about 0.5 LD_{50} of CS, and resulted in high levels of cyanide in blood that were reached quickly, with peak levels 4 to 16 min after injection. Equitoxic doses of malononitrile and cyanide were also evaluated for generating blood cyanide. It should be noted that CS and malononitrile possess two nitrile residues and may in theory give rise to two cyanide ions per molecule of the parent compound. This has been investigated and current evidence indicates that *in vivo,* only one cyanide radical is converted to cyanide; thus, the total amount of cyanide generated may be minimal.[24] Studies to ascertain cyanide production, measured as plasma thiocyanate levels, among human volunteers exposed to CS have been conducted.[53,63] Negligible levels of plasma thiocyanate were detected in both studies.

Sulfhydryl-containing enzymes such as lactic dehydrogenase, glutamic dehydrogenase, and pyruvic decarboxylase are alkylated by CS.[64] CS also reacts with a number of nucleophilic moieties such as glutathione. Based on studies on the effect of CS on lactic dehydrogenase, Cucinell et al. postulated that the toxic effects of CS were the result of CS inhibition of sulfhydryl-containing enzymes.[57] For example, CS is known to react with the SH group of lipoic acid, a coenzyme in the pyruvate decarboxylase system. Regarding the mechanism of action of CS, it is theorized that the irritant and painful effect of CS may be due to bradykinin release.[31,57] Ballantyne and Swanston have reported that both CS and CN are SN_2 alkylating agents, indicating that they react directly with nucleophilic sites.[38] Many of the toxic effects of these irritants may be due to alkylation of nucleophilic sites, including SH-containing enzymes.[57,65] Interactions of electrophilic metabolites with nucleophilic moieties of biological material with potential consequences are highlighted in Figure 11.4. The conversion of CS to cyanide with malononitrile as an intermediate has led Jones and Israel "To postulate that the toxic effects attributed to CS may arise from the conversion of CS *in vivo* to cyanide."[66] Representative of detoxification is the NADPH-dependent reduction of the benzylidene double bond in CS to yield 2-chlorobenzaldehyde—a metabolic conversion that leads to a decrease in the lethal potency and peripheral sensory irritancy.

6. Human Toxicology

When exposed to CS, humans experience immediate signs and symptoms that disappear in minutes on cessation of exposure. CS causes only transient effects on the eye and irritation and blistering of the skin at high concentrations. Healthy individuals repeatedly exposed to CS do not manifest ill effects. Human volunteers have been exposed to CS under varying conditions and concentrations to determine the

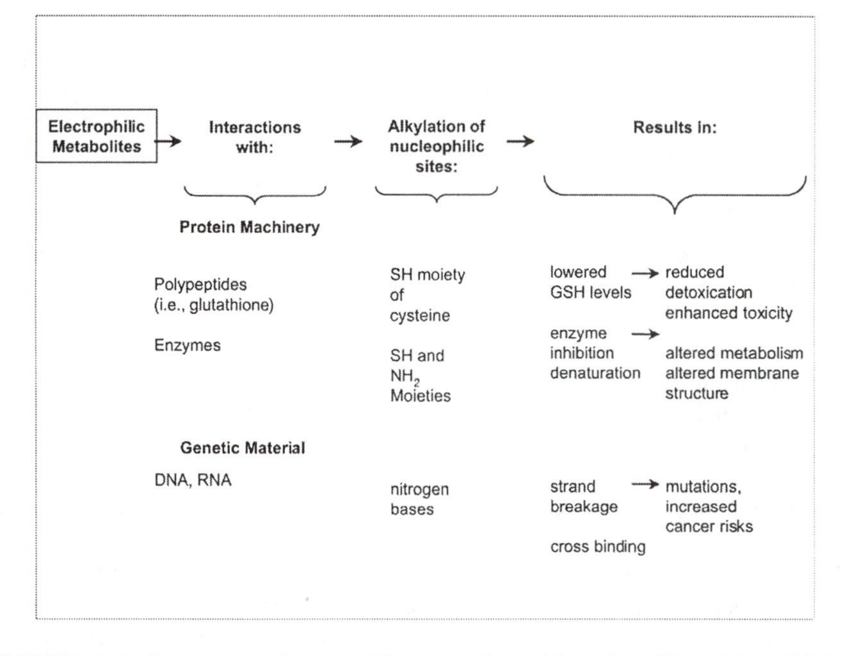

FIGURE 11.4 Interactions of electrophilic metabolites with nucleophilic moities of biological material.

ICt_{50} value, defined as the concentration that will incapacitate 50% of the exposed population in one minute. The incapacitating signs and symptoms include intense burning of the eyes, nose, and respiratory tract; profuse lacrimation; excessive salivation; blepharospasm; tightness in the chest; and a feeling of suffocation.[31] The time to incapacitation did not appear to differ among the test subjects exposed to CS via the different dispersion techniques, reduced ambient temperatures, or subjects with medical histories indicating respiratory, cardiovascular, or hepatic dysfunction.

However, at whole-body exposures at elevated temperatures (i.e., 95°F) and 35–97% relative humidity, the time to incapacitation was shortened. McNamara et al. reported that men may work without any signs of discomfort in an atmosphere where CS gradually accumulates, whereas these concentrations were intolerable to individuals entering the contaminated area from unexposed areas.[31] Thus, it appears that adaptation develops gradually as the CS concentration increases. When the "tolerant" individual left the contaminated area for short periods of 10 to 30 min, the tolerance was lost, and re-entry into the contaminated areas resulted in intolerable irritation. Moreover, additional studies on human volunteers have documented the development of tolerance to CS.[34]

Rose and Smith have reported alleged toxic reactions in human beings exposed to agent CS.[20] The Himsworth et al. report (Parts I and II) as to whether CS could produce serious toxic effects was the focus of an in-depth inquiry following the use of CS in Londonderry Northern Ireland in 1969.[67,68] In their inquiry into the adverse health and toxicological effects of CS, Himsworth et al. found no evidence, even among the most heavily exposed individuals of incapacitation that prevented their egress from a CS-contaminated environment. No evidence was found that previously healthy persons exposed 3 weeks before had developed any illness. Attention focused on susceptible subpopulations such as the elderly, the very young, pregnant women, and those with pre-existing cardiopulmonary dysfunction. Infants exposed to CS promptly recovered from the irritating effects of CS when removed to fresh air. There was no indication that CS exposure markedly altered the pre-existing pulmonary function of individuals with cardiopulmonary compromise. Pertaining to reproductive function and pregnancy, Himsworth et al. concluded that CS exposure had not significantly affected reproductive function.[68]

Park and Giammona have described the effects of CS in a 4-month old infant following prolonged exposure.[69] The infant manifested severe respiratory distress and symptoms including copious nasal and oral secretions, sneezing and coughing, and upper airway obstruction. The patient was released from the hospital but was rehospitalized within 24 h with a diagnosis of pneumonitis. The patient was treated and released following a 28-day hospitalization.

Another report pertaining to the health effects of CS following its application in riot-control and law enforcement situations is that of Anderson et al.[70] They described the findings of a review of case studies of detainees presented for medical treatment following exposure to CS in a riot-control situation. The complaints were consistent with CS toxicity. During this civil disturbance, large quantities of CS were used in a confined area and under humid conditions. Two months after the incident, when the patients were asymptotic, the case notes of all patients who had presented to the clinic within 21 days after exposure with possible CS-related symptoms were reviewed. The most common complaint was coughing. Although the majority of patients had completely recovered within 2 weeks of exposure, one asthmatic child 10 years of age had sore throat and shortness of breath that persisted for 38 days following exposure. Additionally, a 3-month old infant with confirmed hematemesis was admitted to the hospital for observation. Since there was a 6 to 8 h delay from exposure to presentation at the clinic, the immediate and transient effects of lacrimation and rhinorrhea were not reported. There was a high incidence of skin burns, usually areas covered by clothing, and many of which healed with scarring and disfigurement. There was no clinical evidence of serious sequelae to CS exposure in the patients examined. However, the high incidence of burns due to the large amounts of CS generated in a confined area and a high humidity was a cause for concern. Anderson et al. confirmed the findings of Himsworth et al., at least with respect to the transient nature of riot-control agent-induced effects involving the eye and upper respiratory tract.[67,68]

CS was adopted by the United States Army as its standard riot-control agent in 1959, and has largely replaced CN as the riot-control agent of choice worldwide. This was based on the low mammalian toxicity and the high sensory irritant potency of CS. It was used in the United Kingdom in 1969 to quell riots in Northern Ireland. In spite of its extensive use, there have been no verified causes of death in humans following CS application.[6,24,31] There have been several alleged reports of death following CS exposure; however, these were non-verifiable and/or incorrect. Hu reported that a middle-aged adult was overcome by CS and suffered heart failure and hepatic damage and eventually succumbed.[10] A review of the original report of Krapf and Thalmann indicated that the subject did indeed suffer heart failure and hepatic insult.[71] This individual was hospitalized, treated, and discharged 3 months after the exposure in a condition capable of work.

Additionally, a report focused on allegations that the Israeli Defense Forces had misused United-States-manufactured CS.[72] The alleged misuse was reported to have caused numerous deaths, principally among the elderly and ill. The majority of casualties purported to be less than a year old or over 55 years of age. The Physicians for Human Rights reported in 1988 that they could not confirm that deaths were linked to tear gas exposure.[73] The United States Department of State did not have any medical evidence to support a direct causation between CS inhalation and the number of deaths reported. It was concluded that only four deaths might have been attributable to CS use by the Israeli Defense Forces. It also appears that at the time, Israel was utilizing two types of tear gas, but generally employed CN. Thus, the allegations of death following the use of CS in the West Bank and Gaza were unsubstantiated.

There are no authenticated reports of death from CS smokes.[24] Published estimates of the human acute lethal inhalation dosage of CS vary between 25,000 and 150,000 mg · min/m^3. A widely quoted estimate of the human LCt_{50} for CS, from United States sources, is 61,000 mg · min/m^3. Estimates of lethal levels for man can be derived only by extrapolation from animal data since humans can withstand only minute dosages of riot-control agent. Also, in light of the array in lethal dose response noted in various animal species, conservative values should be adopted. Furthermore, estimates of lethal amounts on the basis of deaths occurring in law enforcement operations can be quite imprecise. The Himsworth et al. report (Part II) concluded that the physical properties of CS smoke and the unpleasant nature of the symptoms produced exposures that were self-limiting and short.[68] For irritants such as CS, a person is considered incapacitated when the exposed individual will no longer remain in the contaminated atmosphere. Motivated persons may remain in a cloud of irritant for longer periods of time, since a condition of adaptation occurs, and the irritant effects are diminished. The irritant ICt_{50} for CS, which is considered intolerable for 1 minute is 0.1 to 10 mg/m^3. However, the exact concentration depends on the individual's degree of motivation.[31]

B. Dibenz (b,f)-1:4-Oxazepine (CR)

Dibenz (b,f)1:4-oxazepine (CR), a more recent addition to the riot-control family of compounds, was first synthesized in 1962. It is a potent sensory irritant of low toxicity. The overt signs (e.g., eye and skin irritation) of exposure are more transitory than those of other riot-control agents such as CS. CR does not induce vesication or contact sensitization.

1. Mammalian Toxicology

The low toxicity of CR has been demonstrated in various animal species following exposure via different routes—the data having been summarized by Ballantyne.[74] LD_{50} and LCt_{50} values for CR and the commonly used riot-control agents are summarized in Table 4. Comparison of the acute toxicity of CR to that of CN (1-chloroacetophenone) and CS (2-chlorobenzylidene malononitrile) indicates that CR is less toxic by all routes of exposure. CR-dosed animals exhibit ataxia (incoordination), spasms, convulsions, and rapid breathing. These effects gradually subside over a 15- to 60-min period, after which time the animals either appear normal, or there is increasing respiratory distress and death. Pathologic changes found in animals after i.v. and oral dosing included congestion of liver sinusoids and alveolar capillaries. Surviving animals manifest no gross abnormalities at necropsy and no histological abnormalities. After i.p. administration, toxic signs include muscle weakness and heightened sensitivity to handling. Toxic effects persist through the first day after exposure, and some animals exhibit CNS effects. Animals surviving the post-exposure period exhibit no gross or histologic abnormalities at necropsy. Ballantyne[74] also studied the effects of CR in various animal species following inhalation exposure. Animals were acutely exposed to varying exposure periods and test article concentrations of CR aerosol or CR smoke. Rats exposed to CR aerosol (13,050 to 428,400 mg · min/m^3) manifested nasal secretions and blepharospasm (uncontrollable closure of the eyelids), which subsided within 1 h of cessation of exposure. There were no mortalities among the CR-exposed rats. In rabbits, guinea pigs, and mice exposed to CR aerosol , no deaths occurred at Ct of up to 68,400 mg · min/m^3. Animals exposed to pyrotechnically generated CR manifested alveolar capillary congestion and intra-alveolar hemorrhage. Congestion of the liver and kidneys was also noted.

Pattle and co-workers[75] evaluated the potential of CR aerosol to produce physiological and ultrastructural changes of the lung. Rats were exposed to high dosages of CR aerosol (Ct = 115,000 mg · min/m^3). Electron microscopy revealed that organelles (e.g., lamellated osmiophilic bodies) were not affected following exposure to CR. In studies by Colgrave et al., lungs were evaluated following exposure to CR aerosol at dosages of 78,200, 140,900, and 161,300 mg · min/m^3.[76] The lungs appeared normal on gross examination; however, on microscopic examination, there were indications of mild congestion, hemorrhage, and emphysema. Electron microscopy revealed isolated swelling and thickening of the epithelium and early capillary damage, as evidenced by ballooning of the endothelium. The authors concluded that very high doses of CR aerosol produced only minimal pulmonary damage.

Lundy and McKay studied the effects of CR on the cardiovascular system.[77] These studies examined the effects of CR, administered via the i.v. route, on cardiovascular activity. A dose-dependent increase in blood pressure of short duration was observed. Stimulation of the heart rate and increased arterial catecholamine content were also noted following treatment with CR. The authors postulated that the CR-induced cardiovascular effect was related to sympathetic nervous system effects as evidenced by the abolition of CR-induced pressor effect by phentolamine and 6-hydroxydopamine.

Marrs and co-workers reported findings on the repeated-dose inhalation toxicity of aerosolized CR in mice and hamsters.[78] Animals were exposed to CR at test article concentrations of 204, 236, and 267 mg/m^3 for 5 days per week for 18 weeks. Follow-up observations were conducted to detect recovery from or persistence of toxic effects. High concentrations of CR affected the survival of both species, and no single cause of death could be ascertained, although pneumonitis was evident in many cases. CR exposure produced minimal organ toxicity; however, chronic inflammation of the larynx was noted in mice. No significant pulmonary lesions were manifested; however, the occurrence of alveologenic carcinoma in a single low-dose group mouse and in a single high-dose group mouse was seen. The validity of these findings, as well as interpretations/conclusions, may be questioned since the spontaneous frequency of alveologenic carcinoma is high in many mouse strains.[79,80] Further, this tumor type is dissimilar in many respects from human types of lung tumors. No lung tumors were noted in hamsters exposed to CR. Likewise, no lesions were present in the larynx of hamsters exposed to CR aerosol. Histopathologic evaluation of the liver revealed hepatic lesions in mice; however, these were of infective origin and not test article related. In general, Marrs et al. concluded that CR exposure at high concentrations reduced survivability and that CR produced minimal organ-specific toxicity at levels many times the intolerable human dose ($IC_{50} = 0.7$ mg/m^3, within 1 min[24]; $IC_{50} = 0.15$ mg/m^3, within 1 min).[78,81]

A number of studies have been reported on the repeated-dose toxicity of CR following dermal administration.[82–84] Owens and co-workers studied the effects of CR following multiple dermal application in rabbits and monkeys.[82] In the study by Marrs and co-workers, CR in acetone was applied to the skin of mice (5 days/wk for 12 wk).[84] The animals were kept for an additional 80 weeks following the end of the application period. No abnormalities were noted that could be attributed to CR, but a high incidence of fatty infiltration of the liver was noted in one strain of mice, which was most likely due to acetone. These investigators concluded that the repeated dermal application of CR had little effect on the skin. They further postulated that in view of the absence of any specific organ toxicity, absorption of even substantial amounts of CR would have little effect.

2. Ocular and Cutaneous Effects

CR was initially noted by Higginbottom and Suschitzky to cause intense lacrimation and skin irritation.[85] Studies on the irritancy of CR, CN, and CS in a number of species have been conducted.[82,83,86–89] Owens et al. evaluated the ocular effects of 1% CR solutions in rabbits and monkeys following single- or multiple-dose application.[82] Mild and transitory eye effects (mild redness, mild chemosis) were observed in rabbits and monkeys after a single dose of 1% CR solution. Multiple applications over a 5-day period of 1% CR solution to the eye produced only minimal ocular effects. Rengstorff et al. reported moderate conjunctivitis following the application of a 5% CR solution to the eyes of rabbits.[86] Histological examination revealed normal corneal and eyelid tissues. Biskup et al.[83] reported no signs of eye irritation in animals following single- or multiple-dose applications of 1% CR solution. Ballantyne and Swanston[88] also studied the ocular irritancy of CR and arrived at a threshold concentration (TC_{50}) for blepharospasm in several species. Ballantyne et al.[89] conducted extensive studies on the ocular effects of CR as a solid (0.1 to 5 mg), as aerosol (360 to 571 mg/m^3, 30 min exposure), and in solution (1 to 10% in polyethylene glycol). Measurements of intraocular tension and corneal thickness were conducted as well as histological examination of the eyes. CR in solution resulted in mild to moderate concentration-related ocular effects usually of several days duration—transient even at the higher concentrations. Solid CR resulted in lacrimation and minor irritation of the conjunctivae and eyelids. Exposure to CR aerosol (Ct = 10,800 mg · min/m^3; Ct = 17,130 mg · min/m^3) resulted in mild lacrimation and conjunctival injection with clearing in 1 h. CR solutions produced reversible dose-related increases in corneal thickness. Ballantyne et al. concluded that CR produced considerably less damage to the eye than CN and that there was a much greater degree of safety for CR than CN.[89]

With regard to skin toxicity, CR produces a transient erythema; however, it does not induce vesication and contact sensitization or delay the healing of skin injuries.[24,90] The burning sensation on exposure to CR persists for 15 to 30 min, and erythema may last for 1 to 2 h.

3. Reproductive Toxicity and Developmental Effects

The effects of CR on rabbit and rat embryonic development were studied by Upshall.[91] Animals were exposed to aerosolized CR at concentrations of 2, 20, and 200 mg/m^3 for 5- to 7-min exposures. Additionally, some rats were dosed intragastrically at 2, 20, and 100 mg/kg on days 6, 8, 10, 12, and 14 of pregnancy, and others were dosed intragastrically with 400 mg/kg on days 7, 10, and 13 of pregnancy. Rabbits were dosed intragastrically with CR (0.2, 2, and 20 mg/kg) on days 6, 8, 10, 12, 14, 16, and 18 of pregnancy.

Recorded data included the number of litters, litter size and weight, number of abnormal litters, number of live fetuses, and placental weight. Pregnant female rats exposed to CR aerosol did not manifest toxic effects. There were no dose-related effects of CR on the parameters measured or the number or type of fetal malformations. Predominant abnormalities observed in all groups were skeletal in nature (e.g., poorly ossified sternebrae, extra ribs). Fetuses from female rats dosed intragastrically with CR exhibited skeletal anomalies in all groups. Pregnant rabbits exposed to CR aerosol did not manifest overt signs of toxicity. There were no dose-related effects of CR on any of the parameters measured and the numbers or types of malformation. No externally visible malformations were seen in any group. No dose-related effects of CR were noted in the fetuses in any group. Based on the overall observations, the authors concluded that CR was neither teratogenic nor embryotoxic to rats and rabbits.

4. Genotoxicity and Carcinogenicity

There is a paucity of data addressing the subject of genotoxicity with respect to CR exposure. A review of the database has identified a single study in the mainstream medical literature.[92] In the above-cited study, the mutagenic potential of technical grade CR and its precursor (2-aminodiphenyl ether) was evaluated. Various strains of *S. typhimurium* served as the microbial test for predicting mutagenic response. Mammalian assay systems for the detection of mutations consisted of the following: Chinese hamster cell mutagenesis (V79/HGPRT system); mouse lymphoma cell mutagenesis (L5178Y/TK+/TK−); and the micronucleus test (erythrocytes). CR and its precursor were negative in all assays. The results suggest that CR is not mutagenic; however, conclusions as to the genetic threat of CR to humans must await further genotoxicity testing utilizing additional genotoxicity assays. With regard to carcinogenicity, very little research has been conducted on the carcinogenic potential of CR or its ability to cause other chronic health effects.

5. Clinical Chemistry

Husain et al. studied the effects of CR and CN aerosols on clinical chemistry parameters (e.g., plasma glutamic-oxaloacetic tansaminase (GOT), plasma glutamic-pyruvic transaminase (GPT), acid phosphatase, and alkaline phosphatase).[93] Rats were exposed via inhalation to aerosols of CR or CN. Animals exposed to CR aerosol exhibited no significant changes in plasma GOT and GPT activities or in acid and alkaline phosphatase activities. In contrast, CN-exposed animals manifested significant increases in GOT, GPT, acid phosphatase, and alkaline phosphatase activities. The conclusion drawn from the study was that exposure to CN aerosol could lead to tissue damage.

6. Metabolism, Metabolic Fate, and Mechanisms

The bioconversion and metabolic fate of CR has been studied in some detail.[94–98] The subject of CR metabolism has been reviewed by Upshall.[99] Aerosols of CR are very quickly absorbed from the respiratory tract. Plasma half-life ($T_{1/2}$) of CR after inhalation exposure to CR aerosol is about 5 min. The plasma half-life of CR following i.v. administration is also about 5 min. Balfour studied the uptake and metabolic fate of CR in intact cornea and corneal homogenates.[94] The data indicated that these tissues readily took up CR and metabolized CR to a lactam derivative. The metabolism and fate of CR have been investigated in a series of *in vivo* and *in vitro* studies by French and co-workers and by Furnival et al.[96–98] French et al. studied the *in vivo* metabolism and metabolic fate of CR in rats, guinea pigs, and monkeys after i.v. and intragastric dosing of CR.[96] In the rat, CR is converted to the lactam, followed by the subsequent hydroxylation to yield monohydroxylated derivatives (i.e., 4-, 7-, and 9-hydroxylactams) and the eventual formation of sulfate conjugates (see Figure 11.5). The pathway leading to sulfate conjugate formation of CR metabolites represents the major metabolic pathway in the rat irrespective of dose and the route of administration. The bile contained only small levels of sulfate conjugates. It should be noted that in his review on the metabolism of CR, Upshall discussed the formation of glucuronide conjugates of CR metabolic intermediates that are eventually excreted in the urine as sulfate conjugates.[99] French and co-workers also indicated that similar metabolic products and excretory pathways exist in the guinea pig and monkey; however, only free hydroxylactams were isolated from monkey urine.[96] In the same study,

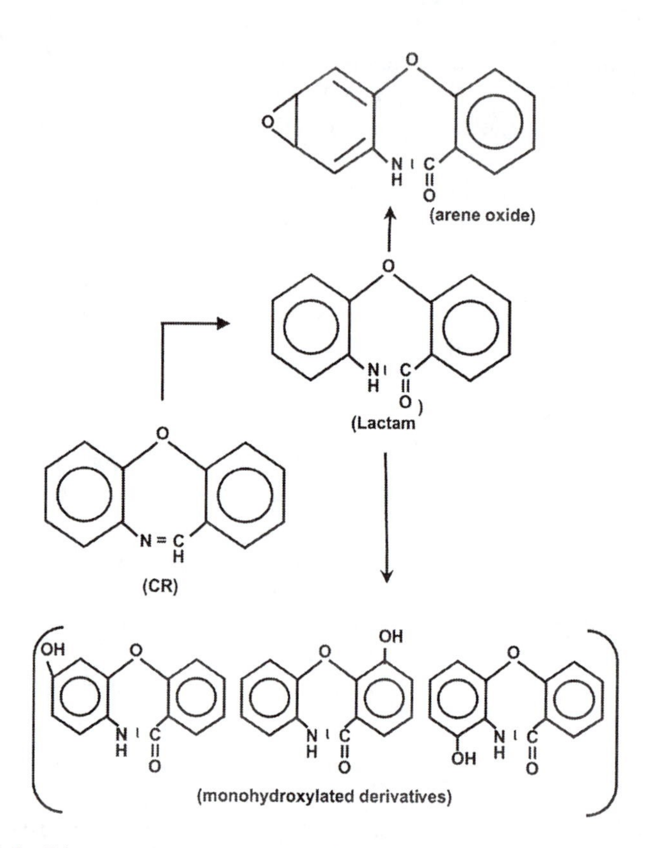

FIGURE 11.5 Bioconversion pathways for CR.

the authors reported the findings of whole-body autoradiography studies in CR-dosed mice, which demonstrated the rapid disappearance of CR from blood into other compartments (liver, kidney, and small intestine). These findings are consistent with the rat studies indicating the rapid absorption, hepatic metabolism, biliary secretion, interohepatic recirculation, and renal excretion. Furnival et al., on the basis of *in vitro* metabolic studies using rat liver preparations, demonstrated that CR was metabolized via the following major pathways: (a) ring opening and reduction and (b) oxidation to lactams.[98] French et al. performed additional *in vitro* and *in vivo* metabolic studies.[97] Their findings supported previous conclusions that the major metabolic fate of CR in the rat is the oxidation to the lactam, subsequent ring hydroxylation, sulfate conjugation, and urinary excretion.

7. Human Toxicology

The effects of CR on human subjects following aerosol exposures, "drenches" with dilute solutions, and local application have been reported by a number of investigators.[37,88,90,100–105] The estimated human LCt_{50} of CR is over 100,000 mg · min/m^3.

Human studies to determine the effects of CR after aerosol or cutaneous exposures were conducted, and the findings have been summarized in a National Academy of Sciences report.[6] Human subjects manifested mostly ocular and respiratory effects after acute exposure to CR aerosol. Ocular effects consisted of lacrimation, irritation, and conjunctivitis; and respiratory effects included upper respiratory tract irritation, with choking and dyspnea. Ballantyne et al. described the effects of dilute CR solutions following "splash contamination" on the face.[101] In addition to the classical effects on the eye, CR facial "drenches" also resulted in an immediate increase in blood pressure concomitant with decreases in heart rate. Subsequent studies were conducted by Ballantyne and co-workers on the effects of CR after whole-body "drenches."[37] Immediate increases in blood pressure were noted, as in the previous study; however, Ballantyne et al. concluded that the cardiovascular effects described in both studies were not due to absorbed CR.[37] They theorized that there was insufficient CR uptake to cause the systemic effects on the heart, and the cardiovascular effects were due to the sensory-irritant-induced stress. However, Lundy and McKay suggested that the cardiovascular effects described by Ballantyne et al.[101] were the result of CR-induced effects on the heart via the sympathetic nervous system.[77] Ashton et al. also studied the effects of CR aerosol on the respiratory physiology of humans.[102] Test subjects were exposed to CR aerosol at a mean concentration of 0.25 mg/m^3 (particle size of 1–2 micron) for 1 h. Expiratory flow rate was decreased about 20 min after onset of exposure. The authors theorized that CR stimulated the pulmonary irritant receptors to produce bronchoconstriction and increasing pulmonary blood volume by augmenting sympathetic tone.

8. Ocular and Cutaneous Effects (Human)

Utilizing procedures developed for CS, Ballantyne and Swanston conducted a comparative study including human subjects to assess the irritant potency of CR.[88] Dilute solutions of CR (CR in saline) were applied to the eyes to ascertain the threshold concentration for producing blepharospasm (uncontrollable closure of the eyelids). The median threshold concentration (TC_{50}) to produce blepharospasm for man is 8.6×10^{-7} M. Comparative TC_{50} values in several animal species are: $TC_{50} = 7.9 \times 10^{-5}$ M, rabbit and $TC_{50} = 3.5 \times 10^{-5}$, guinea pig. The TC_{50} to produce sensation on the human eye is 4.9×10^{-7}M (9.1×10^{-2} mg/l solution). The authors suggested that CR at a concentration of 3.3×10^{-6} M would be incapacitating based on extrapolation from human eye data on sensation. In general, the data also indicate that the molar concentration required to elicit threshold effects on the human eye is smaller for CR than for CS. Ballantyne and Swanston also postulated that a CR concentration of less than 0.25 M (5% solution) would not produce structural damage to the eye when applied to the conjunctiva.[88] They also cited data by Hogg on the threshold irritant response (burning sensation) of the human eye to CR aerosol.[106] A TC_{50} for sensation of 4.0×10^{-3} mg/m^3 (4.0×10^{-6} mg/l) was calculated for CR aerosol. Thus, the human eye is much more sensitive to CR aerosol ($TC_{50} = 4.0 \times 10^{-6}$ mg/l) than to CR in solution ($TC_{50} = 9.1 \times 10^{-2}$ mg/l). This is consistent with the report that the human eye is more sensitive to CS aerosol than to

CS in solution.[87] Other studies by these researchers included a study to ascertain the effect of CR solution (1% CR) splashed on the face and the effects of very dilute solutions (0.0025 to 0.001%) of CR on volunteer subjects exposed (whole body spray or showers) to test material.[101] After a 15-s, individual drench with CR, subjects exhibited intense stinging of the eyes, injection of the conjunctivae, profuse lachrymation, and blepharospasm. The stinging of the eyes was rapid in onset, which occurred within seconds. Additionally noted was a very rapid onset of stinging of the skin around the eyes, which rapidly intensified to a strong, burning sensation. Group drenches of 1 min duration were also conducted. The ocular effects noted were similar to those observed in the individual 15-s drenches. Compared to CR, the effects elicited by CS were of shorter duration, less severe, and more variable. It should also be noted that following CS exposure, stinging of the eyes was the first biological effect seen. From the data, it was concluded that even very dilute solutions of CR (0.0025 to 0.001%) produced sensory ocular effects.

Over the years there has been considerable interest in the cutaneous effects of sensory irritant compounds, and several studies on the dermal effects of CR in humans have been published.[90,100,101] Weigand and Mershon studied the dermal effects of dilute CR and CS solutions (CR in propylene glycol and CS in trioctyl phosphate).[100] Test subjects were patch tested on various anatomical sites with concentrations of test article ranging between 0.01 to 1.0%, and exposure duration was for 5 and 30 min. Stinging sensation was evident following exposure to both compounds, with CR eliciting a response of greater intensity. The onset of stinging was more prompt at higher ambient temperatures. Transient erythema of varying degrees was evident and subsided within 4 h.

Holland evaluated skin reactions to CR in humans following application of CR in graded quantities as CR powder or as dry material moistened with saline.[90] Erythema was noted in 10 min and faded on removal of test article. When moistened, CR resulted in marked irritation. No swelling or vesication was evident, even under adverse conditions. It was concluded that CR is capable of producing acute cutaneous discomfort. In comparing the results with similar studies on CS and CN, Holland concluded that all reactions to CR are mild and transient compared to that of CS, which resulted in an erythema of greater duration and to that of CN, which produced blistering.[90]

Ballantyne and co-workers drenched volunteer subjects with very dilute solutions of CR and CS for 15- and 60-s durations.[37] In the studies comprising subjects that were exposed individually, stinging of the skin around the eyes was of rapid onset and spread to other parts of the face. The burning sensation involving facial skin was the next pronounced feature for about the first minute. Scalp and ears were not usually affected. During the second minute, stinging was associated with the back of the neck and irritation of the genital area. Stinging of the shoulder and back followed in 3 to 4 min, and the burning sensation was intense in approximately 5 min. After about 5 to 6 min, other anatomical sites (e.g., chest, abdomen, thighs, and buttocks) were affected. At 10 min, the burning sensation of the skin was intense, primarily affecting the trunk and back. Within 15 min, the skin sensation had subsided. By 20 min, skin sensations were reduced to mild tingling or had disappeared. Erythema of the skin was produced within several minutes and persisted for 1 to 2 h. No other skin changes were noted. Many areas of the skin were somewhat resistant to irritation, which included such sites as the ears, nose, scalp, palms of the hands, knees, and the lower legs. In general, a more intense response was elicited by CR at higher concentrations; however, it should be noted that individual variations were more marked than the differences between CR concentrations. In the group drenching studies, burning of the skin was the most prominent symptom. As with the individual drenches, considerable variation in the severity of the symptoms was manifested. Compared with CR, the effects elicited by CS were less severe, of shorter duration, and more variable. Stinging of the skin followed a similar progression (face, neck, genital areas, shoulders and back, chest, abdomen, and thighs) as seen with the CR drenches. The studies by Ballantyne et al. demonstrated that very dilutc solutions of CR and CS produce a strong stimulation of sensory receptors in the skin and mucous membranes.[37] The burning sensation was more intense and of longer duration on exposure to CR than with CS. Skin irritation and erythema were evident following exposure to either CR or CS, and the signs were more pronounced with CS than with CR. No individual drenched with CR or CS manifested edema, vesication, or desquamation.

C. Chloroacetophenone (CN)

Chloroacetophenone or CN, a white crystalline solid with an apple blossom odor, is commonly known as tear gas or Mace®. Chloroacetophenone was first synthesized in 1871 and was studied for its use as a tear gas shortly after World War I. CN acts directly on the mucous membranes to produce irritation, burning, and pain of the eyes, nose, throat, and respiratory tract. Ocular effects include lacrimation, blepharospasm, and conjunctivitis. Irritation of the respiratory tract produces sneezing, coughing, secretions, nasal congestion, and a sense of suffocation. The onset of some or all of these symptoms is immediate and persists from 5 to 20 min after removal from the contaminated atmosphere.

1. Mammalian Toxicology

Inhalation studies consisting of acute and repeated-dose have been conducted in various animals to ascertain the comparative toxicity of CN. The toxicology of CN has been reviewed and summarized in a National Academy of Sciences report, by McNamara et al., and by Hu.[6,31,107] The early toxicity studies on CN were highly variable, and subsequent studies conducted in the mid-1960s using an array of animal species were designed to provide more quantitative data. In these studies, CN was dispersed in acetone or from commercially available thermal grenades. Sublethal effects observed on exposure to CN consisted of the following: lacrimation, conjunctivitis, copious nasal secretions, salivation, hyperactivity, dyspnea, and lethargy. Cutaneous effects seen in the exposed animals consisted mainly of erythema. Dyspnea was the salient biological finding on post-exposure, which was exhibited in all animals. Ocular (i.e., conjunctivitis) and dermal effects (i.e., erythema) persisted for 3 to 7 days after exposure. The primary cause of death following CN inhalation was from pulmonary damage. Lethality (LCt_{50}) values for CN in various species are as follows: rat, 8,878 mg · min/m^3; guinea pig, 7,984 mg · min/m^3; and dog, 7,033 mg · min/m^3. Pathological findings in animals that died after CN aerosol exposures consisted of pulmonary congestion, edema, emphysema, tracheitis, bronchitis, and bronchopneumonia in dogs, and pulmonary congestion, edema, and bronchopneumonia in rats, mice, and guinea pigs. The pathological findings reported by Ballantyne and Swanston in animals that died after CN inhalation included congestion of the alveolar capillaries, alveolar hemorrhage, and excessive secretion in the bronchi and bronchioles.[38] There were also areas of acute inflammatory cell infiltration of the trachea, bronchi, and bronchioles.

McNamara et al. cited the findings of a repeated-dose study on the effects of thermally generated CN on guinea pigs, dogs, and monkeys.[31] In one series of experiments, guinea pigs and monkeys were exposed on 10 consecutive days to CN at Ct from 2,300 to 4,000 mg · min/m^3 for a total exposure dosage of 31,445 mg · min/m^3. This dosage would be expected to be lethal to about 75% of the guinea pigs and 100% lethal to monkeys if given in a single exposure. This exposure regimen resulted in the death of five guinea pigs and no deaths in the exposed monkeys. The toxicity of CN is considerably less when administered in divided dosages. These findings were confirmed in further studies using dogs that were exposed on 10 consecutive days to Ct of CN ranging from 3,000–7,000 mg · min/m^3 for a total dosage of 60,000 mg · min/m^3. A subsequent repeated-dose study was also conducted in guinea pigs, dogs, and monkeys exposed daily for 10 days to Ct ranging from 4,200 to 13,000 mg · min/m^3 for a total exposure of 88,000 mg · min/m^3. This dosage was lethal in the majority of animals for all species tested. Overall, these studies demonstrated the lack of cumulative toxicity of CN when administered as repeat doses.

In a recent study, Kumar and co-workers reported findings on the effects of multiple exposure to CN and CR in mice.[108] Animals were exposed to test article at concentrations equivalent to the 0.05 LC_{50} (CN [87 mg/m^3] or CR 1008 mg/m^3) for 15 min/day for 5 and 10 days. Biochemical endpoints measured included blood glucose, plasma urea, transaminase enzymes (SGOT,SGPT), liver alkaline phosphatase (ALP), liver acid phosphatase (ACP), liver glutathione (GSH) levels, and hepatic lipid peroxidation (malondialdehyde [MDA] formation). Clinical parameters affected following repeated exposure included decreased hepatic glutathione and increased lipid peroxidation. The hepatic acid phosphatase increased after the 5-day exposure to CN and the glutathione levels decreased after the 10-day CN exposure. CN-induced elevation in acid phosphatase levels reflected the release of lysosomal enzyme from the liver, indicative of tissue injury. CR exposure did not produce significant alteration in hepatic biochemical parameters. Additionally, hyperglycemia was observed after exposure to CN, an effect previously reported by Husain.[93] It was suggested that the hyperglycemia was induced by stress-mediated release of epinephrine, which is known to elevate glucose levels. Significant decreases in body weight gain were also noted on exposure to these compounds, with CN having a more prominent effect on body weight. Overall, these findings were consistent with results reported by Ballantyne on the repeated-dose effects of orally administered CR in various animals.[74] Histopathologic changes following CN exposure included hemorrhage, perivascular edema, congestion of the alveolar capillaries, occluded bronchioles, and alveolitis. Renal histopathology demonstrated congestion and coagulative necrosis in the cortical renal tubules in CN-exposed mice. Hepatic histopathology consisted of cloudy swelling and lobular and centrolobular necrosis of hepatocytes following CN exposure. The National Research Council cited a subchronic study that was conducted under the National Toxicology Program.[6] Mice and rats were exposed to CN aerosol for 13 weeks, and the findings indicated no gross clinical signs in rats or mice, except irritation of the eyes including opacity. No microscopic lesions were noted compared with controls.

2. Ocular and Cutaneous Effects

CN is a potent irritant and is more likely to cause serious eye effects than CS. The ocular irritation caused by CN signals avoidance and the intense lacrimation and blepharospasm initiates a defense mechanism. High concentrations of CN may result in chemical injury to the eye with corneal and conjunctival edema, corneal edema, erosion or ulceration, chemosis, and focal hemorrhages.[109–111] CN-induced ocular effects on the rabbit eye following treatment with various CN formulations has been investigated by Ballantyne et al.[89] and Gaskins et al.[112] Ocular effects included lacrimation, chemosis, iritis, blepharitis, and keratitis; severity was dependent on the formulation.

Exposure to CN has been associated with primary irritation and allergic contact dermatitis.[113–115] CN is a potent skin irritant and is more likely to cause serious effects to the skin than CS induces. Severe exposure to CN results in skin injury that may consist of severe generalized itching, a diffuse and intense erythema, severe edema, and vesication. In addition to being a more potent skin irritant than CS, CN is considered a more potent skin sensitizer.[114]

3. Genotoxicity and Carcinogenicity

Carcinogenicity bioassays have been conducted in rats and mice.[116] There was no indication of carcinogenic activity of chloroacetophenone in male rats exposed to test article. Equivocal evidence of carcinogenicity of chloroacetophenone was based on findings in female rats, indicating an increase in fibroadenomas of the mammary gland. The findings of a 2-year inhalation bioassay in mice suggested no carcinogenic activity in male or female mice exposed to CN.

4. Metabolism, Metabolic Fate, and Mechanisms

Little is known about the metabolism and full metabolic fate of CN. Chloroacetophenone is converted to an electrophilic metabolite. It is a SN_2 alkylating agent that reacts with SH groups and other nucleophilic sites of biomolecules. Alkylation of SH-containing enzymes leads to enzyme inhibition with disruption of cellular processes. Based on the potential to disrupt enzyme function, Castro investigated the effects of alkylating agents, including CN, on human plasma cholinesterase.[117] CN was found to inhibit ChE via a non-SH interaction. It is thought that some of the toxic effects of CN may be due to alkylation of SH-containing enzymes.

5. Human Toxicology

The initial LCt_{50} estimate for humans, based on extrapolation from animal data, was 7,000 mg · min/m^3, which was subsequently revised and established as 14,000 mg · min/m^3. In human volunteer studies, the immediate effects on exposure to CN were a burning sensation or stinging in the eyes, nose, throat, and exposed skin. This was followed by lacrimation, salivation, rhinorrhea, and dyspnea. Lacrimation persisted for about 20 min post-exposure, while conjunctivitis and blepharospasm persisted for up to 24 h. High levels of CN may produce chemical injury to the eye characterized as corneal and conjunctival edema, chemosis, and loss of corneal epithelium.[111] Physical injuries may also occur following dispersion via grenade-type tear gas devices.[109,111] Punte et al. studied the effects of CN on human subjects.[118] Individuals were exposed to CN aerosol at a Ct below 350 mg · min/m^3, which is considered the maximum safe inhaled dosage for humans. Common symptoms include rhinorrhea, lacrimation, blurred vision, conjunctivitis, and burning of the throat. Less frequent, but more severe symptoms included difficulty in breathing, nausea, and burning in the chest. Persistence of effects was negligible with no overt clinical signs noted approximately 10 min after cessation of exposure.

Varying values for the incapacitating dosage (ICt_{50}) of CN have ranged from 20 to 50 mg · min/m^3. The ICt_{50} of CN is comparable to adamsite (DM), an early riot-control agent that replaced CN; however, it is considerably greater than the ICt_{50} of CS, which replaced CN in turn. The estimate for the human LCt_{50} of CN dispersed from solvent is 7,000 mg · min/m^3 dispersed from grenades.[31] Other reported estimates are in the range between 8,500 to 25,000 mg · min/m^3. According to Punte et al.,[118] the maximum safe inhaled dose of CN for man is estimated at 500 mg · min/m^3. Pulmonary lesions may occur at the inhalation dosages, and the effects of CN exposure in confined spaces can be severe, as reported by Thornburn.[119] Exposed individuals manifested lacrimation, conjunctivitis, conjunctival edema, upper respiratory tract irritation, cough, dyspnea, and skin burns. Death from high concentrations of CN may occur; the post-mortem examination may reveal edema and congestion of the lungs, alveolar hemorrhage, necrosis of the mucosal lining of the lungs, and bronchopneumonia.[120,121] Lethal exposures to CN have been reported.[119–121]

D. Oleoresin Capsicum (OC)

Oleoresin capsicum (OC), an extract of pepper plants, is a mixture containing many substances and capsaicinoids, including the active ingredient capsaicin (8-methyl-N-vanillyl-6-nonenamide or [N-(4-hydroxy-3-methoxybenzyl)-8-methyl-trans-6-noneamide]). OC is a highly effective irritant that has received much attention as a less-than-lethal agent in civilian, governmental, and military sectors, and OC spray ("pepper spray") has gained popularity as a police weapon in recent years. Since OC is a natural product (capsicum fruits are established food additives), it is considered safe—a viewpoint that is not necessarily accurate. OC has been incorporated into a variety of formulations and marketed as "pepper gas," "pepper mace," and "pepper spray" for self-defense, criminal incapacitation, law-enforcement, and riot-control purposes. Used as a spray, OC rapidly induces involuntary closure of the eyes and lacrimation. It also causes respiratory-related effects such as severe coughing and sneezing, nasal irritation, bronchoconstriction, and shortness of breath. OC also causes burning sensation of the skin and loss of motor control. Consequently, exposed individuals can in most instances be easily subdued. Acute effects of capsaicin and capsaicinoids are primarily associated with the respiratory system (e.g., bronchospasm, respiratory arrest, pulmonary edema) but may also include hypertensive crisis and hypothermia. There have been numerous reports concerning deaths related to OC use. Although a causal relationship has not been established, most of the reported deaths have occurred within 1 h following exposure. Literature can be obtained regarding the chemistry, physiology, and toxicology of OC, capsaicin, and capsaicinoids.[122–130]

1. Mammalian Toxicology

Capsaicin, the active ingredient of oleoresin capsicum (OC), was prepared and evaluated for physiological effects in humans as early as the 1920s. Interest in the development of capsaicin as a riot-control agent decreased as research efforts were directed to the development of the newly synthesized agent, CS. Unlike other riot-control agents such as CS, CR, and CN, which have definite chemical compositions, oleoresin capsicum is a mixture of compounds containing capsaicinoids (capsaicin and structural analogs), various acids and esters, alcohols, aldehydes, ketones, and carotenoid pigments.[131–134] Keller et al. have identified numerous compounds in OC by gas chromatography-mass spectrometry (GC-MS).[135] The capsaicinoid content of the dried fruit has been reported to range from 0.1 to 1%.[126] The capsaicinoid content of the oleoresin is as follows: capsaicin (~70%), dihydrocapsaicin (~20%), norhydrocapsaicin (~7%), homocapsaicin (~1%), and homodihydrocapsaicin (~1%). Capsaicin is considered to be the active ingredient of oleoresin capsicum, and little consideration has been given to the other capsaicinoids regarding biological effects and mechanisms. Generally, the capsaicin analogs have similar effects, although with different potencies.[134]

Toxicological studies have been conducted on both OC and capsaicin; however, not much is known regarding the toxicity of OC. Because OC is a much-utilized food component, OC is widely regarded as safe, with a low-order of toxicity.[136] Data on the toxicology of OC is extant, particularly when administered by inhalation. However, recent inhalation studies on OC have indicated the low inhalation toxicity of OC.[137,138]

The toxicology of capsaicin is much better characterized. Toxicological and pharmacological data on capsaicin have been derived from both animal and human studies to include inhalation exposures. The pharmacologic actions of capsaicin and capsaicinoids were characterized in the 1950s.[139,140] More recent studies revisited the toxicology of these substances and LD_{50} values (0.56 mg/kg, i.v.; 7.6 mg/kg, i.p.; 7.8 mg/kg, i.m.; 9.0 mg/kg, s.c.; 190 mg/kg, intragastric; 512 mg/kg, dermal; and 1.6 mg/kg, intratracheal) for capsaicinoids and capsicum extracts were reported by Glinsukon et al.[141] It was reported by the same authors that the toxicity of capsaicin in the capsicum extract was about four-fold greater than that of pure capsaicin administered intraperitoneally. Guinea pigs appear to be more susceptible than mice or rats, whereas hamster and rabbits are less vulnerable to the toxic actions of capsaicin. The probable cause of death is respiratory paralysis.

Capsaicin has profound, acute effects on respiratory function, and the pulmonary toxicology of capsaicin has been studied in some detail. Capsaicin may induce the Kratschmer reflex, which on inhalation of an irritant causes apnea, bradycardia, and a biphasic fall and rise in aortic blood pressure. Animals and humans exposed to capsaicin manifest bronchoconstriction, the release of the neuropeptide substance P from sensory nerve terminals, and airway mucosal edema.[142–146] Capsaicin administration produces species-related pulmonary effects. In the guinea pig, intravenous and intra-arterial dosing induces bronchoconstriction.[147] In the dog and cat, i.v. administration of capsaicin results in bronchoconstriction that is dependent on a vagal cholinergic reflex. Aerosol exposure of cats to capsaicin also evokes a vagal-mediated cholinergic reflex bronchoconstriction.[148] A study designed to elucidate the mechanism by which capsaicin via aerosol exposure produces bronchoconstriction in guinea pigs suggests a vagal/cholinergic and non-cholinergic local axon reflex contributes to this effect.[149]

The cardiorespiratory effects of capsaicin have been studied following i.v. administration, which resulted in a triphasic effect on blood pressure and altered cardiac parameters.[150,151] Capsaicin induces complex effects on the cardiovascular system consisting of tachypnea, hypotension (seen in the Bezold-Jarrish reflex), bradycardia, and apnea.

The neurotoxic action of capsaicin on C-fiber afferents is well-known, and capsaicin has been used as a selective probe to study the role of nociceptors and neurogenic inflammation.[152–154] In animals, exposure to high doses of capsaicin and its analogs results in long-lasting insensitivity for stimuli such as pain, irritants, and temperature.[126] Capsaicin-induced desensitization may be manifested for weeks and associated with structural changes that are reversible. Long-term effects involving the respiratory tract are characterized by desensitization of the airways to chemical irritants and the marked inhibition of vagal bronchoconstriction effects.[155] Capsaicin-induced desensitization is caused by acute and excessive depletion of the neurotansmitter substance P, which is expressed as a lack of normal physiological response to stimuli such as heat and cold. High doses of systemic capsaicin induce a permanent or long-lasting desensitization of capsaicin-sensitive afferent nerves in newborn rats. In adult rats, the same doses provoke a long-lasting but temporaneous block of the nerves. In both cases, transmission of pain in response to various noxious stimuli was inhibited or abolished in capsaicin-treated animals. The effect is postulated to be capsaicin-induced with the consequent neurodegeneration of C-fiber receptors.[142] More recently, reports suggested that this effect can be dissociated by using lower doses.[156]

Alteration in thermoregulation can result on exposure to capsaicin and capsaicinoids. Weiss has reviewed some of the studies on capsaicin's effect on body temperature control and corroborated that capsaicin has been used for the last 25 years as the tool of choice in elucidation of physiological body temperature and pain control.[157] Pretreatment of rats and guinea pigs with capsaicin induced an irreversible impairment in thermoregulation. On exposure to heat, body temperature rose concomitant with an inability to discriminate and seek cooler environments.[158] The capsaicin-treated animals consumed less water and became dehydrated. Dermal blood vessels failed to dilate, and the animals did not take appropriate behavior to prevent heat stroke. The same investigators reported that s.c. injections of capsaicin reduced body temperatures and that the dosing regimen resulted in a tolerance to thermal regulation. Frens reported that s.c. injections of capsaicin reduced body temperature in goats.[159] Topical treatment of human skin with 1% capsaicin and capsaicinoids lowered the threshold to thermal pain.[160] Weiss postulated that capsaicin and capsaicinoids might have potentially deleterious physiological consequences in individuals exposed to these substances at elevated temperatures, especially in repeated-exposure scenarios.[157]

The effects of capsaicin and capsaicinoids on the gastrointestinal tract and nutritional impacts have been studied. The duodenal mucosal response to capsaicinoids and altered fat uptake by damaged duodenal epithelium, reported by Nopanitaya and Nopanitaya and Nye, led to further investigations concerning alteration of nutrient absorption and metabolism by capsaicinoids.[161,162] Studies by Sambaiah et al.,[163] Sambaiaha et al.,[164] and Kawada et al.[165] indicated that capsaicinoids had no adverse effect on fat intake or absorption. The lipotropic and hypolidemic effects of capsaicinoids has been examined in some detail.[165–167] It was postulated by Sambaiah and Satayanarayana that capsaicinoids counteract the accumulation of fat in the liver by the reduction of hepatic lipogenesis and/or increased oxidation of lipids in tissues.[168–170] Repeated dosing of capsaicin and capsicum in the rabbit produced pathological changes in several organ systems.[171,172] In the study reported by Lee, capsaicin resulted in hepatic necrosis following multiple-dose administration.[171] Mice, fed a diet containing capsicum extract for 4 weeks, did not exhibit signs of toxicity.[169] Intragastric administration of capsaicin (50 mg/kg/day) or crude extract of capsicum (0.5 mg/kg/day) for 60 days was conducted in rats by Monsereenusorn.[170] The findings are in concordance with those reported by Nopanitaya.[168] Biochemical parameters affected by capsaicin and crude extract included significant reductions in plasma urea nitrogen, glucose, phospholipids, triglyceride, transaminase, and alkaline phosphatase.

2. Ocular and Cutaneous Effects

Typical ocular symptoms associated with exposure to OC aerosol exposure include lacrimation, conjunctival inflammation, redness, severe burning pain, swelling, and blepharospasm. The application of capsaicin to the eye causes neurogenic inflammation and unresponsiveness to chemical and mechanical stimuli. Topical application of capsaicin eliminates the blink reflex for up to 5 days following dosing.[125] Systemic administration of capsaicin is associated with trigeminal nerve fiber degeneration in the cornea.[173] In humans, exposure to OC can cause loss of the blink reflex.

Dermal exposure to aerosolized OC results in intense burning pain, tingling, edema, erythema, and occasionally blistering. Topical application of capsaicin has been reported to deplete the skin of substance P, somatostatin, prostaglandin, and acetylcholine.[125] Studies by Wallengren demonstrated that topical pre-treatment with capsaicin enhances different experimental inflammations including allergic dermatitis.[174] Multiple exposures of the skin over a period of minutes exaggerate the response. It is postulated that capsaicin amplifies inflammation via the release of substance P from the skin.

3. Mutagenicity and Carcinogenicity

There is wide concern regarding the mutagenic and carcinogenic potential of capsaicinoids. The mutagenic potential of capsaicinoids has been evaluated in both microbial and mammalian mutation assays. The mutagenicity of capsaicinoids has been extensively tested in the Ames (*S. typhimurium*) assay.[175–179] Buchanan and co-workers evaluated the mutagenicity of chili pepper oleoresins and capsaicinoids.[175] Neither the oleoresin nor the purified capsaicin produced mutations in *S. typhimurium.* Toth et al. reported that purified capsaicinoids exhibited mutagenic activity in the presence of liver-activating enzymes; however, the material was non-mutagenic when mouse liver fractions were used.[176] Damhoeri and co-workers studied the mutagenic potential of capsicum pepper (oleoresins) using *S. typhimurium* in the absence of metabolic activation.[177] Under the conditions of the assay, the oleoresins were found to be mutagenic. Nagabhushan and Bhide studied the mutagenicity of capsaicin in *S. typhimurium* strains with and without the S-9 liver fraction.[178] In mutagenicity studies by Gannett et al., capsicum and the ethanol extract of red pepper were evaluated using the TA 98 and TA 1535 strains of *S. typhimurium* in the absence and presence of metabolic activation.[179] Findings suggested that capsaicin and the pepper extract were not mutagenic. In the rec+/rec^{-} assay, capsaicinoids were non-mutagenic for *B. subtilis.*[180] A number of studies have also been conducted using mammalian cells (V79 cell line) to ascertain the mutagenic potential for capsaicinoids.[178,179,181] In the V79 mammalian test system, Nagabhushan and Bhide reported that capsaicin was non-mutagenic.[178] However, studies by Gannett et al. and Lawson and Gannett using the V79 cell line suggested that capsaicin and capsaicinoids were genotoxic.[179,181] Naghabhusahan and Bhide studied the mutagenic potential of capsaicin using the Micronucleus Mutation Assay and found capsaicin positive for mutagenicity.[178] The mutagenic potential of capsaicin was assessed in the Dominant Lethal Assay by Narasimhamurthy and Narasimhamurthy and found not to be mutagenic.[182] Overall, in spite of equivocal findings regarding the mutagenic potential, capsaicin and capsaicinoids should be regarded as genotoxic.

Capsaicin has been reported to induce mucous fibrosis in the oral cavity and may play a role in the development of esophageal cancer.[183,184] When administered in the diet, capsaicin induced cancer in the mouse duodenum.[176] A rodent carcinogenesis bioassay to assess the carcinogenic potential of capsaicin was conducted by Toth and Gannet.[185] Increases were noted in the incidence of benign tumors (polyploid adenomas) in the cecum of treated animals. An increased rate of malignant tumors, however, was not evident. Chronic treatment with capsaicin appeared not to alter the general health of the animals, influence growth rate, or alter body weight. The effect of capsaicin on 12-O-tetradecanoylpholrol-13-acetate (TPA), widely used in tumor promotion studies, was examined by LaHann[186] and Sasajima et al.[187]

LaHann[186] concluded that capsaicin appeared to facilitate the onset of TPA-induced tumorigenesis and that capsaicin might increase the risk of skin cancer. In the studies of Sasajima et al.,[187] capsaicin induced ornithine decarboxylase (ODC) activity, an enzyme used as an index of tumor promoting capability. There appears to be sufficient evidence that capsaicin may pose a tumorigenic threat.

4. Metabolism, Metabolic Fate, and Mechanisms

Capsaicin and capsaicinoids undergo bioconversion, which involves oxidative and non-oxidative pathways. The highest enzymatic activity is found in the liver, followed by extrahepatic tissues (e.g., kidney, lung, and small intestine). Saria et al. studied the distribution of capsaicin in tissues of rats following systemic administration.[188] Uptake in the CNS was rapid, and high levels of capsaicin were detected following i.v. dosing. Slow diffusion from the site of application was noted on s.c. administration; however, detectable levels of capsaicin were found in various tissues. Kawada and Iwai studied the *in vivo* and *in vitro* metabolism of the capsaicin analog, dihydrocapsaicin, in rats.[189] The parent compound was metabolized to metabolic products that were excreted in the urine mostly as glucuronides. The metabolic processes involved in the bioconversion of capsaicin and analogs were initially studied by Lee and Kumar.[190] They demonstrated the conversion to catechol metabolites via hydroxylation on the vanillyl ring moiety—findings later confirmed by Miller et al.[191] The conversion of capsaicin by the liver mixed-function oxidase system to arene oxide is one example of metabolism to an electrophilic metabolite. Another pathway leading to highly reactive intermediates involves the formation of a phenoxy radical, which in turn may undergo subsequent conversion to a quinone type product.[192] In addition to the above oxidative pathways, the alkyl side chain of capsaicin is also susceptible to enzymatic oxidation (oxidative deamination).[193] Capsaicin may also undergo non-oxidative metabolism via the hydrolysis of the acid-amide bond to yield vanillylamine and fatty acyl moieties[189,194,195] (see Figure 11.6).

Presently, no definitive mechanism can solely account for capsaicin-mediated toxicity. A number of metabolic pathways are involved in the bioconversion to electrophilic metabolites (e.g., arene oxide, phenoxy radical, quinone). These moieties can interact with nucleophilic sites of macromolecules such as proteins, DNA, and RNA and are thought to be critical in the etiology of capsaicin-induced toxicity, mutagenicity, and carcinogenicity. The potential of covalent binding with microsomal protein, for example, may account for the impact of capsaicin on xenobiotic metabolizing enzymes and liver toxicity. Concerning mitochondrial energy metabolism, Yagi postulated that capsaicin and dihydrocapsaicin produce repression of NADH-quinone oxidoreductase activity, which confirms findings suggesting capsaicin induces inhibitory effects on hepatic mitochondrial bioenergetics.[196]

5. Human Toxicology

Studies have been published concerning the human response to inhaled capsaicin.[145,197–201] The human pharmacology of capsaicin has been reviewed by Fuller.[128] Watson et al. described the clinical effects in individuals exposed to oleoresin

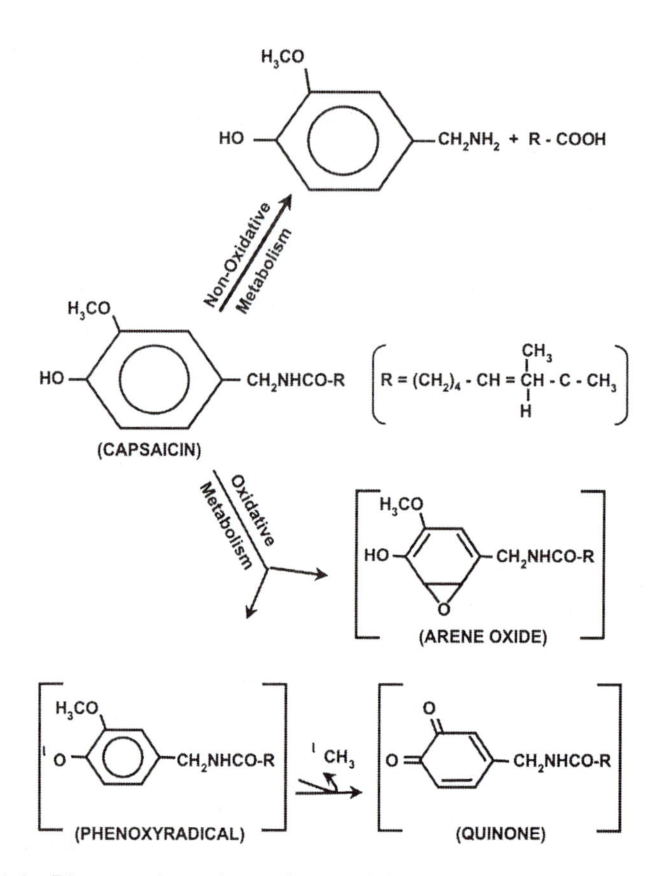

FIGURE 11.6 Bioconversion pathways for capsaicin.

capsicum.[202] The probable lethal oral dose of capsaicin for humans is considered to be 0.5 to 5.0 g/kg.[203] The upper respiratory tract effects have been described.[199,204] Healthy young human adult subjects who were challenged intra-nasally with capsaicin manifested rhinorrhea, sneezing, nasal burning, and congestion.[199] Capsaicin application to the nasal mucosa produced a painful sensation and copious secretion of nasal fluid, and these effects undergo desensitization after repeated application.[204] Studies by Geppetti et al. support the hypothesis that the therapeutic effectiveness of capsaicin treatments in painful diseases might not be linked to nerve fiber degeneration due to the neurotoxic effect of capsaicin, but might rely on desensitization of the mechanism activated by capsaicin on the nerve terminal.[204]

The larynx may represent the primary site of stimulation of inhaled capsaicin.[205] Bronchoconstriction has been the subject of a number of human studies on capsaicin.[145,146,197] Fuller and co-workers demonstrated that when inhaled by humans, capsaicin caused a dose-dependent bronchoconstriction that was the same as in asthmatics and smokers.[145] The majority of subjects manifested coughing and all reported retrosternal discomfort. The studies by Fuller and colleagues confirmed that the bronchoconstrictor reflex following capsaicin stimulation in animals is also present in humans.[145,197] The capsaicin-induced bronchoconstriction and the release of substance P, a neuropeptide, are caused by stimulation of the C-fibers of the non-myelinated afferent fibers. These studies and those using isolated human airway preparations showed that repeated dosing causes tachyphylaxis. In humans, the mechanism of bronchoconstriction following inhalation of capsaicin is uncertain, but possible mechanisms can be inferred from animal studies. Capsaicin has been shown to release substance P, which can cause bronchoconstriction directly by activation of specific receptors or by release of histamine and other mediators. Capsaicin may also cause reflex bronchoconstriction by stimulating C fibers in both pulmonary and bronchial circulation. Therefore, bronchoconstriction could be secondary to substance P release or to a vagal reflex. In addition to the altered neurophysiology of sensory neurons in the airway mucosa, capsaicin also induces the release of tachykinins and neurokinin A. These biologically active substances in turn induce neuromediated inflammation of the epithelium, airway blood vessels, glands, and smooth muscle, which lead to bronchoconstriction, mucous secretion, enhanced vascular permeability, and neutrophil chemotaxis.[206–208]

E. Diphenylaminochloroarsine (Adamsite)

1. Toxicology and Physiological Effects

As previously stated, riot-control agents may be classified as to type (e.g., lacrimators, vomiting agents), based on a salient physiological effect. Diphenylaminochlorarsine (DM) is one of several compounds that include diphenylchloroarsine (DA), diphenylcyanoarsine (DC), and chloropicrin, which are classified militarily as vomiting agents. DM has been characterized as both a vomiting agent and sternutator and was known as adamsite during World War I. DM, which can cause great discomfort, has also been used as a riot-control agent, only by the United States according to Cookson and Nottingham.[12] DM is more toxic than other riot-control compounds—it is considered a potentially dangerous agent in view of a warning per Field Manual 3-l0 that DM may not be used " . . . In any operation where deaths are not acceptable." The estimated human LCt_{50} is 11,000 mg · min/m^3, as reported by Sidell.[15] DM produces symptoms of slightly delayed onset and a relatively long recovery period. DM-related effects do not appear immediately as in the case of CN, CS, and CR. DM-induced effects occur in about 3 min after exposure begins and, depending on the severity of the exposure effects, may last for several hours.[24,209] Unlike the other tear agents, DM is more likely to cause prolonged systemic effects. Signs and symptoms include eye irritation, upper respiratory tract irritation, uncontrolled sneezing and coughing, choking, headache, acute pain, tightness in the chest, nausea, and vomiting. Additionally, DM can cause unsteady gait, weakness in the limbs, and trembling. Ballantyne indicated mental depression as a prominent symptom following exposure to DM.[24] Exposure to high concentrations can result in serious illness as a result of pulmonary damage and edema or death.[209]

A number of investigations on the physiological effects of DM in various species of animals, including monkeys, have been conducted and are summarized in a National Academy of Sciences Report,[6] by McNamara et al.[31] and by Owens et al.[210] Following acute exposure to DM, animals exhibited ocular and nasal irritation, hyperactivity, salivation, labored breathing, ataxia, and convulsions. Punte et al. have reported the results of acute inhalation toxicity studies in several species following exposure (5 to 90 min) to high aerosol concentrations of irritant compounds which included DM.[211] Toxic signs observed in animals were hyperactivity, ocular and nasal irritation, lacrimation, salivation, respiratory distress, and lethargy. Histopathologic examination revealed no abnormalities below an inhaled dose of 500 mg · min/m^3 of DM. LCt_{50} estimates were as follows: rat, 3,700 mg · min/m^3; mouse, 22,400 mg-min/m^3; and guinea pig, 7,900 mg · min/m^3. The theoretical dose received can be calculated from the respiratory volume, the LCt_{50}, and the estimated percent retention. The computed inhaled LD_{50s} for DM are as follows: rat, 14.1 mg/kg; mouse, 17.9 mg/kg; and guinea pig, 2.4 mg/kg. Animals exposed to DM at a dosage of 500 mg · min/m^3 did not exhibit pathologic changes. Animals sacrificed or dying after exposure to DM manifested hyperemia of the trachea, pulmonary congestion and edema, and pneumonia. The clinical and pathologic findings parallel that observed on exposure to pulmonary irritants. Striker and co-workers studied the effects of DM in monkeys exposed to test article at varying concentrations and exposure periods (855 mg/m^3 [3 min]; 1,708 mg/m^3 [5 min]; and 2,615 mg/m^3 [11 mins]).[212] Toxic effects noted at the lowest exposure parameters were limited to a single animal exhibiting some oral and nasal discharge and with diminished response to stimuli. Exposure to a Ct of 8,540 mg · min/m^3 resulted in ocular and nasal irritation, conjunctival congestion, facial erythema, and decreased responses—all signs abated within 24 h. Exposure to a Ct of 28,765 mg · min/m^3 resulted in hyperactivity, copious nasal discharge, conjunctival congestion, marked respiratory distress, and gasping and gagging in all exposed monkeys. Eight deaths had occurred within 24 h in the high-exposure group. Necropsy of the high-dose group revealed congested and extremely edematous lungs. Microscopic examination revealed ulceration of the tracheobronchial tree and pulmonary edema. Additional studies in monkeys were also conducted by Striker et al.[213] The effects of "low" concentrations of DM were evaluated. Animals were exposed to DM at target concentrations of 100 and 300 mg/m^3 for exposure periods of 2 to 60 min and 2 to 40 min, respectively. A progression of toxic signs, characteristic of irritant gases, was noted as the exposure times were increased. At the maximum Ct of 13,200 mg · min/m^3, animals exhibited nausea and vomiting, oral and nasal discharge, and conjunctival congestion. At Cts below 1,296 mg · min/m^3, signs were restricted to blinking. Serious effects involving the eyes have been characterized as necrosis of the corneal epithelium on exposure to DM.[30]

2. Human Toxicology

The human toxicology of DM has been reviewed by Ballantyne,[24] McNamara et al.,[31] and Owens et al.[210] The earliest human study describing the effects following inhalation exposure to adamsite was that of Lawson and Temple.[214] Punter et al.[118] and

Gongwer et al.[215] investigated the effects of varying concentrations of DM on human subjects. Punte and co-workers investigated the onset and persistency of effects following exposure to aerosolized DM and other irritant compounds in a small group of human subjects. The dosage had not exceeded 100 mg · min/m^3, which was considered the maximum "safe" inhaled dose for man. Many of the experiments were terminated so as not to exceed the safe dosage. Subjects reported experiencing a burning sensation of the nose, throat, and chest, coughing and sneezing, and salivation. Several of the symptoms persisted for up to 2 h after termination of exposure. Based on their findings, the estimated ECt_{50} for irritation (3-min exposure) was 19 mg · min/m^3. The dosage (Ct) required to elicit vomiting and nausea, however, could not be established. Additional human toxicological data were also examined by McNamara et al.[31] McNamara cited a dosage of 49 mg · min/m^3 necessary to cause vomiting and nausea, based on studies in humans exposed to DM at Ct between 7 and 236 mg · min/m^3. However, high confidence in the above estimate is lacking since the estimate was founded on highly variable data. Ballantyne estimated a dosage of 370 mg · min/m^3 was required to cause nausea and vomiting.[24] Inhalation of high concentrations of DM has resulted in severe pulmonary damage and death.[209]

VIII. SUMMARY

The desired effect of all riot-control agents is the temporary incapacitation of individuals via irritation of the mucous membranes and skin. Generally, riot-control agents can produce acute, site-specific toxicity where sensory irritation occurs (e.g., eyes, respiratory tract, and skin). The early riot-control compounds such as chloroacetophenone (CN) and chlorodihydrophenarsazine (DM) have been replaced with "safer" compounds such as CS and oleoresin of capsicum (OC). As much is known of the toxicity of riot-control agents such as chlorobenzylidene malononitrile (CS) as for many regulated chemicals such as pesticides. However, the widespread use of riot-control agents raises questions and concerns regarding their health effects and safety. For modern riot-control agents (e.g., CS and CR), there exists a large margin between dosages that produce harassment and dosages likely to cause adverse health effects. Yet, despite the low toxicity of modern riot-control agents, these compounds are not entirely without risk. The risk of toxicity increases with higher exposure doses and prolonged exposure durations. Pulmonary, dermal, and ocular damage may occur on exposure to high concentrations of these substances, particularly on exposure to DM or CN. Furthermore, it is best recognized that exposure to riot-control agents in enclosed spaces may produce significant toxic effects irrespective of the riot-control agent in question. Also, misuse of riot-control agents has resulted in varying degrees of eye and/or skin damage. Additionally, it is important to note that the intense physical discomfort and anxiety associated with riot-control chemicals may elicit cardiovascular changes that may have significant implications for individuals with pre-existing disease. Reported lethalities are few involving riot-control agents, and then only under conditions of prolonged exposure and high concentrations. Recently, concern has focused on the deaths resulting from law enforcement use of OC, a riot-control agent generally regarded as safe since it is a natural product. As with other xenobiotics, not enough is known concerning the long-term/chronic effects of riot-control agents. Repeated-dose studies have been conducted for some of the riot-control agents; however, additional studies are needed to address concerns magnified by the potential of multiple exposures during situations of civil unrest. Clearly, there is considerable need for additional studies, both applied and basic, to define and delineate the biological and toxicological actions of riot-control agents.

REFERENCES

1. Prentiss, A.M., *Chemicals in War: A Treatise on Chemical Warfare,* McGraw-Hill Book Co., Inc., New York, 1937.
2. Jacobs, M.B., *War Gases, Their Identification and Decontamination,* Interscience Publishers, Inc., New York, 1942.
3. Waitt, A.H., *Gas Warfare,* Duell, Sloan and Pearce, New York, 1942.
4. Sartori, M., *The War Gases,* D. VanNostrand, Co., Inc., New York, 1943.
5. Sidell, F.R. and Goldwyn, P.M., Chemical and biological weapons: A primer, *N. Engl. J. Med.,* 274, 21, 1966.
6. National Academy of Sciences, *Possible Long-Term Health Effects of Short-Term Exposure to Chemical Agents,* Vol 2, *Cholinesterase Reactivators, Psychochemicals, and Irritants & Vesicants,* National Academy Press, Washington, DC, 1984.
7. Robinson, J., Chemical weapons, in *Chemical and Biological Warfare,* Rose, S., Ed., G. Harrap, London, 1968.
8. Volmer, K., La querre chimique, *Genie,* 10, 1969.
9. World Health Organization (WHO), *Health Aspects of Chemical and Biological Weapons,* World Health Organization, Geneva, Switzerland, 1970.
10. Hu, H., Toxicodynamics of riot-control agents (lacrimators), in *Chemical Warfare Agents,* Somani, S.M., Ed., Academic Press Inc., New York, 1992, chap. 11.
11. National Research Council (NRC), *Chemical and Biological Terrorism,* National Academy Press, Washington, DC, 1999.
12. Cookson, J. and Nottingham, J., *A Survey of Chemical and Biological Warfare,* Monthly Review Press, New York, 1969, chap. 5.
13. Verwey, W.D., *Riot-Control Agents and Herbicides in War,* A.W. Sijthoff, Leyden, 1977.
14. Departments of the Army, The Navy, and The Air Force, Field Manual (Army FM 8–285, Navy Navmed P-5041, Air Force AFM 160-11), *Treatment of Chemical Agent Casualties and Conventional Military Chemical Injuries,* Depts. of the Army, Navy, AirForce, Washington, DC, 1990.
15. Sidell, F.R., Riot-control agents, in *Textbook of Military Medicine, Medical Aspects of Chemical and Biological Warfare,* Office of the Surgeon General, U.S. Army, TMM Publications, Borden Institute, Washington, DC, 1997, chap. 12.
16. Takafuji, E.T. and Kok, A.B., The chemical warfare threat and the military health care provider, in *Textbook of Military Medicine, Medical Aspects of Chemical and Biological Warfare,* Office of the Surgeon General, U.S. Army, TMM Publications, Borden Institute, Washington, DC, 1997, chap. 4.
17. Office of the Press Secretary (The White House), *Press Release dated June 23, 1994 (Subject: Riot Control Agents),* Washington, DC.

18. Swearengen, T.F., *Tear Gas Munitions,* Charles C Thomas, Springfield, IL, 1966.
19. Royer, J. and Gainet, F., Ocular effects of ethyl bromoacetate tear gas, *Bull. Soc. Ophthalmol.* (France), 73, 1165, 1973.
20. Rose, S. and Smith, R., CS—a case for concern, *New Sci.,* September 1969.
21. Jones, G.R.N., CS in the balance, *New Sci. Sci. J.,* 50, 690, 1971.
22. Rothschild, J.H., *Tomorrow's Weapons,* McGraw-Hill Book Co., New York, 1964.
23. Corson, B.B. and Stoughton, R.W., Reactions of alpha, beta unsaturated dinitriles, *J. Am. Chem. Soc.* 50, 2825, 1928.
24. Ballantyne, B., Riot-control agents (biomedical and health aspects of the use of chemicals in civil disturbances), in *Medical Annual,* Scott, R.B. and Frazer, J. Eds., Wright and Sons, Bristol, 1977, 7.
25. MacLeod, I.F., Chemical mace: ocular effects in rabbits and monkeys, *J. Foren. Sci.,* 14, 34, 1969.
26. Midtbo, A., Eye injury from tear gas, *Acta Ophthalmol.,* 42, 672, 1964.
27. Hoffman, D.H., Injuries of the eyes from short-range discharge of tear gas weapons, *Klin. Monstsbl. Augenheilkd,* 147, 625, 1965.
28. Bregeat, P., Ocular injuries by lacrimogenic agents, *Bull. Soc. Ophthalmol.* (France), 68, 531, 1968.
29. Levine, R.A. and Stall, C.J., Eye injury caused by tear-gas weapons, *Am. J. Ophthalmol.,* 65, 497, 1968.
30. Grant, W.M., *Toxicology of the Eye,* 3rd ed., Charles C Thomas Publishers, Springfield, IL, 1986, 882.
31. McNamara, B.P., Owens, E.J., Weimer, J.T., Ballard, T.A., and Vocci, F.J., Toxicology of riot control chemicals CS, CN, and DM, *Edgewood Arsenal Technical Report, EATR—4309, (Nov. 1969),* Dept. of the Army, Edgewood Arsenal Medical Research Laboratory, Edgewood Arsenal, MD.
32. Punte, C.L., Weimer, J.T., Ballard, T.A., and Wilding, J.L., Toxicologic studies on o-chlorobenzylidene malononitrile, *Toxicol. Appl. Pharmacol.,* 4, 656, 1962.
33. Punte, C.L., Owens, E.J., and Gutentag, P.J., Exposures to ortho-chlorobenzylidene malononitrile, *Arch. Environ. Hlth.,* 6, 72, 1963.
34. Owens, E.J. and Punte, C.L., Human respiratory and ocular irritation studies utlizing o-chlorobenzylidene malononitrile aerosols, *Am. Ind. Hyg. Assoc. J.,* 24, 262, 1963.
35. Beswick, F.W., Holland, P., and Kemp, K.H., Acute effects of exposure to ortho-chlorobenzylidene malononitrile (CS) and the development of tolerance, *Br. J. Ind. Med.,* 29, 298, 1972.
36. Ballantyne, B. and Callaway, S., Inhalation toxicology and pathology of animals exposed to o-chlorobenzylidene malononitrile (CS), *Med. Sci. Law,* 12, 43, 1972.
37. Ballantyne, B., Gall, D., and Robson, D.C., Effects on man of drenching with dilute solutions of o-chlorobenzylidene malononitrile (CS) and dibenz (b,f)-1:4-oxazepine (CR), *Med. Sci. Law,* 16, 159, 1976.
38. Ballantyne, B. and Swanston, D.W., The comparative acute mammalian toxicity of 1-chloroacetophenone (CN) and 2-chlorobenzylidene malononitrile (CS), *Arch. Toxicol.,* 40, 75, 1978.
39. U.S. Army Edgewood Arsenal, Chemical Research and Development Laboratories, *Special Summary Report on the Toxicology of CN, CS, and DM,* Directorate of Medical Research, September 1965, Edgewood Arsenal, MD, 1965.
40. Marrs, T.C., Colgrave, H.V., Cross, N.L., Gazzard, M.F., and Brown, R.F.R., A repeated dose study of the toxicity of inhaled 2-chlorobenzylidene malononitrile (CS) aerosol in three species of laboratory animals, *Arch. Toxicol.* 52, 183, 1983.

41. Gutentag, P.J., Hart, J., Owens, E.J., and Punte, C.L., The evaluation of CS aerosol as a riot control agent in man, *Technical Report CWLR 2365,* April 1960, U.S. Army Chemical Warfare Laboratories, Army Chemical Center, 1960.
42. Bowers, M.B., Owens, E.J., and Punte, C.L., Interim report of CS exposures in plant workers, *CWL Technical Memorandum 24–50,* June 1960, U.S. Army Chemical Warfare Laboratories, Army Chemical Center, MD, 1960.
43. Hellerich, A., Goldman, R.H., Bottiglieri, N.G., and Weimer, J.T., The effects of thermally-generated CS aerosols on human skin, *Technical Report EATR 4075,* Jan. 1967. U.S. Army Medical Research Laboratory, Edgewood Arsenal, MD, 1967.
44. Rothberg, S., Skin sensitization potential of the riot-control agents BBC, DM, CN and CS in guinea pigs, *Mil. Med.,* 135, 552, 1970.
45. Upshall, D.G., Effects of o-chlorobenzylidene malononitrile (CS) and the stress of aerosol inhalation upon rat and rabbit embryonic development, *Toxicol. Appl. Pharmacol.,* 24, 45, 1973.
46. Von Daniken, A. , Friederich, U., Lutz, W.K., and Schlatter, C., Tests for mutagenicity in salmonella and covalent binding to DNA and protein in the rat of the riot control agent o-chlorobenzylidene malononitrile (CS), *Arch. Toxicol.,* 49, 15, 198l.
47. Zeiger, E., Anderson, B., Haworth, S., Lawlor, T., Mortelmans, K., and Speck,W., Salmonella mutagenicity tests. III. Results from the testing of 255 chemicals, *Environ. Mutagen,* 9 (Suppl. 9), 1, 1987.
48. Reitveld, E.C., Delbressine, L.P., Waegemaekers, T.H., and Seutter-Berlage, F., 2-Chlorobenzylmercapturic acid, a metabolite of the riot control agent 2-chlorobenzylidene malononitrile (CS) in the rat, *Arch. Toxicol.,* 54, 139, 1983.
49. Wild, D., Eckhardt, K., Harnasch, D., and King, M.T., Genotoxicity study of CS (ortho-chlorobenzylidene malononitrile) in salmonella, drosophila, and mice, *Arch. Toxicol.* 54, 167, 1983.
50. National Institutes of Health (National Toxicology Program), *Toxicology and Carcinogenesis Studies of CS2 (94% o-chlorobenzalmalononirile) (CAS No. 2698-41-1)* in *F344/N Rats and B6C3Fl Mice (Inhalation Studies),* NTP TR 377, March 1990, National Toxicology Program, Research Triangle Park, NC.
51. McGregor, D.B., Brown, A., Cattanach, P., Edwards, I., Mcbride, D., and Caspary, W.J., Responses of the L5l78Y tk + /tk- mouse lymphoma cell forward mutation assay II: 18 coded chemicals, *Environ. Mol. Mut.,* 11, 91, 1988.
52. Schmid, E., Bauchinger, M., Ziegler-Skylakaksis, K., and Andrae, U., 2-Chlorobenzylidene malononitrile (CS) causes spindle disturbances in V79 Chinese hamster cells, *Mutat. Res.,* 226, 133, 1989.
53. Leadbeater, L., The absorption of ortho-chlorobenzylidenemalononitrile (CS) by the respiratory tract, *Toxicol. Appl. Pharmacol.,* 25, 101, 1973.
54. Patai, S. and Rappaport, Z., Nucleophilic attacks on carbon-carbon double bonds. Part II. Cleavage of arylmethylenemaloninitriles by water in 95% ethanol, *J. Chem. Soc.* (London), Part 1 (71), 383, 1962.
55. Nash, J.B., Doucet, B.M., Ewing, P.L., and Emerson, G.A., Effects of cyanide antidotes and inanition on acute lethal toxicity of malononitrile in mice, *Arch. Int. Pharmacodyn.,* 84, 385, 1950.
56. Stern, J., Weil-Malherbe, H., and Green, R.H., The effects and the fate of malononitrile and related compounds in animal tissues, *Biochem. J.,* 52, 114, 1952.
57. Cucinell, S.A., Swentzel, K.C., Biskup, R., Snodgrass, H., Lovre, S., Stark, W., Feinsilver, L., and Vocci, R., Biochemical interactions and metabolic fate of riot-control agents, *Fed. Proc.,* 30, 86, 1971.

58. Feinsilver, L., Chambers, H.A., Vocci, F.J., Daasch, L., and Berkowitz, L.M., Some metabolites of CS from rats, *Edgewood Arsenal Technical Report, Series No. 4521,* (May 1971), 1971.
59. Leadbeater, L., Sainsbury, G.L., and Utley, D., o-Chlorobenzyl malononitrile, a metabolite formed from o-chlorobenzylidene malononitrile (CS), *Toxicol. Appl. Pharmacol.,* 25, 111, 1973.
60. Paradowski, M., Metabolism of toxic doses of o-chlorobenzylidene malononitrile (CS) in rabbits, *Polish J. Pharmcacol. Pharm.,* 31, 563, 1979.
61. Brewster, K., Harrison, J.M., Leadbeater, L., Newman, J., and Upshall, D.G., The fate of 2-chlorobenzylidene malononitrile (CS) in rats, *Xenobiotica,* 17, 911, 1987.
62. Frankenberg, L. and Sorbo, B., Formation of cyanide from o-chlorobenzylidene malononitrile and its toxicological significance, *Arch. Toxikol.,* 31, 99, 1973.
63. Swentzel, K.C., Merkey, R.P., Cucinell, S.A., Weimer, J.T., and Vocci, F.J., Unchanged thiocyanate levels in human subjects following exposure to CS aerosol, *Edgewood Arsenal Technical Memorandum, EATM 100–8,* June 1970, Edgewood Arsenal MD, 1970.
64. Lovre, S.C. and Cucinell, S.A., Some biological reactions of riot-control agents, *Technical Report EATR, 4399,* U.S. Army Medical Research Laboratory, Edgewood Arsenal, MD, 1970.
65. Mackworth, J.F., The inhibition of thiol enzymes by lachrymators, *Biochem. J.,* 42, 82, 1948.
66. Jones, G.R.N. and Israel, M.S., Mechanism of toxicity of injected CS gas, *Nature,* 228, 1315, 1970.
67. Himsworth et al., *Report of the Enquiry into the Medical and Toxicological Aspects of CS (Orthochlorobenzylidene Malononitrile). Part I. Enquiry into the Medical Situation following the Use of CS in Londonderry on 13th and 14th August, 1969,* Cmnd 4173, H.M.S.O, London, 1969.
68. Himsworth et al., *Report of the Enquiry into the Medical and Toxicological Aspects of CS (Orthochlorobenzylidene Malononitrile). Part II. Enquiry into Toxicological Aspects of CS and Its Use for Civil Purposes,* Cmnd 4775, H.M.S.O., London, 1971.
69. Park, S. and Giammona, S.T., Toxic effects of tear gas on an infant following prolonged exposure, *Am. J. Dis. Child.,* 123, 245, 1972.
70. Anderson, P.J., Lau, G.S.N., Taylor, W.R.J., and Critchley, J.A., Acute effects of the potent lacrimator o-chlorobenzylidene malononitrile (CS) tear gas, *Hum. Exper. Toxicol.,* 15, 461, 1996.
71. Krapf, R. and Thalmann, H., Akute exposition durch CS-rauchgas und linische beobachtungen, *Schweiz. Med. Wochenschr.,* 11, 2056, 1981.
72. United States General Accounting Office (GAO), *Use of U.S. Manufactured Tear Gas in the Occupied Territories, GAO/NSIAD–89–128,* Washington, DC, 1989.
73. Physicians for Human Rights, *The Casualties of Conflict: Medical Care and Human Rights in the West Bank and Gaza Strip,* Physicians for Human Rights, Somerville, MA, 1988.
74. Ballantyne, B., The acute mammalian toxicology of dibenz (b,f)-1:4-oxazepine, *Toxicology,* 8, 347, 1977.
75. Pattle, R.E., Schock, C., Dirnhuber, P., and Creasy, J.M., Lung surfactant and organelles after an exposure to dibenzoxazepine (CR), *Br. J. Exper. Pathol.,* 55, 213, 1974.
76. Colgrave, H.F., Brown, R., and Cox, R.A., Ultrastructure of rat lungs following exposure to aerosols of dibenzoxazepine (CR), *Br. J. Exper. Pathol.,* 60, 130, 1979.
77. Lundy, P.M. and McKay, D.H., Mechanism of the cardiovascular activity of dibenz [b,f][1,4] oxazepine (CR) in cats, *Suffield Technical Paper No. 438,* Defence Research Establishment, Ralston, Alberta, Canada, 1975.

78. Marrs, T.C., Colgrave, H.F., and Cross, N.L., A repeated dose study of the toxicity of technical grade dibenz-(b,f)-1,4 oxazepine in mice and hamsters, *Toxicol. Lett.,* 17, 13, 1983.
79. Grady, H.G. and Stewart, H.L., Histogenesis of induced pulmonary tumors in strain A mice, *Am. J. Pathol.,* 16, 417, 1940.
80. Stewart, H.L., Dunn, T.B., Snell, K.C., and Deringer, M.K., Tumors of the respiratory tract, in *WHO Monographs on the Pathology of Laboratory Animals,* Vol I.: *The Mouse,* Turusov, S.U., ed., International Association for Research on Cancer (IARC), Lyons, 1979, 251.
81. McNamara, B.P., Vocci, F.J., Owens, E.J., Ward, D.M., and Anson, N.M., The search for an effective riot-control agent solvent system for large volume dispersers, *Edgewood Arsenal Technical Report, 4675 (October 1972),* 1972.
82. Owens, E.J., Weimer, J.T., Ballard, T.A., Ford, D.F., Samuel, J. B., Hopcus, M.W., Merkey, R.P., and Olson, J.S., Ocular, cutaneous, respiratory, intratracheal toxicity of solutions of CS and EA 3547 in glycol and glycol ether in animals, *Technical Report No. 4446, October 1970,* Edgewood Arsenal, Aberdeen Proving Ground, MD, 1970.
83. Biskup, R.C., Swentel, K.C., Lochner, M.A., and Fairchild, D.G., Toxicity of 1% CR in propylene glycol/water (80/20), *Technical Report EB-TR-75009, May 1975,* Edgewood Arsenal, Aberdeen Proving Ground, MD, 1975.
84. Marrs, T.C., Gray, M.I., Colgrave, H.F., and Gall, D., A repeated dose study of the toxicity of CR applied to the skin of mice, *Toxicol. Lett.,* 13, 259, 1982.
85. Higginbottom, R. and Suschitzky, H., Synthesis of heterocyclic compounds II. Cyclization of O-nitrophenyl oxygen, *J. Chem. Soc.,* 2367, 1962.
86. Rengstorff, R.H., Petrali, J.P., Mershon, M., and Sim, V.M., The effect of the riot control agent dibenz (b,f)-1:4-oxazepine (CR) in the rabbit eye, *Toxicol. Appl. Pharmacol.,* 34, 45, 1975.
87. Ballantyne, B. and Swanston, D.W., The irritant potential of dilute solutions of ortho-chlorobenzylidene malononitrile (CS) on the eye and tongue, *Acta Pharmacol. Toxicol.,* 32, 266, 1973.
88. Ballantyne, B. and Swanston, D.W., The irritant effects of dilute solutions of dibenzoxazepine (CR) on the eye and tongue, *Acta Pharmacol. Toxicol.,* 35, 412, 1974.
89. Ballantyne, B., Gazzard, M.F., Swanston, D.W., and Williams, P., The comparative ophthalmic toxicology of l-chloroacetophenone (CN) and dibenz (b,f)- 1:4-oxazepine (CR), *Arch. Toxicol.,* 34, 183, 1975.
90. Holland, P., The cutaneous reactions produced by dibenzoxazepine (CR), *Br. J. Derm.,* 90, 657, 1974.
91. Upshall, D.G., The effects of dibenz (b,f)-1,4-oxazepine (CR) upon rat and rabbit embryonic development, *Toxicol. Appl. Pharmacol.,* 29, 301, 1974.
92. Colgrave, H.F., Lee, C.G., Marrs, T.C., and Morris, B., Repeated-dose inhalation toxicity and mutagenicity status of CR (dibenz (b.f.)-1,4, oxazepine), *Br. J. Pharmacol.,* 78, 169p, 1983.
93. Husain, K., Kumar, P., and Malhotra, R.C., A comparative study of biochemical changes induced by inhalation of aerosols of o-chloroacetophenone and dibenz (b,f)-1,4-oxazepine in rats, *Indian J. Med. Res.,* 94, 76, 1991.
94. Balfour, D.J., Studies on the uptake and metabolism of dibenz (b,f)-1:4-oxazepine (CR) by guinea pig cornea, *Toxicology,* 9, 11, 1978.
95. Harrison, J.M., Clarke, R.J., Inch, T.D., and Upshall, D.G., The metabolism of dibenz (b,f)-1,4-oxazepine (CR): in vivo hydroxylation of 10, 11-dihydrodibenz (b,f)-1,4-oxazepin-11-(1 OH) {nd}one and the NIH shift, *Experentia,* 34, 698, 1978.

96. French, M.C., Harrison, J.M., Inch, T.D., Leadbeater, L., Newman, J., and Upshall, D.G., The fate of dibenz (b,f)-1,4-oxazepine (CR) in the rat, rhesus monkey, and guinea pig. Part I, metabolism *in vivo, Xenobiotica,* 13, 345, 1983.
97. French, M.C., Harrison, J.M., Newman, J., Upshall, D.G., and Powell, G.M., The fate of dibenz (b,f)-1,4-oxazepine (CR) in the rat. Part III, the intermediary metabolites, *Xenobiotica,* 13, 373, 1983.
98. Furnival, B., Harrison, J.M., Newman, J., and Upshall, D.G., The fate of dibenz (b,f)-1,4-oxazepine in the rat. Part II, metabolism *in vitro, Xenobiotica,* 13, 361, 1983.
99. Upshall, D.G., Riot control smokes: Lung absorption and metabolism of peripheral sensory irritants, in *Clinical Toxicology,* Duncan W.A. and Leonard, B.J. Eds., Excerpta Medica, Amsterdam, The Netherlands, 1977, 121.
100. Weigand, D.A. and Mershon, M.M., Cutaneous reaction to EA 3547 in propylene glycol, *Edgewood Arsenal Technical Report No. EATR-4413,* July 1970, U.S. Army Medical Research Laboratory, Edgewood Arsenal, 1970.
101. Ballantyne, B., Beswick, F.W., and Thomas, D., The presentation and management of individuals contaminated with solutions of dibenzoxazepine (CR), *Med. Sci. Law,* 13, 265, 1973.
102. Ashton, I., Cotes, J.E., Holland, P., Johnson, G.R., Legg, S.J., Saunders, M.J., and White, R.G., Acute effect of dibenz (b,f)- 1:4-oxazepine upon the lung function of healthy young men, *J. Physiol.,* 275, 85p, 1978.
103. Bandman, A.L. and Savateyev, N.V., Toxicology of CR, *Voen. Med. Zh.* 3, 84, 1977.
104. Choi, D.D. and Landis, W.G., A review of toxicologic studies on dibenz-l,4-oxazepine (CR), U.S. Army Chemical Systems Laboratory, Edgewood Arsenal, MD, 1982.
105. Beswick, F.W., Chemical agents used in riot control warfare, *Hum. Toxicol.,* 2, 247, 1983.
106. Hogg, M.A.P., (personal communication, as reported by Ballantyne, B. and Swanston, D.W., 1974–see Reference 88).
107. Hu, H., Tear gas—harassing agent or toxic chemical weapon?, *JAMA,* 262, 660, 1989.
108. Kumar, P., Flora, S.J.S., Pant, S.C., Sachan, A.S., Saxena, S. P., and Das Gupta, S., Toxicological evaluation of 1-chloroacetophenone and dibenz(b,f)-1,4-oxazepine after repeated inhalation exposure in mice, *J. Appl. Toxcicol.,* 14, 411, 1994.
109. Hoffman, D.H., Eye burns caused by tear gas, *Br. J. Ophthalmol.,* 51, 265, 1967.
110. Macrae, W.G., Willinsky, M.D., and Basu, P.K., Corneal injury caused by aerosol irritant projectors, *Can. J. Ophthalmol.* 5, 3, 1970.
111. Leopold, I.H. and Lieberman, T.W., Chemical injuries of the cornea, *Fed. Proc. Am. Soc. Exp. Biol.,* 30, 92, 1971.
112. Gaskins, J.R., Hehir, R.M., McCaulley, D.F., and Ligon, E.W., Lacrimating agents (CS and CN) in rats and rabbits—acute effects on mouth, eyes and skin, *Arch. Environ. Hlth.,* 24, 449, 1972.
113. Penneys, N.S., Israel, R.M., and Indgin, S.M., Contact dermatitis due to l-chloroacetophenone and chemical mace, *New Engl. J. Med.,* 281, 1969.
114. Chung, C.W. and Giles, A.I., Sensitization of guinea pigs to alpha-chloroacetophenone (CN) and ortho-chlorobenzylidene malononitrile (CS) tear gas chemicals, *J. Immunol.,* 109, 284, 1972.
115. Holland, P. and White, R.G., The cutaneous reactions produced by o-chlorobenzylidene-malononitrile and ω-chloroacetophenone when applied directly to the skin of human subjects, *Br. J. Derm.,* 86, 150, 1972.
116. National Institutes of Health (National Toxicology Program), *Toxicology and Carcinogenesis Studies of 2-Chloroacetophenone (CAS No. 532–27–4) in F344/N Rats and B6C3Fl Mice (Inhalation Studies), NTP TR 379,* March 1990, National Toxicology Program, Research Triangle Park, NC.

117. Castro, J.A., Effects of alkylating agents on human plasma cholinesterase, *Biochem. Pharmacol.,* 17, 295, 1968.
118. Punte, C.L., Gutentag, P.J., Owens, E.J., and Gongwer, L.E., Inhalation studies with chloroacetophenone, diphenylaminochloroarsine, and pelargonic morpholide—II. human exposures, *Am. Ind. Hyg. Assoc. J.,* 23, 199, 1962.
119. Thorburn, K.M., Injuries after use of the lacrimatory agent chloroacetophenone in a confined space, *Arch. Environ. Hlth.,* 37, 182, 1982.
120. Stein, A.A. and Kirwan, W.E., Chlolroacetophenone (tear gas) poisoning: A clinicopathologic report, *J. Foren. Sci.* 9, 374, 1964.
121. Chapman, A.J. and White, C., Death resulting from lacrimatory agents, *J. Foren. Sci.,* 23, 527, 1978.
122. Maga, J.A., Capsicum, *Crit. Rev. Food Sci. Nutr.,* 6, 177, 1975.
123. Monsereenusorn, Y., Kongsamut, S., and Pezalla, P.D., Capsaicin—a literature survey, *CRC Crit. Rev. Toxicol.,* 321, 1982.
124. Suzuki, T. and Iwai, K., Constituents of red pepper species: Chemistry, biochemistry, pharmacology and food science of the pungent principal of capsicum species, in *The Alkaloids,* 23, Brossi, A., Ed., Academic Press, Orlando, FL, 1984, chap. 4.
125. Buck, S.H. and Burks, T.F., The neuropharmacology of capsaicin: Review of some recent observations, *Pharmacol. Rev.,* 38, 179, 1986.
126. Govindarajan,V.S. and Sathyanarayana, M.N., Capsicum—production, technology, chemistry, and quality. Part V. Impact on physiology, pharmacology, nutrition, and metabolism; structure, pungency, pain and desensitization sequences, *Critical Review of Fd. Sci. Nutri.,* 29, 435, 1991.
127. O'Neill, T.P., Mechanism of capsaicin action: Recent learnings, *Resp. Med.,* 85, 35, 1991.
128. Fuller, R.W., The human pharmacology of capsaicin, *Arch. Int. Pharmacodyn,* 303, 147, 1990.
129. Holzer, P., Capsaicin: Selective toxicity for thin primary sensory neurons, in *Handbook of Exper. Pharmcaol.,* 102, 420, 1992.
130. Busker, R.W. and van Helden, H.P.M., Toxicologic evaluation of pepper spray as a possible weapon for the Dutch police forces. Risk assessment and efficacy, *Am. J. Foren. Med. Pathol.,* 19, 509, 1998.
131. Teranishni, R., Keller, U., Flath, R.A., and Mon, T.R., Comparison of batchwise and continuous steam distillation—solvent extraction recovery of volatiles from oleoresin capsicum, African type (capsicum frutesens), *J. Agric. Food Chem.,* 28, 156, 1980.
132. Games, D.E., Alcock, N.J., van der Greef, J., Nyssen , L.M., Maarse, H., and Ten, M.C., Analysis of pepper and capsicum oleoresins by high performance liquid chromatography—mass spectrometry and field desorption mass spectrometry, *J. Chromatog.,* 294, 269, 1984.
133. Govindarajan, V.S., Capsicum—production, technology, chemistry, and quality. II. Processed products, standards, world production and trade, *Critical Rev. Fd. Sci. Nutr.,* 23, 207, 1986.
134. Cordell, G.A. and Araujo, D.E., Capsaicin: Identification, nomenclature, and pharmacology, *Ann. Pharmacother.,* 27, 330, 1993.
135. Keller, U., Flath, R.A., Mon, R.A., and Teranishi, R., Volatiles from red pepper (capsicum spp.), in Teranishi, R. and Berrera-Beniez, H., Eds., *Quality of Selected Fruits and Vegetables in North America,* Symposium Series 1070, American Chemical Society, Washington, DC, 137, 1980/1981.
136. Clede, B., Oleoresin capsicum, *Law Order,* March, 63, 1993.
137. Zarc International, personal communication, 1993.

138. Mioduszewski, R., unpublished data, 1997.
139. Issekutz, B., Lichtneckert, I., and Nagy, H., Effect of capsaicin and histamine on heat regulation, *Arch. Int. Pharmacodyn.,* 81, 35, 1950.
140. Toh, C.C., Lee, T.S., and Kiang, A.K., The pharmacological actions of capsaicin and analogues, *Br. J. Pharmacol,* 10, 175, 1955.
141. Glinsukon, T., Stitmunnaithum, V., Toskulkao, C., Buranawuti, T., and Tangkrisanavinont, V., Acute toxicity of capsaicin in several animal species, *Toxicon,* 18, 215, 1980.
142. Jansco, G., Kiraly, E., and Jansco-Gabor, A., Pharmacologically-induced selective degeneration of chemosensitive primary sensory neurons, *Nature,* 270, 741, 1977.
143. Russel, J.A. and Lai-Fook, S.J., Reflex bronchoconstriction induced by capsaicin in the dog, *J. Appl. Physiol.,* 47, 961, 1979.
144. Davis, B., Roberts, A.M., Coleridge, H.M., and Coleridge, J.C.B., Reflex tracheal gland secretion evolved by stimulation of bronchal c-fiber in dogs, *J. Appl. Physiol.,* 53, 985, 1982.
145. Fuller, R.W., Dixon, C.M.S., and Barnes, P.J., Bronchoconstrictor response to inhaled capsaicin in humans, *J. Appl. Physiol.,* 58, 1080, 1985.
146. Hathaway, T.J., Higenbottam, T.W., Morrison, J.F., Clelland, C. A., and Wallwork, J., Effects of inhaled capsaicin in heart-lung transplant patients and asthmatic subjects, *Am. Rev. Respir. Dis.,* 148, 1233, 1993.
147. Biggs, D.F. and Goel, V., Does capsaicin cause bronchospasm in guinea-pigs?, *Eur. J. Pharmacol.,* 115, 71, 1985.
148. Adcock, J.J. and Smith, T.W., Inhibition of reflex bronchoconstriction by the opioid peptide BW 443C8l in the anaesthetized cat, *Br. J. Pharmacol.,* 96, 596, 1989.
149. Buchan, P. and Adcock, J.J., Capsaicin-induced bronchoconstriction in the guinea pig: Contribution of vagal cholinergic reflexes, local axon reflexes and their modulation by BW 443C81, *Br. J. Pharmacol.,* 1 105, 448, 1992.
150. Chahl, L.A. and Lynch, A.M., The acute effects of capsaicin on the cardiovascular system, *Acta Physiol. Hung.,* 69, 413, 1987.
151. Porszasz, R. and Szolesanyi, J., Circulatory and respiratory effects of capsaicin and resiniferatoxin on guinea pigs, *Acta Biochim. Biophys. Hung.,* 26, 131, 1991/1992.
152. Szolcsanyi, J., Effect of pain-producing chemical agents on the activity of slowly-conducting afferent fibres, *Acta Physiol. Acad. Sci. Hung.,* 56, 86, 1980.
153. Jansco, G., Kiraly, E., Such, G., Joo, F., and Nagy, A., Neurotoxic effect of capsaicins in mammals, *Acta Physiol. Hung.,* 69, 295, 1987.
154. Lynn, B., The immediate and long-term effects of applying capsaicin to cutaneous nerves, *Acta Physiol. Hung.,* 69, 287, 1987.
155. Lundberg, J.M. and Saria, A., bronchial smooth muscle contraction induced by stimulation of capsaicin-sensitive sensory neurons, *Acta Physiol. Scand.,* 116, 473, 1982.
156. Dray, A., Bettaney, J., and Forster, P., Capsaicin desensitization of peripheral nociceptive fibers does not impact sensitivity to other noxious stimuli, *Neurosci. Lett.,* 99, 50, 1989.
157. Weiss, R., personal communication, 1993.
158. Szolcsanyi, J., Disturbances of thermoregulation induced by capsaicin, *J. Therm. Biol.,* 8, 207, 1983.
159. Frens, J., Pharmacological evidence for a set-point mechanism in thermoregulation, in *Drugs, Biogenic Amines and Body Temperature,* 3rd Symposium on the Pharmacology of Thermoregulation (1976), Banff, Alta, Karger, Basel, 1977, 20.
160. Konietzny, F. and Hensel, H., The effect of capsaicin on the response characteristics of human c-polymodal nociceptors, *J. Therm. Biol.,* 8 , 213, 1983.
161. Nopanitaya, W., Effects of capsaicin in combination with diets of varying protein content on the duodenal absorbtive cells of the rat, *Am. J. Dig. Dis.,* 19, 439, 1974.

162. Nopanitaya, W. and Nye, S.W., Deuodenal mucosal response to the pungent principle of hot pepper (capsicum) in the rat: Light and electron microscopic study, *Toxicol. Appl. Pharmacol.,* 30, 149, 1974.
163. Sambaiah, K., Satyanarayana, M.N., and Rao, M.V., Effect of red pepper (chillies) and capsaicin on fat absorption and liver fat in rats, *Nutr. Rep. Int.,* 18, 521, 1978.
164. Sambaiaha, K., Srinivasan, M.R., Satyanarayana, M.N., and Chandrasekhara, N., Influence of capsaicin on the absorption of amino acids and fat in rats, *J. Fd. Sci. Technol.,* 21, 155, 1984.
165. Kawada, T., Hagihara, K.I., and Iwai, K., Effect of capsaicin on lipid metabolism in rats fed a high fat diet, *J. Nutr.,* 116, 1272, 1986.
166. Sambaiah, K. and Satyanarayana, M.N., Lipotropic-like activity of red pepper, *J. Fd. Sci. Technol.,* 19, 30, 1982.
167. Sambaiah, K. and Satyanarayana, M.N., Influence of red pepper and capsaicin on body composition and lipogenesis in rats, *J. Biosci.,* 4, 425, 1982.
168. Nopanitaya, W., Long-term effects of capsaicin on fat absorption and the growth of the rat, *Growth,* 37, 269, 1973.
169. Jang, J.J., Devor, D.E., Logsdon, D.L., and Ward, J.M., A 4-weeek feeding study of ground red chili (capsicum annuum) in male B6C3Fl mice, *Fd. Chem. Toxicol.,* 30, 783, 1992.
170. Monsereenusorn, Y., Subchronic toxicity studies of capsaicin and capsicum in rats, *Res. Comun. Chem. Path. Pharmacol.,* 41, 95, 1983.
171. Lee, S.O., Studies on the influence of diets and lipotropic substances upon the various organ and metabolic changes in rabbits on long-term feeding with red pepper (I) histopathological changes of the liver and spleen, *Kor. J. Int. Med.,* 6, 383, 1963.
172. Lee, S.O., Studies on the influence of diet and lipotrophic substances upon the various organs and metabolic changes in rabbits on long-term feeding with red pepper. II. Histopathologic changes of various organs except the liver and spleen, *Kor. J. Int. Med.,* 8, 471, 1963.
173. Shimizzu, T., Fujita, S., Izumi, K., Koja, T., Ohba, N., and Fukuda, T., Corneal lesions induced by the administration of capsaicin in neonatal mice and rats, *Arch. Pharmacol.,* 326, 347, 1984.
174. Wallengren, J., Ekman, R., and Miller, H., Capsaicin enhances allergic contact dermatitis in the guinea pig, *Cont. Derm.,* 24, 30, 1991.
175. Buchanan, R.L., Goldstein, S., and Budroe, J.D., Examination of chili pepper and nutmeg oleoresins using the salmonella/mammalian microsome mutagenicity assay, *J. Food Sci.,* 47, 330, 1981.
176. Toth, B., Rogan, E., and Walker, B., Tumorigenicity and mutagenicity studies with capsaicin of hot peppers, *Anticancer. Res.,* 4, 117, 1984.
177. Damhoeri, A., Hezoron, T.I., and Matsuyana, A., *In vitro* mutagenicity tests on capsicum pepper, shallot, nutmeg, oleoresins, *Agric. Biol. Chem.,* 49, 1519, 1985.
178. Nagabhushan, M. and Bhide, S.V., Mutagenicity of chili extract and capsaicin in short-term tests, *Environ. Mutag.,* 7, 881, 1985.
179. Gannett, P.M., Nagel, D.L., Reilly, J., Lawson, T., Sharpe, J., and Toth, B., The capsacinoids: Their separation, synthesis and mutagenicity, *J. Org. Chem.,* 53, 1064, 1988.
180. Ungsurungsu, M., Suthienkal, O., and Paovalo, C., Mutagenicity screening of popular Thai spices, *Fd. Chem. Toxicol.,* 20, 527, 1982.
181. Lawson, T. and Gannett, P., The mutagenicity of capsaicin and dihydro-capsaicin in V79 cells, *Cancer Lett.,* 48, 109, 1989.
182. Narasimhamurthy, M. and Narasimhamurthy, K., Non-mutagenicity of capsaicin in albino mice, *Fd. Chem. Toxicol.,* 26, 955, 1988.

183. Sirsat, S.M. and Khanolkar, V.R., Submucous fibrosis of the palate and pillars of the fauces, *Ind. J. Med. Sci.,* 16, 189, 1962.
184. Pindborg, J.J., Poulsen, H.E., and Zachariah, J., Oral epithelial changes in thirty Indians with oral cancer and submucous fibrosis, *Cancer,* 20, 1141, 1967.
185. Toth, B. and Gannet, P., Carcinogenicity of lifelong administration of capsaicin of hot pepper in mice, *In Vivo,* 6, 59, 1992.
186. LaHann, T.R., Effect of capsaicin on croton oil and TPA-induced tumorigenesis and inflammation, *Proc. W. Pharmacol. Soc.,* 29, 145, 1986.
187. Sasajima, K., Willey, J.C., Banks-Schlegel, S.P., and Harris, C.C., Effects of tumor promoters and carcinogens on growth and differentiation of cultured human esophageal epithelial cells, *J. Natl. Cancer Inst.,* 78, 419, 1987.
188. Saria, A., Skofitsch, G., and Lembeck, F., Distribution of capsaicin in rat tissues after systemic administration, *J. Pharm. Pharmacol.,* 34, 273, 1982.
189. Kawada, T. and Iwai, K., *In vivo* and *in vitro* metabolism of dihydro capsaicin, a pungent principle of hot pepper in rats, *Agric. Biol. Chem.,* 49, 441, 1985.
190. Lee, S.S. and Kumar, S., Metabolism *in vitro* of capsaicin, a pungent principle of red pepper, with rat liver homogenates, in *Microsomes, Drug Oxidation and Chemical Carcinogenesis,* Vol 2, Coon, M.J., Conney, A.H., Estabrook, R.W., Gelboin, H.V., Gillette, J.R., and O'Brien P.J., Eds., Academic Press, New York, 1980, 1009.
191. Miller, M.S., Brendel, K., Burks, T.F., and Sipes, G., Induction of capsaicinoids with drug metabolism enzymes, *Biochem. Pharmacol.,* 32, 547, 1983.
192. Surh, Y.J. and Lee, S.S., Capsaicin, a double-edged sword: Toxicity, metabolism, and disease preventive potential, *Life Sci.,* 56, 1845, 1995.
193. Wehmeyer, K.R., Kasting, G.B., Powell, J.H., Kuhlenbeck, D.L., Underwood, R.A., and Bowman, L.A., Application of liquid chromatography with on-line radiochemical detection to metabolism studies on a novel class of analgesics, *J. Pharm. Biomed. Anal.,* 8, 177, 1990.
194. Kawada, T., Suzuki, T., Takahashi, M., and Iwai, K., Gastrointestinal absorption and metabolism of capsaicin and dihydrocapsaicin, *Toxicol. Appl. Pharmacol.,* 72, 449, 1984.
195. Oi, Y., Kawada, T., Wantanabe, T., and Iwai, K.J., Induction of capsaicin-hydrolyzing enzyme activity in rat liver by continuous oral administration of capsaicin, *J. Agric. Fd. Chem.,* 40, 467, 1992.
196. Yagi, T., Inhibition by capsaicin of NADH-quinone oxidoreductases is correlated with the presence of energy-coupling site 1 in various organisms, *Arch. Biochem. Biophys.,* 281, 305, 1990.
197. Fuller, R.W., Pharmacology of inhaled capsaicin in humans, *Resp. Med.,* 85 (Suppl. A), 31, 1991.
198. Nichol, G., Nix, A., Barnes, P.J., and Chung, K.F., Prostaglandin F2 alpha enhancement of capsaicin induced cough in man: Modulation by beta 2 adrenergic and anticholinergic drugs, *Thorax,* 45, 694, 1990.
199. Bascom, R., Lageu-Sobotka, A., and Proud, D., Effect of intranasal capsaicin on symptoms and mediator release, *J. Pharmacol. Exp. Ther.,* 259, 1323, 1991.
200. Barros, M.J., Zammattio, S.L., and Rees, P.J., Effect of changes in inspiratory flow rate on cough responses to inhaled capsaicin, *Clin. Sci.,* 81, 539, 1991.
201. Philip, G., Baroody, F.M., Proud, D., Naclerio, R.M., and Togias, A.G., The human nasal response to capsaicin, *J. Allerg. Clin. Immunol.,* 94, 1035, 1994.
202. Watson, W.A., Stremel, K.R., and Westdorp, E.J., Oleoresin capsicum (cap-stun) toxicity from aerosol exposure, *Ann. Pharmacothera.,* 30, 733, 1996.
203. Gosselin, R.E., Hodge, H.C., Smith, R.P., and Gleason, M.N., *Clinical Toxicology of Commercial Products,* 4th ed., Williams and Wilkins, Baltimore, MD, 1976.

204. Geppetti, P., Tramontana, M., DelBianco, E., and Fusco, B.M., Capsaicin-desensitization to the human nasal mucosa selectively reduces pain evoked by citric acid, *Br. J. Clin. Pharmacol.,* 35, 178, 1993.
205. Collier, J.G. and Fuller, R.W., Capsaicin inhalation in man and the effects of sodium cromoglycate, *Br. J. Pharmacol.,* 81, 113, 1984.
206. Tominack, R.L. and Spyker, D.A., Capsicum and capsaicin—a review: Case report of the use of hot peppers in child abuse, *Clin. Toxicol.,* 25, 591, 1987.
207. Blanc, P., Liu, D., Juarez, C., and Beushey, H.A., Cough in hot pepper washers, *Chest,* 99, 27, 1991.
208. McDonald, M., Infections intensify neurogenic plasma extravasation in the airway mucosa, *Am. Rev. Respir. Dis.,* 146, S 40, 1992.
209. (British) Ministry of Defence, *Medical Manual of Defence against Chemical Agents,* JSP 312, HMSO, London, 1972.
210. Owens, E.J., McNamara, B.P., Weimer, J.T., Ballard, T.A., Thomas, W.U., Hess, T.L., Farrand, R.L., Ryan, S.G., Merkey, R.P., Olson, J.S., and Vocci, F.J., *The Toxicology of DM, Technical Report 4108,* U.S. Army Medical Research Laboratory, Edgewood Arsenal, MD, 1967.
211. Punte, C.L., Ballard, T.A., and Weimer, J.T., Inhalation studies with chloroacetophenone, diphenylaminochloroarsine and pelargonic morpholide—I. Animal exposures, *Am. Ind. Hyg. Assoc. J.,* 23, 194, 1962.
212. Striker, G.E., Streett, C.S., Ford, D.F., Herman, L.H., and Helland, D.R., A clinicopathological study of the effects of riot-control agents in monkeys. III. Diphenylaminochloroarsine (DM) greande, *Technical Report 4070,* U.S. Army Medical Research Laboratory, Edgewood Arsenal, MD, 1967.
213. Striker, G.E., Streett, C.S., Ford, D.F., Herman, L.H., and Welland, D.R., A clinicopathological study of the effects of riot-control agents on monkeys. V. Low concentrations of diphenylaminochloroarsine (DM) or o-chlorobenzylidene malononitrile (CS) for extended periods, *Technical Report 4072,* U.S. Army Medical Research Laboratory, Edgewood Arsenal, MD, 1967.
214. Lawson, W.E. and Temple, J.W., Report on the relation between concentration limit of tolerance for diphenylamine chloroarsine and the development of a continuous flow apparatus for testing, *EACD 92 (Jan 1922),* 1922.
215. Gongwer, L.E., Ballard, T.A., Gutentag, P.J., Punte, C.L., Owens, E.J., Wilding, J.L., and Hart, J.W., The comparative effectiveness of four riot-control agents, *Technical Memorandum 24–18,* U.S. Army Chemical Warfare Laboratories, Army Chemical Center, MD, 1958.

12 Pharmacological Countermeasures for Botulinum Intoxication*

Michael Adler, George A. Oyler, James E. Keller, and Frank J. Lebeda

CONTENTS

I. Introduction 373
A. Characteristics of Botulinum Neurotoxin (BoNT) Intoxication 375
B. Symptomology 375
C. Functional Domains of BoNT 376
II. Manifestations of Botulism 377
A. Foodborne Botulism 377
B. Wound Botulism 378
C. Infant Botulism 378
III. Treatment Options 379
A. Pharmacological Intervention 380
1. Potassium Channel Blockers 380
2. Specific Strategies for Therapeutic Intervention 382
a. Inhibitors of Binding 382
b. Inhibitors of Internalization 383
c. Inhibitors of Metalloprotease Activity 384
IV. Conclusions and Future Research 386
Acknowledgments 387
References 387

I. INTRODUCTION

The botulinum neurotoxins (BoNTs)** comprise a family of seven distinct neurotoxic proteins (A–G) produced by immunologically discrete strains of the anaerobic bacterium, *Clostridium botulinum.*[1,2] These toxins act on peripheral cholinergic

---

*The opinions or assertions contained herein are the private views of the authors, and are not to be construed as reflecting the view of the Department of the Army or the Department of Defense.

**In this chapter, BoNT is used to designate both pure botulinum neurotoxin as well as the neurotoxin complex. Some authors prefer to use BoTx for the latter.

0-8493-0872-0/01/$0.00+$.50

synapses to inhibit spontaneous and impulse-dependent release of acetycholine (ACh).[3,4] Intoxication by BoNT results in muscle weakness, which can be fatal when the diaphragm and intercostal muscles become sufficiently compromised to impair ventilation.[5] BoNTs are the most potent substances in nature and ingestion of as little as 1 ng/kg is sufficient to cause human lethality.[6]

The purpose of this chapter is to use the insights gained in our understanding of the mechanism of BoNT action during the past decade to establish a conceptual framework within which to develop effective treatment strategies for intoxication. The chapter is organized into three major topics consisting of (1) an overview of BoNT action, (2) a description of foodborne, wound, and infant botulism, and (3) a discussion of possible treatment options.

From the first description of botulism in 1793 until the mid 1950s, the primary role of BoNT was that of a public health problem due to its association with food poisoning.[6,7] Although implicated in only a small fraction of all foodborne illnesses (<0.1%), the severity of the clinical syndrome produced by BoNT and the potential for multiple casualties led each outbreak to be considered as a potential health crisis.[7–9] During and shortly after World War II, BoNT was developed as a biological weapon, due to its ability to cause extensive morbidity and mortality.[10] BoNT reverted once again to this role with the Iraqi stockpiling of BoNT prior to the Persian Gulf War (1991), and with the rise of terrorist organizations, such as the Japanese Aum Shinrikyo cult, which has not only produced but reportedly deployed biological and chemical agents.[11–13]

Systematic research on the mechanism of action of BoNT began with Emile Pierre van Ermengem's historic isolation and characterization of *Clostridium botulinum* following a large outbreak in Ellezelles, Belgium, in 1895 and has continued with increasing interest and excitement to the present time.[2,7,9] Early work on BoNT intoxication established the existence of multiple serotypes, localized the site of action to peripheral cholinergic synapses, proposed the mechanism of impaired ACh release, and ruled out non-cholinergic, sensory, and central nervous system involvement.[1–5] Most of these findings were firmly established by 1949.[14,15]

The remarkable specificity for peripheral cholinergic synapses and long duration of action led to the use of BoNT/A for a growing number of focal dystonias and movement disorders following its approval in 1989 as an "orphan drug" by the Food and Drug Administration for the treatment of strabismus, blepharospasm, and hemifacial spasm.[16] The attributes that render BoNT a dreaded poison also make the neurotoxin an ideal therapeutic agent to treat diseases of muscle hyperactivity. In addition to its original indications, some BoNT serotypes are also used for treatment of spasticity from brain and spinal cord injuries, stroke, multiple sclerosis, cerebral palsy, and numerous other disorders. Expansion and refinement in its clinical use constitute the most active focus of current BoNT research, and a number of excellent reviews have been published recently.[16–18]

In the last decade, enormous progress has been made in understanding the action of BoNT at the molecular level. This was spurred by three crucial developments: (1) elucidation of the amino acid sequence leading to recognition of the zinc binding motif, (2) demonstration of zinc metalloprotease activity with identification of substrates and cleavage sites, and (3) solution of the crystal structure for BoNT/A and /B.[19–21] These developments opened the possibility for rational studies of mechanism and active-site inhibitor design.

A. Characteristics of Botulinum Neurotoxin (BoNT) Intoxication

The typical manifestation of botulism is a flaccid paralysis that is bilateral and descending, involving primarily skeletal muscle but also structures innervated by autonomic parasympathetic fibers.[1–9] Human intoxication is caused by serotypes A, B, E, and, to a much lesser extent, by serotype F and is manifested as foodborne, wound, and infant botulism.[1,2,6] Wound and infant botulism, however, are usually confined to serotypes A and B.[22,23] Data compiled for foodborne botulism during the past 50 years in the United States indicate that serotype A was responsible for 37.6% of all outbreaks while sertoypes B, E, and F accounted for 13.7%, 15.1%, and 0.7%, respectively, of intoxications in which serotype involvement was established.[7]

Clostridium botulinum spores are widely distributed in soils, sea sediments, decaying vegetation, animal carcasses, and sewage.[7] The intestinal tracts of birds, mammals, and fish may also acquire *Clostridium botulinum* as a transient member of their intestinal flora[6,7] The hosts do not exhibit botulism since growth of *Clostridium botulinum* is suppressed when there is competition from other organisms and a functional immune system.[7–9] The resistance of clostridial spores to harsh environmental conditions enables their dissemination by air currents and dust particles leading to surface contamination of exposed food products.[7–9] Botulism is not contagious, however, and contact with spores does not usually lead to disease except in young infants or following germination in wounds (see Sections IIB and IIC).[7–9]

B. Symptomology

The clinical syndrome of botulism reflects toxin-induced blockade of ACh release from neuromuscular and neuroeffector junctions.[1,2] The basic syndrome of BoNT intoxication is similar for foodborne, infant, and wound botulism and does not vary appreciably among toxin serotypes.[7,9] The earliest symptoms generally include visual disturbances (diplopia, blurred vision) and xerostomia.[7] With low-level exposure these symptoms may gradually resolve, even in the absence of medical intervention.[7] In more severe cases, the initial symptoms are followed by dysphasia, dysphonia, and dysarthria, reflecting an especially high susceptibility of cranial efferent terminals to BoNT action.[1,2] A descending generalized skeletal muscle weakness may then develop, progressing from the upper to the lower extremities.[7,9,17] Involvement of the diaphragm and intercostal muscles can lead to ventilatory failure and death unless appropriate supportive care is provided.[7,17] Although motor function is severely impaired, there is little or no sensory alteration or central nervous system (CNS) involvement in botulinum intoxication.[2,5,7]

Symptoms are usually observed 12–36 h after exposure, although onset times as short as 4 h or as long as 8 days have been reported.[7] The preponderance of symptoms, including the potentially lethal respiratory collapse, stems from inhibition of neuromuscular transmission.[2,5,14,15] Parasympathetic dysfunction is responsible for blurred vision, xerostomia, constipation, and urinary retention.[5,7,24] The absence of more extensive autonomic involvement in BoNT intoxication reflects a lower overall sensitivity of the autonomic nervous system to the actions of the neurotoxin.[25]

C. Functional Domains of BoNT

BoNT are synthesized as ~150 kDa single-chain protoxins (range, 140–167 kDa). They are proteolytically activated (nicked) to form dichain molecules consisting of an ~50 kDa light chain (L-chain) and an ~100 kDa heavy chain (H-chain).[1,2] The two chains are coupled by a single disulfide bond and by non-covalent forces.

In addition to the neurotoxin, all serotypes of *Clostridium botulinum* synthesize a group of non-toxic proteins, designated as neurotoxin-associated proteins (NAP); some of these possess hemagglutinin activity. These proteins associate with BoNT in the bacterial culture medium by noncovalent interactions and protect the neurotoxin from proteolytic and low pH-mediated inactivation. They have also been suggested to facilitate absorption of BoNT from the gastrointestinal tract into the bloodstream.[26] The ability of BoNT to manifest oral toxicity has generally been attributed to the presence of these proteins; conversely, the inability of tetanus neurotoxin (TeNT) to produce foodborne intoxication has been ascribed to the absence of such NAP.[27] Recent data from Maksymowych et al.[28] has raised some questions of the importance of NAP in BoNT toxicity, especially with regard to their role in transcytosis of the neurotoxin. These investigators demonstrated that pure BoNT/A lacking NAP was still toxic to mice, although to a lesser extent than the toxin complex. When examined at elevated concentrations, the differences in potency between pure and NAP-containing neurotoxin were progressively reduced. These results indicate that pure neurotoxin does not require accessory proteins for absorption. Moreover, even though the NAP are clearly protective, sufficient pure neurotoxin can survive the inhospitable environment of the gastrointestinal tract to produce lethality.

In conformity with the sequential processing of bacterial protein toxins such as diphtheria or cholera toxin, the action of BoNT involves multiple discrete steps: binding to surface receptors, internalization via receptor-mediated endocytosis, transport from endosome to cytosol, and cleavage of target proteins in the cytosol.[1,2,17,19] Binding and internalization are mediated by the C- and N-terminal domains of the BoNT H-chain, respectively.[1,2] The L-chains have zinc metalloprotease activity, targeted selectively to one of three proteins that are required for the docking and fusion of synaptic vesicles with active zones at the cytoplasmic surface of the nerve terminal.[16–20]

Serotypes B, D, F, and G cleave different sites on the synaptic vesicle protein, synaptobrevin (VAMP), whereas serotypes A and E cleave the presynaptic membrane-associated protein SNAP-25.[17] Serotype C1 is unique in that it cleaves two cytoplasmic proteins, syntaxin and SNAP-25.[29] Interaction of these proteins (designated as SNARE) on the surface of synaptic vesicles and active zone membranes is required for voltage- and Ca^{2+}-dependent release of neurotransmitter; cleavage by

BoNT inhibits this process, leading to muscle weakness and paralysis.[30] Cleavage of SNARE proteins appears to be sufficient to account for all actions of the BoNT. The SNARE hypothesis has received near-universal acceptance since its introduction in the early 1990s; an alternative hypothesis has been advanced based on results obtained in PC12 pheochromocytoma cells,[31] however, its generality has not yet been established.

For each BoNT serotype, the dichain form constitutes the active configuration of the neurotoxin; the isolated L- and H-chains are devoid of systemic toxicity. The absence of toxicity is consistent with findings that the L-chain cannot ordinarily gain access to the cytosol unless it is coupled to the H-chain and that, on its own, the H-chain lacks the ability to inhibit neurotransmitter release.[1,2] The isolated L-chain does, however, remain enzymatically active as evidenced by its ability to cleave SNARE proteins in cell-free assays,[32] and by its capacity to inhibit ACh release in skeletal muscle when delivered by liposomes.[33] It is not clear whether any portion of the H-chain is internalized along with the L-chain and, if so, whether it exerts a role in enhancing the catalytic activity or stability of the L-chain.

All BoNT serotypes suppress ACh release, show high specificity for cholinergic synapses, and share the same overall mode of action; they differ, however, in potency and in duration of action. Type A neurotoxin exhibits the highest potency,[6] and types A and C1 produce the longest intoxication times.[34,35] Other differences include targeting of different functional surface receptors at the motor nerve terminal[36] and cleaving unique peptide bonds in the appropriate SNARE proteins.[17]

II. MANIFESTATIONS OF BOTULISM

In adults, botulinum intoxication generally results from ingestion of preformed toxin elaborated in contaminated foods (foodborne), or from colonization by *Clostridium botulinum* of deep wounds with subsequent production of toxin (wound botulism).[8,37] A third form, termed infant botulism, is observed in young infants and originates from colonization of the large intestine by *Clostridium botulinum* with subsequent production and absorption of toxin.[22,23] Rarely, adults also exhibit a syndrome resembling infant botulism and some authors regard this as the fourth manifestation of botulism.[23]

A. Foodborne Botulism

Elaboration of BoNT in foods requires contact with *Clostridium botulinum* spores under conditions that allow bacterial cell proliferation and toxin production. These consist of an anaerobic environment, temperatures between 4 and 40°C, pH above 4.6, water activity greater than 0.94 (<10% NaCl), and lack of adequate preservatives.[7–9,38] The requirements for growth of *Clostridium botulinum* are stringent, especially anaerobiosis, making outbreaks relatively rare; nevertheless, episodes of foodborne botulism constitute a persistent public health threat.[9] In fact, food-related botulism outbreaks in the United States have shown no significant reduction during the past century, with an average of approximately 24 cases per year.[7] The primary vehicle for foodborne botulism presently and during most of the 20th century has been improperly prepared home-preserved food products, often involving vegetables with a low acid content.[7–9] Other sources are restaurants that use unsafe food handling procedures and contaminated commercially canned food products; the latter has become rare since the introduction of modern methods.[7,39]

Although the numbers of outbreaks have been relatively constant, the case-to-fatality ratio has improved markedly. From 1899 to 1950, foodborne botulism was associated with 60% mortality; from 1950 to 1996, the average annual mortality fell to 15.5%, and decreased even further to under 10% during the last decade.[7] These advances in survival have come primarily from improvements in critical care.[7,39] Further reductions in morbidity and mortality from botulinum intoxication will require better methods for detection of BoNT and availability of specific pharmacological treatments.[11,12]

B. Wound Botulism

Wound botulism is relatively rare, accounting for only 5% of all outbreaks. The majority of these are caused by serotype A, and the remainder by serotype B.[7] The neurological symptoms of wound botulism differ little from that of foodborne botulism except for the general absence of gastrointestinal symptoms. Historically, this form of botulism was so uncommon that it was not even recognized until the last half of the 20th century. From its discovery in 1943 until 1985, only 33 incidents of wound botulism were documented.[7,37] An examination of these cases indicated that wounds susceptible to *Clostridium botulinum* are generally deep, with avascular areas, but need not appear obviously infected or necrotic. Additional risk factors include compound fractures and extensive crush injuries.[7,37] Contamination of wounds with *Clostridium botulinum* spores leads to germination and colonization at the injection site. Localized weakness results from production of toxin at the wound site, and systemic botulism can occur from toxin transmitted via the bloodstream to distant targets.[37]

From 1980 to the present time, wound botulism has been observed predominantly in illicit drug users following repeated subcutaneous administration of narcotics or in individuals with nasal or sinus lesions from chronic cocaine abuse.[7–9,37] Recent increases in subcutaneous and intranasal routes for drug abuse have led to a greater incidence of wound botulism. During the last decade alone, wound botulism in the above population has exceeded the total reported in the preceding 40 years by a factor of almost three.[7]

For reasons that are not completely understood, wounds are much more likely to be contaminated by *Clostridium tetani* than with *Clostridium botulinum.* Although an aggressive vaccination program has nearly eliminated tetanus in developed nations, the absence of universal tetanus vaccination in many "third world" countries results in over 300,000 cases annually.[8] A large number of these occur in neonates, often by infection of the umbilical stump.[8]

C. Infant Botulism

Infant botulism is a consequence of intoxication by BoNT following ingestion or inhalation of clostridial spores that colonize the large intestine; young infants, especially those between 2 and 4 months of age, are susceptible to this form of botulism.[22,23] Germination of spores and growth of vegetative cells leads to production of BoNT; the neurotoxin thus elaborated crosses the intestinal wall and reaches susceptible targets such as skeletal muscle via the bloodstream.[23] The characteristic symptoms are poor sucking, constipation, generalized weakness, and respiratory insufficiency. The risk factors are not completely understood, but the incidence drops off sharply after 28 weeks of age, which is likely to be related to development of a more diversified intestinal flora. The latter has been shown to suppress germination and growth of *Clostridium botulinum* spores in mice.[40] Of all food products that may be contaminated with *Clostridium botulinum* spores, honey has been the one most often implicated in infant botulism; it is therefore recommended that honey not be given to young infants.[41]

Although infant botulism was not recognized until a large outbreak occurred in California in 1976,[22] it is currently the most prevalent form of botulism in the United States, accounting for approximately 70% of all cases.[7,23] Because infant botulism results from a continual elaboration of BoNT, it is more effectively treated by antitoxin than is foodborne botulism. Recently concluded clinical trials carried out with a human botulinum immune globulin (BIG) has revealed a greater than two-fold reduction in the mean duration of hospitalization in infants treated with BIG; treatment was effective even when infusion was initiated several days after the onset of symptoms (Arnon, personal communication).

Under rare conditions, adults may manifest a syndrome similar to infant botulism. Such cases generally occur in hospitalized patients treated with a long course of multiple antibiotics that eliminate the normally suppressive intestinal flora; other predisposing factors include inflammatory bowel disease and surgical alterations of the bowel.[23]

III. TREATMENT OPTIONS

BoNT are the most potent toxins known to mankind and exposure to as little as 1 ng/kg by ingestion or 3 ng/kg by inhalation can result in human fatality.[6] Treatment consists of intensive care and infusion of trivalent equine antitoxin.[7] Over 80% of adults with botulism receive antitoxin, although this passive immunization is only effective if administered early during the course of illness.[7] This temporal limitation of antitoxin treatment was appreciated as early as 1929.[42] Recovery, especially from type A intoxication, is slow and residual symptoms may persist for years after exposure. Vaccination provides a high degree of protection and is commonly administered to laboratory investigators who are at risk of exposure. The current vaccine is pentavalent (A–E), and has been available from the Centers for Disease Control and Prevention (CDC) as an Investigational New Drug (IND) for the past 40 years.[43] The vaccine is administered at 0, 2, and 12 weeks and requires a booster at 1 year to generate long-term protection. A heptavalent vaccine (A–G), consisting of conventional formaldehyde-treated toxin but of higher purity than the IND product, is currently under development by the United States Army.[10,11] In addition, a vaccine made from the recombinant binding domain of BoNT H-chain (C-fragment) is also under development.[11]

Although these vaccines are highly effective, all require multiple injections and as much as 1 year from onset to generate adequate protection. In addition, since the BoNT antibodies remain elevated for an unknown period of time after the 1-year booster, vaccinated individuals may be precluded from use of local BoNT administration for treatment of spasticity or movement disorders that may develop during their lifetime.[17] These limitations argue strongly in favor of a supplementary pharmacological approach for the management of botulism.

A. Pharmacological Intervention

From the time that inhibition of ACh release was established as the mechanism of BoNT action, attempts were made to antagonize the neurotoxin by measures that enhance ACh release. Until recently, however, development of a treatment for BoNT intoxication had low priority, in part because early efforts were generally unsuccessful and in part because an effective vaccine and antitoxin were already available. Currently, there is an increased impetus to develop pharmacological treatments following recognition of the potential for overdose with the expanding clinical use of BoNT.[17] In addition, the experience gained in preparation for a potential BoNT threat during the Persian Gulf War made it clear that delays in generating adequate protection by the BoNT vaccine were incompatible with the requirement for rapid deployment of military personnel.[10–12]

Some of the earliest putative BoNT antagonists were cholinesterase inhibitors, selected for their ability to prolong the actions of ACh. Carbamate anticholinesterase agents such as neostigmine and physostigmine were investigated as early as 1924 in animals[44] and in 1947 on nerve muscle preparations,[14] but they were unable to antagonize the effect of BoNT. More recent studies have tended to confirm earlier findings,[45] although there have been occasional reports of human botulism responding to the short-acting cholinesterase inhibitor, edrophonium.[46]

Other potential antagonists of BoNT such as elevated calcium, calcium ionophores, lanthanum, black widow spider venom, 2,4-dinitrophenol, and agents that raise cyclic AMP levels were examined for their ability to reverse BoNT toxicity. Addition of the above compounds to BoNT-intoxicated preparations led to increases in the frequency of spontaneous miniature endplate potentials (MEPP) but resulted in little or no enhancement in the amplitude of evoked endplate potentials (EPP).[47,48] Since these compounds generally increased spontaneous but not evoked activity, they were not considered to be of practical value for treatment of BoNT intoxication.

1. Potassium Channel Blockers

Potassium channel blockers were found to be more effective in antagonizing the paralytic action of BoNT than were the former group of compounds. Their higher efficacy comes from their ability to prolong the duration of the presynaptic action potential, leading to a greater influx of calcium during nerve stimulation. Coupling of increased calcium influx to nerve impulses enables the potassium blockers to produce striking increases in the amplitude of EPP and of nerve-evoked twitch tensions.[49]

A number of potassium channel blockers have been evaluated for their ability to antagonize the actions of BoNT including guanidine, 4-aminopyridine, 3,4-diaminopyridine (3,4-DAP), and tetraethylammonium.[50–52] Of these, the most promising candidate was 3,4-DAP; 4-aminopyridine exhibited undesirable CNS side effects, and tetraethylammonium caused a marked postsynaptic depression of end-plate potentials and muscle contractions that actually exacerbated BoNT-mediated inhibition.[45–49]

When added to nerve-muscle preparations prior to BoNT, 3,4-DAP produced a marked delay in the time-to-block of nerve-evoked muscle contractions.[52] When applied after BoNT paralysis, 3,4-DAP was able to restore tensions to near control values.[45,50–53] Unlike many candidate antagonists, 3,4-DAP could restore tension even several days after total paralysis was established.[53] In spite of these successes with 3,4-DAP, two fundamental limitations were noted: its efficacy was largely limited to serotype A,[50–52] and it had a brief *in vivo* lifetime relative to that of BoNT.[53] Of the two, the latter is less critical since the short lifetime can be compensated by use of an infusion delivery as shown recently by Adler et al.[54]

The basis for the lack of response to 3,4-DAP by the other serotypes is not well understood. Recent identification of the cleavage sites suggests that some characteristic of the target proteins may provide an answer: serotypes B, D, F, and G cleave synaptobrevin on the synaptic vesicle membrane, while serotype A cleaves SNAP-25 at the active zone.[17] However, serotype E also cleaves SNAP-25, but muscles intoxicated by BoNT/E do not respond appreciably better to 3,4-DAP than do those exposed to serotypes that cleave synaptobrevin.[52,53]

At a functional level, serotype A-intoxicated neuromuscular junctions undergo an attenuated but synchronous release of ACh following stimulation; preparations intoxicated by serotypes B, D, and F produce asynchronous release where the ACh quanta are dispersed and cannot summate to produce suprathreshold EPP.[48,50,51] It is readily apparent that the lack of synchrony would prevent 3,4-DAP from restoring transmitter release, however, the precise relationship between cleavage of synaptobrevin and loss of synchrony is not readily apparent.

An additional concern with potassium blockers comes from human case reports. These indicate that while the potassium blockers guanidine and 3,4-DAP produced a moderate increase in muscle strength, their use did not lead to the return of spontaneous ventilation in BoNT-intoxicated individuals.[55,56] It is not clear if human diaphragm or intercostal muscles are refractory to the potassium blockers or whether the doses used were insufficient to reverse muscle paralysis.[55,56] The latter may be the case, since BoNT/A-paralyzed rat or mouse diaphragm muscles respond vigorously to the actions of 3,4-DAP.[45,52] Higher doses of 3,4-DAP were not attempted in these patients to avoid the risk of seizures and other potential side-effects. At the present time, the potassium blockers hold promise as potential therapeutic agents, but development of more selective compounds or targeting of the inhibitors to neuromuscular synapses will be required to exploit their full potential.

2. Specific Strategies for Therapeutic Intervention

The above examples of treatment strategies are based on antagonizing the actions of BoNT after the neurotoxin has undergone internalization and subsequent cleavage of some or all of its target protein. In addition, since these approaches were developed before the intracellular targets were identified, they do not specifically address the basic mechanism of toxin action. Rather, they act by elevating intracellular calcium levels in an attempt to compensate for the toxin-mediated inhibition of ACh release. The discrete stages of clostridial neurotoxin action of binding, internalization, and catalysis suggest that there are multiple sites for direct pharmacological intervention. These stages are mediated by different domains of BoNT and, in principle, each can be specifically inhibited.[1,2] Three areas where significant progress has been made are discussed below.

a. Inhibitors of binding

A reasonable starting point for developing pharmacological countermeasures for BoNT intoxication is the use of receptor antagonists to reduce or prevent the binding of toxin to the nerve terminal. Complications with this approach are that each serotype may have a unique receptor and that the receptor appears to be a complex of a polysialoganglioside and protein.[57] Evidence for involvement of gangliosides is extensive.[58–60] However, as pointed out by Middlebrook, there are numerous findings that are at variance with the proposal that gangliosides constitute productive BoNT receptors.[61] The most obvious is that polysialogangliosides are distributed throughout the nervous system, while the neurotoxins show specificity for peripheral cholinergic nerve terminals.[62]

The protein component of the BoNT receptor is less well characterized. In recent years, the synaptic vesicle protein, synaptotagmin has been suggested as a receptor candidate for BoNT serotypes A, B, and E.[63,64] Sharing of a single protein receptor by these serotypes is inconsistent, however, with the absence of significant competition among them for binding to motor nerve terminals.[65] A possible explanation for these apparent inconsistencies is that the ganglioside and protein receptor components may be arranged in distinct geometric patterns on the cholinergic nerve terminal, and that each serotype recognizes a unique arrangement. Alternatively, instead of a dual receptor with polysialoganglioside and protein components, the receptor may be a serotype-selective glycoprotein in which the common feature is a sialic acid in the carbohydrate moiety.[61,66]

Regardless of the actual identity of the BoNT receptor, pronounced antagonism of neurotoxin binding has been achieved with lectins from *Triticum vulgaris* (TVL) and *Limax flavus*.[66] Pretreatment by these lectins led to a concentration-dependent inhibition in the binding of BoNT/B and TeNT to preparations of rat brain membranes, approaching total inhibition at the highest concentration. The most effective lectins were those that had an affinity for N-acetyl-α-sialic acid; six lectins with specificities for other carbohydrates were ineffective.[66] In complementary experiments on mouse phrenic nerve-hemidiaphragm preparations, TVL delayed the time-to-block of nerve-elicited muscle contractions with all BoNT serotypes examined.

If one defines the time-to-block in the presence and absence of BoNT antagonist as a protective index, the values for the different serotypes ranged from 1.3 to 1.9. Although the physiological actions of TVL appear less striking than its antagonism of binding, it must be borne in mind that a 10-fold decrease in bound neurotoxin can only be expected to produce a 2-fold slowing in the time-to-block.[66]

The major advantage of the lectins is that they are effective against all clostridial toxin serotypes, although to different degrees. Disadvantages of lectins include (1) they must be administered as pretreatments, and (2) they only slow the time course but do not prevent muscle paralysis. The first limitation is inherent in the basic mechanism of BoNT action; thus no antagonist of surface receptor binding would be expected to be protective once BoNT is internalized and symptoms are manifested.

b. *Inhibitors of internalization*

Following binding of the clostridial neurotoxins to surface receptors on cholinergic nerve terminals, the toxins undergo internalization prior to reaching their ultimate intracellular targets.[1,2] Internalization is thought to involve endocytosis of the BoNT-receptor complex, acidification of the resulting endocytotic vesicle, dissociation of the L- and H-chains, and release of the L-chain into the cytosol.[1,2] The most direct evidence for internalization comes from experiments in which colloidal gold-BoNT conjugates have been visualized inside cholinergic motor axon terminals[62] and torpedo electric organ synaptosomes.[67]

Internalization affords the next opportunity to ameliorate the toxic actions of BoNT. A number of pharmacological agents have been examined for inhibition of this process with various degrees of success. Simpson demonstrated that pretreatment of phrenic-nerve hemidiaphragm preparations with the lysosomotropic agents ammonium chloride or methylamine hydrochloride delayed the time-to-block of nerve-evoked contractions after exposure to BoNT serotypes A, B, C1, and TeNT.[67] The maximum protective index was approximately 2, making this strategy somewhat more effective than use of putative receptor blockers.[66]

Incubation of nerve-muscle preparations with ammonium chloride and methylamine hydrochloride was effective if applied before, concurrently, or 10–20 min after toxin exposure. The efficacy of the lysosomotropic agents was reduced rapidly with further delays in addition, such that no effect was observed if they were administered 30–35 min after toxin exposure.

Other candidates examined for inhibiting BoNT-mediated internalization were the antimalarial agents choloroquine and hydroxychloroquine.[69] These drugs were selected on the basis of interfering with the actions of a large group of peptide hormones and protein toxins that exert their actions following internalization.[70] The maximal efficacies of the above 4-aminoquinolines were similar to those of ammonium chloride and methylamine hydrochloride, and both groups exhibited a comparable therapeutic window. They differed in that effective concentrations of the 4-aminoquinolines also produced a reversible depression of neuromuscular transmission.

Work on antimalarial agents was extended by Deshpande et al.[71] to identify candidates that did not block neuromuscular transmission, had a longer therapeutic window, and could delay the time-to-block to a greater degree. These investigators examined a large group of 4- and 8-aminoquinoline compounds as well as analogous acridines for their efficacy against BoNT in mouse diaphragm preparations. The most effective compounds were quinacrine, amodiaquine, and chloroquine; 8-aminoquinolines such as primaquine were ineffective. The highest protective index, 3.9, was obtained with 20 μM amodiaquine. This was achieved with no deleterious effects on neuromuscular transmission, and thus defines the present limit for inhibitors of internalization. Unfortunately, the therapeutic window could not be extended; no protection was observed if the antimalarial agents were added ≥ 40 min after exposure to BoNT/A or /B.

A somewhat different approach for attempting to prevent or reduce the internalization of BoNT was to treat nerve-muscle preparations with the proton ionophores monensin and nigericin.[72,73] These ionophores act by depleting vesicular pH gradients, thereby interfering with several stages in the delivery of active L-chain in the cytosol. These ionophores were found approximately as effective as the other inhibitors of internalization. They were more toxic, however, and high concentrations led to a depression of neuromuscular transmission.[72,73] Toxicity is difficult to avoid with this group of agents since proton gradients are required for a number of cellular reactions such as the synthesis of ATP and filling of synaptic vesicles.

c. *Inhibitors of metalloprotease activity*

The third area for therapeutic intervention is inhibition of the metalloprotease activity of the BoNT light chains. This approach is potentially the most promising, especially since the crystal structures of BoNT/A[21] and /B[74] have been solved, and studies on crystallographic data for the other serotypes are currently in progress. The presence of a zinc binding motif in the L-chain of all clostridial neurotoxins and findings that zinc is required for neurotoxin-mediated proteolysis of SNARE proteins[17–20] suggest that two classes of potential inhibitors may be effective in antagonizing the toxic actions of BoNT L-chain: transition metal chelators and zinc metalloprotease inhibitors. Simpson et al.[75] demonstrated that the zinc chelator N,N,N′,N′-tetrakis(2-pyridylmethyl)ethylenediamine (TPEN) caused a marked slowing in the time-to-block of nerve evoked muscle contractions when added to phrenic nerve-hemidiaphragm preparations prior to BoNT. The maximum efficacy was equivalent to or greater than that achieved with TVL, ammonium chloride, methylamine hydrochloride, or the antimalarial agents. In common with the above inhibitors, TPEN was effective against all BoNT serotypes examined. In addition, when co-applied with TVL or the lysosomotropic agents, the protection observed with TPEN was approximately additive with that of the former compounds. These results are encouraging since they demonstrate that, in principle, concerted inhibition of the different stages in the production of toxicity is a viable strategy for managing BoNT intoxication.

Sheridan and Deshpande[76] examined a number of additional chelators on nerve-evoked twitch tensions and concluded that both a high affinity for zinc and membrane permeability are required for antagonism of BoNT. Interestingly, although TPEN works by an entirely different mechanism, it was subject to the same brief therapeutic window as were the lectins and lysosomotropic agents. This was not entirely unexpected since the deficit in ACh release resulting from BoNT-mediated proteolysis of SNARE proteins cannot be immediately overcome by inhibition of BoNT.

If zinc chelators are administered after substrate proteolysis is complete, functional recovery will involve removal of the cleaved fragments followed by incorporation of intact SNARE, both of which are unlikely to occur during the course of an acute *in vitro* experiment. Moreover, inhibition of BoNT would have to be sustained for perhaps several weeks in order to prevent cleavage of newly synthesized and incorporated SNARE proteins.[35]

The results with TPEN in the isolated nerve-muscle preparations were sufficiently encouraging to test this chelator for protection against challenge by BoNT; unfortunately, these results were less encouraging. First, TPEN was found to be highly toxic *in vivo,* producing rapid lethality at doses above 20 mg/kg in mice.[77] Second, at the highest tolerated dose, TPEN only increased survival by 2–3 h following a 20 LD_{50} challenge of BoNT/A or /B. Toxicity of TPEN was also observed with primary and clonal cells in culture. TPEN concentrations $\geq$ 10 μM produced morphological damage with characteristics of apoptosis.[78,79]

Studies with ion replacement indicated that chelation of zinc was the proximal cause of cytotoxicity, and examination of a variety of chelators suggested that those with high membrane permeability were especially apt to produce cell death.[78] Based on these findings, metal chelators may have a limited use in the therapy of botulinum intoxication since the requirements for efficacy against BoNT are the same ones that promote cellular toxicity.

A more promising approach is the development of metalloprotease inhibitors to target the catalytic activity of BoNT L-chain. This endeavor was made possible by the discovery of the zinc metalloprotease activity of the clostridial toxins that began in 1989 when the HEXXH signature sequence of zinc binding proteins was noted in the TeNT L-chain by Jongeneel et al.[80] Their finding suggested that clostridial neurotoxins possessed zinc-dependent protease activity. During the next 4 years, the SNARE protein substrates and serotype-specific cleavage sites were identified and correlated with intoxication in a series of elegant studies carried out primarily by Montecucco, Schiavo, and colleagues.[81,82] It is noteworthy that an enzymatic activity for BoNT/A was suspected as far back as 1947 by Guyton and Marshall[14] in their seminal study on botulinum intoxication:

> . . . this minute quantity of toxin necessary to produce poisoning, the duration of poisoning and the physical properties of the toxin all tend to characterize the toxin as a destructive enzyme . . .

Early work with zinc metalloprotease inhibitors focused on the well-characterized agents captopril and phosphoramidon. These were found to have little or no inhibitory activity against any BoNT serotype.[72,83] Phosphoramidon analogs in which Leu-Trp was replaced by Phe-Glu to resemble the cleavage site of synaptobrevin exhibited little increase in inhibitory activity; one analog was slightly more potent and two were significantly less potent than the parent compound.

Using a somewhat different strategy, Schmidt and colleagues made substitutions near the cleavage site of a 17-mer SNAP-25 peptide that was slowly cleaved by BoNT/A.[84] Substitution of Cys in the P1 or P2 position of the peptide transformed it from substrate to competitive inhibitor. The best compounds had K_i values of 2 μM. These peptide inhibitors cannot immediately be used as therapeutic agents since they would be unstable *in vivo* and would have difficulty gaining access to the nerve terminal cytosol to inhibit internalized BoNT L-chain. They can serve, however, as templates for synthesis of organic drug candidates.

Adler et al.[32] tested a series of isocoumarin compounds that were originally designed as elastase inhibitors. Molecular modeling studies suggested that these compounds may interact favorably with the BoNT/B active site, and several candidates were able to inhibit BoNT/B L-chain activity. The most effective compound in this series was 7-N-phenylcarbamoylamino-4-chloro-3-propyloxyisocoumarin (ICD 1578), which had an IC_{50} of 28 μM when tested in a cleavage assay using a 50-mer synaptobrevin peptide. Although the potency of the isocoumarins is somewhat low, they have the advantage of greater stability and higher lipid solubility relative to the peptide inhibitors.

IV. CONCLUSIONS AND FUTURE RESEARCH

Efforts to develop pharmacological inhibitors of BoNT have increased substantially during the last decade. The major focus of the current research is the design and synthesis of specific metalloprotease inhibitors. Most of the ongoing drug discovery efforts were initiated prior to publication of the crystal structure for BoNT and will be aided enormously by the availability of precise structural information. The crystal structure of BoNT/A has been available since 1998[21] and that of BoNT/B was recently described;[74] it is reasonable to expect the crystal structures of all BoNT serotypes to be resolved during the next few years.

Results to date indicate that a number of low molecular weight inhibitors and small peptides are effective against BoNT in cell-free *in vitro* systems. Development of safe and effective metalloprotease inhibitors with *in vivo* efficacy will no doubt be difficult. Some of the challenges involve targeting of drugs to the nerve terminal, ensuring their access to the intracellular compartment, and increasing the bioavailability of the drugs to match the duration of the toxin. In addition, different inhibitors may be needed for each serotype, requiring at least four parallel efforts. A more complete characterization of the BoNT receptor and a better understanding of the internalization process will aid in accomplishing these objectives by refining the drug delivery methodologies.

In addition, it may be necessary to accelerate the removal of cleavage products from the nerve terminal[85] and to introduce non-cleavable SNARE analogs for a more rapid recovery.[86] The latter is especially relevant for treatment of persistent serotypes such as BoNT/A.[35] The progress in understanding the mechanism of action of the BoNT and detailed structural information gained during the last decade suggest that pharmacological treatments for BoNT intoxication will soon be a reality.

ACKNOWLEDGMENTS

The authors are grateful for the many valuable discussions that have so greatly benefited us over the years with Drs. Edward Brown, Michael Byrne, Sharad Deshpande, Oliver Dolly, Daniel Drachman, Ernst Habermann, Brennie Hackley, Michael Goodnough, Jane Halpern, Harry Hines, Eric Johnson, John Middlebrook, Maurice Montal, Cesare Montecucco, Elaine Neale, James Nicholson, Mark Olson, Gerald Parker, James Powers, James Schmidt, Robert Sheridan, Lance Simpson, Bal Ram Singh, Leonard Smith, Raymond Stevens, Subramanyam Swaminathan, Timothy Umland, and Lura Williamson.

REFERENCES

1. Habermann, E. and Dreyer, F., Clostridial neurotoxins: Handling and action at the cellular and molecular level, *Curr. Top. Microbiol. Immunol.,* 129, 93, 1986.
2. Simpson, L.L., Peripheral actions of the botulinum toxins, in *Botulinum Neurotoxins and Tetanus Toxin,* Simpson, L.L., Ed., Academic Press, New York, 1989, chap. 7.
3. Brooks, V.B., An intracellular study of the action of repetitive nerve volleys and of botulinum toxin on miniature end-plate potentials, *J. Physiol.* (Lond.), 134, 264, 1956.
4. Kao, I., Drachman, D.B., and Price, D.L., Botulinum toxin: Mechanism of presynaptic blockade, *Science* (Wash.), 193, 1256, 1976.
5. Dickson, E.C. and Shevky, E., Botulism: Studies on the manner in which the toxin of *Clostridium botulinum* act upon the body, *J. Expl. Med.* 38, 327, 1923.
6. Gill, D.M., Bacterial toxins: A table of lethal amounts, *Microbiol. Rev.,* 46, 86, 1982.
7. Shapiro, R.L., Hatheway, C., and Swerdlow, D.L., Botulism in the United States: A clinical and epidemiological review, *Ann. Inter. Med.,* 129, 221, 1998.
8. Snydman, D.R., Foodborne diseases, in *Mechanism of Microbial Diseases,* Schaechter, M., Medoff, G., and Schlessinger, D., Eds., Williams and Wilkins, Baltimore, MD, 1989, chap. 58.
9 Smith, L.D.S., *Botulism: The Organism, Its Toxins, The Disease,* Charles C Thomas, Springfield, 1977, chap. 6.
10. Franz, D.R., Defense against toxin weapons, in *Textbook of Military Medicine: Medical Aspects of Chemical and Biological Warfare.* Part I. Zajtchuk, R. and Bellamy, R.F., Eds., Office of the Surgeon General, Department of the Army, at TMM Publications Borden Institute, Washington, DC, 1997, chap. 30.
11. Middlebrook, J.L. and Franz, D.R., Botulinum toxins, in *Textbook of Military Medicine: Medical Aspects of Chemical and Biological Warfare.* Part I., Zajtchuk, R. and Bellamy, R.F., Eds., Office of the Surgeon General, Department of the Army, at TMM Publications Borden Institute, Washington, DC, 1997, chap. 33.
12. Franz, D.R., Jahrling, P.B., Friedlander, A.M., McClain, D.J., Hoover, D.L., and Bryne, W.R., Clinical recognition and management of patients exposed to biological warfare agents, *JAMA,* 278, 399.
13. Atlas, R.M., The threat of biological weapons, *Crit. Rev. Microbiol.,* 24, 157, 1998.
14. Guyton, A.C. and MacDonald, A., Physiology of botulinus toxin, *Arch. Neurol. Psych.,* 57, 578, 1947.
15. Burgen, A.S.V., Dickens, F., and Zatman, L.J., The action of botulinum toxin on the neuromuscular junction, *J. Physiol.* (Lond.), 109, 10, 1949.

16. Schantz, E.J. and Johnson, E.A., Botulinum toxin: The story of its development for the treatment of human disease, *Perspectives in Biology and Medicine,* 40, 317, 1997.
17. Jankovic, J. and Brin, M.F., Botulinum toxin: Historical perspective and potential new indications, *Muscle Nerve,* 6, S129, 1997.
18. Johnson, E.A., Clostridial toxins as therapeutic agents: Benefits of nature's most toxic proteins, *Annu. Rev. Microbiol.,* 53, 551, 1999.
19. Montecucco, C., Papini, E., and Schiavo, G., Bacterial protein toxins penetrate cells via a four-step mechanism, *FEBS Lett.,* 346, 92, 1994.
20. Schiavo, G., Rossetto, O., Santucci, A., DasGupta, B.R., and Motecucco, C., Botulinum neurotoxins are zinc proteins, *J. Biol. Chem.,* 267, 23479, 1992.
21. Lacy, D.B., Tepp, W., Cohen, A.C., DasGupta, B.R., and Stevens, R.C., Crystal structure of botulinum neurotoxin type A and implications for toxicity, *Nat. Struct. Biol.,* 5, 898, 1998.
22. Pickett, J., Berg, B., Chaplin, E., and Brunstetter-Shafer, M., Syndrome of botulism in infancy: Clinical and electrophysiologic study, *N. Engl. J. Med.,* 295, 770, 1976.
23. Arnon, S.S., Botulism as an intestinal toxaemia, in *Infections of the Gastrointestinal Tract,* Blaser, M.J., Smith, P.D., Ravdin, J.I., Greenberg, H.B., and Guerrant, R.L., Eds., Raven Press, New York, 1995, 257.
24. Ambache, N., A further survey of the action of Clostridium botulinum toxin upon different type of autonomic nerve fibre, *J. Physiol.* (London), 113, 1, 1951.
25. MacKenzie, I., Burnstock, G., and Dolly, J.O., The effects of purified botulinum neurotoxin type A on cholinergic, adrenergic and non-adrenergic, atropine-resistant autonomic neuromuscular transmission, *Neuroscience,* 7, 997, 1982.
26. Sakaguchi, G., Clostridium botulinum toxins, *Pharmacol. Ther.,* 19, 165, 1982.
27. Singh, B.R., Li, B., and Read, D., Botulinum versus tetanus neurotoxins: Why is botulinum neurotoxin but not tetanus neurotoxin a food poison?, *Toxicon,* 33, 1541, 1995.
28. Maksymowych, A.B., Reinhard, M., Malizio, C.J., Goodnough, M.C., Johnson, E.A., and Simpson, L.L., Pure botulinum neurotoxin is absorbed from the stomach and small intestine and produces peripheral neuromuscular blockade, *Infect. Immun.,* 67, 4708, 1999.
29. Williamson, L.C., Halpern, J.L, Montecucco, C., Brown, J.E., and Neale, E.A., Clostridial neurotoxins and substrate proteolysis in intact neurons: Botulinum neurotoxin C acts on synaptosomal-associated protein of 25 kDa, *J. Biol. Chem.,* 271, 7694, 1996.
30. Sutton, R.B., Fasshauer, D., Jahn, R., and Brunger, A.T., Crystal structure of a SNARE complex involved in a synaptic exocytosis at 2.4 Å resolution, *Nature,* 395, 347, 1998.
31. Ray, P., Berman, J.D., Middleton, W., and Brendle, J., Botulinum toxin inhibits arachidonic acid release associated with acetylcholine release from PC12 cells. *J. Biol. Chem.,* 268, 11057, 1993.
32. Adler, M., Nicholson, J.D., Cornille, F., and Hackley, Jr., B.E., Efficacy of a novel metalloprotease inhibitor on botulinum neurotoxin B activity, *FEBS Lett.,* 429, 234, 1998.
33. de Pavia, A. and Dolly, J.O., Light chain of botulinum neurotoxin is active in mammalian motor nerve terminals when delivered via liposomes, *FEBS Lett.,* 277, 171, 1990.
34. Eleopra, R., Tugnoli, V., Rossetto, O., De Grandis, D., and Montecucco, C., Different time courses of recovery after poisoning with botulinum neurotoxin serotypes A and E in humans, *Neurosci. Lett.,* 256, 135, 1998
35. Keller, J.E., Neale, E.A., Oyler, G., and Adler, M., Persistence of botulinum neurotoxin action in cultured spinal cord cells, *FEBS Lett.,* 456, 137, 1999.
36. Daniels-Holgate, P.U. and Dolly, J.O., Productive and non-productive binding of botulinum neurotoxin A to motor nerve endings are distinguished by its heavy chain, *J. Neurosci. Res.,* 44, 263, 1996.

37. Mershon, M.H. and Dowell, V.R., Epidemiologic, clinical, and laboratory aspects of wound botulism, *N. Engl. J. Med.,* 289, 1005, 1973.
38. Baird-Parker, A.C. and Freame, B., Combined effect of water activity, pH and temperature on the growth of Clostridium botulinum spore inocula, *J. Appl. Bacteriol.,* 30, 420, 1967.
39. O'Mahoney, M., Mitchell, E., et al., An outbreak of foodborne botulism associated with contaminated hazelnut yoghurt, *Epidemiol. Infect.,* 104, 389, 1990.
40. Sugiyama, H. and Mills, D.C., Intraintestinal toxin in infant mice challenged intragastrically with Clostridium botulinum spores, *Infect. Immun.,* 21, 59, 1978.
41. Arnon, S.S., Midura, T.F., and Damus, K., Honey and other environmental risk factors for infant botulism, *J. Pediatr.,* 94, 331, 1979.
42. Hewlett, R.T., Bacillus Botulinus, in *A System of Bacteriology in Relation to Medicine,* Vol. III, His Majesty's Stationary Office, London, 1929, 373.
43. Fiock, M. A., Cardella, M. A., and Gearinger, N.F., Immunologic response of man to purified petavalent ABCDE botulinum toxoid, *J. Immunol.,* 90, 697, 1963.
44. Edmunds, C.W. and Keiper, G.F., Jr., The action of botulinus toxin, *JAMA,* 83, 495, 1924.
45. Adler, M., Scovill, J., Parker, G., Lebeda, F.J., Piotrowski, J., and Deshpande, S.S., Antagonism of botulinum toxin-induced muscle weakness by 3,4-diaminopyridine in rat phrenic nerve-hemidiaphragm preparations, *Toxicon,* 33, 527, 1995.
46. Cherington, M., Clinical spectrum of botulism, *Muscle Nerve,* 21, 701, 1998.
47. Simpson, L., Use of pharmacologic antagonists to deduce commonalities of biologic activity among clostridial neurotoxins, *J. Pharmacol. Exp. Ther.,* 245, 867, 1988.
48. Thesleff, S., Botulinal neurotoxins as tools in studies of synaptic mechanisms, *Quart. J. Exp. Physiol.,* 74, 1003, 1989.
49. Adler, M., Oliveira, A.C., Albuquerque, E.X., Mansour, N.A., and Eldefrawi, A.T., Reaction of tetraethylammonium with the open and closed conformation of the acetylcholine receptor ionic channel complex, *J. Gen. Physiol.,* 74, 129, 1979.
50. Lundh, H., Leander, S., and Thesleff, S., Antagonism of the paralysis produced by botulinum toxin in the rat, *J. Neurol Sci.,* 32, 29, 1977.
51. Molgo, J., Lundh, H., and Thesleff, S., Potency of 3,4-diaminopyridine (3,4-DAP) and 4-aminopyridine on mammalian neuromuscular transmission and the effect of pH changes, *Eur. J. Pharmacol.,* 61, 25, 1980.
52. Simpson, L., A preclinical evaluation of aminopyridines as putative therapeutic agents in the treatment of botulism, *Infect. Immun.,* 52, 858, 1986.
53. Adler, M., MacDonald, D.A., Sellin, L.C., and Parker, G.W., Effect of 3,4-diaminopyridine on rat extensor digitorum longus muscle paralyzed by local injection of botulinum neurotoxin, *Toxicon,* 34, 237, 1996.
54. Adler, M., Capacio, B., and Deshpande, S.S., Antagonism of botulinum toxin A-mediated muscle paralysis by 3,4-diaminopyridine delivered via osmotic minipumps, *Toxicon,* 38, 1381, 2000.
55. Molgo, J., DasGupta, B.R., and Thesleff, S., Characterization of the actions of botulinum neurotoxin type E at the rat neuromuscular junction, *Acta Physiol. Scand.,* 137, 497, 1989.
56. Cherington, M. and Schultz, D., Effect of guanidine, germine and steroids in a case of botulism, *Clin. Toxicol,* 11, 19, 1977.
57. Davis, L.E., Johnson, J.K., Bicknell, J.M., Levy, H., and McEnvoy, K.M., Human type A botulism and treatment with 3,4-diaminopyridine, *Electromyogr. Clin. Neurophysiol.,* 32, 379, 1992.
58. Montecucco, C., How do tetanus and botulinum toxins bind to neuronal membranes?, *Trends Biochem. Sci.,* 11, 315, 1986.

59. Van Heyningen, W.E. and Miller, P.A., The fixation of tetanus toxin by ganglioside, *J. Gen. Microbiol.,* 24, 107, 1961.
60. Shapiro R.E., Specht, C.D., Collins, B.E., Woods, A.S., Cotter, R.J., and Schnaar, R.L., Identification of a ganglioside recognition domain of tetanus toxin using a novel ganglioside photoaffinity ligand, *J. Biol. Chem.,* 272, 30380, 1997.
61. Middlebrook, J.L., Cell surface receptors for protein toxins, in *Botulinum Neurotoxins and Tetanus Toxin,* Simpson, L.L., Ed., Academic Press, New York, 1989, chap. 5.
62. Dolly, J.O., Black, J., Williams, R.S., and Melling, J., Acceptors for botulinum neurotoxin reside on motor nerve terminals and mediate its internalization, *Nature* (Lond.), 307, 457, 1984.
63. Nishiki, T., Kamata, Y., Nemoto, Y., Omori, A., Ito, T., Takahashi, M., and Kozaki, S., Identification of protein receptor for Clostridium botulinum type B neurotoxin in rat brain synaptosomes, *J. Biol. Chem.,* 269, 10498, 1994.
64. Li, L. and Singh, B.R., Isolation of synaptotagmin as a receptor for types A and E botulinum neurotoxin and analysis of their comparative binding using a new microtiter plate assay, *J. Nat. Toxins,* 7, 215, 1998.
65. Lalli, G., Herreros, J., Osborne, S.L., Montecucco, C., Rosetto, O., and Schiavo, G., Functional characterisation of tetanus and botulinum neurotoxins binding domains, *J. Cell Sci.,* 112, 2715, 1999.
66. Bakry, N., Kamata, Y., and Simpson, L.L., Lectins from Triticum vulgaris and Limax flavus are universal antagonists of botulinum neurotoxin and tetanus toxin, *J. Pharmacol. Exp. Ther.,* 258, 830, 1991.
67. Blasi, J., Egea, G., Castiella, M.J., Arribas, M., Solsona, C., Richardson, P.J., and Marsal, J., Binding of botulinum neurotoxin to pure cholinergic nerve terminals isolated from the electric organ of Torpedo, *J. Neural. Transm.,* 90, 87, 1992.
68. Simpson, L.L., Ammonium chloride and methylamine hydrochloride antagonize clostridial neurotoxins, *J. Pharmacol. Exp. Ther.,* 225, 546, 1983.
69. Simpson, L.L., The interaction between aminoquinolines and presynaptically acting neurotoxins, *J. Pharmacol. Exp. Ther.,* 222, 43, 1982.
70. Goldstein, J.L., Anderson, R.G.W., and Brown, M.S., Coated pits, coated vesicles, and receptor-mediated endocytosis, *Nature* (Lond.), 279, 679, 1979.
71. Deshpande, S.S., Sheridan, R.E., and Adler, M., Efficacy of certain quinolines as pharmacological antagonists in botulinum neurotoxin poisoning, *Toxicon,* 35, 433, 1997.
72. Adler, M., Deshpande, S.S., Sheridan, R.E., and Lebeda, F.J., Evaluation of captopril and other potential therapeutic compounds in antagonizing botulinum toxin-induced muscle paralysis, in *Therapy with Botulinum Toxin,* Jankovic, J. and Hallett, M., Eds., Marcel Dekker, Inc., New York, 1994, chap. 5.
73. Sheridan, R.E., Protonophore antagonism of botulinum toxin in mouse muscle, *Toxicon,* 34, 849, 1996.
74. Swaminathan, S. and Eswaramoothy, S., Structural analysis of the catalytic and binding sites of *Clostridium botulinum* neurotoxin B, *Nature Struct. Biol.,* 3, 693, 2000.
75. Simpson, L.L., Coffield, J.A., and Bakry, N., Chelation of zinc antagonizes the neuromuscular blocking properties of the seven serotypes of botulinum neurotoxin as well as tetanus toxin, *J. Pharmacol. Exp. Ther.,* 267, 720, 1993.
76. Sheridan, R.E. and Deshpande, S.S., Interaction between heavy metal chelators and botulinum neurotoxin at the neuromuscular junction, *Toxicon,* 33, 539, 1995.
77. Adler, M., Dinterman, R., and Wannemacher, R.W., Protection by the heavy metal chelator N, N, N′, N′-tetrakis(2-pyridylmethyl)ethylediamine (TPEN) against the lethal action of botulinum neurotoxin A and B, *Toxicon,* 35, 1089, 1997.

78. Sheridan, R.E. and Deshpande, S.S., Cytotoxicity induced by intracellular zinc chelation in rat cortical neurons, *In vitro Molec. Toxicol.,* 33, 539, 1995.
79. Adler, M., Shafer, H., Hamilton, T., and Petrali, J.P., Cytotoxic actions of the heavy metal chelator TPEN on NG108–15 neuroblastoma-glioma cells, *NeuroToxicology,* 20, 571, 1999.
80. Jongeneel, C.V., Bouvier, J., and Bairoch, A., A unique signature identifies a family of zinc-dependent metallopeptidases, *FEBS Lett.,* 242, 211, 1989.
81. Montecucco, C. and Schiavo, G., Tetanus and botulism neurotoxins: A new group of zinc proteases, *Trends Biochem. Sci.,* 18, 324, 1993.
82. Rossetto, O., Deloye, F., Poulain, B., Pellizzari, R., Schiavo, G., and Montecucco, C., The metallo-proteinase activity of tetanus and botulinum neurotoxins, *J. Physiol.* (Paris), 89, 43, 1995.
83. Adler, M., Nicholson, J.D., Starks, D.F., Kane, C.T., Cornille, F., and Hackley, Jr., B.E., Evaluation of phosphoramidon and three synthetic phosphonates for inhibition of botulinum neurotoxin B catalytic activity. *J. Appl. Toxicol.,* 19, S5, 1999.
84. Schmidt, J.J., Stafford, R.G., and Bostian, K.A., Type A botulinum neurotoxin proteolytic activity: Development of competitive inhibitors and implication for substrate specificity at the S1′ binding site, *FEBS Lett.,* 435, 61, 1998.
85. Ferrer-Montiel, A.V., Gutierrez, L.M., Apland, J.P., Canaves, J.M., Gil, A., Viniegra, S., Biser, J.A., Adler, M., and Montal, M., The 26-mer peptide released from SNAP-25 cleavage by botulinum neurotoxin E inhibits vesicle docking, *FEBS Lett.,* 435, 84, 1998.
86. O'Sullivan, G.A., Mohammed, N., Foran, P.G., Lawrence, G.W., and Dolly, J. O., Rescue of exocytosis in botulinum toxin A-poisoned chromaffin cells by expression of cleavage resistant SNAP-25, *J. Biol. Chem.,* 274, 36897, 1999.

13 Psychological Factors in Chemical Warfare and Terrorism*

James A. Romano, Jr. and James M. King

CONTENTS

I. Introduction 393
II. Chemical Warfare (CW) Agents and Performance..................... 394
III. Countermeasures and Performance 397
IV. Summary.. 402
Acknowledgments.. 403
References... 403

I. INTRODUCTION

The early 21st century is faced with a significant health threat in the form of chemical weapons. These have been used more frequently and in more diverse settings since they were first used on World War I (WWI) battlefields. The potential to cause large numbers of serious casualties among deployed and deploying military forces and among civilian populations provides a stark reminder to medical planners of the limits of both military and civilian medicine. However, medical countermeasures to these chemical warfare (CW) agents have been, and continue to be, developed. These CW agents, their countermeasures, and the health care implications of their use are described in this chapter. We suggest likely psychological, physiological, and neurological effects that will be encountered should these agents be employed against, or their countermeasures used by, U.S. forces and citizens on the battlefield or in the homeland. We further suggest that these pharmacologic and toxicologic actions will occur in the broad context of a nuclear, biological, chemical (NBC) environment with its attendant confounding variables. For example, recent studies point to potential increased difficulty in the differential diagnosis of stress reaction vis-à-vis organophosphate (OP)-induced organic brain syndromes. Knowledge of the

*The opinions or assertions contained herein are the private views of the authors, and are not to be construed as reflecting the view of the Department of the Army or the Department of Defense.

behavioral effects of the CW agents and of their medical countermeasures is imperative for military and civilian medical and mental health planners as they prepare to deal with possible incidents involving battlefield CW agent use or chemical terrorism.

II. CHEMICAL WARFARE (CW) AGENTS AND PERFORMANCE

The early 21st century is faced with a significant health threat in the form of chemical weapons. These have been used more frequently and in more diverse settings since they were first used on World War I (WWI) battlefields. Traditionally, U.S. Armed Forces have been concerned with four classes of CW agents: (a) choking (e.g., phosgene [CG] and chlorine, which were employed simultaneously during WWI), (b) blood (e.g., cyanide [CN]), (c) blister (e.g., sulfur mustard [HD]), and (d) nerve agents (e.g., sarin [GB]). These agents differ in terms of their rapidity of action, lethality, and the requirement for prompt and/or sustained medical care. The accompanying tables provide a summary of the major CW agents of concern, including their historical mortality/morbidity, principal target tissue, current/proposed countermeasures, and principal behavioral effects of the CW agents and their countermeasures. The CW agents' potential to cause large numbers of serious casualties provides a stark reminder to medical planners of the limits of both military and civilian medicine. The purpose of this review is to describe likely psychological, physiological, or neurological effects that will be encountered should these agents be used on the integrated battlefield or against homeland facilities and personnel, with emphasis on their psychological or behavioral effects. Defense of the homeland against potential terrorist use of CW agents is clearly an emerging requirement for both our military and civilian medical response systems.[1–5]

Choking and blood agents were first used on the battlefield early in WWI. Although several choking agents (e.g., phosgene and chlorine) were employed simultaneously, it has been determined that the choking agent phosgene produced a large number of casualties requiring extensive hospitalization.[6] The primary clinical effect of phosgene is a pulmonary edema following a clinical latent period of variable length. The latent period is dependent primarily on the intensity of exposure (viz., the Ct, or concentration $\times$ time). The latency period is also partly dependent on the physical activity of the exposed individual with higher activity levels found to be more detrimental. For a review of phosgene's effects, see Sciuto.[6] He suggests that the primary effects of phosgene upon military performance would undoubtedly be the result of its capacity to produce deep lung injury. In addition, Sciuto cites evidence that phosgene may produce a toxic encephalopathy in man.[6] Phosgene exposure and subsequent intoxication almost certainly would cause decreased O_2 delivery to the central nervous system (CNS) and to other body systems, with accompanying behavioral and functional deficits under conditions in which continuous mental and/or physical performance is demanded.

Blood agents appeared on WWI battlefields within weeks after the initial use of choking agents and were employed in an effort to rapidly produce lethal casualties.

The principal blood agent is cyanide (CN), a biochemical poison that has a marked CNS toxicity.[7] Acute, sublethal poisoning is associated with a Parkinson-like syndrome, to include a selective neurodegeneration in experimental animals.[8] In man, Utti et al. reported a delayed onset Parkinsonism developing over a 2-year period post-intoxication.[9] Gradients of CNS effects can be observed either after exposure to lower concentrations or following exposure to lethal amounts via the oral or percutaneous routes.[10] Initial signs of CNS excitement, including anxiety and agitation, may progress to signs of CNS depression, such as coma and dilated, unresponsive pupils. Correspondingly, in animals, pathological studies from WWI and WWII indicate that residual cyanide lesions are significant only in the case of animals receiving a narrow range of exposures just below the lethal dose. As reported by Moore and Gates, recovering animals show residual neurological damage, principally in the cerebrum and cerebellum.[7] Kanthasamy et al. demonstrated a correlation between loss of dopaminergic neurons and locomotor dysfunction following CN intoxication.[11] Thus, cyanide should be regarded as a CNS-active CW agent with potential to produce residual CNS or neurobehavioral effects. However, no studies or casualty figures exist to suggest that cyanide is effective in producing militarily significant casualty rates in a trained and protected force. It is believed by some to have been used against unprotected populations and remains a threat to armed forces and civilians in both conventional and unconventional conflicts.

The most effective agent at producing casualties in WWI was the so-called "mustard gas," or HD. Smith points out that mustard produced hundreds of thousands of casualties, many of whom required extensive hospitalization, even more casualties than were seen with phosgene.[12] The pernicious nature of mustard was reinforced in the mid-1980s in the Iran-Iraq War, in which it produced an estimated 45,000 chemical casualties. The major clinical effects of exposure to HD are significant skin, eye, and pulmonary lesions, which are usually nonfatal. The length of hospitalization following these injuries was estimated to be 42 days per casualty during WWI and up to 70 days during the Iran-Iraq War.[13–15] Reports of neuropsychiatric effects, such as severe apathy, impaired concentration, and diminished libido have appeared in the literature, generally in the form of case reports rather than experimental studies.[16,17] However, in general, the psychological and behavioral effects observed after HD exposure are attributable to post-traumatic stress disorder (PTSD).[18] Lohs reported that chemical munitions workers manifested specific neurological sequelae such as impaired concentration, diminished libido, and sensory hypersensitivity.[17] Unfortunately, for many of these workers, additional or co-exposures were noted, which confound our ability to identify relationships between HD exposure and effects. Examination of the literature also reveals a paucity of long-term follow-up in the area of psychiatric sequelae linked to HD exposure.

Other potential blister agents include lewisite and phosgene oxime (CG). Lewisite was synthesized during the late stage of WWI, but there are no reports of its battlefield use. Its antidote, British Anti-lewisite, finds medicinal use today as a heavy metal chelator. Although classified as a vesicant, CG is a corrosive urticant for which, like lewisite, there are no reports of battlefield use. There are also no reports of CNS effects following intoxication with these agents.

Following WWI, work in Germany to develop organophosphorus (OP) insecticides led to the identification of a new class of CW agent compounds of extreme toxicity, the nerve agents (GA [Tabun], GB [Sarin], GD [Soman], etc.). As highly active CNS agents, sublethal exposures to this type of CW agent can be expected to produce prominent deficits in behavior and performance. McDonough provided a review emphasizing human reports of the toxicological and neurobehavioral effects of exposure to nerve agents and their medical countermeasures.[19] A considerable body of literature exists on this topic, and the interested reader is encouraged to use McDonough's review as well as Longo's, Karczmar's, and McDonough and Shih's as useful starting points.[19–22]

Earlier studies using nerve agents were often carried out in doses that produced depressed cholinesterase levels or overt signs of intoxication in animals or symptoms of cholinergic poisoning in man. Animal studies have explored the effects of nerve agents on learning, long-term or short-term memory, nociception, general activity levels, or propioception.[23–28] Occasionally, these studies were designed to demonstrate pharmacologic and behavioral antagonism with anticholinergic drugs such as atropine or antidotal combinations of atropine and oximes.[13,21,29,30] Some human experimental studies have appeared in the literature, with the U.S. Army military volunteer test program ending in the 1970s.[13] These human studies are supplemented by reports of behavioral effects of exposures to other OP compounds, those in widespread use as pesticides. According to McDonough, the behavioral effects of OP in man can be divided into three classes: (1) effects on cognitive processes, (2) effects on mood/affect, and (3) disturbances of sleep/wakefulness.[19] The intensity and duration of these behavioral effects of acute OP exposure are generally related to the magnitude of blood ChE inhibition.[29] In fact, based on his review, McDonough concluded that behavioral CNS symptoms of OP exposure have not been seen at levels of ChE inhibition less than 50% and are more typically seen when inhibition reaches 70–80% of control levels.[19] Somewhat less effort has been given to characterizing the effects of "low-level" exposures to nerve CW agent, perhaps due to the steepness of their toxicity curves and the resultant difficulty in determination of low-level dosages.[26] Recent concern over the possible sublethal exposure of members of the U.S. Armed Forces to the nerve agent GB during and following the Gulf War has led to a renewed study of this problem. The reader can identify pertinent on-going studies and review their annual reports and abstracts on the Internet at http://www.va.gov/resdev/pgrpt97.htm.[31]

Much of the data regarding long-term neurological sequelae to exposures to cholinesterase inhibitors in man have been gathered following accidental exposures to OP pesticides. In one of the major studies, Savage et al. found significant deficits on several cognitive tests of memory and abstraction, without apparent effect on the EEG or neurological examinations.[32] Steenland et al. found deficits in vibrotactile sensitivity and sustained attention among previously intoxicated subjects vs. controls; however, as with the previous study, nerve conduction tests and neurologic examinations were negative.[33] In a follow-up of humans suffering moderate to severe exposure to the CW agent GB, a deficit in the Digit Symbol Substitution Test unrelated to post-traumatic stress was observed by Yokoyama et al.[34] The Digit Symbol

Substitution Test of the WAIS (Webster Adult Intelligence Scale; Japanese version) was described as measuring "motor persistence, sustained attention, response speed, and visuomotor coordination." The decrease in Digit Symbol test scores was found 6 to 8 months after acute poisoning, at a time when ChE levels had returned to normal. The WAIS deficit was also unrelated to the level of ChE inhibition on the day of poisoning. The authors suggested that acute ChE inhibition might not be a good predictor of the chronic effects of GB on psychomotor performance.

For chronic, subclinical exposures to OP, the data are less consistent, but some authors have reported subclinical effects on the CNS and peripheral nervous system (PNS). On the other hand, Ames et al. surveyed 45 pesticide applicators, each of whom had at least one documented episode of asymptomatic AChE exposures.[35] He reported no CNS or PNS effects. For example, Stephens et al., studying a population of 146 sheep dippers with an average of 15 years of potential exposure to several OP (diazinon, propetamphos, chlorfenvinphos) found significant changes in sustained attention and speed of information processing.[36] These authors found no effects on memory or learning. Duffy et al. reported EEG changes in CW agent chemical plant workers that persisted for more than 1 year after accidental exposure to GB, but they suggested that the functional significance of these changes was unclear.[37] A review of the literature by a National Academy of Sciences (NAS) expert panel also indicated they were unable to clearly identify a functional correlate of the observed EEG changes.[38]

We believe that studies to determine the potential long-term psychologic/neurologic sequelae following OP intoxication are confounded by factors such as low response rates, possible selection and follow-up biases (loss to follow-up of the most severe cases), compensatory psychologic response, possible co-exposures, and the like. For this reason, the recent national investment into additional research in this area is well founded. The findings discussed in this section are summarized in Table 1.

III. COUNTERMEASURES AND PERFORMANCE

Pharmacological or medical countermeasures to these CW agents can produce CNS sequelae, and these neurobehavioral effects will also be discussed briefly in this chapter. This section focuses primarily on currently available countermeasures, with one exception, that of scavenging enzymes. Of course, CW agents, which have high toxicity in CNS tissue, like the nerve agents and blood agents, require medical countermeasures that may be expected to produce significant CNS and performance effects, and those effects would need to be evaluated. In the case of choking agents (e.g., phosgene) and blister agents (e.g., sulfur mustard), current treatments are symptomatic and supportive, and have few performance impacts beyond those produced by the agents themselves. Two areas of interest regarding countermeasures are highlighted. First, the rapid action of CN and the need for current therapies to be administered intravenously (a difficult requirement on the battlefield or in a mass casualty situation) have led to programs to develop prophylactics or pretreatments for this CW agent. The most promising approach to this point has been pretreatment by methemoglobin-forming compounds. In this strategy, methemoglobin acts as a scavenger

TABLE 13.1
Morbidity of Four Classes of Chemical Warfare Agents and Their Principal Targets and Medical Countermeasures

Class of CW Agent	Historic Lethality/Morbidity in Warfare	Principal Target Tissue	Physiological/ Performance Effects
Choking (e.g., phosgene)	1%[a]	Deep lung compartment such as pulmonary capillary	Pulmonary edema, hypoxia
Blood (e.g., cyanide)	Unknown[b]	Cellular respiratory enzymes	Depression of cortical function, unconsciousness, convulsions
Blister (e.g., "mustard gas")	2–4%[c]	Skin, airway, eyes, GI tract, bone marrow	Loss of function due to skin, lung, ocular lesions, recovery over time
Nerve (e.g., sarin)	Unknown[d]	CNS, neuromuscular junction; cholinergic synapse	GI tract, miosis, nausea, weakness, loss of consciousness, convulsions

Note: Estimates derived from U.S. published sources.

[a]WWI figures for the U.S. are estimates because phosgene was often mixed with chlorine; however, a total of 6834 injured (average hospitalization = 49 days) have been directly attributed to phosgene with 66 fatalities.

[b]No data from wartime use; however, wartime experiences suggest difficulty in achieving militarily effective concentrations unless confined to closed spaces.

[c]WWI, 2% with 27,711 U.S. injured; Iran-Iraq War, 4% with 45,000 estimated injured.

[d]No data from wartime use; however, on 20 March 1995, using a primitive method of dispersal, sarin was released on Tokyo subways with 5,500 people seeking medical care; approximately 1500 had defined symptoms of exposure, and 12 casualties died. Less well known is the fact that on June 27, 1999, sarin was released in Matsumoto, Japan, with estimates of 471 subjects exposed to sarin and 7 deaths.

with a much higher affinity for CN than the terminal mitochondrial cytochrome oxidase a_3 protein. Displacement of CN from cytochrome a_3 permits the return to normal respiration.[39] Research studies in animals have led to an estimate that methemoglobin levels of 5–15% will be protective against up to 2 × the LD_{50} of cyanide.[40,41] Putting this in perspective, the level of methemoglobin in smokers' blood is 2%. The effects of a level of 5–15%-induced methemoglobin on performance need to be evaluated. However, chronic methemoglobinemia is seen by some as a toxic condition with potential for long-term hematotoxicity.[42] Thus, other pretreatment strategies have been explored as part of a drug discovery program. A strategy of providing sulfur-donating compounds to possibly speed the conversion of cyanide to thiocyanate was one of the alternatives pursued. A number of promising compounds, some without any apparent disruptive effects on behavior (in mice), have been identified and the authors suggested that those newly synthesized compounds provide a rationale for a new class of anti-CN pretreatment compounds.[43] Although some limited data were reported by Baskin et al. concerning the ability of anti-CN compounds to reverse the motor (performance) impairment caused by a near-lethal dose of CN, a significant data gap exists in the area of recovery of function after mild to moderate CN intoxication.[43]

Second, medical protection from the effects of nerve agents may involve pretreatment with a reversible acetylcholinesterase inhibitor like pyridostigmine bromide or with biological scavengers targeted at these CW agent compounds. The drug pyridostigmine bromide (PB) has recently been given a considerable amount of scrutiny. This attention stems from its use as a pretreatment to protect U.S. Armed Forces against the potential use of nerve agents by the forces of Saddam Hussein during the Gulf War. Several studies of PB effects on military performance have been reviewed.[44] These studies include effects of PB upon a number of components of pilot performance such as tolerance to altitude changes or acceleration tolerance.[45,46] Additionally, Kolka and Stephenson examined the effects of PB on endurance and parameters of exercise physiology under several simulated environments.[47] In general, there were no significant effects of PB found on these performance parameters. Laboratory studies using more traditional psychological tests also support the lack of PB effects on performance.[48] Similarly, following chronic PB administration, soldiers of the Israeli Defense Force (IDF) reported infrequent and mild effects.[49] Additionally, PB's potential health consequences, when taken by otherwise healthy U.S. services members, were reviewed by Dunn et al.[43] These authors suggested that PB has a "good safety record over the years of its administration to patients with myasthenia gravis." Moreover, at the prescribed dose of 30 mg every 8 h, no significant decrements have been found in performance of a variety of military tasks. It is not within the scope of this paper to discuss the linkage, if any, between PB and Persian Gulf War Illness. A comprehensive review of the literature regarding PB with emphasis on how it might pertain to Persian Gulf War Illness can be found at http://www.gulflink.osd.mil/library/randrep/pb_paper.[50] However, the reader is again directed to http://www.va.gov/resdev/pgrpt97.htm.[30] This website provides summaries and abstract reports of more than a score of research projects examining the health effects of PB either alone or in combination with a variety of other compounds. In general, these studies explore the possible linkages and complex chemical interactions of PB and other compounds in relation to Persian Gulf War Illness.

Protection from nerve agent poisoning may also involve pretreatment with a biological scavenger. Lenz and Cerasoli suggest that the biological scavenger approach avoids the side effects associated with the current nerve agent antidotes and, from the limited amount of animal data available, appears to prevent or significantly alleviate the neurobehavioral effects of the nerve agents.[51–53] Moreover, the level of protection from the neurobehavioral effects of nerve agents is greater than that seen following the use of "conventional therapy" for nerve agents. Several approaches to the scavenger concept have been explored—antibodies, enzymes, and catalytic bioscavengers. The latter, in general, have been developed by introducing amino acid changes into stoichiometric enzymes, such as human butyrylcholinesterase and human carboxylesterase. Promising results have been obtained, and this research program remains active.[5]

The demonstrations of bioscavenger protection from the behavioral deficits that normally follow exposure to nerve agents have been made in several animal species. For mice, there is a well-characterized deficit in ability to respond on the inverted screen test.[54,55] Mice pretreated with fetal bovine serum acetylcholinesterase (FBS-AChE), equine butyrylcholineserase (Eq-BuChE) or human BuChE were completely protected from these deficits.[52,53,56] An elegant study by Brandeis et al. characterized the effects of soman on naïve rats or rats pretreated with human BuChE.[57] The Morris Water Maze (MWM) was employed in these studies, as maze learning or spatial memory tests had previously been demonstrated to be sensitive to the effects of cholinergic drugs to include OP nerve agents such as soman.[24] In Brandeis' study, naïve rats given a toxic but sublethal dose of soman had significant impairments in MWM performance that persisted for several weeks. Pretreatment with human BuChE provided significant protection from those effects of nerve agent. In fact, performance of the BuChE-pretreated animals was indistinguishable from untreated controls. Furthermore, the BuChE enzyme itself was without effect on MWM performance.

In non-human primates, a series of studies using the Serial Probe Recognition (SPR) task have yielded similar results. The scavenger enzyme was without effect on performance in the SPR task.[58] The exposure of rhesus monkeys to soman produced pronounced deficits in this task, and the animals were protected by pretreatment with BuChE (even after exposure to 4–5 $\times$ LD_{50} of nerve agent).[25] Using the Primate Equilibrium Platform, investigators at the School of Aerospace Medicine demonstrated the effects of soman on performance.[59] They found complete protection of animal performance with either FBS-AChE or Eq-BuChE, even when up to a cumulative 5 $\times$ LD_{50} of nerve agent were administered.[59] When compared to similar studies employing conventional therapies such as atropine and 2-PAM, the scavenger enzymes approach provided a remarkable improvement. The findings on effects of current medical countermeasures to CW agents discussed in this section are summarized in Table 2.

One final point about the response to chemical agents on the battlefield or in the homeland concerns the possible presence of stress casualties. These have been variously labeled as cases of "gas hysteria," "gas mania," or "gas neurosis." In a recent review of this literature, Stokes and Banderet reported that the official U.S. Army Medical Department history of WWI notes that two such cases occurred for each actual chemical injury.[60] Their analysis suggested several origins for these cases: (a) conversion disorders, (b) mistaking normal physiological stress symptoms for exposure to CW agents (despite significant efforts to train soldiers in proper recognition of signs of poisoning), (c) mistaking or magnifying the symptoms of minor illnesses, and (d) deliberate faking or malingering. One might add the possibility of an additional type of "self-inflicted wound" to this list; by this, we mean the inadvertent or misguided use of antidotal compounds, e.g., atropine and diazepam.

TABLE 13.2
Potential Performance Effects of Current Countermeasures

Agent Type	Countermeasure	Potential Physiological/Performance Effects
Choking	Symptomatic	N/A
Blood	Sodium nitrate (i.v.)	Reduces blood oxygen capacity, hypotension[a]
	Sodium thiosulfate (i.v.)	
Blister	Symptomatic	N/A
Nerve	Atropine	Mydriasis, reduced sweating, dry mouth,[b] mild sedation, delirium, at 10 mg or higher
	2-PAM Cl	Minimal, except at high i.v. doses[2]
	Pyridostigmine	Mild GI and urinary discomfort in ~50%, severe in ~1% (see also http://www.gulflink.osd.mil/library/rand/rep/pb_paper)
	Diazepam	Sedation, muscle relaxation, drowsiness, ataxia

[a]Baskin, S.I. and Brewer, T., in *Textbook of Military Medicine—Medical Aspects of Chemical and Biological Warfare,* Zajtchuk, R. and Bellamy, R.F., eds., Office of the Surgeon General, Department of the Army, Washington, DC, 1997, chap. 10.

[b]Sidell, F.R., in *Textbook of Military Medicine—Medical Aspects of Chemical and Biological Warfare,* Zajtchuk, R. and Bellamy, R.F., eds., Office of the Surgeon General, Department of the Army, Washington, DC, 1997, chap. 2.

Self-administration of two nerve agent antidote autoinjectors can produce headache, restlessness, and fatigue, symptoms that can be aggravated in a tired, dehydrated, or stressed individual. The possibility of widespread mental health disorders that may result from fear or actual use of chemical weapons has also been recently reviewed by DiGiovanni.[61] He pointed to potential difficulties in sorting out the physical effects of CW agents from the psychological response of the exposed person. Confusion about these various phenomena may result in behaviors that are difficult to differentiate from a backdrop of generalized war syndromes. In their review of poorly understood war syndromes from the U.S. Civil War to the Persian Gulf War, Hyams et al. analyzed symptoms to identify possible unifying factors.[62] These authors concluded that there was little evidence of a single, unique war syndrome that is unrelated to psychological stress. It is also certainly true that rumors of CW agent use or of its actual use can contribute to high rates of acute combat stress reactions. Ursano suggests that CW agents stir fear in military personnel for several reasons: (1) the particular, personal psychological fears of CW agents, (2) a sense of the need to continue to operate after a CW agent attack, and (3) due to the indiscriminate nature of CW agents, a fear for the safety of family members.[63]

Insights into the likelihood and consequences of such reactions can be found in a series of reports describing effects of SCUD missile attacks on Israeli civilian populations during the Persian Gulf War. One such report described the use of "anti-chemical warfare kits," to include protective masks, during the Gulf War.[64] The authors reported that the unfortunate misuse of such equipment led to the death by suffocation of 13 individuals. These deaths occurred despite detailed instructions and demonstrations provided by civil defense personnel. It appears that even intensive education efforts may not completely prevent inappropriate actions under the stress of anticipated CW agent attacks. Bleich et al. interviewed 773 Israeli civilians who were treated at 12 hospital emergency departments after either a missile attack or a false alarm.[65] Their report was revealing: 43% suffered from a stress reaction, and 27% had mistakenly injected themselves with atropine, an antidote to nerve agents. The authors surmised that anxiety might also have caused the death of several citizens who panicked during the missile attacks. Furthermore, the authors made several interesting suggestions as to the staffing of emergency centers and the processing of stress casualties. These suggestions included establishment of "stress reaction centers" co-located with the receiving hospital, the types of interventions to be found at these centers, and the optimal time frames necessary to achieve the patient's return to his home or family. Finally, they pointed to potential increased difficulty in the differential diagnosis of stress reaction vis-à-vis "OP-induced organic brain syndromes."[65] The relationships between PTSD and structural changes in brain related to sarin exposure, and the clinical implications of that relationship as it pertains to the survivors of the Tokyo subway poisoning are described by Yokoyama et al.[66] These authors did extensive evaluations of the neurobehavioral, CNS, and PNS effects in 18 exposed, hospital-treated patients. In general, Yokoyama et al. supported Bleich's contention of confounding neurobehavioral and psychological symptoms. Incidents of this type will present an enduring challenge to emergency medicine and psychology.

IV. SUMMARY

Had this review been developed in the late 1980s, its application and focus would have generally been limited to the protection of deployed U.S. Armed Forces. We believe that the world has changed. The CW agents discussed in this review should be considered as threats to deployed and deploying forces, fixed military installations either overseas or in the continental U.S., and to the American homeland and its population. Moreover, in the last decade the unthinkable has happened—CW agents have been employed against unprotected civilians by terrorists. These events have challenged the civilian health care and emergency response systems charged with their protection. The magnitude and impact of this challenge to both the military and civilian health care systems are carefully delineated by Stokes and Banderet, Sidell et al., and Ohtomi.[60,67,68] These works provide a detailed look at the current approaches to management of CW agent casualties. Therefore, the current world situation requires that we properly prepare by ensuring that accurate, up-to-date information on the behavioral, functional, and neurological impacts of CW agents, as well as their medical countermeasures, are available. A number of credible Internet sites provide general toxicologic and first-aid information and, on occasion, medical training. See http://www.nbc-med.org, http://www.cbiac.apgea.army.mil, and http://chemdef.apgea.army.mil as useful starting points.

Exposure to CW agents may produce immediate and/or long-lasting effects on behavior and psychology of exposed individuals. The greater the initial exposure, the greater the time for recovery of normal function. Medical countermeasures are available to mitigate, or even prevent, these effects. Occasionally exposure to the countermeasures may produce transient effects on behavior and/or psychological performance. Research promises to provide, at least for some agents, countermeasures that completely block these effects of CW agents, while being innocuous in their own right.

Nevertheless, these pharmacologic and toxicologic actions will occur in the broad context of the so-called "nuclear, biological, and chemical (NBC)" combat environment (and terrorist activity) with its attendant confounding variables. These phenomena will remain among the most challenging and enduring areas of military and emergency medicine and psychology.

ACKNOWLEDGMENTS

The authors wish to thank Ms. Patricia D. Little for her skillful editorial assistance in preparation of this chapter. Her excellence in secretarial skills, assistance in compiling accurate tables, and methodical approach to presentation of the reference citations enabled the publication of this material. The authors would also like to express their thanks to the following reviewers who contributed greatly to the shape of the final document: Drs. Irwin Koplovitz, Tsung-Ming Shih, and John Skvorak. The following also provided fruitful discussion and comment which influenced the paper: Drs. David Lenz, John McDonough, and Bill Smith, each of whom provided additional, separate contributions to this book.

REFERENCES

1. Chandler, R.W. and Backschies, J.R., *The New Face of War,* AMCODA Press, McLean, VA, 1998.
2. Chandler, R.W. and Trees, R.J., *Tomorrow's War, Today's Decisions,* AMCODA Press, McLean, VA, 1996.
3. *Presidential Decision Directive #62—Combating Terrorism,* Office of the Press Secretary, the White House, Washington, DC, 1998.
4. *Presidential Decision Directive #63—Critical Infrastructure Protection,* Office of the Press Secretary, the White House, Washington, DC, 1998.
5. National Domestic Preparedness Office, Department of Justice, Federal Bureau of Investigation, URL http://www.ndpo.com.
6. Sciuto, A.M., Disruption of gas exchange in mice following exposure to the chemical threat agent phosgene, *J. Mil. Psychol.,* in press.

7. Moore, S. and Gates, M., Hydrogen cyanide and cyanogen chloride, in *Chemical Warfare Agents and Related Chemical Problems.* Parts I and II, Summary technical report of Division 9, National Defense Research Committee, vol. 1., Washington, DC, 2, 7, 1946.
8. Mills, E.G., Gunasekar, P.G., Li, I., Borowitz, J.L., and Isom, G.E., Differential susceptibility of brain areas to cyanide involves different modes of cell death, *Toxicol. Appl. Pharmacol.,* 156, 6, 1999.
9. Utti, R.J., Jajput, A.H., Ashenhurst, E.M., and Rozdilksky, B., Cyanide-induced Parkinsonism: A clinicopathologic report, *Neurology,* 35, 921, 1985.
10. Baskin, S.I. and Brewer, T., Cyanide poisoning, in *Textbook of Military Medicine—Medical Aspects of Chemical and Biological Warfare,* Zajtchuk, R. and Bellamy R.F., Eds., Office of the Surgeon General, Department of the Army, Washington, DC, 1997, chap. 10.
11. Kanthasamy, A.G., Borowitz, J.L., Pavlokovik, G., and Isom, G.E., Dopaminergic neurotoxicity of cyanide: Neurochemical, histological, and behavioral characterization, *Toxicol. Appl. Pharmacol.,* 126, 156, 1994.
12. Vedder, E.B., The Vesicants, in: *The Medical Aspects of Chemical Warfare,* Vedder, E.B., Ed., Williams and Wilkins Co., Baltimore, 1925.
13. Sidell, F.R. and Hurst, C.G., Long-term health effects of nerve agents and mustard, in *Textbook of Military Medicine—Medical Aspects of Chemical and Biological Warfare,* Zajtchuk, R. and Bellamy, R.F., Eds., Office of the Surgeon General, Department of the Army, Washington, DC, 1997, chap. 8.
14. Willems, J., Clinical management of mustard gas casualties, *Ann. Med. Mil. Belgicae,* (Supplement)1–6, 1989.
15. Sidell, F.R. and Hurst, C.G., in *Textbook of Military Medicine—Medical Aspects of Chemical and Biological Warfare,* Zajtchuk, R. and Bellamy, R.F., Eds., Office of the Surgeon General, Department of the Army, Washington, DC, 1997, chap. 7.
16. Eisenmenger, W., Drasch, G., von Clarsmann, M., Kietschner, E., and Roider G., Clinical and morphological findings on mustard gas [bis(2-chloroethyl) sulfide] poisoning, *J. Foren. Sci.,* 36, 1688, 1991.
17. Lohs, K., *Delayed Toxic Effects of Chemical Warfare Agents,* Stockholm: Almquist and Wiltsell, SIPRI Monograph, 1975.
18. Balali-Mood, M., First reports of delayed toxic effects of yperite poisoning in Iranian fighters, *Proc. Int. Assoc. Foren. Toxicol.,* 1986.
19. McDonough, J.H., Performance impacts of nerve agents and their pharmacologic countermeasures, *J. Mil. Psychol.,* in press.
20. Longo, V.G., Behavioral and electroencephalographic effects of atropine and related compounds, *Pharmacol. Rev.,* 18, 965, 1996.
21. Karczmar, A., Acute and long-lasting concentrations of organophosphorus agents, *Fund. Appl. Toxicol.,* 4(2), 51, 1984.
22. McDonough, J.H. and Shih, T.A., Neuropharmacological mechanisms of nerve agent-induced seizures and neuropathology, *Neurosci. Biobehav. Rev.,* 21(5), 559, 1997.
23. McDonough, J.H., Smith, R.F., and Smith, C.D., Behavioral correlates of soman-induced neuropathology: Deficits in DRL acquisition, *Neurobehav. Toxicol. Teratol.,* 8, 179, 1986.
24. Raffaele, K., Hughey, D., Wenk, G., Olton, D., Modrow, H., and McDonough, J., Long-term behavioral changes in rats following organophosphonate exposure, *Pharmacol. Biochem. Behav.,* 27, 407, 1987.
25. Castro, C.A., Gresham, V.C., Finger, A.V., Maxwell, D.M., Solana, R.P., Lenz, D.E., and Broomfield, C.A., Behavioral decrements in rhesus monkeys trained on a serial probe recognition task despite protection again soman lethality by butyrylcholinesterase, *Neurotoxicol. Teratol.,* 16, 145, 1994.

26. Romano, J.A., Penetar, D.M., and King, J.M., A comparison of physostigmine and soman using taste aversion and nociception, *Neurobehav. Toxicol. Teratol.,* 7, 243, 1985.
27. Romano, J.A. and Landauer, M.R., Effects of the organophosphorus o-ethyl-n-dimethyl phosphorametocyanidate (tabun) on flavor aversions, locomotor activity, and rotarod performance, *Fund. Appl. Toxicol.,* 6, 62, 1986.
28. Blick, D.W., Murphy, M.R., Brown, G.C., Yochmowitz, M.G., Fanton, J.W., and Hartgraves, S.L., Acute behavioral toxicity of pyridostigmine or soman in primates, *Toxicol. Appl. Pharmacol.,* 126, 311, 1994.
29. Grob, D. and Harvey, A.M., Effects in man of the anticholinesterase compound sarin isopropyl methyl-phosphonofluoridate, *J. Clin. Invest.,* 37, 350, 368, 1958.
30. Romano, J.A., Terry, M., Murrow, M., and Mays, M., Protection from soman-induced lethality and incapacitation by atropine and 2-PAM chloride in the guinea pig, Cavia porcellus, *Drug Chem. Toxicol.,* 14, 21, 1991.
31. *Annual Report to Congress: Federally Sponsored Research on Gulf War Veterans' Illnesses for 1977,* Department of Veterans Affairs, URL http://www.va.gov/resdev/pgrpt97.htm.
32. Savage, E., Keefe, T., Mounce, L., Heaton, R., Lewis, J., and Burcar, P., Chronic sequelae of acute organophosphate pesticide poisoning, *Arch. Environ. Health,* 43, 38, 1990.
33. Steenland, K., Jenkins, B., Ames, R.G., O'Malley, M., Chrislop, D., and Russo, J., Chronic neurological sequelae in organophosphate pesticide poisoning, *Am. J. Public Health,* 84, 731, 1995.
34. Yokoyama, K., Araki, S., Kaysuyutci, M., Nishihitani, M., Okumura, T., Ishimatsu, S., Takasu, N., and White, R.F. Chronic neurobehavioral effects of Tokyo subway sarin poisoning in relation to post-traumatic stress disorders, *Arch. Environ. Health,* 53, 245, 1998.
35. Ames, R., Steenland, K., Jenkins, B., Chrislop, D., and Russo, J. Chronic neurologic sequelae to cholinesterase inhibition among agriculture pesticide applicators, *Arch. Environ. Health,* 50, 440, 1995.
36. Stephens, R., Spurgeon, A., Calvert, I.A., Beach, J., Levy, L.S., Berry, H., and Harrington, J.M., Neuropsychological effects of long-term exposure to organophosphates in sheep dip, *Lancet,* 345, 1135, 1995.
37. Duffy, F.H., Burchfiel, J.L., Bartels, P.H., Gaon, M., and Sim, V.M., Long-term effects of an organophosphate upon the human encephalogram, *Toxicol. Appl. Pharmacol.,* 47, 161, 1979.
38. National Academy of Science, Committee on Toxicology, *Possible Long-Term Health Effects of Short-Term Exposure to Chemical Agents.* Vol. I. *Anticholinesterases and Anticholinergics,* prepared by Panel on Anticholinesterase Chemicals, Panel on Anticholinergic Chemicals, Committee on Toxicology, National Academy of Science, National Academy Press, Washington, DC, 1982.
39. Baskin, S.I. and Fricke, R.E., The pharmacology of p-aminopropiophenone in the detoxification of cyanide, *Cardio. Drug Rev.,* 10(13), 358, 1992.
40. Johnson, W.D. and Becci, P., Effects of methemoglobin versus potassium cyanide intoxication, U.S. Army Medical Research and Materiel Command, Fort Detrick, MD, Contract #DAMD17–83-C-3083, *DTIC #AD B108718L.*
41. Bright, J.E. and Marrs, T.C., Effect of p-aminopropiophenone (PAPP), a cyanide antidote, on cyanide given by intravenous infusion, *Hum. Toxicol.,* 6, 133, 1987.
42. Smith, R.P., Toxic responses of the blood, in *Casarett and Doull's Toxicology: The Basic Science of Poisons* (4th ed.), Amdur, M.O., Doull, J., and Klaasen, C.D., Eds., Pergamon Press, New York, 1991, chap. 8.

43. Baskin, S.I., Porton, D.W., Rockwood, G.A., Romano, J.A., Patel, H.C., Kiser R.C., Cook, C.M., and Ternay, A.L. *In vitro* and *in vivo* comparison of sulfur donors as antidotes to acute cyanide intoxication, *J. Appl. Toxicol.,* 19, 173, 1999.
44. Dunn, M.A., Hackley, B.E., and Sidell, F.R., Pretreatment for nerve agent exposure. in *Textbook of Military Medicine—Medical Aspects of Chemical and Biological Warfare,* Zajtchuk, R. and Bellamy, R.F., Eds., Office of the Surgeon General, Department of the Army, Washington, DC, 1997, chap. 6.
45. Schifflett, S.G., Stranges, S.F., Slater, T., and Jackson, M.K., Interactive effects of PB and attitude on performance, in *Proceedings of the 6th Medical Chemical Defense Bioscience Review,* Aberdeen Proving Ground, MD, 605, 1987.
46. Forster, E.M., Barber, J.A., Parker, F.R., Whinnery, J.E., Burton, R.R., and Bell I., Effects of pyridostigmine bromide on acceleration tolerance and performance, *Aviat. Space Environ. Med.,* 65, 110, 1994.
47. Kolka, M.A. and Stephenson, L.A., Human temperature regulation during exercise after oral pyridostigmine administration, *Aviat. Space Environ. Med.,* 6, 220, 1990.
48. Gall, D., The use of therapeutic mixtures in treatment of cholinesterase inhibition, *Fundam. Appl. Toxicol.,* 1, 214, 1981.
49. Sharabi, Y., Danon, Y.L., Berkenstadt, H., Almog, S., Mimouni-Block, A., Zisman, A., Dani, S., and Atsmon, J., Survey of symptoms following intake of PB during the Persian Gulf War, *Isr. J. Med. Sci.,* 27, 656, 1991.
50. Golomb, B., *A Review of the Scientific Literature as It Pertains to Gulf War Illnesses.* Vol. II. Pyridostigmine Bromide. URL http//:www.gulflink.osd.mil/library/randrep/pb_paper.
51. Lenz, D. and Cerasoli, D., Nerve agent bioscavengers: Protection with reduced behavioral deficits, *J. Mil. Psychol.,* in press.
52. Raveh, L., Greenwald, J., Marcus, P., Papier, Y., Cohen, E., and Ashani, Y., Human butyrylcholinesterase as a general prophylactic antidote for nerve agent toxicity, *Biochem. Pharmacol.,* 45, 2465, 1993.
53. Raveh, L., Grauer, E., Greenwald, J., Cohen, E., and Ashani, Y., The stoichiometry of protection against soman and VX toxicity in monkeys pretreated with human butyrylcholinesterase, *Toxicol. Appl. Pharmacol.,* 145, 45, 1997.
54. Wolfe, A.D., Rush, R.S., Doctor, B.P., Koplovitz, I. and Jones, D., Acetylcholinesterase prophylaxis against organophosphate toxicity, *Fundam. Appl. Toxicol.,* 9, 266, 1987.
55. Koplovitz, I., Romano, J.A., and Stewart, J.R., Rapid assessment of motor performance decrement following soman poisoning in mice, *Drug Chem. Toxicol.,* 22, 221, 1987.
56. Maxwell, D.M., Brecht, K.M., Doctor, B.P., and Wolfe, A.D., Comparison of antidote protection against soman by pyridostigmine, HI-6 and acetylcholinesterase, *J. Pharmacol. Exper. Ther.* 264, 1085, 1993.
57. Brandeis, R., Raveh, L., Grunwald, J., Cohen, E., and Ashani, Y., Prevention of soman-induced cognitive deficits by pretreatment with human butyrylcholinesterase in rats, *Pharmacol. Biochem. Behav.,* 126, 311, 1993.
58. Broomfield, C.A., Maxwell, D.M., Solana, R.P., Castro, C.A., Finger, A.V., and Lenz, D.E. Protection of butyrylcholinesterase against organophosphorus poisoning in nonhuman primates, *J. Pharmacol Exp. Ther.,* 259, 633, 1991.
59. Wolfe, A.D., Blick, D.W., Murphy, M.R., Miller, S.A., Gentry, M.K., Hartgraves, S.L., and Doctor, B.P., Use of cholinesterase as pretreatment drugs for the protection of rhesus monkeys against soman toxicity, *Toxicol. Appl. Pharmacol.,* 117, 189, 1992.
60. Stokes, J.W. and Banderet, L.E., Psychological aspects of chemical defense and warfare, *Mil. Psychol.,* 9, 395, 1997.

61. DiGiovanni, C., Domestic terrorism with chemical or biological agents: Psychiatric aspects, *Am. J. Psychiatry,* 156, 1500, 1999.
62. Hyams, K.C., Wignall, F.S., and Roswell, R., War syndromes and their evaluation: From the U.S. Civilian War to the Persian Gulf War, *Ann. Intern. Med.,* 125, 398, 1996.
63. Ursano, R.J., Combat stress in the chemical and biological warfare environment, *Aviat. Space Environ. Med.,* 59(12), 1123, 1988.
64. Arenburg, B. and Hiss, J., Suffocation from misuse of gas masks during the Gulf War, *Br. Med. J.,* 3018, 92, 1992.
65. Bleich, A., Dycian, A., Koslowsky, M., Solomon, Z., and Wiener, M., Psychic implications of missile attacks on a civilian population: Israeli lessons from the Persian Gulf War, *JAMA,* 268(5), 613, 1992.
66. Yokoyama, K., Araki, S., Murata, K., Nishihitani, M., Okumura, T., Ishimatsu, S., and Takasu, N., Chronic neurobehavioral and central and autonomic nervous system effects of Tokyo subway sarin poisoning, *J. Physiol.,* 92, 317, 1998.
67. Sidell, F.R., Nerve agents, in *Textbook of Military Medicine—Medical Aspects of Chemical and Biological Warfare,* Zajtchuk, R. and Bellamy, R.F., Eds., Office of the Surgeon General, Department of the Army, Washington, DC, 1997, chap. 5.
68. Ohtomi, S., Medical experience with sarin casualties in Japan, in *1996 Medical Defense Bioscience Review Proceedings,* King, J.M., Ed., U.S. Army Medical Research Institute of Chemical Defense Aberdeen Proving Ground, MD, AD A 321842. Vol 3, 1182, 1996.

14 Emergency Response to a Chemical Warfare Agent Incident: Domestic Preparedness, First Response, and Public Health Considerations

David H. Moore and Steve M. Alexander

CONTENTS

I. Introduction 410
A. Executive Initiatives 411
B. Legistlative Initiatives 412
C. The Federal Response Plan 413
II. Chemical Warfare (CW) Agents 414
A. Means of Intoxication 414
B. Clinical Effects 416
1. Nerve Agents 416
2. Vesicants 416
3. Blood Agents 417
4. Pulmonary Agents 417
C. Emergency Medical Treatment 417
1. Nerve Agents 417
2. Vesicants 418
3. Blood Agents 419
4. Pulmonary Agents 419
III. Programs to Protect Civilian Populations 419
A. Civilian vs. Military Response Considerations 419
B. The Israeli Model 420
C. Chemical Stockpile Emergency Preparedness Program 421

0-8493-0872-0/01/$0.00+$.50

IV. Elements of the Response. 422
A. Clues to the Presence of Chemical Warfare Agents 422
B. HAZMAT Response 422
C. Currently Available Detection Technology. 423
D. Laboratory Analysis 424
1. Nerve Agents 425
2. Vesicants. 426
3. Cyanide 426
V. Protection of First Responders and Victims 426
A. Programs to Improve the Response 426
1. Training 426
2. The Chemical and Biological Hotline 427
3. Chemical Weapons Improved Response Program 427
4. Applied Research. 427
5. Mass Casualty Decontamination 428
VI. Issues Related to Low-Dose Exposure to Chemical Agents. 430
VII. Summary. 431
Acknowledgment. 431
References 431

I. INTRODUCTION

The past 20 years have evidenced a steady, upward trend in the number of nations having chemical warfare (CW) capability. The proliferation and use of CW agents within unstable sectors of the world is an additional cause for grave concern regarding the possible future use of such agents, both in open conflict and in the hands of terrorists. The televised images shown of the reactions to potential CW missile attacks by Iraq against targets in Saudi Arabia and Israel during the Gulf War are still vivid in our minds. The bombings of the World Trade Center in New York in 1993 and the Alfred P. Murrah Federal Building in Oklahoma City in 1995 brought the face of terrorism into every American home with startling clarity. Terrorism is no longer something that happens overseas, occurs at an American embassy, or is encountered only by travelers. In fact, the face of the terrorist can be that of someone's neighbor or one-time classmate. Terrorism can derive from clandestine, state-directed initiatives or from small splinter groups with special interests or agendas. The Aum Shinrikyo cult in Japan successfully manufactured the CW agent sarin and combined psychological manipulation and religious zeal to support the terrorist actions of its organization. Access to chemical manufacturing facilities and suppliers around the globe provides terrorists the availability of precursors and chemical reagents, while faltering economic conditions can pave the way for theft or sale of the chemical agents themselves. A particularly troubling phenomenon is the availability of accurate information about the chemical properties, uses, and effects of CW agents on the Internet. This medium makes once highly sensitive information available to virtually every person on earth.

On a more optimistic note, the increased worldwide awareness of the need to control the production, storage, and use of CW agents has led to the formulation and signing of a treaty, outlined in the Chemical Weapons Convention, calling for the prohibition and elimination of CW agent production. The Russian Federation's chemical weapons arsenal includes 40,000 metric tons of toxic agents at seven storage sites. The United States had about 31,000 metric agent tons prior to the start of its Stockpile Disposal Program and over 25,000 tons of these toxic agents still remain.[1] While such prohibition is a noble endeavor, implementation of aggressive projects to destroy these enormous stockpiles of CW agents poses significant public health problems for civilian populations living near storage sites, destruction facilities, or transportation routes. Additionally, the manufacture of large quantities of toxic industrial chemicals that is prevalent in most industrialized countries can present an accident threat to civilian communities surrounding such an area. Facilities of this nature are also not immune to sabotage or terrorist attack. The targeting of an industrial complex producing or utilizing toxic industrial chemicals as a means of waging war was the topic of a recent international conference.[2]

Over the past 5 years, subsequent to the publication of the last edition of this text, enormous emphasis has been placed on domestic preparedness for possible use of weapons of mass destruction (WMD). Chemical warfare agents, along with nuclear weapons and biological warfare agents, are included in this category. The reader is referred to the previous edition where much of the information on medical and public health considerations of CW agents remains accurate.[3] This chapter is designed to expand on the previous work and to put this information into a more current context.

A. Executive Initiatives

In 1986, National Security Decision Directive 207 was released, which highlighted the need for a coordinated, centrally managed approach to combat terrorism. It also reaffirmed federal agencies' roles and responsibilities under the auspices of the National Security Council. The Department of State was responsible for coordinating the national response to international terrorism while the Federal Bureau of Investigation (FBI) through the Department of Justice was responsible for domestic terrorism.[4]

In June 1995, following the bombing of the Oklahoma City Federal building, Presidential Decision Directive 39 was issued. This directive detailed federal agency roles and responsibilities and defined three anti-terrorism strategies as follows: (1) to proactively reduce vulnerabilities to attack and deter such attacks before they occur, (2) to conduct crisis management of terrorist attacks in responding to such acts and conduct activities to apprehend and punish terrorists, and (3) to manage the consequences of these attacks.[5] Furthermore, the highest priority was given to developing effective capabilities to respond to the threat posed by weapons of mass destruction in the hands of terrorists. The FBI was charged with taking the lead in crisis management, while the Federal Emergency Management Agency (FEMA) was tasked with ensuring national preparedness for a terrorist attack. Additionally, the Attorney General was directed to study the threat terrorism posed to critical U.S. infrastructure, including the water supply. The President's Commission on Critical Infrastructure Protection published a report in October 1997 recognizing several shortcomings in the capabilities of first responders—firefighters, paramedics, and police—to effectively deal with the challenges of a WMD attack.[4] The Commission recommended responders receive additional assets, to include adequate equipment and training.

Presidential decision directive 62, signed in May 1998, created a more systematic approach to the problem by reinforcing and clarifying the current counter-Terrorism mission of the more than 40 federal agencies involved.[6] The Directive also established the Office of the National Coordinator for Security, Infrastructure Protection and Counter-Terrorism. The Coordinator oversees and reports to the President on the policies and programs relevant to preparedness and consequence management for WMD. Presidential Decision Directive 63, signed on the same day, required more effective interaction between government agencies and the private sector.[7] As a result, the National Infrastructure Protection Center at the FBI was tasked to increase the capabilities of federal agencies by increasing information sharing. The Center also serves as the principal facilitator of a coordinated federal response to a WMD incident.

B. Legislative Initiatives

The U.S. Congress has acknowledged the threat of terrorism and has enacted legislation to address the issue. Certain acts of terrorism are now federal crimes, regardless of where they are committed.[4,8] The Foreign Crisis Act of 1961 prohibits U.S. assistance to foreign countries whose governments supported terrorism. The Act to Combat International Terrorism (PL 98–533), enacted in 1984, offers a bounty to persons providing information leading to the arrest of a terrorist in any country if the target was a U.S. citizen or property. Furthermore, Congress has delineated agency roles and responsibilities and appropriated funds to enhance federal, state, and local response capabilities to terrorism, including those involving chemicals. The National Defense Authorization Act of 1994 (PL 103–160) by Congress was designed to strengthen federal response planning for potential terrorist use of chemical or biological (CB) weapons. The Anti-Terrorism and Effective Death Penalty Act of 1996 (PL 104–132) prohibits terrorist fund-raising, financial transactions, and other assistance to terrorists. It prescribes procedures for removing alien terrorists from the U.S. and expands and strengthens criminal prohibitions and penalties pertaining to terrorism.

Title XIV of the Defense Authorization Act of 1997 (PL 104–201), championed by Senators Nunn, Lugar, and Domenici, directed the Defense Department to assist federal, state, and local officials with training, technical advice, equipment, and other actions necessary to increase the local response capabilities to respond to and manage the consequences of a WMD terrorist incident. Specifically, the Defense Department was directed to assist civilian officials in developing chemical and biological defensive programs, and to help the Public Health Service to organize Metropolitan Medical Strike Teams.

Public Law 105–119, the Departments of Commerce, Justice, and State, the Judiciary and Related Agencies Appropriations Act of 1998 directed the federal government to provide grants to state and local governments to procure detection, decontamination, personal protective, and communications equipment. The Act also directed the Attorney General to fund the operation of large training facilities at Fort McClellan, Alabama and The New Mexico Institute of Mining and Technology.

House Report 105–825, the Conference Report Accompanying Department of Justice (DOJ) Fiscal Year 1999 Appropriations Act, provided more than $100 million to the Office of Justice Programs. This money is to be used to buy equipment, conduct training, provide technical assistance, and fund research and development. This appropriation provides for the equipment and training required by first responders, fire and emergency services and law enforcement, for a WMD incident.

C. The Federal Response Plan

The Federal Response Plan establishes the process and implementing structure for federal agencies to lend assistance in any declared disaster or emergency, including terrorists' use of chemical agents.[9] The plan organizes the federal response under 12 Emergency Support Functions, with a designated agency assigned primary responsibility for each, and others given responsibility in support roles.

Crisis management is primarily a law enforcement function. It refers to activities undertaken to pre-empt terrorist attack. The Department of Justice has assigned responsibility for these duties to the FBI. The Bureau acts as the on-scene manager for the federal government and controls access to the incident.[10] Once an attack has occurred, consequence management responsibilities predominate. State and local officials have primary authority for this. The Federal Emergency Management Agency, as the lead federal agency for consequence management, coordinates and structures the federal response to supplement state and local assets.[11] During an event, there would be both crisis and consequence management activities occurring simultaneously.

The Department of Health and Human Services directs and resources the federal response under Emergency Support Function #8, Health and Medical Services. Response actions under this function are grouped into four general categories: prevention, medical services, mental health services, and environmental health. The Centers for Disease Control and Prevention serve as the lead in this mission to assess the health and medical effects of exposure, conduct field investigations, collect samples, provide advice on protection from the hazard, and lend technical assistance for treatment and decontamination of victims.

Special teams from within the federal government can rapidly respond to assess an incident and help locate and examine an unknown WMD device. More than two dozen Weapons of Mass Destruction Civil Support Detachments, formerly called Rapid Assessment and Initial Detection (RAID) Teams have been formed.[12] An early assessment will determine the type of agent used and the location of downwind hazard. This information is critical for making appropriate decisions regarding areas currently contaminated and allowing for evacuation of those potentially in danger. The Department of Defense has established the Chemical and Biological Incident Response Force to respond rapidly in the event of a chemical or biological incident. This dedicated force under the control of the U.S. Marine Corps is equipped with the most current detection equipment and trained for mass casualty decontamination and consequence management.

The Office of Emergency Preparedness provides other emergency response teams, such as Disaster Mortuary Teams. Additionally, it has established special National Medical Response Teams to provide treatment, decontamination, and special pharmaceuticals to treat up to 1,000 patients. The Metropolitan Medical Strike Teams, also established under this office, ensure the continued viability of a jurisdiction's existing health system given the added burden of a WMD incident. The Metropolitan Medical Response System consists of parts of existing local systems that can be called in to provide triage, treatment, and patient decontamination. This system transports patients who have been decontaminated at the scene to other facilities as appropriate for continued care. The System also assists medical facilities in developing procedures that ensure patients are decontaminated before they enter a facility.[13]

II. CHEMICAL WARFARE (CW) AGENTS

Medical professionals and emergency response personnel seldom see mass casualties that resemble CW agent casualties. With the increased threat of terrorism world-wide, a focus on the management of CW agent casualties is timely and appropriate. A goal of any WMD response plan should be to train teams of professionals to understand chemical agent threats and how to respond to them efficiently. A more complete description of the agents, their effects, and the medical management of casualties are presented in other texts and are only summarized in this section for general information.

Chemical warfare agents are either lethal in their effects or incapacitating, depending upon the class of agent, the concentration, and the period of exposure. The lethal agents, nerve, blood and pulmonary, or choking agents and the incapacitating vesicant agents will be covered below. Excluded from this discussion are other incapacitants and riot-control agents.

Chemical warfare agents are also classified as "persistent" and "non-persistent." The former includes the vesicants such as sulfur mustard (HD) and Lewisite (L) and the nerve agent VX. Non-persistent agents are more volatile and do not remain in an open environment for more than a few hours. Among these are phosgene, cyanide, and the nerve agents, tabun (GA), sarin (GB), soman (GD), and cyclosarin (GF). Toxicity follows exposure to chemical agents dispersed as solids, liquids, aerosols, or vapor (see Table 14.1). Chemical warfare agents have characteristics that make them uniquely suited to warfare. In addition to their extreme toxicity, their chemical structures are simple, and the manufacturing processes for most are relatively uncomplicated and inexpensive. Cyanide and phosgene represent particularly significant hazards because they are manufactured in large quantities for use in industry and are shipped in bulk by truck or train.

A. Means of Intoxication

Most CW agents were designed to be volatile and non-persistent and are encountered as vapor or gas. These agents can also be dispersed as an aerosol following a

TABLE 14.1
Comparison of Potencies of Chemical Warfare Agents

CW Agent	ECt_{50}	LCt_{50}
Nerve agents	3–5	10–200
Mustard	50–100	1500
Cyanide	>1000	2500–5000
Phosgene	>1000	3000

Note: Ct is concentration of vapor (mg/m^3) × time (minutes of exposure); ECt_{50} is the Ct producing clinical symptoms in 50% of the exposed population; LCt_{50} is the Ct that is lethal for 50% of the exposed population.

detonation. The persistence of the agent is dependent on factors such as temperature, pressure, and wind speed. Thus, for some of the nerve agents such as GA, GB, and GD, as well as phosgene and chlorine, the primary route of intoxication is through the respiratory tract. The nerve agents VX and thickened GD and the vesicant agent sulfur mustard are three of the most persistent CW agents and pose a threat from dermal absorption as liquids or droplets. These agents can pose vapor hazards as well.

Contamination of foodstuffs by chemical agents may occur from contact with vapor, aerosol, or droplets. The effects of the chemical agents on food depend on the nature of the agent, as well as the nature of the food. For example, foods having a low water content and a high fat content such as butter, oils, fatty meats, and fish absorb vesicant and nerve agents so readily that removal of the agents is virtually impossible. Chemical agents can cause the food to become highly toxic without changing the appearance of the food. Unprotected foodstuffs may be so contaminated that their consumption will produce gastrointestinal irritation or systemic poisoning. Protected foodstuffs in cans and bottles or food wrapped in heavy plastic are not affected by agent vapor and can be salvaged following decontamination.[14]

Few environmental factors impact community and individual well being more than the ready availability of adequate and safe, potable water. Surface water sources in the area of a chemical release could become contaminated. The contamination of water, whether intentional or inadvertent, may reach concentrations that could produce casualties. Deep ground water reservoirs and protected water storage tanks are regarded as safe sources of drinking water following a vapor release of chemical agents. While avoiding any possibly contaminated water source should be a goal, methods such as reverse osmosis are available to treat large volumes of potentially contaminated water for emergency drinking. However, these techniques may not eliminate low-dose exposure to the contaminating agent. The fate and distribution of CW agents in the environment have been recently reviewed.[15]

B. Clinical Effects

1. Nerve Agents

Nerve agents exert their effects by inhibition of the enzyme acetylcholinesterase (AChE), leading to accumulation of excess levels of the neurotransmitter acetylcholine (ACh) at cholinergic synapses. Enzyme inhibition is both rapid and irreversible, thus making organophosphorus (OP) nerve agents highly toxic and extremely dangerous chemicals (see Table 1). These agents were designed to kill or incapacitate enemy forces, disrupt military operations, and deny terrain to the adversary. However, in unscrupulous hands, they have been shown to be effective weapons of terror.

Nerve agents gain entry by absorption through the lungs or skin and impair the activity of cholinergic synapses, including those of smooth and skeletal muscle, autonomic ganglia, and the central nervous system. Acute toxic effects of nerve agents can be elicited at very low concentrations, while lethal effects are observed at somewhat higher concentrations. Threshold symptoms for vapor exposure are commonly stated to be miosis, rhinorrhea, and airway constriction,[16] generally appearing at a Ct (vapor concentration $\times$ exposure time) of 2–3 mg $\cdot$ min/m^3. Low-to-moderate exposure of skin to liquid nerve agent causes localized sweating, nausea, vomiting, and a feeling of weakness.[17] Lethal amounts of vapor or liquid cause a rapid cascade of events, culminating within a minute or two in convulsion, loss of consciousness, apnea, paralysis, and death.[18] Toxicity is thus concentration-dependent, requiring a defined minimal concentration of agent; recovery generally occurs by synthesis of new AChE.

Additionally, after both vapor and liquid agent exposure, there are CNS effects that vary in intensity and duration. After mild to moderate exposure to nerve agent, there may be forgetfulness, an inability to concentrate, insomnia, impaired judgment, nightmares, irritability, and depression. These effects may be present for 4 to 6 weeks. They may also occur upon recovery from acute, severe effects of exposure. Long-term and low-dose effects of the nerve agents are the topic of Chapter 1 of this text.

2. Vesicants

Sulfur mustard (H, HD) has been a major military threat agent since World War I. Lewisite (L) also falls into this class. Sulfur mustard constitutes both a vapor and a liquid threat to all exposed skin and mucous membranes. The effects are delayed, appearing 2–24 h after exposure. Mustard reacts with tissues within minutes of absorption. In extracellular water, it rapidly forms a highly reactive cyclic compound that binds to enzymes, proteins, and other substances. Mustard is a strong alkylating agent that causes cross-linking in DNA strands leading to cell death.[19,20] Blood, tissue, and blister fluid do not contain mustard and cannot cause further toxicity. Typical effects occur in the eye, ranging from mild to severe conjunctivitis, blepharitis, and damage to the cornea. Airways react with initial irritation, progressing to severe damage of lower airways with higher concentrations; respiratory failure and pneumonia, in addition to bone marrow suppression, may lead to death in sulfur mustard poisoning. Skin injury initially shows erythema, followed by formation of vesicles that later coalesce to form bullae.[19]

3. Blood Agents

Chemical warfare blood agents are hydrogen cyanide (hydrocyanic acid, AC) and cyanogen chloride (CK). Cyanide is a rapidly acting lethal agent that causes death within 6 to 8 min after inhalation of a high concentration (Table 1). However, few toxic effects are seen below a lethal concentration. Once absorbed, the cyanide ion rapidly combines with the active site of the enzyme cytochrome oxidase interfering with aerobic metabolism, creating excess lactic acid and metabolic acidosis.[21] Cell death is the final outcome. The organs most susceptible to cyanide are the CNS and the heart. The onset and progression of signs and symptoms are slower after ingestion of cyanide or after inhalation of a low concentration of vapor. There may be an asymptomatic period of several minutes, followed by the initial transient hyperpnea. This may be followed by feelings of anxiety, agitation, vertigo, weakness, nausea, vomiting, and trembling. Later, there is a loss of consciousness, decrease in respiration, convulsions, apnea, and cardiac arrest.

4. Pulmonary Agents

Phosgene is a simple, highly volatile molecule ($COCl_2$) known as carbonyl chloride (CG). The odor of phosgene has been described as an odor of newly mown hay. Upon inhalation, CG chemically induces acute lung injury because of a reaction of its carbonyl group with groups affecting cell membrane stability.[22] This allows plasma to leak into the alveoli and produces pulmonary edema. There is a symptom-free period (10 min to 24 h) that varies with the amount of CG inhaled. Substantial toxicity can occur from levels of 1.5–2 ppm. Its aroma is detectable at 2–3 ppm, hence toxicity may occur without subject awareness.[17] In cases where individuals are exposed to high concentrations of phosgene, there may be initial symptoms of mucous membrane irritation followed by pulmonary edema, hypoxia, hypotension, bronchospasm, right heart failure, and death.

C. Emergency Medical Treatment

1. Nerve Agents

The principles of care for a casualty with nerve agent intoxication include termination of exposure, maintenance of ventilation, administration of antidotes, and supportive therapy. For successful medical management, early and intense therapy after severe exposure to nerve agents is necessary to prevent death. The condition of the patient will dictate the need for specific treatment procedures and the order of administration. It is of utmost importance that medical care providers are protected from contamination by use of appropriate protective clothing, otherwise they may become additional casualties.[16]

Decontamination of the casualty is not necessary if exposure is due only to vapor. However, clothing that may trap vapor needs to be removed. The patient should be removed from the source of the vapor, if possible, or fitted with a protective mask if this is a practical alternative. Removal of affected clothing and decontamination of the underlying skin will help terminate liquid agent exposure. Visible droplets of agent can be wiped or blotted off, followed by flushing or rinsing with copious amounts of water. The agents can be neutralized with soap and water or by 0.5% hypochlorite solution, followed by rinsing with water. It should be noted that neutralization of the agent is not immediate and can take minutes to hours to complete. Agent absorbed into the skin cannot be removed, so the casualty from liquid exposure may continue to worsen because of continued absorption of the agent from the dermis and subcutaneous tissues. As a rule, the longer the asymptomatic period after exposure, the milder will be the eventual symptoms.[17,23]

Endotracheal intubation and assisted ventilation with oxygen is essential for apneic casualties. This is necessary for survival and may be needed for an extended period. Airway resistance is initially high because of bronchoconstriction as well as copious secretions. If respiratory impairment is only mild to moderate, it can be reversed by antidote treatment without the need for ventilation.[16]

Antidotal therapy includes the use of atropine to block the effects of excess ACh primarily at peripheral muscarinic receptor sites. Following atropine use, secretions are reduced and constriction of smooth muscle is reversed. Since it has little effect at nicotinic sites, skeletal muscle fasciculation will continue. Similarly, miosis will not be reversed. Pralidoxime chloride (2-PAM Cl) is the oxime of choice in the U.S. for reactivation of nerve agent inhibited ChE. By breaking the OP-enzyme bond, oximes restore normal activity of the enzyme. Clinically, this is noticeable in those organs with nicotinic receptors. Abnormal activity in skeletal muscle decreases and normal strength returns. Diazepam is an anticonvulsant drug used to reduce brain damage caused by prolonged seizure activity, as seen in severe poisoning from nerve agents. Diazepam therapy is indicated with other therapy at the onset of severe effects from a nerve agent whether convulsions are present or not. The use of atropine, 2-PAM, and diazepam therapy affords considerable benefit for most cases of nerve agent exposure. For a more complete discussion of medical management of nerve agent casualties, see references.[16,17]

2. Vesicants

Sulfur mustard causes tissue damage within several minutes after contact without showing corresponding clinical signs. Decontamination must be performed immediately after contact to prevent injury. However, even slightly delayed decontamination can reduce the severity of lesions. Irrigation with water, use of soap and water with no scrubbing, or 0.5% hypochlorite are all effective measures.[17] Medical management of a patient exposed to mustard can range from symptomatic care for erythema to total management for a severely ill patient with burns, immunosuppression, and multi-system involvement.[18,19] Following decontamination to include the flushing of eyes, medical management of the casualty should be administered at a fixed medical facility as symptoms are delayed. Erythema is treated with soothing lotions and topical steroids, while large blisters can be drained. Denuded areas need irrigation with saline or sodium hypochlorite, topical antibiotics, and observation for infection. Eye injuries need initial saline irrigation, followed by use of atropine eyedrops, antibiotic ointments, and sterile petrolatum to prevent lid adhesions. Early pulmonary symptoms in the upper airway respond to steam inhalation and cough suppressants, but later infection needs specific antibiotic therapy. Mechanical ventilation and bronchodilators can help. Death can occur because of pulmonary insufficiency and infection complicated by a comprised immune response caused by mustard-induced bone marrow suppression. Some general precautions for care of casualties include maintenance of fluid balance and adequate nutritional support. Additionally, close monitoring of white blood cell counts and liberal use of analgesics and antipruritics aid in treatment.[19] Historical reports indicate that most casualties from mustard will require therapy for their skin and eye injuries, and a few additional ones will need attention for pulmonary lesions. Only a very few will require intensive care.[18]

3. Blood Agents

Exposure is terminated by evacuation to fresh air or by masking. Skin decontamination is not necessary because of the highly volatile nature of the agents. However, wet, contaminated clothing should be removed and the underlying skin decontaminated with water. Detoxification of cyanide is preceded by its removal from the cytochrome oxidase complex by intravenous injection of sodium nitrite. This treatment forms methemoglobin to which cyanide preferentially binds.[21] Intravenous injection of sodium thiosulfate follows. This sulfate combines with cyanide to produce thiocyanate, which is excreted by the kidneys. Supportive care consists of providing oxygen and correcting any metabolic acidosis. Full recovery is relatively rapid following cyanide intoxication if the antidotes are given before cessation of cardiac activity.

4. Pulmonary Agents

Termination of exposure by physical removal of the casualty or use of a mask is a vital first measure. Rest is important as any physical exertion shortens the latent period and increases the severity of symptoms.[17,22] The airways need to be kept clear and the circulation checked for hypotension. Bronchospasm is treated with bronchodilators. Positive airway pressure may be required. Oxygen therapy is indicated to treat hypoxia and intubation may also be required. In the absence of a bacterial infection, the toxic effects of low to moderate exposures to phosgene will be relatively short-lived with proper respiratory monitoring in place.[22] Exposure to moderate to high concentrations of phosgene may result in acute respiratory distress and death.

III. PROGRAMS TO PROTECT CIVILIAN POPULATIONS

A. Civilian vs. Military Response Considerations

Prior to plans for the destruction of the CW stockpiles of most countries, and preceding the sarin attacks in Matsumoto and Tokyo, the development and implementation of defensive measures against CW agents was primarily centered on the military use of these chemicals. However, military and civil defense planners face very different situations when planning for a potential chemical threat, mainly with respect to prior knowledge about the identity of the enemy and the time, place, and means of attack. The value of deployment of a chemical detection system and the use of highly specific antidotes, therapeutics, or pretreatment drugs diminishes considerably in the most probable civilian terrorism situation, in which the enemy, the agent, the time, and the place of attack are unknown. For civil defense purposes, it is therefore, more appropriate to emphasize prior planning, medical treatment, and consequence management over prevention. The responsibility for prevention in this case is left to intelligence and law enforcement agencies. However, it would be advisable to include the medical community in the distribution of pre-incident intelligence in order to maximize the medical response in dealing with chemical incidents. One significant difference between military and civilian response planning for a CW incident is that the populations to be protected are fundamentally different. In military chemical defense planning, the population of interest is primarily healthy young males between the ages of 18 to 26 years, while the civilian community at risk includes this prior population, as a minority, as well as both genders, the very young, the very old, and the sick. Furthermore, acceptable levels of exposure for military personnel operating in a CW environment differ significantly from those of first responders and medical personnel who might find themselves involved in a civilian emergency situation. In the first case, military standards prevail while NIOSH exposure criteria should be considered when CW responses are planned or when appropriate protective equipment is designed for utilization in civilian emergencies.

B. The Israeli Model

While the Japanese have actually experienced the consequences of a CW agent incident, there is no population more aware of the present day threat of CW than that of the nation of Israel. Early in this country's history, and well ahead of the invasion of Kuwait by Iraq in 1990, Israel had implemented an aggressive program of domestic preparedness. Their program of "Homeland Defense" includes extensive preparations for the widespread use of CW agents against the population. Doctrine was established for the protection of the civilian population by the Ministry of Defense and is based on the concept of "protected space." Protected space is a readily accessible space capable of providing occupants with protection against both conventional and non-conventional weapons for several hours. Since 1992, all new buildings, as well as additions to existing buildings, have been required to be equipped with a protective space that meets specific engineering specifications. Expedient measures to provide a protective space within a home or office are taught in civil defense training classes. Examples of such measures include the use of wet towels at the bottom of a doorway or the use of tape around door and window openings. Furthermore, every Israeli citizen is issued a protective mask and atropine injectors to be used in the event of a chemical attack. As mentioned earlier, a civilian population is not homogeneous and the Israelis have adopted measures to meet the needs of most of the citizenry. The government has provided standard protective masks of various sizes including a protective mask with a blower unit for those who cannot use the standard mask, a protective hood-kit designed for children ages 3 to 8 years old, an infant protective suit with a blower designed for infants 0 to 3 years old, and a medical hood-kit designed for people with impaired respiratory systems. Distribution of protective kits is administered by the Israeli Defense Force's Home Front Command through a system of 21 exchange centers located throughout the country.[24] Hospitals and medical centers prepare for the use of chemical agents by stockpiling needed medications and by routinely conducting realistic mass chemical casualty exercises. Their experience during the Gulf War demonstrated the need for mass decontamination facilities and efficient triage. Following missile attacks on Israel during the Gulf War, there were large numbers of panicked "worried well." The presence of large numbers of patients, who think they have been exposed to a chemical agent, or who are presenting with psychosomatic symptoms, could severely limit a medical facility's ability to respond rapidly and effectively to the needs of critical casualties. Planning, as well as training and education of the populace, should minimize these effects in a future incident.

C. Chemical Stockpile Emergency Preparedness Program

In the U.S., the Chemical Stockpile Emergency Preparedness Program, or CSEPP, is a partnership among state, local, and federal governments. It was created as a result of a directive from Congress that chemical weapons stockpiled at eight U.S. Army installations in the U.S. be destroyed over the course of several years. A total of 39 counties in 10 states participate in CSEPP. The slight but real threat of an emergency involving chemical agents at these sites necessitates that local officials and responders remain ready for such an emergency and involve the community in their efforts. The U.S. Army is custodian of the stockpiles, while the Federal Emergency Management Agency is the source of long-standing experience in planning for contingencies in civilian areas. Partnering with the U.S. Environmental Protection Agency and the U.S. Department of Health and Human Services, the combined effort of CSEPP allows for the funding, guidance, resources, and training needed to effectively provide protection to communities surrounding stockpile sites. Protective measures are determined for each community based on its unique needs and considerations. The plans and procedures are appropriate for the specific agents stored at the nearby Army installation. The most common emergency protective measures are evacuation and shelter-in-place and are based on two planning zones, the Immediate Response Zone and the Protective Action Zone.[25] The distance from the stockpile for each zone varies and is based on risk analyses. In the case of a stockpile accident, the community is informed and instructed through radio and television Emergency Broadcast/Alert Systems and/or over loudspeakers. Sirens and tone alert radios serve to alert and warn residents in Immediate Response Zones. Evacuation routes are designated and fully equipped shelters are identified. This provide the right atmosphere for people to stay calm and follow the recommendations of local officials and emergency managers. Public information facilities in neighbor-hoods serve to disseminate information and function as response command centers for questions from the media and concerned citizens.

IV. ELEMENTS OF THE RESPONSE

A. CLUES TO THE PRESENCE OF CHEMICAL WARFARE AGENTS

If CW agents were employed in a civilian situation, real-time detection and monitoring are not currently widely available for many of the agents. With the exception of a public announcement by a terrorist group of the employment of a toxic chemical agent or a public announcement of an accidental release, signs and symptoms of persons exposed to the toxicant will most likely be the first indication of the presence of the agent. For this reason, it is essential that first responders and medical personnel be familiar with the clinical aspects of CW agent intoxication. This is not only critical for treating the casualties, but just as important to protect the responders themselves and to limit the spread of contamination. In a previous section, we examined the clinical signs of CW agent intoxication and how CW agents could be dispersed in the environment. Because of their chemical characteristics, CW agent use in a domestic terrorist incident may not be associated with a high explosive event, and these agents are likely to be dispersed in such a manner that would primarily involve a vapor hazard. In the immediate vicinity of the incident, where there may be a continuing source of agent vapor, the probability of detecting the CW agent is greatest. However, due to the chemical properties (persistent vs. non-persistent) of many of the agents, detection may not be possible at the time emergency medical personnel arrive at an incident. Once casualties of a CW agent incident are removed from the area of the attack and become accessible to medical personnel, the signs and symptoms of the patients may be the only detection method available to guide incident commanders and law enforcement personnel. Thus, improved detection and improved diagnostic technologies are two sides of the improved response coin. More will be said about efforts to enhance the ability to rapidly and specifically diagnose exposure to chemical agents.

Effective and efficient incident response depends on the rapid and accurate identification of the chemical agents involved. The protection of first responders and emergency medical personnel at local medical facilities, as well as the effective treatment of casualties, hinges on this critical capability. Various devices capable of detecting chemical agents in the environment are available to civilian communities. Many devices were designed for military applications but have now been adapted for civilian use.

B. HAZMAT RESPONSE

An emergency response to an incident that involves the accidental or intentional release of toxic chemicals or materials will typically be categorized as a hazardous materials (HAZMAT) incident. With the greater emphasis placed on this type of response, HAZMAT incident response plans have become increasingly standardized across the country. Specialized HAZMAT teams are routinely activated to respond in such situations. HAZMAT teams are typically part of the fire services and will possess chemical detection equipment. The first responders, typically the police or fire department, must be capable of determining that a HAZMAT incident has occurred. Unfortunately, most emergency response vehicles do not have any chemical detection equipment, and the first responder must make a quick judgement call whether or not to call in HAZMAT units.

The equipment needs of early responders to a domestic incident in which CW agents may be involved are significantly different than those for military personnel. The military has the advantage of intelligence information that enables the users of the equipment to predict a probable threat agent and the likely area of impact from the chemical agent. In the case of first responders to a domestic terrorist incident, there are currently no such benefits of intelligence. The medical personnel on site will require equipment capable of detecting the widest range of chemical agents. HAZMAT teams are routinely equipped with chemical detection devices and detection kits, but these are usually chemical-specific tests indicating only the presence or absence of a single suspected toxic industrial chemical or class of chemicals. Chemical detection equipment currently used by HAZMAT teams varies considerably by locality, with many large metropolitan areas having significant technology available. Today, most local response units in the U.S. have limited capability for CW agent detection. A critical review of the abilities of emergency responders to detect chemical agents has recently been made available.[26]

C. Currently Available Detection Technology

A wide variety of commercial equipment is available for detection of hazardous chemicals, including a number of CW agents. A listing of the technologies and the manufacturers of the technology have been compiled by various organizations.[27–29] The following is a list of the principal technologies employed in currently available detection equipment that can be employed by HAZMAT teams and medical units. The Metropolitan Medical Strike Teams organized and equipped by the U.S. Public Health Service, for example, have purchased detection paper, three different detection kits, and portable chemical agent detectors and monitors.

Ion Mobility Spectrometry (IMS) technology is used to detect nerve, vesicant, and blood agents. The Chemical Agent Monitor (CAM) uses ion mobility spectrometry to provide a portable, hand-held point detection instrument for monitoring nerve or vesicant agent vapors. Minimum levels detectable are about 100 times the acceptable exposure limit (AEL) for the nerve agents and about 50 times the AEL for vesicants. This insensitivity to low concentrations limits the utility of this instrument to check the efficacy of decontamination efforts or in occupational exposure measurements.

Acoustic wave sensors are also used to detect nerve and vesicant agents. The Surface Acoustic Wave Chemical Agent Detector (SAW Mini-CAD) is a commercially available, pocket-sized instrument that can automatically monitor for trace levels of toxic vapors of both sulfur mustard and the G nerve agents with a high degree of specificity.

Color change chemistry detectors can detect nerve, vesicant, and blood agents. Colorimetric tubes are the most common detection technology used by HAZMAT teams. There are several hundred different types of colorimetric tubes available that can detect a variety of chemicals. A HAZMAT team's analytical capabilities usually include tests for chlorine, cyanide, phosgene gas, and organophosphate pesticides. Direct reading detector tubes can be used for both short-term and long-term measurements. Recently, one manufacturer of detection tubes has developed a kit specifically for OP nerve agents. Routine HAZMAT tests rarely include a capability to detect vesicant agents. This is particularly unfortunate, as the clinical signs of exposure to sulfur mustard can be delayed, and a continued exposure of personnel can result until signs appear or until the agent is detected.

The M18 detection kit and the M256A1 kit are military items. The M18 is a colorimetric device for measuring the concentration of selected airborne chemicals. The M18 comes with detector tubes for cyanide, phosgene, Lewisite, sulfur mustard, and nerve agents GA, GB, GD, and VX.

The M256A1 kit includes detector tickets that detect low concentrations of cyanide, vesicant, and nerve agents in vapor form. The tests take approximately 15 min. Sensitivity of this kit is such that the tests may provide a negative reading at concentrations below that immediately dangerous to life and health (IDLH). Occupational Safety and Health Administration (OSHA) rules call for the use of maximum personal protection until concentrations can be shown to be less than 50 times the AEL. The IDLH is the maximum concentration of a contaminant to which a person could be exposed for 30 min without experiencing any escape-impairing or irreversible health effects. The AEL is a general term indicating a level of exposure that is unlikely to result in adverse health effects.

M8 and M9 detection papers provide a rapid, inexpensive test for the presence of liquid mustard or nerve agents. Use of the paper should be as a screening test only because of the paper's propensity to show false positive results. False positive results for the presence of a CW agent in a civilian community could produce hysteria and panic.

Additional detectors currently available use other technology such as electrochemical detectors for vesicant, nerve, blood, and choking agents, infrared spectroscopy detectors for vesicant and nerve agents, and photo ionization detectors for the detection of nerve and blister agents.

Mass spectrometry (MS), gas chromatography (GC), and Fourier transform infrared (FTIR) spectrometry technologies are the basis for fixed facility CW agent detection and can provide definitive identification of the agent. Furthermore, they possess the sensitivity to detect low levels of the agents in the range of occupational exposure levels. Where these new technologies can be of benefit to the medical community is in fixed medical facilities for monitoring air samples for low levels of agents that may cause an occupational hazard or in predicting the medical impact of a chemical event. However, few local governments, organizations, or medical facilities have invested in such equipment to date.

D. Laboratory Analysis

As mentioned earlier, clinical signs and symptoms displayed by casualties will be the earliest and often an accurate method for determining the presence of a specific class of chemical agent. This rapid diagnosis is essential in saving lives and preventing further exposure. Furthermore, the signs and symptoms of the patient will provide the most important information on which to base emergency treatment. In emergency situations, medical personnel should have available analytical methods and tools for monitoring patients and confirming their diagnosis. Additionally, forensic scientists must have techniques to retrospectively piece together the clinical course of a patient or to examine aspects of the incident. The following is a brief discussion of some of the techniques available for laboratory analysis.

1. Nerve Agents

The requirement for direct measurements of nerve agents and/or their metabolites in clinical samples has given rise to the development of sensitive techniques utilizing gas chromatography-mass spectrometry (GC-MS),[30] gas chromatography-tandem mass spectrometry (GC-MS-MS), and capillary gas chromatography.[31] Extremely sensitive retrospective detection of organophosphorus nerve agents has been recently described by Polhuijs et al.[32] The research group at TNO Prins Maurits Laboratory in the Netherlands has applied this method in analyzing blood samples obtained from victims of the Tokyo sarin incident. The technique reactivates sarin-inhibited enzyme by treating the inhibited enzyme with fluoride ions, thus converting the OP leaving group into the corresponding phosphofluoridate. The U.S. Army Medical Research Institute of Chemical Defense at Aberdeen Proving Ground, Maryland proposes several GC-MS methods for measuring nerve agent metabolites in urine in their Technical Bulletin, TB MED 296.[33] These techniques are based on extensive animal studies.[34–36] In these studies, almost total recoveries of the given doses for GB and GF, in metabolite form, were obtained from the exposed animal's urine. Sensitive methods like those just described can be used in following the course of medical treatment, in health surveillance programs, and as forensic tools.

Since normal serum cholinesterase activity ranges from 182 to 804 IU/L, the determination of a single cholinesterase (ChE) inhibition level can only be an indirect indication of toxicity resulting from exposure to nerve agents. The following information exemplifies this rather well. In a report of Japanese victims of the Tokyo sarin attack, patients who exhibited moderate symptoms of intoxication had serum ChE values ranging from 300–750 IU/L.[37] Additionally, these patients had red blood cell (RBC) ChE activity ranging between 0.3 and 2.0 IU vs. 1.2–2.0 IU for asymptomatic patients.

The basis for the standard method for determining blood (ChE) inhibition is the measurement of the enzymatic products derived when either acetylcholine or acetylthiocholine are used as substrates. The rate of formation of acetate is measured by changes in pH, while the formation of thiocholine is determined colorimetrically.[38] A portable device utilizing this method, the Test-Mate OP Kit (EQM Research Incorporated, 2585 Montana Avenue, Cincinnati, OH 45211), provides a rapid, reasonably sensitive assay for ChE inhibition following OP exposure. The kit should be appropriate for emergency contingencies, and only very small blood samples are required. Military forces are viewing such a portable kit as a means to screen and monitor chemical casualties and have adopted it for Theater Army Medical Laboratories. Such a rapid detection device could also be useful in a triage or hospital setting.

2. Vesicants

The last documented use of sulfur mustard against military forces was during the Iran-Iraq conflict. Several sensitive GC-MS and GC-MS-MS were developed and have been used to demonstrate exposure to sulfur mustard in samples from casualties of this conflict.[39–44] In cases of suspected exposure to sulfur mustard, U.S. Army laboratories utilize methods described in TB MED 296.[33] In this method, thiodiglycol (2,2-thiodiethanol), one of the *in vivo* degradation products of sulfur mustard, is used to confirm an exposure.[45–47]

3. Cyanide

Several assays for detecting the presence of cyanide in blood samples have been used in the past.[48,49] However, a more efficient gas chromatography and mass spectrometry method for determining cyanide and its major metabolite, thiocyanate, in blood has been described.[50] Another gas chromatographic procedure in which cyanide is converted to cyanogen chloride has also proved to be specific, sensitive, and rapid, thus permitting measurements in emergency situations.[51] Alternatively, there is an automated fluorometric measurement described in TB MED 296.[33,52] With this method, tests for both plasma free CN- and total blood CN- are accomplished directly by a completely automated method requiring less than 30 min.

V. PROTECTION OF FIRST RESPONDERS AND VICTIMS

A. Programs to Improve the Response

1. Training

The U.S. has placed strong emphasis on effective training for civilian response to a chemical incident within the city response structure. Under federal law, training assistance is to be provided to 120 U.S. cities. Recent legislation has directed that this assistance be extended to 37 additional metropolitan areas.

Accordingly, a federal interagency team coordinates the training and exercise programs with the target cities. Courses are offered in six subjects: Awareness, Operations, Technician-HAZMAT, Technician-Emergency Medical Services, Hospital Provider, and Incident Command. City officials determine which training best fits their needs, and teams consisting of CB experts from the Department of Defense (DoD), as well as professional civilian first responders, present the training. Those receiving the training later promulgate it on a sustaining basis in their own communities. The effectiveness of the training is assessed in subsequent "tabletop" and functional exercises that provide direct feedback to the city for future activities and programs the city may pursue in order to enhance the response to chemical terrorism.

2. The Chemical and Biological Hotline

Congressional legislation has specifically directed the establishment of a designated telephonic link to a designated source of relevant data and expert advice for the use of state or local officials responding to emergencies involving a weapon of mass destruction. This link, the Chemical and Biological Hotline, (1-800-424-8802) is operated 7 days a week, 24 h a day.[53] It provides emergency technical assistance from a variety of federal agencies, or, if warranted, an actual federal response to assist first responders during incidents. The hotline is intended for use by first responders as well as state emergency operations centers and medical facilities. Assistance is provided on a wide array of subjects, which include personal protective equipment, decontamination systems and methods, toxicology information, and medical symptoms and treatment for exposure to CB agents.

3. Chemical Weapons Improved Response Program

To address the issue of identifying response capability shortfalls, the DoD established the Chemical Weapons Improved Response Program. The program has formed an alliance between federal, state, and local government offices as well as various industry organizations. Through an iterative process of workshops, conferences, exercises, and technical studies, the program has structured its work around four functional working groups: Health and Safety; Emergency Management; Emergency Response; and Law Enforcement. The responsibility of each group is to identify, prioritize, and arrive at solutions pertinent to first responders.

A major focus of this program is on issues that impact the well being of the community at large and the challenges associated with maximizing the impact of local and regional public health and medical services. In examining the impact of a chemical terrorist attack, three priorities were identified:[54] timely and accurate communication between responder groups, a system to identify and distribute available resources including pharmaceuticals, and operating procedures for all responsible agencies.

To address the myriad of issues a chemical attack would present to a community, a response template plan with national applicability, referred to as the Off-Site Triage, Treatment and Transport Center, was formulated.[55] The concept was designed to address the non-critical and non-exposed patients who can be expected to seek medical help. This concept requires the set-up of an ancillary medical facility to handle the large number of walking wounded and "worried well" casualties that are expected from a terrorist attack. The facility would be a temporary site that supplements existing assets, since many of these centers can be set-up as are deemed necessary. The care envisioned to be provided in such a center includes decontamination, initial entry into the medical care system for patients not processed at the scene, care for non-critical patients, transportation to medical facilities for patients in need, and mental health care.

4. Applied Research

This DoD program has initiated a number of research studies that are of interest to first responders. Examples include studies on the protection afforded by standard firefighting turnout gear from CW agents, methods for mass casualty decontamination, and the effect of positive or negative pressure ventilation on vapor concentrations inside structures.

Firefighters respond to incidents wearing what is referred to as turnout gear and self-contained breathing apparatus (SCBA). The protection afforded by these ensembles was assessed in test chambers using a chemical simulant. The equipment was found to provide the wearer sufficient protection against nerve and blister agents to allow for the reconnaissance and rescue of victims. Furthermore, the protective efficacy of the gear can be increased simply by using common, heavy-duty duct tape.[56] Guidelines established from these studies are meant to assist Incident Commanders in making decisions to enter chemical agent vapor environments to perform rescue, reconnaissance, mitigation, or detection operations and to establish minimum criteria for entry for first responders. In a parallel study addressing law enforcement and emergency medical services issues, six commercial, level C chemical protective suits and the standard police duty uniforms were assessed to determine what protection they afforded the wearer.[57] Testing revealed protective factors ranging from 2 for the standard duty uniform to 42 for commercial suits. The commercial ensembles consisting of the respirator, gloves, and over-garment were felt to provide adequate protection to responders in areas of low concentration as might be found at the perimeter of the incident, but not for activities in areas where the threat is expected to be much greater.

Another tool fire fighters bring to an incident are fans used to blow smoke out of the building in order to facilitate evacuation and rescue. These fans were assessed for use in reducing chemical vapor hazards.[58] Using simulants, it was determined that dramatic reductions in vapor concentrations can be attained with these fans by creating positive pressure ventilation in structures. Fans can reduce the vapor concentration 50–70% within the first 10 min of use. This reduction significantly increases the first responders' protection above and beyond the adequate protection provided by standard turnout gear with SCBA.

5. Mass Casualty Decontamination

Following a chemical terrorist attack, it can be assumed that many of the civilians that are present are not contaminated, but during the crucial minutes immediately following the event, there is currently no way of determining with certainty those who are contaminated. HAZMAT teams that respond to accidents or spills of industrial chemicals are well prepared to decontaminate themselves but have limited capability to care for mass casualties. However, a terrorist attack could well involve hundreds, if not thousands of victims, so the scale of the response must be expanded. An underlying assumption has been that decontamination should occur as soon as possible after exposure, so a recent study considered methods using equipment fire fighters already bring to every incident. Fire fighters have access to large amounts of water, so three water-based decontamination methods were assessed: water alone, soap and water, and bleach and water.[59] Water alone was found to be a very effective method via physical removal. The shear forces and dilution achieved by using high-volume, low-pressure (60 pounds per square inch) showering was found to be the most practical method of mass casualty decontamination. Various shower applications were examined. Some included the use of tarps and establishing decontamination corridors for privacy concerns. The use of soap provides a slight improvement via ionic degradation of the chemical agent, however a supply of it has to be on hand. This requires that fire fighters either bring it on their trucks, creating a logistical burden, or otherwise procure it on scene, which would take time. The use of bleach (sodium hypochlorite) and water was not recommended. Although these solutions react with most chemical agents, the preparation and application of the solution would take time, a distinct disadvantage where speed is critical.

Based on their findings, investigators proposed the following principles to guide the decontamination of large numbers of civilians at chemical incidents:

- Expect a 5:1 ratio of unaffected/affected casualties.
- Decontaminate as soon as possible.
- Disrobing is decontamination; top to bottom, the more the better.
- Water flushing generally is the best mass decontamination method.
- After known exposure to liquid agent, first responders must self-decontaminate as soon as possible to avoid serious effects.

Triaging casualties at an incident site may exceed emergency responders' capabilities. There may be too many people to rescue, decontaminate, and treat, regardless of exposure to the chemical agent. Victims must be prioritized into ambulatory and non-ambulatory groups, and further grouped based on agent signs and symptoms, or the likelihood of exposure. Emergency care providers will have to decide when, for example, to perform only hasty decontamination, if at all, on a severely injured casualty who is not clearly a chemical casualty. Conversely, casualties displaying symptoms of severe chemical exposure may require antidotes and other aid before decontamination is possible. Toxic exposure issues emergency personnel could encounter on the perimeter of an incident have recently been addressed.[57] The threat to responders who perform activities on the perimeter, be they police or emergency medical service, should be minimal. As a routine matter, no significant vapor, aerosol, or liquid danger is expected. The most likely threat will come as a result of a wind shift or off-gassing from people exiting the incident scene.

Recent advances made by scientists working in the U.S. Army Medical Chemical Defense Research Program include two potential products that may aid in casualty decontamination. A skin exposure reduction paste (formerly known as Topical Skin Protectant, TSP) was just approved for a New Drug Application by the FDA. The formulation consists of a mixture of a perfluorinated polyether base oil to which is added fine polytetrafluoroethylene particulate as thickener. Animal studies have shown that this formulation is effective for at least 4 h of continuous contact with CW agents. An enzymatic sponge for medical decontamination of CW agents has recently been tested. These polyurethane sponges will have cholinesterase and organophosphate hydrolase enzymes immobilized to the polymer. The enzymes retain their activity for long periods of time and under harsh environmental conditions. The sponge itself is wetted with a solution that absorbs the agent from the skin. Once removed from the skin, the sponge could be reactivated with an oxime.

VI. ISSUES RELATED TO LOW-DOSE EXPOSURE TO CHEMICAL AGENTS

The health effects of exposure to low doses of CW agents have been of considerable interest for several decades. During the period of large-scale production of CW agents in the U.S., this subject was a particularly important occupational health issue for workers in production plants. New attention to this issue was raised when the results of human testing involving chemical agents, conducted by the U.S. Army, was the topic of National Research Council reports.[60,61] Interest in this issue peaked again when risk assessment and public health programs were initiated in response to the chemical demilitarization of the stockpiles of these same weapons. It was at this time that the U.S. Department of Health and Human Services published their conclusions in the Federal Register regarding the risk of adverse health effects to exposure to low doses of nerve agent. The most recent interest in this subject has been generated as a result of "Gulf War Syndrome." One of the suggested causes of this malady is that soldiers were exposed to low levels of CW agents during their period of service in the Gulf. Several panels of experts have reviewed these suggestions extensively.[62,63] Current knowledge of the health effects of exposure to low doses of nerve agents has been reviewed and is the subject of a chapter in this text.[64–66] However, this issue remains critical and is of significant importance to civilian responders to a CW agent incident. The issues are basic. To respond rapidly and effectively to a chemical incident and to respond in such a way as to save lives, first responders must subject themselves to levels of the agents that may exceed current occupational exposure limits (Table 2). Currently available detection technology for use at the scene of an incident cannot measure chemical agents at these occupational exposure levels. Additionally,

TABLE 14.2
Human Exposure Values for Sarin

Agent GB

- LCt_{50} = 100 mg min/m^3
- No death dose = 10 mg min/m^3
- NNM effect dose = 4 mg min/m^3
- ECt_{50} (miosis) = 2–4 mg min/m^3
- NOAEL = 0.5 mg min/m^3
- MSC (1 h) = 0.001 mg/m^3
- MSC (8 h) = 0.0003 mg/m^3
- Safety factor of 0.1 for general population

0.0001 mg/m^3 (1 h)
0.00003 mg/m^3 (8 h)
0.000003 mg/m^3 (72 h)
IDLH 0.2 mg/m^3 (air-supplied respirator required)
Hand held detector threshold 0.01 mg/m^3

Note: NNM = no neuromuscular effect level; MSC = maximum safe concentration; and IDLH = immediately dangerous to life and health.

many of the protective ensembles that will be used by first responders will not protect down to these levels or have not been definitively tested for their protective efficacy. While the data appears to point to no adverse health effects from an acute, low dose of nerve agent, studies are continuing that may provide additional support to these conclusions or may find effects that have previously gone undetected.

VII. SUMMARY

The accidental or intentional release of chemical agents is similar to the hazardous materials incidents that metropolitan public safety personnel contend with routinely. The emergency response to a release of CW agent can use the existing framework of response to toxic chemical incidents and can be modified and enhanced for maximum effectiveness. Additionally, poison control centers located throughout the country deal with chemical poisonings on a daily basis and can serve as the initial focus of efforts to improve the response of the medical community to dangerous CW agents. Information in this chapter has described the adverse health effects of CW agents and the threat they pose from the standpoint of protecting civilian communities in the event of their use. Future advances should include improved methods of detection or laboratory analysis and better knowledge of the potential health consequence of such exposures in a general population.

Information in other chapters in this text describes areas of ongoing work that will expand our knowledge base, thus allowing for greater ability to deal with chemical agent exposure in mass casualty situations.

ACKNOWLEDGMENT

The authors would like to acknowledge the exceptional contributions of Ms. Suman Adler in the preparation of this chapter. Her efforts to pursue the most recent references related to this rapidly changing area and her organizational and editing skills were critical to the completion of this work.

REFERENCES

1. *United States—Russian Federation Cooperative Efforts in the Area of Chemical Weapons Destruction,* URL http://www-pmcd.apgea.army.mil/text/CTR/ctr_dr_smith.html.
2. *Proceedings of the Chemical and Biological Medical Treatment Symposium* (CB MTS-Industry I), ISSN 1092-7255, Zagreb, Croatia, 1999.
3. *Chemical Warfare Agents,* Somani, S.M., Ed., Academic Press, San Diego, 1992.
4. Government Accounting Office, *Combating Terrorism: Federal Agencies' Efforts to Implement National Policy and Strategy,* GAO/NSIAD-97–254, 1997.
5. Federal Emergency Management Agency Abstract, Memorandum for Mr. John F. Sopko, Senate Government Affairs Committee, Unclassified FEMA Abstract on Presidential Decision Directive 39, National Security Council, Washington, DC, 1996.
6. The White House, Fact Sheet, *Combating Terrorism: Presidential Decision Directive 62,* Office of the Press Secretary, Washington, DC, 1998.

7. The White House, Fact Sheet, *Protecting America's Critical Infrastructures: Presidential Decision Directive 63,* Office of the Press Secretary, Washington, DC, 1998.
8. Department of Justice, Office of Justice Programs, Office for State and Local Domestic Preparedness Support Report to Congress, *Responding to Incidents of Domestic Terrorism: Assessing the Needs of State and Local Jurisdictions,* Washington DC, 1999.
9. *Introduction to the Basic Plan of the Federal Response Plan, April 1999,* Federal Emergency Management Agency, URL http://www.fema.gov/r-n-r/frp/frpintro.htm.
10. Federal Emergency Management Agency, *The Federal Response Plan, Terrorism Annex,* Figure D-Overview of a Disaster Operation, April 1999, URL http://www.fema.gov/r-n-r/frp/frpfigd.htm.
11. Federal Emergency Management Agency, *The Federal Response Plan, Terrorism Incident Annex,* URL http://www.fema.gov/r-n-r/frp/frpterr.htm.
12. Office of the Assistant Secretary of Defense (Public Affairs), *Integration of Reserve Components Responds to New Threat,* Comments by Deborah Lee, Assistant Secretary of Defense for Reserve Affairs, Robert Walker, Acting Secretary of the Army, and B.G. Roger Schultz, Director of Military Support, Army Operations Center, Washington, DC, 1998.
13. OPHS Testimony: Anti-Terrorism Measures, Statement by Robert Knouss, M.D., before the House Subcommittee on National Security, International Affairs, and Criminal Justice Committee on Government Reform and Oversight, URL http://www.os.dhhs.gov/progorg/asl/testify/y981002c.txt.
14. *Food and Water Operations in a Nuclear, Biological and Chemical Environment,* Moore, D.H., Ed., Euromed Working Group on Food Hygiene and Food Technology, 1989.
15. Munro, N.B., Talmage, S.S., Griffen, G.D., Waters, L.C., Watson, A.P., King, J.F., and Hauschild, V., The sources, fate, and toxicity of chemical warfare agent degradation products, *Environ. Health Persp.,* 107 (12), 933, 1999.
16. Sidell, F.R., Clinical considerations in nerve agent intoxication, in *Chemical Warfare Agents,* Somani, S.M., Ed., Academic Press, San Diego, CA, 1992, 155.
17. *Medical Management of Chemical Casualties Handbook,* 3rd ed., Chemical Casualty Care Division, U.S. Army Medical Research Institute of Chemical Defense, Aberdeen Proving Ground, MD, 1999.
18. Sidell, F.R. and Hurst, C.G., Clinical considerations in mustard poisoning, in *Chemical Warfare Agents,* Somani, S.M., Ed., Academic Press, New York, 1992, 51.
19. Sidell, F.R., Urbanetti, J.S., Smith, W.J., and Hurst, C.G., Vesicants, in *Textbook of Military Medicine—Medical Aspects of Chemical and Biological Warfare,* Zajtchuk, R. and Bellamy, R.F., Eds., Office of the Surgeon General, Washington, DC, 1997, 197.
20. Smith, W.J. and Dunn, M.A., Medical defense against blistering chemical warfare agents, *Arch. Derm.,* 127, 1207, 1991.
21. Baskin, S.I. and Brewer, T.G., Cyanide poisoning, in *Textbook of Military Medicine—Medical Aspects of Chemical and Biological Warfare,* Zajtchuk, R. and Bellamy, R.F., Eds., Office of the Surgeon General, Washington, DC, 1997, 271.
22. Urbanetti, J.S., Toxic inhalational injury, in *Textbook of Military Medicine—Medical Aspects of Chemical and Biological Warfare,* Zajtchuk, R. and Bellamy, R.F., Eds., Office of the Surgeon General, Washington, DC, 1997, 247.
23. Sidell, F.R., Nerve agents, in *Textbook of Military Medicine—Medical Aspects of Chemical and Biological Warfare,* Zajtchuk, R. and Bellamy, R.F., Eds., Office of the Surgeon General, Washington, DC, 1997, 129.
24. *Homefront Command, Important Information Concerning Civil Defense,* URL http://www.idf.il/english/organization/homefront/index.stm.

25. Federal Emergency Management Agency, *Preparedness, Training, and Exercises,* URL http://www.fema.gov/pte/csepp1.htm.
26. Chemical and Biological Defense Information Analysis Center, *The Emergency Responders' Ability to Detect Chemical Agents,* Gunpowder Branch, Aberdeen Proving Ground, MD, 2998.
27. *Chemical and Biological Terrorism: Research and Development to Improve Civilian Medical Response,* National Academy Press, Washington, DC, 1999.
28. Chemical and Biological Defense Information Analysis Center, *Worldwide Chemical Detection Equipment Handbook,* Gunpowder Branch, Aberdeen Proving Ground, MD, 1995.
29. Chemical and Biological Defense Information Analysis Center, *Assessment of Chemical Detection Equipment for Hazmat Responders,* CBIAC Gunpowder Branch, Aberdeen Proving Ground, MD, 1998.
30. Black, R.M., Clarke, R.J., Read, R.W., and Reid, M.T.J., Application of gas chromatography-mass spectrometry and gas chromatography-tandem mass spectrometry to the analysis of chemical warfare samples found to contain residues of the nerve agent sarin, sulphur mustard and their degradation products, *J. Chromatogr. A.,* 662:301, 21, 1994.
31. Bonierbale, E., Debordes, L., and Coppet, L., Application of capillary gas chromatography to the study of hydrolysis of the nerve agent VX in rat plasma, *J. Chromatogr. B: Biomed. Appl.,* 688:255, 64, 1997.
32. Polhuijs, M., Langenberg, J.P., and Benschop, H.P., New method for retrospective detection of exposure to organophosphorous anticholinesterase: Application to alleged sarin victims of Japanese terrorists, *Toxicol. Appl. Pharmacol.,* 146(1) 156, 1997.
33. *Assay Techniques for the Detection of Exposure to Sulfur Mustard, Cholinesterase Inhibitors, Sarin, Soman, GF, and Cyanide,* Technical Bulletin TB MED 296, Headquarters, Department of the Army, 1996.
34. Harris, L.W., Braswell, L.M., Fleisher, J.P., and Cliff, W.J., Metabolites of pinacolylmethylphosphonofluridate (soman) after enzymatic hydrolysis *in vitro, Biochem. Pharmacol.,* 13, 1129, 1964.
35. Reynolds, M.L., Little, P.J., Thomas, B.F., Bagley, R.B., and Martin, B.R., Relationship between the biodisposition of (3h) soman and its pharmacological effects in mice, *Toxicol. Appl. Pharmacol.,* 80, 409, 1985.
36. Lenz, D.E., Boisseau, J., Maxwell, D.M., and Heir, E., Pharmacokinetics of soman and its metabolites in rats, *Proceedings of the 6th Medical Chemical Defense Bioscience Review,* Aberdeen Proving Ground, MD, U.201, AD B121516, 1987.
37. Masuda, N., Takatsu, M., Morinari, H., and Ozawa, T., Sarin poisoning in Tokyo subway, *Lancet,* 345, 1446, 1995.
38. Ellman, G.L., Courtney, K.D., Andres, J., Jr., and Featherstone, R.M., A new and rapid colorimetric determination of acetylcholinesterase activity, *Biochem. Pharmacol.,* 7, 88, 1961.
39. Fidder, A., Noort, D., deJong, A.L., Trap, H.C., deJong, L.P., and Benschop, H.P., Monitoring of *in vitro* and *in vivo* exposure to sulfur mustard by GC/MS determination of the n-terminal valine adduct in hemoglobin after a modified edman degradation, *Chem. Res. Toxicol.,* 9, 788, 1996.
40. Fidder, A., Noort, D., deJong, L.P., Benschop, H.P., and Hulst, A.G., N7-(2-hydroxyethylthythioethyl)-guanine: A novel urinary metabolite following exposure to sulfur mustard, *Arch. Toxicol.,* 10, 854, 1996.
41. Noort, D., Hulst, A.G., Trap, H.C., deJong, L.P., and Benschop, H.P., Synthesis and mass spectrometric identification of the major amino acid adducts formed between sulphur mustard and hemoglobin in human blood, *Arch. Toxicol.,* 71, 171, 1997.

42. Benschop, H.P., van der Schans, G.P., Noort, D., Fidder, A., Mars-Groenendijk, R.H., and deJong, L.P., Verification of exposure to sulfur mustard in two casualties of the Iran-Iraq conflict, *J. Toxicol.,* 21(4), 249, 1997.
43. Black, R.M., Clarke, R.J., Harrison, J.M., and Read, R.W., Biological fate of sulfur mustard: Identification of valine and histidine adducts in hemoglobin from casualties of sulphur mustard poisoning, *Xenobiotica,* 27(5), 499, 1997.
44. Black, R.M. and Read, R.W., Improved methodology for the detection and quantitation of urinary metabolites of sulfur mustard using gas chromatography-tandem mass spectrometry, *J. Chromatog. B: Biomed. Appl.,* 665, 97, 1995.
45. Jakubowski, E.M., Woodard, C.L., Mershon, N.M., and Dolzine, T.W., Quantitation of thiodiglycol urine by electron ionization gas chromatography-mass spectrometry, *J. Chromatog.* 528, 184, 1990.
46. Davison, C.D., Roman, R.S., and Smith, P.K., Metabolism of bis-β-chloroethyl sulphide (sulphur mustard gas), *Biochem. Pharmacol.,* 7, 65, 1961.
47. Roberts, J.J. and Warwick, G.P., Studies of the mode of action of alkylating agents-VI. The metabolism of bis-2-chloroethyl sulphide (mustard gas) and related compounds, *Biochem. Pharmacol.,* 12, 1329, 1963.
48. Feldstein, M. and Klendshoj, N.C., The determination of cyanide in biological fluids by microdiffusion analysis, *J. Lab. Clin. Med.,* 44, 166, 1954.
49. Lundquist, P., Rosling, H., and Sorbo, B., Determination of cyanide in whole blood, erythrocytes, and plasma, *Clin. Chem.,* 31, 591, 1985.
50. Kage, S., Nagata, T., and Kudo, K., Determination of cyanide and thiocyanate in blood by gas chromatography and gas chromatography-mass spectrometry, *J. Chromatog. B: Biomed. Appl.* 675, 27, 1996.
51. Odoul, M., Fouillet, B., Nouri, B., Chambon, R., and Chambon, P., Specific determination of cyanide in blood by headspace gas chromatography, *J. Anal. Toxicol.,*18, 205, 1994.
52. Groff, W.A., Sr., Stemler, F.W., Kaminskis, A., Froehlich, H.R., and Johnson, R.P., Plasma-free cyanide and blood total cyanide: A rapid completely automated microdistillation assay, *Clin. Toxicol.,* 23, 133, 1985.
53. *U.S. Army Soldier and Biological Chemical Command Domestic Preparedness Fact Sheet,* August 1999, URL http://dp.sbccom.army.mil/fs/dp_hotline.html.
54. *Domestic Preparedness 1997–1998 Summary Report,* Chemical Weapons Improved Response Program, U.S. Army Soldier and Biological Chemical Command, Aberdeen Proving Ground, MD, 1999.
55. *Domestic Preparedness 1999 Summary Report,* Chemical Weapons Improved Response Program, U.S. Army Soldier and Biological Chemical Command, Aberdeen Proving Ground, MD, 2000.
56. U.S. Army Soldier and Biological Chemical Command, Domestic Preparedness, Chemical Weapons Improved Response Program, *Guidelines for Incident Commander's Use of Firefighter Protective Equipment (FFPE) with Self-Contained Breathing Apparatus (SCBA) for Rescue Operations during a Terrorist Chemical Agent Incident,* URL http://dp.sbccom.army.mil/cwirp.
57. U.S. Army Soldier and Biological Chemical Command, Domestic Preparedness, Chemical Weapons Improved Response Program, *Interim Summary Report for Law Enforcement and Emergency Services Protective Ensemble Testing,* May 1999, URL http://dp.sbccom.army.mil/cwirp.
58. U.S. Army Soldier and Biological Chemical Command, Domestic Preparedness, Chemical Weapons Improved Response Program, *Use of Positive Pressure Ventilation (PPV) Fans to Reduce the Hazards of Entering Chemically Contaminated Buildings Summary Report,* October 1999, URL http://dp.sbccom.army.mil/cwirp.

59. U.S. Army Soldier and Biological Chemical Command, Domestic Preparedness, Chemical Weapons Improved Response Program, *Guidelines for Mass Casualty Decontamination during a Terrorist Chemical Agent Incident,* January 2000, URL http://dp.sbccom.army.mil/cwirp.
60. Panel on Anticholinesterase Chemicals, *Possible Long-Term Health Effects of Short-Term Exposure to Chemical Agents.* Vol. I. *Anticholinesterases and Anticholinergics,* Committee on Toxicology and Environmental Health Hazards, Assembly of Life Sciences, National Academy Press, Washington, DC, 1982.
61. Coordinating Subcommittee, *Possible Long-Term Health Effects of Short-Term Exposure to Chemical Agents.* Vol III. *Final Report, Current Health Status of Test Subjects,* Committee on Toxicology, Board on Toxicology and Environmental Health Hazards, Commission on Life Sciences, National Research Council, Washington, DC.
62. Presidential Advisory Committee on Gulf War Veterans' Illnesses: Final Report, U.S. Government Printing Office, Washington, DC, 1996.
63. Report of the Defense Science Board, Persian Gulf War Health Effects, Office of the Under Secretary of Defense for Acquisition and Technology, Washington, DC, Ref # 94-S-2248, 1994.
64. Sidell, F.R. and Hurst, C.G., The long-term health effects of nerve agents and mustard, in *Textbook of Military Medicine—Medical Aspects of Chemical and Biological Warfare,* Zajtchuk, R. and Bellamy, R.F., Eds., Office of the Surgeon General, Department of the Army, Washington, DC, 1997, 232.
65. Moore, D.H., Long term health effects of low-dose exposure to nerve agents, *J. Physiol.* (Paris), 92, 325, 1998.
66. Moore, D.H., Health effects of exposure to low doses of nerve agents—A review of present knowledge, *Drug Chem. Toxicol.,* 21(Suppl 1), 123, 1998.

Index

A

Absorption, of pyridostigmine, 147
Acceleration tolerance, pyridostigmine and, 164
Acceptable exposure limit (AEL), 423–424
Acetylcholine (ACh)
  physostigmine and, 175
  in stress, 91–92
Acetylcholinesterase, 220–222, see also Cholinesterases
  blood-brain barrier and, 124–126
  as compared with cholinesterase bioscavengers, 196
Acetylcholinesterase inhibition, 1–2, 11, 92–94
  in vitro studies, 14–15
Acoustic wave sensors, 423
Acrolein, 325, 330, 332
Act to Combat National Terrorism (PL 98–533), 412
Acute exposure, to nerve agents, 5–10
Adamsite (diphenylaminochloroarsine), 324, 325, 359–361
Adenosine antagonists, cyanide and, 309
Adenosine receptors, nerve agents and, 106–107
Adherens junctions, 124
ADP-ribosylation, by cyanide, 312
Adrenergic system, blood-brain barrier and, 129
Age, pyridostigmine and, 165–166
Agent Orange Act of 1991, 262
Aging, of nerve agents, 217
Altermonas sp., 228
Alzheimer's disease, 166, 197
Amblyopia, tobacco, 305
Ames test, of capsaicin, 356–357
4-Aminopyridine, in botulism, 381
Aminoquinolines, in botulism, 384
Ammonium chloride, in botulism, 383, 384
Amodiaquine, in botulism, 384
Amyl nitrite, as cyanide antidote, 313
Anesthesia, pyridostigmine and, 166
Angiotensin, blood-brain barrier penetration and, 133
Antibodies, as bioscavengers, 220
Anticholinesterases, see Cholinesterase inhibition; Nerve agents
Anticonvulsants, 217
Antidotes, see also specific agents
  to cyanide, 313–314
  neurobehavioral effects, 400–401
Antimalarials, in botulism, 383–384
Asthma, 163
Astrocytes, blood-brain barrier and, 128
ATPase inhibition, 106
Atropine, 85, 159
  pretreatment with, 65–66
Autoimmune demyelination diseases, 130, 133–134

B

BCHE gene, 136
Behavioral effects, see Neurobehavioral effects
Benz bromide, 332
Benzyl bromide, 325
Benzyl iodide, 325
Bimakalim, 309
Bioscavengers, 215–237
  advantages of, 218–219
  background and principles, 215–237
  behavioral effects, 233–235
  catalytic, 228–233
  as compared with standard treatments, 217–218
  for nerve agents, cholinesterases, 192–210, 221–228
    carboxylesterase, 205–208
    enhancement of scavenging activity, 202–205
    guinea pig soman studies with HuBChE, 195–196
    huperzine, 208
    immobilized for decontamination of organophosphates, 208–210
    non-human primate studies, 196–202
    rodent studies, 193–195
    soman studies in comparison with other agents, 196
    stability *in vivo,* 193
  stochiometric, 220–228
    antibodies, 220
    enzymes, 220–228, see also Cholinesterases
Birth defects, see also Reproductive effects
  Gulf War exposure and, 271–272, 275–276
Black widow spider venom, as botulism antagonist, 380
Blister agents
  countermeasure effects, 401
  detection and measurement, 426
  emergency response to, 416, 418–419
  neurobehavioral effects, 395, 398

Blood agents, see also Cyanide
countermeasure effects, 401
detection and measurement, 426
emergency response, 417, 419
neurobehavioral effects, 394–395, 398
Blood-brain barrier, 121–144
as complex genetic trait, 136–137
conditions inducing disruption, 134–137
acute insults, 135
pathophysiological, 134–135
psychological and physiological stressors, 135–136
functional characteristics, 128–132
cholinergic involvement, 129
investigative approaches, 130–131
pericellular cell passage, 129–130
physiological considerations in transport, 128–129
transgenic engineering models, 131–132
genetic aspects, 136–137
modulators, 132–134
immunomodulators and multi-drug transport, 132–134
nitric oxide and vasoactive agents, 132
physical basis, 122–128
adherens and tight junctions, 124
astrocytes and, 128
potential involvement of acetyl-cholinesterase, 124–126
signal-transducing elements, 126–128
vascular endothelial cells, 123–124
Botulinum immune globulin, 379
Botulinum neurotoxin, 373–392
general properties, 373–374
manifestations of botulism, 377–379
foodborne, 377–378
infant, 378–379
wound, 378
symptoms and toxic effects, 374–377
treatment options, 379–386
inhibitors of binding, 382–383
inhibitors of internalization, 383–384
inhibitors of metalloprotease activity, 384–386
pharmacologic intervention: potassium channel blockers, 380–381
Bradykinin, blood-brain barrier penetration and, 133
Brain injury, blood-brain barrier and, 135
British Anti-Lewisite, 395
Bromoacetone, 325
Bromobenzyl cyanide, 325, 330, 332
Butyrylcholinesterase, 222, see also Cholinesterases

C

Ca^{2+} ATPase inhibition, 106
Ca^{2+} elevation, cyanide and neuronal, 308
Calabar bean, see Physostigmine
Calcium, elevated, as botulism antagonist, 380
Calcium channel blockers, cyanide and, 309
Calcium ionophores, as botulism antagonist, 380
Capillary gas chromatography, 426
Capsaicin (oleoresin of capsicum), 325, 352–359, see also Oleoresin of capsicum (OC)
Capsicum sp., 328, see also Oleoresin of capsicum (OC)
Captopril, in botulism, 385–386
Carbamates, 145–190, see also specific agents
neostigmine, 177–180
physostigmine, 166–177
pyridostigmine bromide, 147–166
Carbaryl, 11
Carboxylesterase, 205–208, 222–226
Carcinogenicity
of chlorobenzylidene malononitrile, 336
of chloroacetophenone, 351
of riot-control agents, 336, 344–345, 351, 356–357, see also Riot-control agents
of sulfur mustard, 247–252
Cardiac effects, of cyanide, 310–312
Catalytic bioscavengers, 228–233
Caterins, 124
Chelating agents, 395
in botulism, 384–385
Chemical Agent Monitor (CAM), 423
Chemical and Biological Hotline, 427
Chemical Stockpile Emergency Preparedness Program, 421
Chemical warfare agents, as compared with military chemicals, 323
Chemical Weapons Improved Response Program, 427
Chlorfenvinphos, 397
Chlorine, 394
Chloroacetone, 325, 330, 332
Chloroacetophenone (CN), 325, 349–352
chemistry, 327
human toxicology: physiological effects, 352
mammalian toxicology, 349–351
genotoxicity and carcinogenicity, 351
metabolism, fate, and mechanism, 351
ocular and cutaneous effects, 351
Chlorobenzylidene malononitrile (CS), 325, 332–341
carcinogenicity, 336
chemistry, 326–327
genotoxicity, 336

human toxicology, 338–341
mammalian toxicoloty, 333–335
metabolism, metabolic fate, and mechanisms, 336–338
ocular and cutaneous effects, 335
reproductive/developmental toxicity, 335–336
2-Chloro-1-phenylethanone, see Chloroacetophenone
Chloropicrin (PS), 324, 325, 330, 332
Chloroquine, in botulism, 383, 384
Chlorpyrifos, 11
Choking agents, see also specific agents
countermeasure effects, 401
neurobehavioral effects, 394–395, 398
Choline acetyl transferase (ChAT)
blood-brain barrier and, 129
in stress, 90–91
Cholinergic involvement, in blood-brain barrier, 129
Cholinergic toxicity, of nerve agents, 82–102, see also under Toxicity
Cholinesterase activity, in stress, 89–90
Cholinesterase inhibitors, see Carbamates and specific agents
Cholinesterases
as bioscavengers, 21–228, 192–210
carboxylesterase, 205–208
enhancement of scavenging activity, 202–205
guinea pig soman studies with HuBChE, 195–196
huperzine, 208
immobilized for decontamination of organophosphates, 208–210
non-human primate studies, 196–202
rodent studies, 193–195
soman studies in comparison with other agents, 196
stability *in vivo,* 193
enhancement of scavenging activity, 202–205
oxime amplification, 203
by site-specific mutagenesis, 203–205
hydrolyzing enzymes, 205
neurobehavioral effects, 400–402
Chronic fatigue syndrome, 289
Cingulin, 124, 125
Civilian protection, 419–421
Chemical Stockpile Emergency Preparedness Program, 421
as compared with military response, 419–420
Israeli model, 420–421
Clostridium botulinum, 373–392, see also Botulinum neurotoxin
Clostridium tetani, 378
Cluster methodology, 272–273
CN (chloroacetophenone), 325, 349–352, see also Chloroacetophenone
CNS effects, see Neurobehavioral effects
Cobalt diedetate (kelocyanor), as cyanide antidote, 313
Color change chemistry detectors, 423–424
Computed tomography (CT scan), of blood-brain barrier function, 130–131
Contrast agents, for blood-brain barrier studies, 130–131
CR (dibenz (b,f)-1:4-oxazepine), 325, 341–349, see also Dibenz (b,f)-1:4-oxazepine
CS (σ-chlorobenzylidene malononitrile), 325, 332–341, see also Chlorobenzylidene malononitrile
Cutaneous effects
of riot-control agents, 335, 343–344, 347–349, 351, 355–356
of sulfur mustard, 248–249
Cyanide, 301–320, 338, 417, 419
ADP-ribosylation by, 312
antidotes, 313–314
cardiac effects, 310–312
chemical reactivity, 305
detection and measurement, 426
exposure, 302–303
metabolism of, 306–308
neural tissue effects, 309–310
cellular Ca^{2+} elevation, 308
hyperpolarization, 309–310
neuronal activation, 310
neuronal metabolism, 308
oxidative stress and, 308–309
neurobehavioral effects, 395, 397–398
production in neural tissue, 312–313
symptoms, 303–305
Cystine, cyanide metabolism and, 307

D

Decontamination, 428–429
DEET (N,N-diethyl-m-toluamide), 11, 165, 277, 283
Defense Authorization Act of 1997 (PL 104–201), 412
Demyelination disorders, 130, 133–134
Departents of Commerce, Justice, and State, the Judiciary and Related Agencies Appropriations Act of 1998, 412–413
Department of Veterans Affairs Persian Gulf Health Registry, 283–289
Depleted uranium, Gulf War Syndrome and, 278–279
Detection technology, 423–424

Developmental toxicity, of riot control agents, 335–336, 344
DFP, acute exposure, 5–6
3,4-Diaminopyrdine (3,4-DAP), in botulism, 381
Diaphragmatic contractility, cyanide and, 311–312
Diazepam, 85, 217, 418
Diazinon, 397
Dibenz (b,f)-1:4-oxazepine (CR), 325, 341–349
  chemistry, 327
  human toxicology
    ocular and cutaneous effects, 347–349
    physiological effects, 346–347
  mammalian toxicology, 342–343
    carcinogenicity, 344–345
    clinical chemistry, 345
    genotoxicity, 344–345
    metabolism, metabolic fate, and mechanisms, 345–346
    ocular and cutaneous effects, 343–344
    reproductive toxicity and developmental effects, 344
Dichlorvos, 207
N,N-Dimethyl-carbamate, see Carbamates and specific agents
2,4-Dinitrophenol, as botulism antagonist, 380
Diphenylaminochloroarsine (adamsite), 324, 325, 359–361
  human toxicology, 360–361
  toxicology and physiological effects, 359–360
Dipyridamole, cyanide and, 309
Distribution of pyridostigmine, 147–149
DlgA protein, 126
DM (adamsite, diphenylaminochloroarsine), 324, 325, 359–361
DoD Chemical Weapons Improved Response Program, 427
DoD Clinical Evaluation Program, 283–289
Dopaminergic system, blood-brain barrier and, 128–129
Dosage, terminology of, 333
Drosophila studies, of blood-brain barrier, 124–125

E

Ecological fallacy, 265
Elimination pathways, nerve agents, 58–69
  by covalent binding, 59–62
  in exposure detection, 63–64
  by hydrolytic degradation, 58–59
  renal excretion, 62–63
Emergency response, 409–431
  to blood agents, 417, 419
  civilian protection, 419–421
    Chemical Stockpile Emergency Preparedness Program, 421
    as compared with military response, 419–420
    Israeli model, 420–421
  classification of chemical warfare agents for, 414–419
    by clinical effects, 416–417
    by emergency treatment, 417–419
    by means of intoxication, 415
  comparison of potencies, 415
  elements of, 422–426
    clues to presence of CW weapons, 422
    detection technology, 423–424
    HAZMAT incidents and teams, 422–423
    laboratory analysis, 424–426
  first-responder and victim protection, 426–429
    mass casualty decontamination, 428–429
    training, 426–428
  legislative and regulatory
    executive initiatives, 411–412
    federal response plan, 413–414
    legislative initiatives, 412–413
  low-dose exposure issues, 430–431
  to nerve agents, 416, 417–418
  to pulmonary agents, 417, 419
  to vesicants, 416, 418–419
Endothelial cells, of cerebral vasculature, blood-brain barrier and, 123–124
Endurance, see Stress, physical
Environmental exposure, pyridostigmine and, 165
Enzymes, sulfhydryl-containing, 338
Enzyme sponge techniques, 208–209
Epidemiology, Gulf War Syndrome and methodology, 291–295
Epilepsy, 134–135
Eserine, see Physostigmine
Ethanol, pyridostigmine and, 164–165
Ethyl bromoacetate, 325
Ethyl iodoacetate, 325
Excretion, of pyridostigmine, 152–153
Exercise training, see also Heat stress; Stress, physical
  physostigmine and, 173–175
Exposure
  dose and duration, 4–5
  estimation of military, 4
Eye injury, by sulfur mustard, 248, 250

F

Federal Emergency Management Administration (FEMA), 411–412
Federal Response Plan, 413–414
Fibromyalgia, 289–290
First-responder and victim protection, 426–429
Foodborne botulism, 377–378, see also Botulinum neurotoxin

Food poisoning, see Botulinum neurotoxin
Foreign Crisis Act of 1961, 412

G

GA, see Tabun
GABAergic system, nerve agents and, 16–17, 106
Gas chromatography, capillary, 426
Gas chromatography-mass spectrometry, 426
Gas chromatography-tandem mass spectrometry, 426
Gas hysteria, 400–401
Gas masks, misuse of, 402
GB, see Sarin
GD, see Soman
Gender, pyridostigmine and, 165–166
Genetic aspects, of blood-brain barrier, 136–137
Genotoxicity
  of riot control agents, 336
  of riot-control agents, 344–345, 351
Geotrichum candidum studies, of carboxyesterase, 206–207
Gliotactin, 125–126
Glyburide, cyanide and, 309
Guanidine, in botulism, 381
Guanine nucleotides, in blood-brain barrier function, 126–128
Gulf War Syndrome, 261–295, 399
  background and history, 262–265
  case definition problems, 289–290
  common exposure subset, 276–283
    chemical and biological warfare agents, 279
    depleted uranium, 278–279
    immunizations, pesticides, and occupational exposures, 277–278
    infectious diseases, 276–277
    Khamisiyah, Iraq, 280–282
    mixed exposure and synergistic, additive, and other effects, 283
    oil well fires, 279
  emergency response implications, 430–431
  future research directions, 290–291
  methodological lessons, 291–295
  population at risk, 265–272
    morbidity, 269–270
    mortality, 267–269
    reproductive outcomes, 271–272
    symptom prevalence, 270–271
  registry enrollees, 283–289
  symptom-based cluster subset, 272–276
    Air National Guard, 274
    123rd ARCOM, 273
    reproductive effects, 275–276
    Seabees, 274–275

H

HAZMAT teams, 422–423, 424, 428–429
  training levels, 426
Heat stress, pyridostigmine and, 160–163
HI-6
  as compared with cholinesterase bioscavengers, 196
  pretreatment with, 64
House Report 105–825, 413
HuBChE, see Cholinesterases
Huperzine, 208
Hu-Pon enzyme, 228–231
Hydrocyanic acid, see Cyanide
Hydromorphone, cyanide and, 312
Hydroxychloroquine, in botulism, 383
Hyperpolarization, cyanide and, 309–310
Hyperthermia, blood-brain barrier and, 135
Hypothermia, blood-brain barrier and, 135

I

2-ICA, cyanide metabolism and, 307
ICD 1578, in botulism, 386
Immunizations, Gulf War Syndrome and, 277–278
Immunomodulators, blood-brain barrier and, 133–134
Infant botulism, 378–379, see also Botulinum neurotoxin
Infectious diseases, in Gulf War Syndrome, 276–277
Inhalation toxicokinetics, see under Toxicokinetics
Institute of Medicine, 264–265, 291–292
Intermediate syndrome, 5
Intravenous toxicokinetics, see under Toxicokinetics
In vitro studies, nerve agents, subclinical exposures, 14–17
Iodoacetone, 325
Ion Mobility Spectrometry (IMS), 423
Iowa Persian Gulf Study Group, 270
Iran-Iraq War, 395
Irritant chemicals, see Riot control agents
Ischemia, blood-brain barrier and, 135
Isocoumarins, in botulism, 386
Israel, SCUD missile attacks, 402
Israeli model, of civilian protection, 420–421

J

Japanese studies, acute exposure to sarin, 7–8
Junctions
  adherens, 124
  proteins associated with blood-brain barrier, 124.125
  tight, 124

K

Kelocyanor (cobalt diedetate), as cyanide antidote, 313
Khamisiyah, Iraq, Gulf War Syndrome exposure and, 280–282
Kinase cascades, 133

L

Laboratory analysis, in emergency response, 424–426
Lacrimators, see Riot control agents
Lectins, in botulism, 382–383
Lewisite, 395
Limax flavus lectins, in botulism, 382–383
Low-level exposure, emergency response implications, 430–431

M

Magnetic resonance imaging (MRI), of blood-brain barrier function, 131
Mass casualty decontamination, 428–429
Mast cells, in blood-brain barrier penetration, 130
mdr protein, 123–124
M18 detection kit, 424
3-Mercaptopyruvate sulfurtransferase, cyanide metabolism and, 306–307
Mestinon, see Pyridostigmine
Metal chelators, in botulism, 384–385
Metalloprotease inhibitors, in botulism, 385–386
Methylamine hydrochloride, in botulism, 383, 384
Metropolitan Medical Strike Teams, 423
Military chemicals, as compared with chemical warfare agents, 323
Momensin, in botulism, 384
Morbidity, in Gulf War Syndrome, 269–270
Mortality, in Gulf War Syndrome, 266–269
M256A1 detection kit, 424
Multiple chemical synsitivity (MCS), 289.290
Multiple sclerosis, blood-brain barrier and, 133–134
Muscarinic receptor downregulation, 11–12
Muscarinic receptors, nerve agents and, 15–16, 85, 94–100
Mustard gas, 245–260, 395, 398, 416, 418–419, see also Sulfur mustard
Mutagenicity, of riot-control agents, 356–357
Myasthenia gravis, 136, 148–154, see also Pyridostigmine
  neostigmine and, 178–180

N

National Academy of Sciences, 3, 262
National Ambulatory Medical Care Survey, 287
National Defense Authorization Act (PL 103–160), 412
National Institutes of Health Technology Assessment Workshop, 263–264
National Security Decision Directive 207, 411
Neostigmine, 177–180
  general aspects, 177–178
  pharmacodynamics, 179–180
  pharmacokinetics, 178–179
Nerve agents, see also Cyanide, neural tissue effects; Organophosphates and specific agents
  acute exposure, animal studies, 8–9
  acute toxicity, 2
  bioscavengers and, 215–237, see also Bioscavengers and specific agents
    advantages of, 218–219
    background and principles, 215–237
    behavioral effects, 233–235
    catalytic, 228–233
    cholinesterases, 192–210
      carboxylesterase, 205–208
      enhancement of scavenging activity, 202–205
      guinea pig soman studies with HuBChE, 195–196
      huperzine, 208
      immobilized for decontamination of organophosphates, 208–210
      non-human primate studies, 196–202
      rodent studies, 193–195
      soman studies in comparison with other agents, 196
      stability *in vivo,* 193
    as compared with standard treatments, 217–218
    stochiometric, 220–228
      antibodies, 220
      enzymes, 220–228, see also Cholinesterases
  blood-brain barrier and, 121–144, see also Blood-brain barrier
  carbamates and, 145–190, see also Carbamates and specific agents
    neostigmine, 177–180
    physostigmine, 166–177
    pyridostigmine bromide, 147–166
  cholinergic toxicity
    histopathological effects, 101
    under stressful conditions, 101–103

chronic health effects, 1–18
of acute exposure, 5–10
background and history, 3–4
in vitro studies, 14–12
recent literature, 4–5
of repeated low-level exposure, 10–14
as compared with organophosphate insecticides, 5–7
countermeasure effects, 401
detection and measurement, 425–426
mechanism of toxicity, 1–2
neurobehavioral effects, 396–397, 398
non-cholinergic toxicity (OPIDN), 102–108, see also under OPIDN; Toxicity
signs of poisoning by, 2
subclinical exposures
GABAergic system effects, 16–17
muscarinic receptor effects, 15–16
nicotinic receptor effects, 17
toxicity, 83–120
cholinergic, 82–102
delayed neurotoxicity, 85–89
mechanisms of action, 83–85
non-cholinergic (OPIDN), 102–108
stress and, 89–82
toxicokinetics, 25–82
elimination pathways, 57–64, see also Elimination pathways and specific agents
inhalation
sarin, 51–52
soman, 45–51, 195–196
upon low-level exposure, 52–57
intravenous, soman and sarin in various species, 33–42
methods employing stereoisomers, 26–29
prophylaxis and treatment and, 69–72
soman, physiologically based modeling, 64–69
subcutaneous, of soman, 42–45
trace analysis in biological samples, 30–33
of V agents, 72–74
Neural tissue effects, of cyanide, 309–310
cellular Ca^{2+} elevation, 308
hyperpolarization, 309–310
neuronal activation, 310
neuronal metabolism, 308
oxidative stress and, 308–309
Neurexin IV gene, 125
Neurobehavioral effects, 393–408
of chemical warfare agents, 394–397
blister agents, 395, 398
choking and blood agents, 394–395, 398
nerve agents, 396–397, 398
of countermeasures, 397–402
administration problems, 397–399
pretreatment, 399–402
as detection method, 233
nerve agents, subclinical repeated exposures, 12–13
of scavengers, 233–235
Neurobehavioral testing, nerve agents, subclinical repeated exposures, 12–13
Neurological effects, of sulfur mustard, 249–250
Neuromuscular blockage, pyridostigmine and, 166
Neurophysiological studies, nerve agents, acute exposure, 8
Neurotoxic esterase inhibition, 85–80
Neurotoxins, botulinum, 373–392, see also Botulinum neurotoxin
Neurotransmission, blood-brain barrier and, 125–128
Nicotinic receptors, nerve agents and, 17, 100–101
Nigericin, in botulism, 384
Nitric oxide, blood-brain barrier and, 132
NMDA receptors, cyanide and, 305
N,N-dimethyl-carbamate, see Carbamates and specific agents
N,N,N,'N'-tetrakis(2-pyridylmethyl)ethylenediamine (TPEN), in botulism, 384–385
7-N-phenylcarbamoylamino-4-chloro-3-propyloisocoumarin (ICD 1578), in botulism, 386

O

OC (oleoresin of capsicum), 325, 352–359, see also Oleoresin of capsicum
Occludin, 124, 125
Occupational exposures, Gulf War Syndrome and, 277–278
Ocular effects
of riot-control agents, 335, 343–344, 347–349, 351, 355–356
of sulfur mustard, 248, 250
Off-Site Triage, Treatment, and Transport Center, 427
Oil well fires, Gulf War Syndrome and, 279
Oklahoma City bombing, 411
Oleoresin of capsicum (OC), 325, 352–359
chemistry, 328
human toxicology, 357–359
mammalian toxicology, 353–357
carcinogenicity, 356–357
metabolism, metabolic fate, and mechanisms, 357
mutagenicity, 356–357
ocular and cutaneous effects, 355–356

123rd ARCOM population group, 273
Operation Desert Shield/Storm, 3, 262, see also Gulf War Syndrome
Organophosphate-induced delayed neurotoxicity (OPIDN), 12, 85, 102–108
 biochemical effects, 104–107
Organophosphate pesticides
 as compared with nerve agents, 5–7
 in greenhouse workers, 13–14
 subclinical repeated exposures, 12–13
Organophosphates, see also Nerve agents
 cholinesterase decontamination, 208–210
 cholinesterase prophylaxis, 191–207
 neurobehavioral effects, 396–397, 398
Oxidative stress response, pyridostigmine and, 164–165

P

2-PAM, 217
Parkinsonian symptoms, in cyanide toxicity, 305, 395
Parkinson's disease, 197
PB, 11
PDZ proteins, blood-brain barrier and, 126–128
Peptide inhibitors, in botulism, 386
Pericellular cell passage, through blood-brain barrier, 129–130
Permethrin, 277, 283
Persian Gulf War, see Gulf War Syndrome
Persistent and nonpersistent agents, 414–415
Pesticides, Gulf War Syndrome and, 277–278
P glycoprotein, 123
Pharmacodynamics
 of neostigmine, 179–180
 of physostigmine, 170–173
 of pyridostigmine, 154–158
Pharmacokinetics
 of neostigmine, 178–179
 of physostigmine, 166–170
 pyridostigmine, 153–154
Phenacyl chloride, see Chloroacetophenone
Phenarsazine chloride, see Diphenylaminochloroarsine
7-N-Phenylcarbamoylamino-4-chloro-3-propyloisocoumarin (ICD 1578), in botulism, 386
Phenyl carbylamine chloride, 325
Phenyl chloromethyl ketone, see Chloroacetophenone
Phosgene, 324, 417, 419
 neurobehavioral effects, 394–395, 398
Phosphoramidon, in botulism, 385–386
Physiological effects, of riot-control agents, 346–347, 359–360
Physostigmine
 factors influencing activity, 173–177
  physical stress, 173–175
  soman, 175–177
 general aspects, 166
 pharmacodynamics, 170–173
 pharmacokinetics, 166–170
Poly (ADP-ribose) polymerase, cyanide and, 312
Polyurethane foams, as organophosphate absorption materials, 208–209
Post-traumatic stress disorder, 395, 402
Potassium channel blockers, in botulism, 380–381
Pralidoxime chloride (2-PAM Cl), 418
Presidential Advisory Committee on Gulf War Syndrome, 3, 264–265, 285
Presidential Directive 62, 412
Pretreatment, see also specific agents
 neurobehavioral effects, 397–402
 with pyridostigmine, 154–160
Primaquine, in botulism, 384
Primate Equilibrium Platform (PEP) task, 197–200
Primate studies
 of cholinesteraes, 196–202
 of cholinesterases, 226–228
Propetamphos, 397
Prophylaxis
 with atropine sulfate, 65–66
 bioscavengers in, 191–210, see also Bioscavengers; Cholinesterases
 with HI-6, 64
 with pyridostigmine, 65
Prostigmine, see Physostigmine
PSD-95 protein, 126
Pseudomonas diminuta, 228
Psychological factors, 11, 393–408, see also Neurobehavioral effects
Pulmonary agents
 countermeasure effectiveness, 401
 emergency response, 417, 419
 neurobehavioral effects, 394–395, 398
Pulmonary effects, of sulfur mustard, 248, 250
Pyridostigmine, 147–166
 absorption, 147
 BhE protein and, 136
 blood-brain barrier and, 136
 as compared with cholinesterase bioscavengers, 196
 distribution, 147–149
 excretion, 152–153
 factors influencing activity, 160–166
  environmental exposure, 165
  gender and age, 165–166
  stress, 160–165
 general aspects, 147–148

metabolism, 149–152
pharmacodynamics in pretreatment, 154–158
pharmacokinetics, 153–154
pretreatment with, 64

Q

Quantitative trait loci, 136–137
Quinacrine, in botulism, 384
o-Quinone, see Physostigmine

R

Radiation, sulfur mustard and, 252–253
Receptor antagonists, in botulism, 382–383
Registry enrollees, for Gulf War Syndrome, 283–289
Renal excretion
of physostigmine, 167
of pyridostigmine, 140–153
Reproductive outcomes, in Gulf War Syndrome, 271–272, 275–276
Reproductive toxicity
of riot control agents, 335–336
of riot-control agents, 344
Research programs, for emergency response, 427–428
Rhodanese, cyanide metabolism and, 306–307
Rho proteins, 124
blood-brain barrier and, 126–128
Riot-control agents, 321–362, see also specific agents
chemistry of selected, 326–328
chloroacetophenone (CN), 327
chlorobenzylidene malononitrile (CS), 326–327
dibenz (b,f)-1:4-oxazepine (CR), 327
oleoresin of capsicum (OC), 328
chloroacetophenone (CN), 325, 349–352
human toxicology: physiological effects, 352
mammalian toxicology, 349–351
genotoxicity and carcinogenicity, 351
metabolism, fate, and mechanism, 351
ocular and cutaneous effects, 351
o-chlorobenzylidene malononitrile (CS), 332–341
carcinogenicity, 336
genotoxicity, 336
human toxicology, 338–341
mammalian toxicoloty, 333–335
metabolism, metabolic fate, and mechanisms, 336–338
ocular and cutaneous effects, 335
reproductive/developmental toxicity, 335–336
clinical aspects of, 328–329
dibenz (b,f)-1:4-oxazepine (CR), 325, 341–349
human toxicology
ocular and cutaneous effects, 347–349
physiological effects, 346–347
mammalian toxicology, 342–343
carcinogenicity, 344–345
clinical chemistry, 345
genotoxicity, 344–345
metabolism, metabolic fate, and mechanisms, 345–346
ocular and cutaneous effects, 343–344
reproductive toxicity and developmental effects, 344
diphenylaminochloroarsine (adamsite), 324, 325, 359–361
human toxicology, 360–361
toxicology and physiological effects, 359–360
general aspects, 322–323
historical perspectives, 323–325
ocular and cutaneous effects of, 329–332
oleoresin of capsicum (OC), 325, 352–359
human toxicology, 357–359
mammalian toxicology, 353–357
carcinogenicity, 356–357
metabolism, metabolic fate, and mechanisms, 357
mutagenicity, 356–357
ocular and cutaneous effects, 355–356
toxicology of, 329
Risk management assessment, 293–295
Rubreserine, see Physostigmine
Russia, nerve agents, acute exposure studies, 8

S

S. *typhimurium* (Ames) test, of capsaicin, 356–357
Sarin, 1, 396, 419–420
acute exposure
Japanese studies, 7–8
in rhesus monkeys, 6–7
Russian studies, 8
carboxyesterase inhibition, 207
chronic health effects, 10
historical development, 83–84
human exposure values, 430
LOAEL and NOEL of subacute dosages, 11
low-dose exposure, 9–10
neurobehavioral effects, 396–397, 398
subclinical repeated exposures, 12
toxicokinetics
inhalation, 51–52
intravenous in various species, 33–42

United States Air Force School of Aerospace Medicine studies, 10–11
Scarring, by sulfur mustard, 248–249
SCUD missile attacks, 402
Seabees population group, 273–274
Second-messenger systems, nerve agents and, 106
Self-contained breathing apparatus (SCBA), 428
Serial Probe Recognition Task testts, 400
Signal-transducing elements, blood-brain barrier and, 126–128
Skin cancer, sulfur mustard and, 248–249, 252–253
SNAP-25 protein, 376–377
SNARE proteins, in botulism, 384–385
Sodium nitrite, as cyanide antidote, 313
Sodium thiosulfate, as cyanide antidote, 313
Soman, 1, 396
acute exposure, Russian studies, 8
carboxyesterase inhibition, 207
physostigmine and, 171–172, 176–177
physostigmine pharmocokinetics and, 175–177
toxicokinetics
inhalation, 45–51, 195–196
upon low-level exposure, 52–57
intravenous in various species, 33–42
physiologically based modeling, 64–69
Spectrometry, detection, 424
Spinal cord reflexes, toxicity and, 107
Stochiometric bioscavengers, 220–228
antibodies, 220
enzymes, 220–228, see also Cholinesterases
Stress, 89–92, See also Toxicity, non-cholinergic
cholinergic toxicity and, 101–103
cyanide toxicity and oxidative, 308–309
environmental, 102–103
non-cholinergic toxicity and, 108
physical, 101–102
physostigmine and, 173–175
pyridostigmine and, 160–165
psychological, blood-brain barrier and, 135–136
pyridostigmine and, 160–165
Stress casualties, 400–401
Stressors, blood-brain barrier and, 135–136
Succinate dehydrogenase inhibition, 106
Sulfhydryl-containing enzymes, 338
Sulfur mustard, 245–260, 395, 398, 416, 418–419
acute subclinical exposure, 251–253
carcinogenesis and, 251–252
radiomimetic effect, 252–253
clinical effects, 246–250
carcinogenesis, 247–248
central nervous system, 249
chronic eye disease, 248
chronic pulmonary disease, 248
cutaneous effects and cancer formation, 248–249
summary for systemic exposures, 249–250
dose dependency of lesion, 254–255
general properties, 245–246
in vitro toxicity studies, 253–254
neurobehavioral effects, 395, 398
Sulfurtransferases, cyanide metabolism and, 306–307
Surface Acoustical Wave Chemical Agent Detector, 423
Synaptobrevin (VAMP), 376–377
Syntaxin, 376–377
Systemic physiological effects, of riot-control agents, 346–347, 359–360

T

Tabun, 1, 12, 83, 396
Taylor Pharmaceutical Cyanide Kit, 313
Temperature, as stressor, 102–103
Test-Mate OP Kit, 426
Thermoregulation, pyridostigmine and, 160–162
Thiosulfate, as cyanide antidote, 313
Thyrotropin-releasing hormone, 107
Tight junctions, 124
Tobacco amblyopia, 305
Tokyo subway poisoning incident, 7–8
Topical Skin Protectant, 429
Torpedo californica studies, of carboxyesterase, 206–207
Toxicity
nerve agents, 83–120
cholinergic, 82–102
biochemical effects, 94–100
histopathological effects, 101
under stressful conditions, 101–103
delayed neurotoxicity, 85–89
mechanisms of action, 83–85
non-cholinergic (OPIDN), 102–108
stress and, 89–82
Toxicokinetics, nerve agents, 25–82
elimination pathways, 57–64, see also Elimination pathways and specific agents
inhalation
sarin, 51–52
soman, 45–51, 195–196
upon low-level exposure, 52–57
intravenous, soman and sarin in various species, 33–42
methods employing stereoisomers, 26–29
prophylaxis and treatment and, 69–72

soman, physiologically based modeling, 64–69
subcutaneous, of soman, 42–45
trace analysis in biological samples, 30–33
of V agents, 72–74
Transgenic models, of blood-brain barrier, 131–132
Traumatic brain injury, blood-brain barrier and, 135
Triage, 429
Trichloronitromethane (chloropicrin, PS), 324, 325
Triorthotolyl phospate (TOTP)-induced neurotoxicity, stress and, 108
Triticum vulgaris lectins, in botulism, 382–383
Tropilidene, 325

U

United States Air Force School of Aerospace Medicine, sarin studies, 10–11
Uranium, depleted, Gulf War Syndrome and, 278–279
U.S. Army Environmental Hygiene Agency, 279
U.S. Army Medical Chemical Defense Research Program, 429

V

Vaccination, for botulism, 379–380
V agents, toxicokinetics, 72–74
VAMP (synaptobrevin), 376–377
VA National Health Survey of Gulf War Era Veterans and Their Families, 270
VA Persian Gulf Health Registry, 283–289
Vasoactive agents, blood-brain barrier and, 132
Vesicants
countermeasure effects, 401
detection and measurement, 426
emergency response to, 416, 418–419
neurobehavioral effects, 395, 398
Vietnam War, 262
Vitamin deficiencies, tobacco amblyopia and, 305
Vomiting agents, see Riot control agents
VX, 1, 12, 84

W

WAIS deficit, in organophosphate poisoning, 397
Water contamination, 415
Wound botulism, 378, see also Botulinum neurotoxin

X

Xylyl bromide, 325, 330, 332

Z

Zinc chelators, in botulism, 384–385
ZO-1 protein, 126–128